Calculus
Know-It-ALL

Calculus
Know-It-ALL
Beginner to Advanced, and Everything in Between

Stan Gibilisco

McGraw
Graw
Hill

New York   Chicago   San Francisco   Lisbon   London   Madrid
Mexico City   Milan   New Delhi   San Juan   Seoul
Singapore   Sydney   Toronto

The McGraw·Hill Companies

Library of Congress Cataloging-in-Publication Data

Gibilisco, Stan.
    Calculus know-it-all : beginner to advanced, and everything in
between / Stan Gibilisco.
            p.   cm.
    Includes index.
    ISBN 978-0-07-154931-8 (alk. paper)
    1. Calculus.   I. Title.
QA300.G498   2009
515—dc22   2008037246

1 2 3 4 5 6 7 8 9 0   DOC/DOC   0 1 4 3 2 1 0 9 8

ISBN 978-0-07-154931-8
MHID 0-07-154931-5

Sponsoring Editor
 Judy Bass

Production Supervisor
 Pamela A. Pelton

Editing Supervisor
 Stephen M. Smith

Project Manager
 Vasundhara Sawhney, International
 Typesetting and Composition

Copy Editor
 Bhavna Gupta, International Typesetting
 and Composition

Proofreaders
 Nigel O'Brien and Sanjukta Chandra,
 International Typesetting and Composition

Art Director, Cover
 Jeff Weeks

Composition
 International Typesetting and Composition

Printed and bound by RR Donnelley.

McGraw-Hill books are available at special quantity discounts to use as premiums and sales promotions, or for use in corporate training programs. To contact a special sales representative, please visit the Contact Us page at www.mhprofessional.com.

To Tim, Tony, Samuel, and Bill

About the Author

Stan Gibilisco is an electronics engineer, researcher, and mathematician. He is the author of *Algebra Know-It-ALL*, a number of titles for McGraw-Hill's *Demystified* series, more than 30 other technical books and dozens of magazine articles. His work has been published in several languages.

Contents

Part 2   Integration in One Variable

Preface

If you want to improve your understanding of calculus, then this book is for you. It can supplement standard texts at the high-school senior, trade-school, and college undergraduate levels. It can also serve as a self-teaching or home-schooling supplement. Prerequisites include intermediate algebra, geometry, and trigonometry. It will help if you've had some precalculus (sometimes called "analysis") as well.

This book contains three major sections. Part 1 involves differentiation in one variable. Part 2 is devoted to integration in one variable. Part 3 deals with partial differentiation and multiple integration. You'll also get a taste of elementary differential equations.

Chapters 1 through 9, 11 through 19, and 21 through 29 end with practice exercises. You may (and should) refer to the text as you solve these problems. Worked-out solutions appear in Apps. A, B, and C. Often, these solutions do not represent the only way a problem can be figured out. Feel free to try alternatives!

Chapters 10, 20, and 30 contain question-and-answer sets that finish up Parts 1, 2, and 3, respectively. These chapters will help you review the material.

A multiple-choice final exam concludes the course. Don't refer to the text while taking the exam. The questions in the exam are more general (and easier) than the practice exercises at the ends of the chapters. The exam is designed to test your grasp of the concepts, not to see how well you can execute calculations. The correct answers are listed in App. D.

In my opinion, most textbooks place too much importance on "churning out answers," and often fail to explain how and why you get those answers. I wrote this book to address these problems. I've tried to introduce the language gently, so you won't get lost in a wilderness of jargon. Many of the examples and problems are easy, some take work, and a few are designed to make you think hard.

If you complete one chapter per week, you'll get through this course in a school year. But don't hurry. When you've finished this book, I recommend *Calculus Demystified* by Steven G. Krantz and *Advanced Calculus Demystified* by David Bachman for further study. If Chap. 29 of this book gets you interested in differential equations, I recommend *Differential Equations Demystified* by Steven G. Krantz as a first text in that subject.

Stan Gibilisco

Acknowledgment

- -

I extend thanks to my nephew Tony Boutelle. He spent many hours helping me proofread the manuscript, and he offered insights and suggestions from the viewpoint of the intended audience.

Calculus
Know-It-ALL

PART 1

Differentiation in One Variable

1

Single-Variable Functions

Calculus is the mathematics of *functions,* which are relationships between sets consisting of objects called *elements.* The simplest type of function is a *single-variable function,* where the elements of two sets are paired off according to certain rules.

Mappings

Imagine two sets of points defined by the large rectangles in Fig. 1-1. Suppose you're interested in the subsets shown by the hatched ovals. You want to pair off the points in the top oval with those in the bottom oval. When you do this, you create a *mapping* of the elements of one set into the elements of the other set.

Domain, range, and variables

All the points involved in the mapping of Fig. 1-1 are inside the ovals. The top oval is called the *domain.* That's the set of elements that we "go out from." In Fig. 1-1, these elements are *a* through *f.* The bottom oval is called the *range.* That's the set of elements that we "come in toward." In Fig. 1-1, these elements are *v* through *z.*

In any mapping, the elements of the domain and the range can be represented by *variables.* A nonspecific element of the domain is called the *independent variable.* A nonspecific element of the range is called the *dependent variable.* The mapping assigns values of the dependent (or "output") variable to values of the independent (or "input") variable.

Ordered pairs

In Fig. 1-1, the mapping can be defined in terms of *ordered pairs,* which are two-item lists showing how the elements are assigned to each other. The set of ordered pairs defined by the mapping in Fig. 1-1 is

$$\{(a,v), (b,w), (c,v), (c,x), (c,z), (d,y), (e,z), (f,y)\}$$

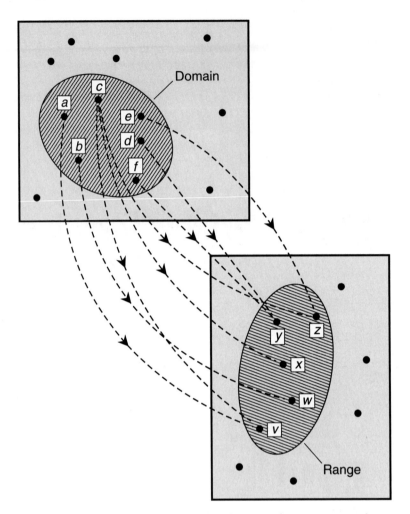

Figure 1-1   A relation defines how the elements of a set are assigned to the elements of another set.

Within each ordered pair, an element of the domain (a value of the independent variable) is written before the comma, and an element of the range (a value of the dependent variable) is written after the comma. Whenever you can express a mapping as a set of ordered pairs, then that mapping is called a *relation*.

- -

Are you confused?

You won't see spaces after the commas inside of the ordered pairs, but you'll see spaces after the commas separating the ordered pairs in the list that make up the set. These aren't typographical errors! That's the way they should be written.

- -

Modifying a relation

A *function* is a relation in which every element in the domain maps to one, but never more than one, element in the range. This is not true of the relation shown in Fig. 1-1. Element *c* in the domain maps to three different elements in the range: *v*, *x*, and *z*.

In a function, it's okay for two or more values of the independent variable to map to a single value of the independent variable. But it is *not* okay for a single value of the independent variable to map to two or more values of the dependent variable. A function can be *many-to-one*, but never *one-to-many*. Sometimes, in order to emphasize the fact that no value of the independent variable maps into more than one value of the dependent variable, we'll talk about this type of relation as a *true function* or a *legitimate function*.

The relation shown in Fig. 1-1 can be modified to make it a function. We must eliminate two of the three pathways from *c* in the domain. It doesn't matter which two we take out. If we remove the pathways represented by (*c,v*) and (*c,z*), we get the function illustrated in Fig. 1-2.

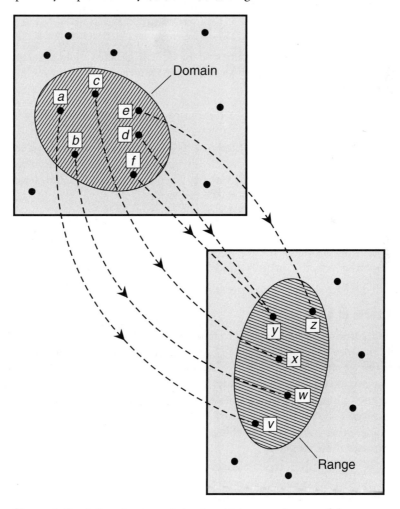

Figure 1-2   A function is a relation in which every element of the
domain is assigned to one, but never more than one,
element of the range.

Here's an informal way to think of the difference between a relation and a function. A relation correlates things in the domain with things in the range. A function operates on things in the domain to produce things in the range. A relation merely sits there. A function does something!

Three physical examples

Let's look at three situations that we might encounter in science. All three of the graphs in Fig. 1-3 represent functions. The changes in the value of the independent variable can be thought of as causative, or at least contributing, factors that affect the value of the dependent variable. We can describe these situations as follows:

- The outdoor air temperature is a function of the time of day.
- The number of daylight hours on June 21 is a function of latitude.
- The time required for a wet rag to dry is a function of the air temperature.

A mathematical example

Imagine a relation in which the independent variable is called x and the dependent variable is called y, and for which the domain and range are both the entire set of *real numbers* (also called the *reals*). Our relation is defined as

$$y = x + 1$$

This is a function between x and y, because there's never more than one value of y for any value of x. Mathematicians name functions by giving them letters of the alphabet such as f, g, and h. In this notation, the dependent variable is replaced by the function letter followed by the independent variable in parentheses. We can write

$$f(x) = x + 1$$

to represent the above equation, and then we can say, "f of x equals x plus 1." When we write a function this way, the quantity inside the parentheses (in this case x) is called the *argument* of the function.

The inverse of a relation

We can transpose the domain and the range of any relation to get its *inverse relation,* also called simply the *inverse* if the context is clear. The inverse of a relation is denoted by writing a superscript –1 after the name of the relation. It looks like an exponent, but it isn't meant to be.

The inverse of a relation is always another relation. But when we transpose the domain and range of a function, we don't always get another true function. If we do, then the function and its inverse reverse, or "undo," each other's work.

Suppose that x and y are variables, f and f^{-1} are functions that are inverses of each other, and we know these two facts:

$$f(x) = y$$

and

$$f^{-1}(y) = x$$

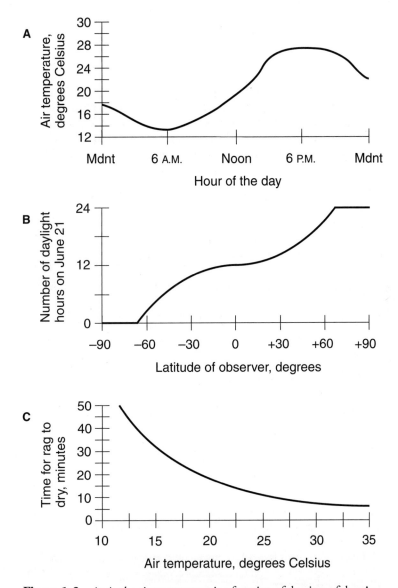

Figure 1-3 At A, the air temperature is a function of the time of day. At B, the number of daylight hours on June 21 is a function of the latitude (positive is north; negative is south). At C, the drying time for a wet rag is a function of the air temperature.

Then the following two facts are also true:

$$f^{-1}\,[f(x)] = x$$

and

$$f\,[f^{-1}\,(y)] = y$$

- -

Are you confused?

It's reasonable to wonder, "Can we tell whether or not a relation is a function by looking at its graph?" The answer is yes. Consider a graph in which the independent variable is represented by the horizontal axis, and the dependent variable is represented by the vertical axis. Imagine a straight, vertical line extending infinitely upward and downward. We move this vertical line to the left and right, so the point where it intersects the independent-variable axis sweeps through every possible argument of the relation. A graph represents a function "if and only if" that graph never crosses a movable vertical line at more than one point. Let's call this method of graph-checking the *vertical-line test*.

Here's a note!

In mathematics, the expression "if and only if" means that logical implication works in both directions. In the above example, we are really saying two things:

- If a graph represents a function, then the graph never crosses a movable vertical line at more than one point.
- If a graph never crosses a movable vertical line at more than one point, then the graph represents a function.

The expression "if and only if" is abbreviated in text as "iff." In logic, it's symbolized by a double-shafted, double-headed arrow pointing to the left and right ($\Leftrightarrow$).

Here's a challenge!

Imagine that the independent and dependent variables of the functions shown in Fig. 1-3 are reversed. This gives us some weird assertions.

- The time of day is a function of the outdoor air temperature.
- Latitude is a function of the number of daylight hours on June 21.
- The air temperature is a function of the time it takes for a wet rag to dry.

Only one of these statements translates into a mathematical function. Which one?

Solution

You can test the graph of a relation to see if its inverse is a function by doing a *horizontal-line test*. It works like the vertical-line test, but the line is parallel to the independent-variable axis, and it moves up and down instead of to the left and right. The inverse of a relation represents a function if and only if the graph of the original relation never intersects a movable horizontal line at more than one point.

When you test the graphs shown in Figs. 1-3A and B, you'll see that they fail the horizontal-line test. That means that the inverses aren't functions. When you transpose the independent and dependent variables in Fig. 1-3C, you get another function, because the graph passes the horizontal-line test.

- -

Linear Functions

When the argument changes in a *linear function,* the value of the dependent variable changes in constant proportion. That proportion can be positive or negative. It can even be zero, in which case we have a *constant function.*

Slope and intercept

In conventional coordinates, linear functions always produce straight-line graphs. Conversely, any straight line represents a linear function, as long as that line isn't parallel to the dependent-variable axis.

The *slope,* also called the *gradient,* of a straight line in *rectangular coordinates* (where the axes are perpendicular to each other and the divisions on each axis are of uniform size) is an expression of the steepness with which the line goes upward or downward as we move to the right. A horizontal line, representing a constant function, has a slope of zero. A line that ramps upward as we move to the right has positive slope. A line that ramps downward as we move to the right has negative slope. Figure 1-4 shows a line with positive slope and another line with negative slope.

To calculate the slope of a line, we must know the coordinates of two points on that line. If we call the independent variable x and the dependent variable y, then the slope of a line, passing through two points, is equal to the difference in the y-values divided by the difference in the x-values. We abbreviate "the difference in" by writing the uppercase Greek letter *delta* (Δ). Let's use a to symbolize the slope. Then

$$a = \Delta y / \Delta x$$

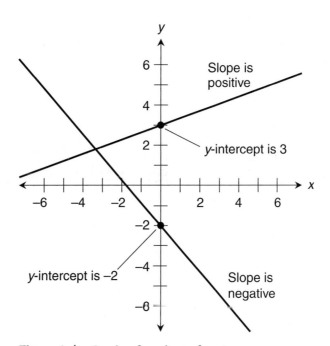

Figure 1-4 Graphs of two linear functions.

We read this as "delta *y* over delta *x*." Sometimes the slope of a straight line is called *rise over run*. This makes sense as long as the independent variable is on the horizontal axis, the dependent variable is on the vertical axis, and we move to the right.

An *intercept* is a point where a graph crosses an axis. We can plug 0 into a linear equation for one of the variables, and solve for the other variable to get its intercept. In a linear function, the term *y-intercept* refers to the value of the dependent variable *y* at the point where the line crosses the *y* axis. In Fig. 1-4, the line with positive slope has a *y*-intercept of 3, and the line with negative slope has a *y*-intercept of –2.

Standard form for a linear function

If we call the dependent variable *x*, then the *standard form for a linear function* is

$$f(x) = ax + b$$

where *a* and *b* are real-number constants, and *f* is the name of the function. As things work out, *a* is the slope of the function's straight-line graph. If we call the dependent variable *y*, then *b* is the *y*-intercept. We can substitute *y* in the equation for $f(x)$, writing

$$y = ax + b$$

Either of these two forms is okay, as long as we keep track of which variable is independent and which one is dependent!

- -

Are you confused?

If the graph of a linear relation is a vertical line, then the slope is undefined, and the relation is not a function. The graph of a linear function can never be parallel to the dependent-variable axis (or perpendicular to the independent-variable axis). In that case, the graph fails the vertical-line test.

Here's a challenge!

Rewrite the following equation as a linear function of *x*, and graph it on that basis:

$$12x + 6y = 18$$

Solution

We must rearrange this equation to get *y* all by itself on the left side of the equals sign, and an expression containing only *x* and one or more constants on the right side. Subtracting 12*x* from both sides gives us:

$$6y = -12x + 18$$

Dividing each side by 6 puts it into the standard form for a linear function:

$$y = (-12x)/6 + 18/6 = -2x + 3$$

If we name the function f, then we can express the function as

$$f(x) = -2x + 3$$

In the graph of this function, the y-intercept is 3. We plot the y-intercept on the y axis at the mark for 3 units, as shown in Fig. 1-5. That gives us the point (0,3). To find the line, we must know the coordinates of one other point. Let's find the x-intercept! To do that, we can plug in 0 for y to get

$$0 = -2x + 3$$

Adding $2x$ to each side and then dividing through by 2 tells us that $x = 3/2$. Therefore, the point (3/2,0) lies on the line. Now that we know (0,3) and (3/2,0) are both on the line, we can draw the line through them.

Here's a twist!

When we move from (0,3) to (3/2,0) in Fig. 1-5, we travel in the negative y direction by 3 units, so $\Delta y = -3$. We also move in the positive x direction by 3/2 units, so $\Delta x = 3/2$. Therefore

$$\Delta y / \Delta x = -3/(3/2) = -2$$

reflecting the fact that the slope of the line is -2. We'll always get this same value for the slope, no matter which two points on the line we choose. Uniformity of slope is characteristic of all linear functions. But there are functions for which it isn't so simple.

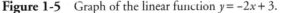

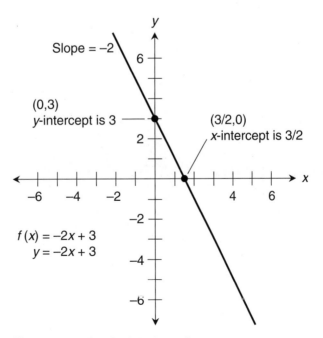

Figure 1-5   Graph of the linear function $y = -2x + 3$.

Nonlinear Functions

When the value of the argument changes in a *nonlinear function,* the value of the dependent variable also changes, but not always in the same proportion. The slope can't be defined for the whole function, although the notion of slope can usually exist at individual points. In rectangular coordinates, the graph of a nonlinear function is always something other than a straight line.

Square the input

Let's look at a simple nonlinear relation. The domain is the entire set of reals, and the range is the set of nonnegative reals. The equation is

$$y = x^2$$

If we call the relation g, we can write

$$g(x) = x^2$$

For every value of x in the domain of g, there is exactly one value of y in the range. Therefore, g is a function. But, as we can see by looking at the graph of g shown in Fig. 1-6, the reverse is not true. For every nonzero value of y in the range of g, there are two values of x in the domain. These two x-values are always negatives of each other. For example, if $y = 49$, then $x = 7$ or $x = -7$. This means that the inverse of g is not a function.

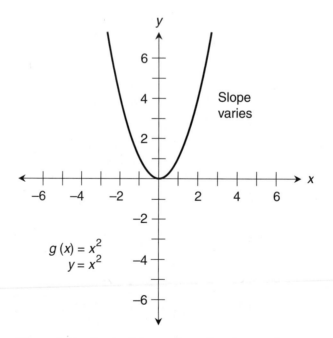

Figure 1-6   Graph of the nonlinear function $y = x^2$.

Cube the input

Here's another nonlinear relation. The domain and range both span the entire set of reals. The equation is

$$y = x^3$$

This is a function. If we call it *h*, then:

$$h(x) = x^3$$

For every value of *x* in the domain of *h*, there is exactly one value of *y* in the range. The reverse is also true. For every value of *y* in the range of *h*, there is exactly one *x* in the domain. This means that the inverse of *h* is also a function. We can see this by looking at the graph of *h* (Fig. 1-7). When a function is one-to-one and its inverse is also one-to-one, then the function is called a *bijection*.

Are you confused?

Have you noticed that in Fig. 1-7, the *y* axis is graduated differently than the *x* axis? There's a reason for this. We want the graph to fit reasonably well on the page. It's okay for the axes in a rectangular coordinate system to have increments of different sizes, as long as each axis maintains a constant increment size all along its length.

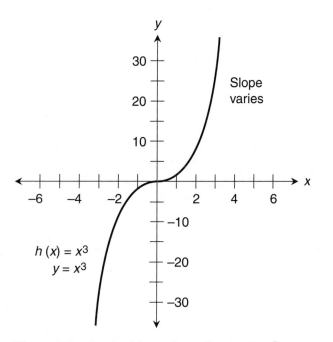

Figure 1-7   Graph of the nonlinear function $y = x^3$.

Here's a challenge!

Look again at the functions g and h described above, and graphed in Figs. 1-6 and 1-7. The inverse of h is a function, but the inverse of g is not. Mathematically, demonstrate the reasons why.

Solution

Here's the function g again. Remember that the domain is the entire set of reals, and the range is the set of nonnegative reals:

$$y = g(x) = x^2$$

If you take the equation $y = x^2$ and transpose the positions of the independent and dependent variables, you get

$$x = y^2$$

This is the same as

$$y = \pm(x^{1/2})$$

The plus-or-minus symbol indicates that for every value of the independent variable x you plug in you'll get two values of y, one positive and the other negative. You can also write

$$g^{-1}(x) = \pm(x^{1/2})$$

The function g is two-to-one (except when $y = 0$), and that's okay. But the inverse relation is one-to-two (except when $y = 0$). So, while g^{-1} is a legitimate relation, it is not a function.

The function h has an inverse that is also a function. Remember from your algebra and set theory courses that the inverse of any bijection is also a bijection. You have

$$h(x) = x^3$$

and

$$h^{-1}(x) = x^{1/3}$$

If a function is one-to-one over a certain domain and range, then you can transpose the values of the independent and dependent variables while leaving their names the same, and you can also transpose the domain and range. That gives you the inverse, and it is a function. However, if a function is many-to-one, then its inverse is one-to-many, so that inverse is not a function.

- -

"Broken" Functions

Relations and functions often show "gaps," "jumps," or "blow-ups" in their graphs. This can happen in countless different ways. Let's look at some examples.

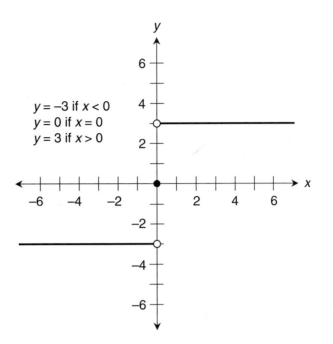

$y = -3$ if $x < 0$
$y = 0$ if $x = 0$
$y = 3$ if $x > 0$

Figure 1-8   Graph of the "broken" function $y = -3$
if $x < 0$, $y = 0$ if $x = 0$, $y = 3$ if $x > 0$.

A three-part function

Figure 1-8 is a graph of a function where the value is −3 if the argument is negative, 0 if the argument is 0, and 3 if the argument is positive. Let's call the function f. Then we can write

$$f(x) = -3 \text{ if } x < 0$$
$$= 0 \text{ if } x = 0$$
$$= 3 \text{ if } x > 0$$

Even though this function takes two jumps, there are no gaps in the domain. The function is defined for every real number x.

The reciprocal function

Figure 1-9 is a graph of the *reciprocal function*. We divide 1 by the argument. If we call this function g, then we can write

$$g(x) = 1/x$$

This graph has a two-part blow-up at $x = 0$. As we approach 0 from the left, the graph blows up negatively. As we approach 0 from the right, it blows up positively. The function is defined for all values of x except 0.

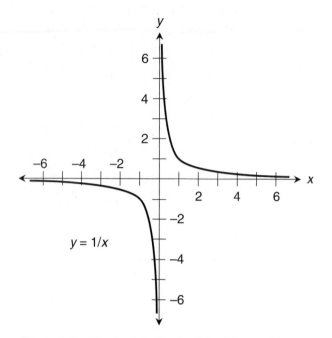

Figure 1-9   Graph of the "broken" function $y = 1/x$.

The tangent function

Figure 1-10 is a graph of the *tangent function* from trigonometry. If we call this function h, then we can write

$$h(x) = \tan x$$

This graph blows up at infinitely many values of the independent variable! It is defined for all values of x except odd-integer multiples of $\pi/2$.

- -

Are you confused?

Does it seem strange that a function can jump abruptly from one value to another, skip over individual points, or even blow up to "infinity" or "negative infinity"? You might find this idea difficult to comprehend if you're the literal-minded sort. But as long as a relation passes the test for a function according to the rules we've defined, it's a legitimate function.

Here's a challenge!

Draw a graph of the relation obtained by rounding off an argument to the nearest integer smaller than or equal to itself. Call the independent variable x and the dependent variable y. Here are some examples to give you the idea:

If $x = 3$, then $y = 3$

If $x = -6$, then $y = -6$

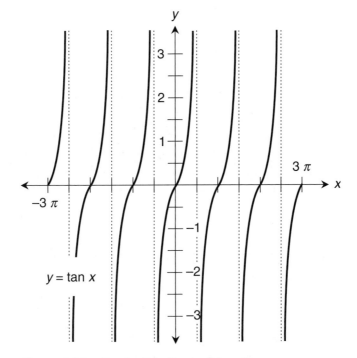

Figure 1-10 Graph of the "broken" function $y = \tan x$.

If $x = \pi$, then $y = 3$

If $x = -\pi$, then $y = -4$

If $x = 4.999$, then $y = 4$

If $x = -5.001$, then $y = -6$

Is the relation represented by this graph a function? How can we tell?

Solution

This graph is shown in Fig. 1-11. It passes the vertical-line test, so it represents a function. We can also tell that this relation is a function by the way it's defined. No matter what the argument, the relation maps it to one, but only one, integer. This type of function is called a *step function* because of the way its graph looks.

- -

Practice Exercises

This is an open-book quiz. You may (and should) refer to the text as you solve these problems. Don't hurry! You'll find worked-out answers in App. A. The solutions in the appendix may not

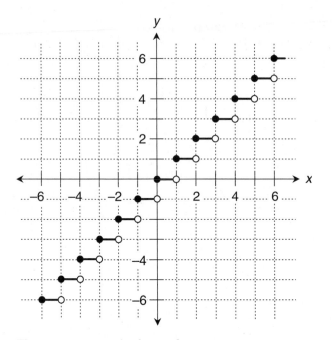

Figure 1-11   Graph of a step function. Every argument
is rounded off to the nearest integer smaller
than or equal to itself.

represent the only way a problem can be figured out. If you think you can solve a particular
problem in a quicker or better way than you see there, by all means try it!

1. Imagine a mapping from the set of all integers onto the set of all nonnegative integers in
    which the set of ordered pairs is

    $$\{(0,0),\ (1,1),\ (-1,2),\ (2,3),\ (-2,4),\ (3,5),\ (-3,6),\ (4,7),\ (-4,8),\ \ldots\}$$

    Is this relation a function? If so, why? If not, why not? Is its inverse a function? If so,
    why? If not, why not?

2. Consider a mapping from the set of all integers onto the set of all nonnegative integers
    in which the set of ordered pairs is

    $$\{(0,0),\ (1,1),\ (-1,1),\ (2,2),\ (-2,2),\ (3,3),\ (-3,3),\ (4,4),\ (-4,4),\ \ldots\}$$

    Is this relation a function? If so, why? If not, why not? Is its inverse a function? If so,
    why? If not, why not?

3. Consider the following linear function:

    $$f(x) = 4x - 5$$

    What is the inverse of this? Is it a function?

4. Consider the following linear function:

$$g(x) = 7$$

   What is the inverse of this? Is it a function?

5. In the Cartesian coordinate xy plane, the equation of a circle with radius 1, centered at the origin (0,0), is

$$x^2 + y^2 = 1$$

   This particular circle is called the *unit circle*. Is its equation a function of x? If so, why? If not, why not?

6. Is the equation of the unit circle, as expressed in Prob. 5, a function of y? If so, why? If not, why not?

7. Consider the nonlinear function we graphed in Fig. 1-6:

$$g(x) = x^2$$

   As we saw, the inverse relation, g^{-1}, is not a function. But it can be modified so it becomes a function of x by restricting its range to the set of positive real numbers. Show with the help of a graph why this is true. Does g^{-1} remain a function if we allow the range to include 0?

8. We can modify the relation g^{-1} from the previous problem, making it into a function of x, by restricting its range to the set of negative real numbers. Show with the help of a graph why this is true. Does g^{-1} remain a function if we allow the range to include 0?

9. Look again at Figs. 1-8 through 1-10. All three of these graphs pass the vertical-line test for a function. This is true even though the relation shown in Fig. 1-9 is not defined when $x = 0$, and the relation shown in Fig. 1-10 is not defined when x is any odd-integer multiple of $\pi/2$. Now suppose that we don't like the gaps in the domains in Figs. 1-9 and 1-10. We want to modify these functions to make their domains cover the entire set of real numbers. We decide to do this by setting $y = 0$ whenever we encounter a value of x for which either of these relations is not defined. Are the relations still functions after we do this to them?

10. Consider again the functions graphed in Figs. 1-8 through 1-10. The inverse of one of these functions is another function. That function also happens to be its own inverse. Which one of the three is this?

Limits and Continuity

While Isaac Newton and Gottfried Wilhelm Leibniz independently developed the differential calculus in the seventeenth century, they both wanted to figure out how to calculate the *instantaneous rate of change* of a nonlinear function at a point in space or time, and then describe the rate of change in general, as a function itself. In the next few chapters, we'll do these things. But first, let's be sure we have all the mathematical tools we need!

Concept of the Limit

As the argument (the independent variable or input) of a function approaches a particular value, the dependent variable approaches some other value called the *limit*. The important word here is *approaches*. When finding a limit, we're interested in what happens to the function as the argument gets closer and closer to a certain value without actually reaching it.

Limit of an infinite sequence

Let's look at an infinite sequence S that starts with 1 and then keeps getting smaller:

$$S = 1, 1/2, 1/3, 1/4, 1/5, \ldots$$

As we move along in S from term to term, we get closer and closer to 0, but we never get all the way there. If we choose some small positive number r, no matter how tiny, we can always find a number in S (if we're willing to go out far enough) smaller than r but larger than 0. Because of this fact, we can say, "If n is a positive integer, then the limit of S, as n gets endlessly larger, is 0." We write this symbolically as

$$\lim_{n \to \infty} S = 0$$

The expression "$n \to \infty$" translates to "as n approaches infinity." When talking about limits, mathematicians sometimes say "approaches infinity" to mean "gets endlessly larger" or "gets arbitrarily large." This expression is a little bit obscure, because we can debate whether large numbers are really closer to "infinity" than small ones. But we'll often hear that expression used, nevertheless.

When talking about this sequence *S*, we can also say, "The limit of $1/n$, as *n* approaches infinity, is equal to 0," and write

$$\operatorname*{Lim}_{n\to\infty} 1/n = 0$$

Limit of a function

Now let's think about what takes place if we don't restrict ourselves to positive-integer arguments. Let's consider the function

$$g(x) = 1/x$$

and allow *x* to be any positive real number. As *x* gets larger, $g(x)$ gets smaller, approaching 0 but never getting there. We can say, "The limit of $g(x)$, as *x* approaches infinity, is 0," and write

$$\operatorname*{Lim}_{n\to\infty} g(x) = 0$$

This is the same as the situation with the infinite sequence of positive integers, except that the function approaches 0 smoothly, rather than in jumps.

- -

Are you a nitpicker?

Let's state the above expression differently. For every positive real number *r*, there exists a positive real number *s* such that

$$0 < g(s) < r$$

Also, if *t* is a real number larger than *s*, then

$$0 < g(t) < g(s) < r$$

Think about this language for awhile. It's a formal way of saying that as we input larger and larger positive real numbers to the function *g*, we get smaller and smaller positive reals that "close in" on 0. This statement also tells us that even if we input huge numbers such as 1,000,000, 1,000,000,000, or 1,000,000,000,000 to the function *g*, we'll never get 0 when we calculate $g(x)$. We can't input "infinity" in an attempt to get 0 out of *g*, either. "Infinity" isn't a real number!

Are you confused?

If the notion of "closing in on 0" confuses you, look at the graph of the function *g* for large values of *x*. As you move out along the *x* axis in the positive direction, the curve gets closer and closer to the *x* axis, where

$g(x) = 0$. No matter how close the curve gets to the axis, you can always get it to come closer by moving out farther in the positive x direction, as shown in Fig. 2-1. But the curve never reaches the x axis.

- -

Sum rule for two limits

Consider two functions $f(x)$ and $g(x)$ with different limits. We can add the functions and take the limit of their sum, and we'll get the same thing as we do if we take the limits of the functions separately and then add them. Let's call this the *sum rule for two limits* and write it symbolically as

$$\lim_{x \to k} [f(x) + g(x)] = \lim_{x \to k} f(x) + \lim_{x \to k} g(x)$$

where k, the value that x approaches, can be a real-number constant, another variable, or "infinity." This rule isn't restricted to functions. It holds for any two expressions with definable limits. It also works for the difference between two expressions. We can write

$$\lim_{x \to k} [f(x) - g(x)] = \lim_{x \to k} f(x) - \lim_{x \to k} g(x)$$

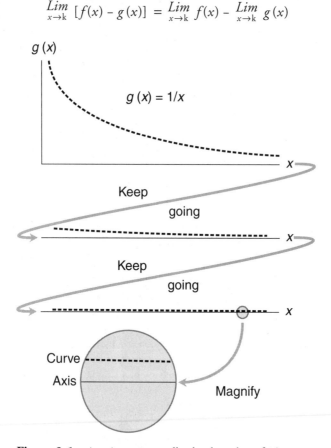

Figure 2-1   As x increases endlessly, the value of $1/x$ approaches 0, but it never actually becomes equal to 0.

In verbal terms, we can say these two things:

* The limit of the sum of two expressions is equal to the sum of the limits of the expressions.
* The limit of the difference between two expressions is equal to the difference between the limits of the expressions (in the same order).

Multiplication-by-constant rule for a limit

Now consider a function with a defined limit. We can multiply that limit by a constant, and we'll get the same thing as we do if we multiply the function by the constant and then take the limit. Let's call this the *multiplication-by-constant rule for a limit*. We write it symbolically as

$$c \lim_{x \to k} f(x) = \lim_{x \to k} c\,[f(x)]$$

where c is a real-number constant, and k is, as before, a real-number constant, another variable, or "infinity." As with the sum rule, this holds for any expressions with definable limits, not only for functions. In verbal terms, we can say this:

* A constant times the limit of an expression is equal to the limit of the expression times the constant.

- -

Here's a challenge!

Determine the limit, as x approaches 0, of a function $h(x)$ that raises x to the fourth power and then takes the reciprocal:

$$h(x) = 1/x^4$$

Symbolically, this is written as

$$\lim_{x \to 0} 1/x^4$$

Solution

As x starts out either positive or negative and approaches 0, the value of $1/x^4$ increases endlessly. No matter how large a number you choose for $h(x)$, you can always find something larger by inputting some x whose absolute value is small enough. In this situation, the limit does not exist. You can also say that it's not defined.

Once in awhile, someone will write the "infinity" symbol, perhaps with a plus sign or a minus sign in front of it, to indicate that a limit blows up (increases without bound) positively or negatively. For example, the solution to this "challenge" could be written as

$$\lim_{x \to 0} 1/x^4 = \infty$$

or as

$$\lim_{x \to 0} 1/x^4 = +\infty$$

- -

Continuity at a Point

When we scrutinize a function or its graph, we might want to talk about its *continuity at a point*. But first, we must know the value of the function at that point, as well as the limit as we approach it from either direction.

Right-hand limit at a point

Consider the following function, which takes the reciprocal of the input value:

$$g(x) = 1/x$$

We can't define the limit of $g(x)$ as x approaches 0 from the positive direction, because the function blows up as x gets smaller positively, approaching 0. To specify that we approach 0 from the positive direction, we can refine the limit notation by placing a plus sign after the 0, like this:

$$\lim_{x \to 0+} g(x)$$

This expression reads, "The limit of $g(x)$ as x approaches 0 from the positive direction." We can also say, "The limit of $g(x)$ as x approaches 0 from the right." (In most graphs where x is on the horizontal axis, the value of x becomes more positive as we move toward the right.) This sort of limit is called a *right-hand limit*.

Right-hand continuity at a point

What about some other point, such as where $x = 1$? As we approach the point where $x = 1$ from the positive direction, g starts out at positive values smaller than 1 and increases, approaching 1. We can see this with the help of Fig. 2-2, which is a graph of g drawn in the vicinity of the point where $x = 1$. (Each division on the axes represents 1/2 unit.) This graph tells us that

$$\lim_{x \to 1+} g(x) = 1$$

We can calculate the actual value of g for $x = 1$, getting

$$g(1) = 1/1 = 1$$

Now suppose that:

- We can define the right-hand limit of a function at a certain point
- We can define the actual value of the function at that point
- The limit and the actual value are the same

When all three of these things are true, we say that the function is *right-hand continuous* at that point. Some texts will tell us that g is *continuous on the right* at the point. In this example, g is right-hand continuous at $x = 1$.

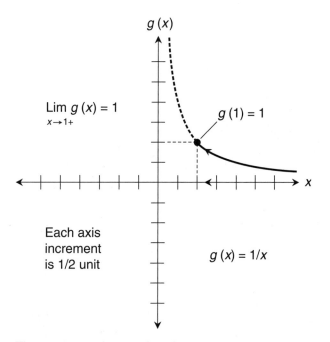

$g(x)$

Lim $g(x) = 1$
$x \to 1+$

$g(1) = 1$

$g(x) = 1/x$

Each axis
increment
is 1/2 unit

x

Figure 2-2   The limit of the function $g(x) = 1/x$ as x
approaches 1 from the right is equal to the
value of the function when $x = 1$. In this
graph, each axis division represents 1/2 unit.

Left-hand limit at a point

Let's expand the domain of g to the entire set of reals except 0, for which g is not defined because 1/0 is not defined. Suppose that we start out with negative real values of x and approach 0 from the left. As we do this, g decreases endlessly, as we can see by looking at Fig. 2-3. (Here, each division on the axes represents 1 unit.) Another way of saying this is that g increases negatively without limit, or that it blows up negatively. Therefore,

$$\underset{x \to 0-}{Lim}\ g(x)$$

is not defined. We read the above symbolic expression as, "The limit of $g(x)$ as x approaches 0 from the negative direction." We can also say, "The limit of $g(x)$ as x approaches 0 from the left." This sort of limit is called a *left-hand limit*.

Left-hand continuity at a point

Now let's look again at the point in the graph g where $x = 1$. Suppose we approach this point from the left, that is, from the negative direction. The value of g starts out at positive values larger than 1. As x increases and approaches 1, g decreases, getting closer and closer to 1. Figure 2-4 illustrates what happens here. (Each division on the axes represents 1/2 unit.) We have

$$\underset{x \to 1-}{Lim}\ g(x) = 1$$

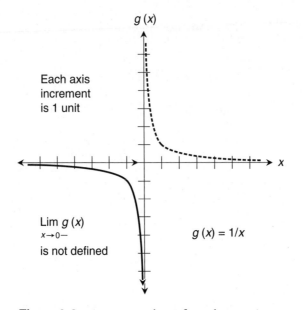

Figure 2-3 As *x* approaches 0 from the negative direction, the value of 1/*x* increases negatively without limit. In this graph, each axis division represents 1 unit.

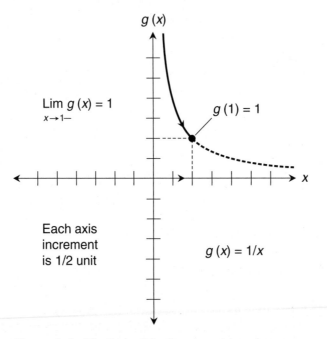

Figure 2-4 The limit of the function *g* (*x*) = 1/*x* as *x* approaches 1 from the left is equal to the value of the function when *x* = 1. In this graph, each axis division represents 1/2 unit.

We already know that $g(1) = 1$. Just as we did when approaching from the right, we can say that g is *left-hand continuous* at the point where $x = 1$. Some texts will say that g is *continuous on the left* at that point.

- -

Are you confused?

It's easy to get mixed up by the meanings of "negative direction" and "positive direction," and how these relate to the notions of "left-hand" and "right-hand." These terms are based on the assumption that we're talking about the horizontal axis in a graph, and that this axis represents the independent variable. In most graphs of this type, the value of the independent variable gets more negative as we move to the left, and more positive as we move to the right. This is true no matter where on the axis we start.

As we travel along the horizontal axis, we might be in positive territory the whole time; we might be in negative territory the whole time; we might cross over from the negative side to the positive side or vice-versa. Whenever we come toward a point from the left, we approach from the negative direction, even if that point corresponds to something like $x = 567$. Whenever we come toward a point from the right, we approach from the positive direction, even if the point is at $x = -53{,}535$. The location of the point doesn't matter. The important thing is the *direction from which we approach*.

- -

"Total" continuity at a point

Now that we've defined right-hand and left-hand continuity at a point, we can define *continuity at a point* in a "total" sense. When a function is both left-hand continuous and right-hand continuous at a point, we say that the function is continuous at that point. Conversely, whenever we say that a function is continuous at a point, we mean that it's continuous as we approach and then reach the point from both the left-hand side and the right-hand side.

- -

Here's a challenge!

Suppose we modify the function $g(x) = 1/x$ by changing the value for $x = 1$. Let's call this new function $g*$. The domain remains the same: all real numbers except 0. The only difference between $g*$ and g is that $g*(1)$ is not equal to 1, but instead is equal to 4, as shown in Fig. 2-5. (In this graph, each axis division is 1 unit.) We have seen that the original function g is both right-hand continuous and left-hand continuous at $(1,1)$, so we know that g is "totally" continuous at $(1,1)$. But what about $g*$? Is this modified function continuous at the point where $x = 1$? Is it left-hand continuous there? Is it right-hand continuous there?

Solution

The answer to each of these three questions is "No." The function $g*$ is not continuous at the point where $x = 1$. It's not right-hand continuous or left-hand continuous there. The limit of $g*$ as we approach $x = 1$ from either direction is equal to 1. That is,

$$\lim_{x \to 1+} g*(x) = 1$$

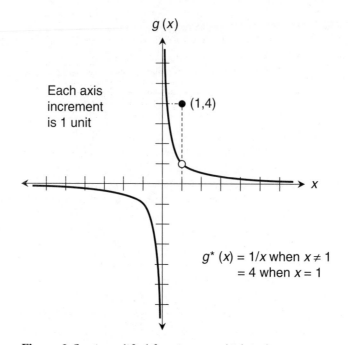

Each axis
increment
is 1 unit

(1,4)

$g^*(x) = 1/x$ when $x \neq 1$
$= 4$ when $x = 1$

Figure 2-5   A modified function g^*, which is the same as
g except that the point (1,1) has been moved
to (1,4).

and

$$\underset{x \to 1-}{Lim}\ g^*(x) = 1$$

As we move toward the point where $x = 1$ from either direction along the curve, it seems as if we should
end up at the point (1,1) when the value of x reaches 1. But when we look at the actual value of g^* at $x = 1$,
we find that it's not equal to 1. There is a *discontinuity* in g^* at the point where $x = 1$. We can also say that
g^* is *not continuous* at the point where $x = 1$, or that g^* is *discontinuous* at the point where $x = 1$.

Here's another challenge!

Look again at the step function we saw in Fig. 1-11 near the end of Chap. 1. Is this function right-hand
continuous at point where $x = 3$? Is it left-hand continuous at that point?

Solution

A portion of that function is reproduced in Fig. 2-6, showing only the values in the vicinity of our point
of interest. Let's call the function $s(x)$. We can see from this drawing that

$$\underset{x \to 3+}{Lim}\ s(x) = 3$$

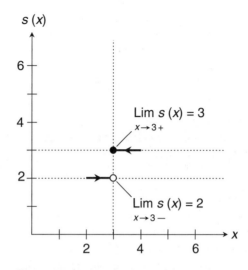

Figure 2-6 Limits of a step function,
$s(x)$, at the point where $x = 3$.

and

$$\underset{x \to 3-}{Lim}\; s(x) = 2$$

The value of the function at the point where $x = 3$ is $s(3) = 3$. The right-hand limit and the actual value are the same, so s is right-hand continuous at the point where $x = 3$. But s is not left-hand continuous at the point where $x = 3$, because the left-hand limit and the actual value are different. The same situation occurs at every integer value of x in the step function illustrated in Fig. 1-11. Because s is not continuous from both the right and the left at any point where x is an integer, s is not continuous at any such point. This function has an infinite number of discontinuities!

- -

Continuity of a Function

A real-number function in one variable is a *continuous function* if and only if it is continuous at every point in its domain. Imagine a line or curve that's smooth everywhere, with no gaps, no jumps, and no blow-ups in its domain. That's what a continuous function looks like when graphed.

Linear functions

All *linear functions* are continuous. A real-number linear function L always has an equation of this form:

$$L(x) = ax + b$$

where x is the independent variable, a is a nonzero real number, and b can be any real number. In rectangular coordinates, the graph of a linear function is a straight line that extends forever in two opposite directions. It never has a discontinuity.

Figure 2-7 shows four generic graphs of linear functions. The independent variable is on the horizontal axis, and the dependent variable is on the vertical axis.

Quadratic functions

All single-variable, real-number *quadratic functions* are continuous. The general form of a quadratic function Q is

$$Q(x) = ax^2 + bx + c$$

where x is the independent variable, a is a nonzero real, and b and c can be any reals. In rectangular coordinates, the graph of a quadratic function is always a *parabola* that opens either straight up or straight down. The domain includes all reals, but the range is restricted to either the set of all reals greater than or equal to a certain *absolute minimum,* or the set of all reals less than or equal to a certain *absolute maximum.* There are never any gaps, jumps, or blow-ups in the graph within the domain.

Figure 2-8 shows four generic graphs of quadratic functions. The independent variable is on the horizontal axis, and the dependent variable is on the vertical axis.

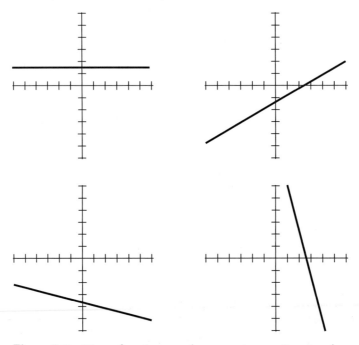

Figure 2-7   Linear functions are always continuous. Imagine the lines extending smoothly forever from both ends.

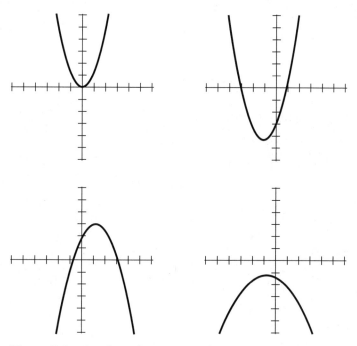

Figure 2-8   Quadratic functions are always continuous. Imagine the curves extending smoothly forever from both ends.

Cubic functions

All single-variable, real-number *cubic functions* are continuous. The general form of a cubic function C is

$$C(x) = ax^3 + bx^2 + cx + d$$

where x is the independent variable, a is a nonzero real, and b, c, and d can be any reals. In rectangular coordinates, the graph of a cubic function looks like a badly distorted letter "S" tipped on its side, perhaps flipped over backward, and then extended forever upward and downward.

Unlike a quadratic function, which has a limited range with an absolute maximum or an absolute minimum, the range of a cubic function always spans the entire set of reals, although the graph can have a *local maximum* and a *local minimum*. The contour of the graph depends on the signs and values of a, b, c, and d. There are no gaps, blow-ups, or jumps within the domain.

Figure 2-9 shows four generic graphs of cubic functions. The independent variable is on the horizontal axis, and the dependent variable is on the vertical axis.

Polynomial functions

All single-variable, real-number *polynomial functions* are continuous. We can write the general form of an *nth-degree polynomial function* (let's call it P_n) as follows, where n is an integer greater than 3, and x is never raised to a negative power:

$$P_n(x) = a_n x^n + a_{n-1} x^{n-1} + a_{n-2} x^{n-2} + \cdots + a_1 x + b$$

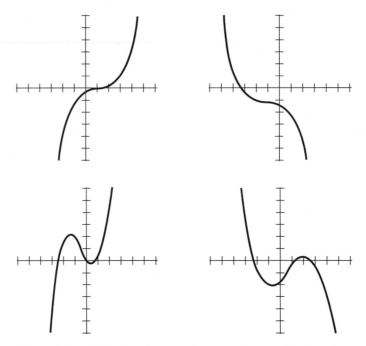

Figure 2-9 Cubic functions are always continuous. Imagine the
curves extending smoothly forever from both ends.

Here, x is the independent variable, a_1, a_2, a_3, . . . , and a_n are called the *coefficients,* and b is
called the *stand-alone constant.* The *leading coefficient* a_n can be any real except 0. All the other
coefficients, and the stand-alone constant, can be any real numbers.

The domain of an nth-degree polynomial function extends over the entire set of real
numbers. If n is even, the range is restricted to either the set of all reals greater than or equal
to a certain absolute minimum, or the set of all reals less than or equal to a certain absolute
maximum. If n is odd, the range spans the entire set of reals, but there may be one or more
local maxima and minima. The contour of the graph can be complicated, but there are never
any gaps, blow-ups, or jumps within the domain.

Other continuous functions

Plenty of other functions are continuous. You can probably think of a few right away, remembering your algebra, trigonometry, and precalculus courses.

Discontinuous functions

A real-number function in one variable is called a *discontinuous function* if and only if it is *not*
continuous at one or more points in its domain. Imagine a function whose graph is a line or
curve with at least one gap, blow-up, or jump. That's what a discontinuous function looks like
when graphed.

Sometimes a discontinuous function can be made continuous by restricting the domain.
We eliminate all portions of the domain that contain discontinuities. For example, the function
$g*$, described earlier in this chapter and graphed in Fig. 2-5, is not continuous over the set of
positive reals, because there's a discontinuity at the point where $x = 1$. But we can make $g*$

continuous if we restrict its domain so that $x > 1$. We can also make it continuous if we restrict the domain so that $0 < x < 1$, or so that $x < 0$. In fact, there are infinitely many ways we can restrict the domain and get a continuous function!

- -

Here's a challenge!

Look again at the function $g(x) = 1/x$ that we worked with earlier in this chapter. If we define the domain of g as the set of all positive reals except 0, is this function continuous? If we include $x = 0$ in the domain and give the function the value 0 there, calling the new function $g^{\#}$, is this function continuous?

Solution

As long as we don't allow x to equal 0, the function $g(x) = 1/x$ is both right-hand continuous and left-hand continuous at every point in the *restricted domain.* That means it's a continuous function, even though its graph takes a huge jump. The blow-up in the graph does not represent a true discontinuity in g, because we don't allow $x = 0$ in the domain.

Now suppose that we modify the function to allow $x = 0$ in the domain, calling the new function $g^{\#}$ and including $(0,0)$. Because neither the right-hand limit nor the left-hand limit is equal to 0 as x approaches 0, the function $g^{\#}$ is neither right-hand continuous nor left-hand continuous at the point where $x = 0$. Because of this single discontinuity, $g^{\#}$ is not a continuous function.

- -

Practice Exercises

This is an open-book quiz. You may (and should) refer to the text as you solve these problems. Don't hurry! You'll find worked-out answers in App. A. The solutions in the appendix may not represent the only way a problem can be figured out. If you think you can solve a particular problem in a quicker or better way than you see there, by all means try it!

1. Find the limit of the infinite sequence

$$1/10, \ 1/10^2, \ 1/10^3, \ 1/10^4, \ 1/10^5, \ldots$$

2. In a series, a *partial sum* is the sum of all the term up to, and including, a certain term. As we include more and more terms in a series, the partial sum usually changes. Find the limit of the partial sum of the infinite series

$$1/10 + 1/10^2 + 1/10^3 + 1/10^4 + 1/10^5 + \cdots$$

as the number of terms in the partial sum approaches infinity.

3. Let x be a positive real number. Does the following limit exist? If so, find it. If not, explain why not.

$$\lim_{x \to \infty} 1/x^2$$

4. Let x be a positive real number. Does the following limit exist? If so, find it. If not, explain why not.

$$\underset{x \to 0+}{Lim}\ 1/x^2$$

5. Consider the base-10 logarithm function (symbolized $\log_{10}$). Sketch a graph of the function $f(x) = \log_{10} x$ for values of x from 0.1 to 10, and for values of f from –1 to 1. Determine

$$\underset{x \to 3-}{Lim}\ \log_{10} x$$

6. Look again at $f(x) = \log_{10} x$. Find

$$\underset{x \to 3+}{Lim}\ \log_{10} x$$

7. Based on the answers to Probs. 5 and 6, can we say that $f(x) = \log_{10} x$ is continuous at the point where $x = 3$? If so, why? If not, why not?

8. Is the function $f(x) = \log_{10} x$ continuous over the set of positive reals? If not, where are the discontinuities? Is this function continuous over the set of nonnegative reals (that is, all the positive reals along with 0)? If not, where are the discontinuities?

9. Sketch a graph of the absolute-value function (symbolized by a vertical line on either side of the independent variable) for values of the domain from approximately –6 to 6. Is this function continuous over the set of all reals? If not, where are the discontinuities?

10. Sketch a graph of the trigonometric cosecant function for values of the domain between, and including, -3π radians and 3π radians. Is this function continuous if we restrict the domain to this closed interval? If not, where are the discontinuities? Remember that the cosecant (symbolized csc) of a quantity is equal to the reciprocal of the sine (symbolized sin). That is, for any x,

$$\csc x = 1\ /\ (\sin x)$$

3

What's a Derivative?

All single-variable linear functions have straight-line graphs. The slope of such a graph can be found easily using ordinary algebra. In this chapter, we'll learn a technique that allows us to find the slope of a graph whether its function is linear or not.

Vanishing Increments

When we want to find the slope of a curve at a point, we're looking for the *instantaneous rate of change* in a function for a specific value of the independent variable. The instantaneous rate of change is the slope of a *tangent line* at that point on the curve.

What is a tangent line?

In the rectangular coordinate plane, a tangent line intersects a curve at a point, and has the same slope as the curve at that point. Figure 3-1 shows two examples of straight lines tangent to curves at certain points (at A and B), and two examples of straight lines that are not tangent to the same curves at those same points (at C and D).

Slope between two points

Imagine a nonlinear function whose graph is a curve in the rectangular xy-plane, and whose equation is

$$y = f(x)$$

Suppose that we want to find the slope of a line tangent to the curve at a specific point (x_0, y_0). Let's call this "mystery slope" M. We can approximate M by choosing some point (x, y) that's near (x_0, y_0) and that is also on the curve, as shown in Fig. 3-2. We construct a line through the two points. The slope of that line is close to M. The difference in the y-values between our two points (x, y) and (x_0, y_0) is

$$\Delta y = y - y_0$$

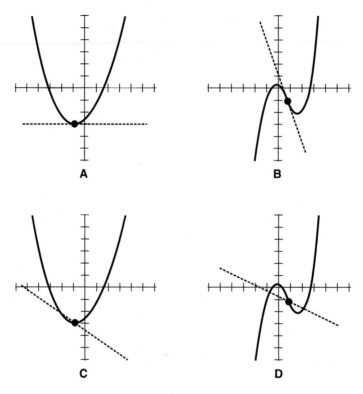

A   **B**

C   **D**

Figure 3-1   At A and B, the dashed lines are tangent to the curves at the points shown by the dots. At C and D, the lines are not tangent to the curves.

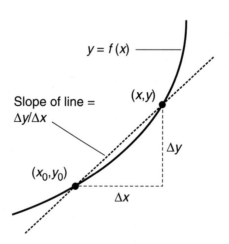

$y = f(x)$

Slope of line $= \Delta y / \Delta x$

(x,y)

Δy

(x_0, y_0)

Δx

Figure 3-2   The slope of a curve at a point (x_0, y_0) can be approximated by constructing a line through that point and a nearby point (x,y), and then finding the slope of the line, $\Delta y / \Delta x$.

and the difference in the *x*-values is

$$\Delta x = x - x_0$$

The slope of the line through the two points is

$$\Delta y / \Delta x = (y - y_0) / (x - x_0)$$

This slope is approximately the same as *M*. The accuracy of our approximation depends on how close together the two points are. We'll get the best results if we choose (x,y) as close to (x_0,y_0) as possible. But we can't make the points identical in an effort to find *M* exactly. If we do that, we get $\Delta y = 0$ and $\Delta x = 0$. Then when we try to calculate the slope, we get 0/0. That's no help!

Converging points

In Chap. 2, we reviewed the theory of limits, which Newton and Leibniz used centuries ago to figure out instantaneous rates of change in the values of functions. Isaac Newton called these rates of change *fluxions*. We call them *derivatives*.

In the situation of Fig. 3-2, we can't put (x,y) directly on top of the point (x_0,y_0); if we do, our problem reduces to nonsense. But we can move (x,y) toward (x_0,y_0) until the two points are *arbitrarily* close together. We get the points as close to each other as we can imagine—and then a little closer! We minimize Δx until it's "too small to see."

As Δx shrinks to almost nothing, Δy does the same. Imagine the point (x,y) getting to within a hair's width, then a bacterium's length, then a proton's diameter of (x_0,y_0). As this occurs, the slope of the line through the two points gets arbitrarily close to *M*, as shown in Fig. 3-3. The actual value of *M* is therefore equal to the limit of $\Delta y / \Delta x$ as Δx approaches 0:

$$M = \lim_{\Delta x \to 0} \Delta y / \Delta x$$

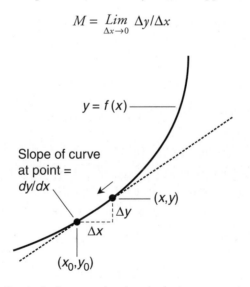

Figure 3-3 As (x,y) approaches (x_0,y_0), the increments Δy and Δx become smaller, and the line approaches the slope of the curve at the point (x_0,y_0). We call this slope *dy* / *dx*.

We now have a way to find the slope of a curve or the instantaneous rate of change in any function, as long as we can find the limit of $\Delta y/\Delta x$ as Δx approaches 0. This limit is the derivative of the function.

Mathematicians call the arbitrarily tiny quantities Δy and Δx *differentials*, and sometimes symbolize them *dy* and *dx*, respectively. They are called *infinitesimals* in some texts. We can express the exact "mystery slope" as

$$M = dy/dx$$

- -

Are you confused?

You might ask, "What's the difference between y and $f(x)$ here? Are they different names for the same thing?" If y is the dependent variable in a function $f(x)$, then they are indeed the same. As you work with calculus in the future, you'll likely see the derivative of a function $y = f(x)$ written in many different ways. Here are some of the variants you should watch for:

$$dy/dx$$
$$y'$$
$$df(x)/dx$$
$$d/dx\, f(x)$$
$$df/dx$$
$$f'(x)$$
$$f'$$

Here's a challenge!

Find the slope M of a line tangent to the graph of the function

$$f(x) = x^2$$

at the point $(x_0, y_0) = (1,1)$ in rectangular xy-coordinates, where $y = f(x)$. What does this slope represent?

Solution

Let's set up a two-point scheme, choosing a movable point (x,y) near $(1,1)$ as shown in Fig. 3-4. The function tells us that $y = x^2$ for all possible values of x and y. The x-value of our movable point is

$$x = 1 + \Delta x$$

The y-value of our movable point is

$$y = 1 + \Delta y = (1 + \Delta x)^2 = 1 + 2\Delta x + (\Delta x)^2$$

The slope of the line passing through our two points is

$$\Delta y/\Delta x = (y - 1)/(x - 1)$$

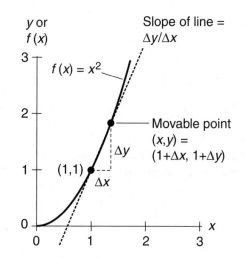

Figure 3-4 Finding the slope of a line tangent to the graph of $f(x) = x^2$ at the point (1,1).

Substituting the quantity $[1 + 2\Delta x + (\Delta x)^2]$ in place of y in the above equation, and also substituting the quantity $(1 + \Delta x)$ in place of x, we get

$$\Delta y / \Delta x = [1 + 2\Delta x + (\Delta x)^2 - 1] / (1 + \Delta x - 1)$$

which can be simplified to

$$\Delta y / \Delta x = [2\Delta x + (\Delta x)^2] / \Delta x$$

and further to

$$\Delta y / \Delta x = 2 + \Delta x$$

To find the slope of the line tangent to the curve at the point (1,1), we must find

$$M = \lim_{\Delta x \to 0} \Delta y / \Delta x$$

In this case, it turns out to be

$$M = \lim_{\Delta x \to 0} 2 + \Delta x$$

It's easy to see that the quantity $(2 + \Delta x)$ approaches 2 as Δx approaches 0. That limit, 2, is the slope of the line tangent to the curve at the point (1,1). It's also the derivative of the function $f(x) = x^2$ at the point where $x = 1$.

Basic Linear Functions

We have just recreated an example of the original "Newton-Leibniz magic," which allows us to find the slope of a line tangent to a curve, or to find the instantaneous rate of change in a nonlinear function, at a specified point. Now let's work out some derivatives over whole domains for some basic linear functions.

Simply a constant

Imagine a *constant function* in which the dependent variable always has the same value, no matter what we input for the independent variable. If we call the function f, then we can describe it simply enough:

$$f(x) = a$$

where x is the independent variable and a is a real number. Let's find the derivative function f' at some unspecified point (x_0, y_0).

We begin by creating a movable point (x,y) near (x_0, y_0), as shown in Fig. 3-5. In the rectangular xy-plane, $y = a$ for all possible values of x. The x-value of our movable point is $x_0 + \Delta x$. The y-value is a. The derivative at (x_0, y_0) can be approximated by calculating the slope of a straight line passing through (x_0, y_0) and (x,y). That slope is

$$\Delta y/\Delta x = (y - y_0) / (x - x_0)$$

If we substitute a for y, a for y_0, and the quantity $(x_0 + \Delta x)$ for x in this equation, we get

$$\Delta y/\Delta x = (a - a) / (x_0 + \Delta x - x_0)$$

which can be simplified to

$$\Delta y/\Delta x = 0/\Delta x = 0$$

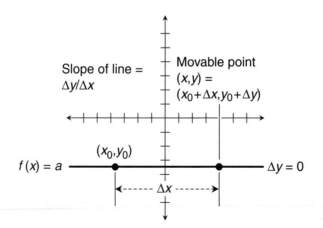

Figure 3-5   Finding the derivative of the constant
function $f(x) = a$ at a point (x_0, y_0).

The derivative $f'(x_0)$ is

$$f'(x_0) = \underset{\Delta x \to 0}{Lim} \ 0$$

No limit can be more straightforward than this! The derivative $f'(x_0)$ is equal to 0, no matter what we choose for x_0. By doing this exercise, we've shown that for any constant function

$$f(x) = a$$

the derivative function is

$$f'(x) = 0$$

which is called, appropriately enough, the *zero function*.

Multiply *x* by a real constant

Now we'll find the derivative of a basic linear function in which the independent variable is multiplied by a real-number constant. Again, let's call the function f. Then

$$f(x) = ax$$

where x is the independent variable and a is the constant. Let's figure out f' at a nonspecific point (x_0, y_0).

We can invent a movable point (x, y) near (x_0, y_0), as shown in Fig. 3-6. In rectangular xy-coordinates, $y = ax$ for all possible values of x and y. The x-value of our movable point is

$$x = x_0 + \Delta x$$

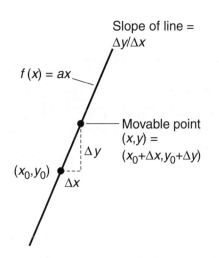

Figure 3-6   Finding the derivative of the basic linear function $f(x) = ax$ at a point (x_0, y_0).

The y-value of the movable point is

$$y = ax = a(x_0 + \Delta x) = ax_0 + a\Delta x$$

We can approximate the derivative by finding the slope of a straight line passing through the points (x_0, y_0) and (x, y). That slope is

$$\Delta y / \Delta x = (y - y_0) / (x - x_0)$$

We know that $y_0 = ax_0$, because f tells us that any x-value must be multiplied by a to get the corresponding y-value. In the above equation, let's substitute the quantity $(ax_0 + a\Delta x)$ for y, substitute ax_0 for y_0, and substitute the quantity $(x_0 + \Delta x)$ for x. That gives us

$$\Delta y / \Delta x = (ax_0 + a\Delta x - ax_0) / (x_0 + \Delta x - x_0)$$

which can be simplified to

$$\Delta y / \Delta x = a(x_0 + \Delta x - x_0) / \Delta x$$

and further to

$$\Delta y / \Delta x = a\Delta x / \Delta x$$

and finally to

$$\Delta y / \Delta x = a$$

The derivative $f'(x_0)$ is therefore

$$f'(x_0) = \underset{\Delta x \to 0}{Lim}\ a$$

This limit here is equal to a. That's all there is to it! Therefore,

$$f'(x_0) = a$$

no matter what value of x_0 we choose. This result tells us that for any basic linear function of the form

$$f(x) = ax$$

the derivative function is

$$f'(x) = a$$

which is a constant function.

- -

Here's a challenge!

Using the calculus techniques we've learned so far (not plain algebra!), find the derivative of the function

$$f(x) = -4x + 5$$

at the general point (x_0, y_0) in rectangular xy-coordinates, where $y = f(x)$. What does this result tell us?

Solution

Once again, let's start out by creating a movable point (x,y) near (x_0, y_0). Figure 3-7 shows the slant of the line. (The coordinate grid is not shown because it would make the illustration needlessly messy. It's the slope we're interested in, anyhow!). In rectangular xy-coordinates, we can graph our function in the slope-intercept form as

$$y = -4x + 5$$

The x-value of our movable point is, as always,

$$x = x_0 + \Delta x$$

The y-value of our movable point is

$$y = -4(x_0 + \Delta x) + 5 = -4x_0 - 4\Delta x + 5$$

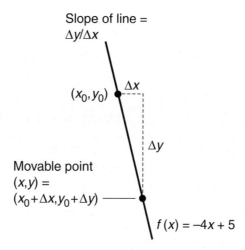

Slope of line =
$\Delta y / \Delta x$

(x_0, y_0) Δx

Δy

Movable point
$(x,y) =$
$(x_0 + \Delta x, y_0 + \Delta y)$ ———

$f(x) = -4x + 5$

Figure 3-7   Finding the derivative of the function $f(x) = -4x + 5$ at a point (x_0, y_0).

We approximate the derivative by finding the slope of a straight line through (x_0, y_0) and (x, y). That slope is

$$\Delta y/\Delta x = (y - y_0) / (x - x_0)$$

Our function f tells us that $y_0 = -4x_0 + 5$. Now, in the above equation, let's substitute the quantity $(-4x_0 - 4\Delta x + 5)$ in place of y, substitute the quantity $(-4x_0 + 5)$ in place of y_0, and substitute the quantity $(x_0 + \Delta x)$ in place of x. That gives us

$$\Delta y/\Delta x = [-4x_0 - 4\Delta x + 5 - (-4x_0 + 5)] / (x_0 + \Delta x - x_0)$$

We had better be careful with the signs here! We can rewrite the above equation as

$$\Delta y/\Delta x = (-4x_0 - 4\Delta x + 5 + 4x_0 - 5) / (x_0 + \Delta x - x_0)$$

which can be simplified to

$$\Delta y/\Delta x = -4\Delta x/\Delta x$$

and finally to

$$\Delta y/\Delta x = -4$$

We got rid of a big mess in a hurry! Now we can say that the derivative $f'(x_0)$ is

$$f'(x_0) = \underset{\Delta x \to 0}{Lim} \ -4$$

This is, again, trivial. The limit is equal to -4. Therefore,

$$f'(x_0) = -4$$

no matter what value of x_0 we choose. This tells us that the derivative of the linear function

$$f(x) = -4x + 5$$

is the function

$$f'(x) = -4$$

The stand-alone constant, 5, in $f(x)$ can be changed to any other real number, and we'll still end up with the same derivative function, $f'(x) = -4$.

- -

Basic Quadratic Functions

The next step in understanding *differentiation* (the process of finding derivatives) is to do it with some basic quadratic functions.

Square *x* and multiply by constant

Let's keep using the name f for our functions. Using the same techniques as we have for simpler functions, we'll now find the derivative of

$$f(x) = ax^2$$

where x is the independent variable and a is a real number. Our mission is to determine f' at a point (x_0, y_0). In rectangular xy-coordinates, we can plot this function as the graph of

$$y = ax^2$$

By now, the routine ought to be getting familiar, and we shouldn't need to draw a graph and label the points, curve, and lines. We create a movable point (x, y) near (x_0, y_0). The x-value of the movable point is, as always,

$$x = x_0 + \Delta x$$

The y-value of the movable point is

$$y = ax^2 = a(x_0 + \Delta x)^2 = a[x_0^2 + 2x_0\Delta x + (\Delta x)^2]$$

$$= ax_0^2 + 2ax_0\Delta x + a(\Delta x)^2$$

The derivative is approximately equal to the slope of a straight line passing through the points (x_0, y_0) and (x, y). That slope is, as always,

$$\Delta y / \Delta x = (y - y_0) / (x - x_0)$$

If we plug the value x_0 into our function, we get $y_0 = ax_0^2$. In the above equation, let's substitute

- The quantity $[ax_0^2 + 2ax_0\Delta x + a(\Delta x)^2]$ in place of y
- The value ax_0^2 in place of y_0
- The quantity $(x_0 + \Delta x)$ in place of x

That gives us

$$\Delta y / \Delta x = [ax_0^2 + 2ax_0\Delta x + a(\Delta x)^2 - ax_0^2] / (x_0 + \Delta x - x_0)$$

which can be simplified to

$$\Delta y / \Delta x = [2ax_0\Delta x + a(\Delta x)^2] / \Delta x$$

and finally to

$$\Delta y / \Delta x = 2ax_0 + a\Delta x$$

The derivative $f'(x_0)$ is therefore

$$f'(x_0) = \lim_{\Delta x \to 0} 2ax_0 + a\Delta x$$

As Δx approaches 0, the second addend in the expression, $a\Delta x$, approaches 0 because it is a constant multiple of Δx. The first addend, $2ax_0$, is not affected by changes in Δx. The limit of the entire expression is therefore equal to $2ax_0$, telling us that

$$f'(x_0) = 2ax_0$$

for any real number x_0. On this basis, we can say that for any function of the form

$$f(x) = ax^2$$

the derivative function is

$$f'(x) = 2ax$$

- -

Are you confused?

In each of these drills, just before we take the limit, we encounter ratios with Δx in the denominator. We also find Δx in each of the terms of the numerator. That lets us cancel out Δx in both the numerator and the denominator, so the expression is no longer a fraction. "But," you might object, "we're forcing Δx down to 0 when we take the limit, aren't we? How can we play around with these little Δx factors if they all ultimately turn out to be 0? Aren't we dividing by 0 when we do this?"

That is an excellent question. But there's no need to worry! The quantity Δx is *never* equal to 0, so it's all right to divide by it, cancel it out, and treat it like any nonzero real number. It can get as close to 0 as we dare to imagine, either on the negative side or the positive side, but it is never true that $\Delta x = 0$. We're seeing what happens to expressions that contain Δx as it gets arbitrarily close to 0, so Δx "thinks it's 0 when it's really not"! This might seem sneaky, almost deceptive—but it works, and it lies at the root of the branch of mathematics that Newton called *infinitesimal calculus,* and that people nowadays call *differential calculus.*

Here's a challenge!

Using the calculus techniques we've learned so far, find the derivative of the function

$$f(x) = -7x^2 + 2x$$

at the general point (x_0, y_0) in rectangular xy-coordinates, where $y = f(x)$.

Solution

We can plot this function as the graph of

$$y = -7x^2 + 2x$$

If it helps you envision the situation, you can go ahead and draw the graph by locating a few points that satisfy the function, and then filling in a smooth curve that passes through those points. We create the usual movable point (x,y) near (x_0, y_0). The x-value of this point is

$$x = x_0 + \Delta x$$

The y-value of the movable point is

$$y = -7x^2 + 2x = -7(x_0 + \Delta x)^2 + 2(x_0 + \Delta x)$$
$$= -7[x_0^2 + 2x_0 \Delta x + (\Delta x)^2] + 2x_0 + 2\Delta x$$
$$= -7x_0^2 - 14x_0 \Delta x - 7(\Delta x)^2 + 2x_0 + 2\Delta x$$

The slope of a straight line passing through (x_0, y_0) and (x,y), which approximates the derivative we're looking for, is

$$\Delta y / \Delta x = (y - y_0) / (x - x_0)$$

The function dictates that $y_0 = -7x_0^2 + 2x_0$. In the above equation, let's substitute

- The quantity $[-7x_0^2 - 14x_0\Delta x - 7(\Delta x)^2 + 2x_0 + 2\Delta x]$ in place of y
- The quantity $(-7x_0^2 + 2x_0)$ in place of y_0
- The quantity $(x_0 + \Delta x)$ in place of x

When we make these substitutions and then find the slope, we get

$$\Delta y / \Delta x = [-7x_0^2 - 14x_0\Delta x - 7(\Delta x)^2 + 2x_0 + 2\Delta x - (-7x_0^2 + 2x_0)] / (x_0 + \Delta x - x_0)$$

The signs are tricky in the numerator here, so we have to be careful! Let's rewrite the above equation as

$$\Delta y / \Delta x = [-7x_0^2 - 14x_0\Delta x - 7(\Delta x)^2 + 2x_0 + 2\Delta x + 7x_0^2 - 2x_0] / (x_0 + \Delta x - x_0)$$

which can be simplified to

$$\Delta y / \Delta x = [-14x_0 \Delta x - 7(\Delta x)^2 + 2\Delta x] / \Delta x$$

and finally to

$$\Delta y / \Delta x = -14x_0 - 7\Delta x + 2$$

The derivative $f'(x_0)$ is

$$f'(x_0) = \lim_{\Delta x \to 0} -14x_0 - 7\Delta x + 2$$

As Δx approaches 0, the second addend, $-7\Delta x$, approaches 0 because it's a constant multiple of Δx. The first addend, $14x_0$, and the third addend, 2, have nothing to do with Δx, so they stay the same as Δx approaches 0. The limit of the entire expression is therefore $-14x_0 + 2$, so we have

$$f'(x_0) = -14x_0 + 2$$

for any real number x_0. We've just shown that

$$f(x) = -7x^2 + 2x$$

has the derivative function

$$f'(x) = -14x + 2$$

- -

Basic Cubic Functions

To differentiate a basic cubic function, we can go through the same process as we did to differentiate the basic linear and quadratic functions. Some of the expressions are a little more complicated this time, but otherwise everything works out in a similar way.

Cube x and multiply by constant

Let's go through our trusty routine to find the derivative of a function f in which we cube the independent variable x and then multiply by the constant a, like this:

$$f(x) = ax^3$$

As in previous situations, we'll determine f' at a nonspecific point (x_0, y_0). In the xy-plane, we can plot the graph of

$$y = ax^3$$

The x-value of a movable point (x,y) in the vicinity of (x_0, y_0) is, once again,

$$x = x_0 + \Delta x$$

The y-value of the movable point is

$$y = ax^3 = a(x_0 + \Delta x)^3 = a[x_0^3 + 3x_0^2\Delta x + 3x_0(\Delta x)^2 + (\Delta x)^3]$$

$$= ax_0^3 + 3ax_0^2\Delta x + 3ax_0(\Delta x)^2 + a(\Delta x)^3$$

The slope of a line passing through (x_0, y_0) and (x,y) is

$$\Delta y/\Delta x = (y - y_0) / (x - x_0)$$

The function says that $y_0 = ax_0^3$. In the above equation, let's substitute

- The quantity $[ax_0^3 + 3ax_0^2\Delta x + 3ax_0(\Delta x)^2 + a(\Delta x)^3]$ for y
- The quantity ax_0^3 for y_0
- The quantity $(x_0 + \Delta x)$ for x

That gives us

$$\Delta y / \Delta x = [ax_0^3 + 3ax_0^2\Delta x + 3ax_0(\Delta x)^2 + a(\Delta x)^3 - ax_0^3] / (x_0 + \Delta x - x_0)$$

which simplifies to

$$\Delta y / \Delta x = [3ax_0^2\Delta x + 3ax_0(\Delta x)^2 + a(\Delta x)^3] / \Delta x$$

and finally to

$$\Delta y / \Delta x = 3ax_0^2 + 3ax_0\Delta x + a(\Delta x)^2$$

The derivative $f'(x_0)$ is therefore

$$f'(x_0) = \lim_{\Delta x \to 0} 3ax_0^2 + 3ax_0\Delta x + a(\Delta x)^2$$

As Δx approaches 0, the second addend, $3ax_0\Delta x$, approaches 0 because it is a constant multiple of Δx. The third addend, $a(\Delta x)^2$, also approaches 0. It's a constant multiple of $(\Delta x)^2$, which must approach 0 as Δx approaches 0. The first addend, $3ax_0^2$, stays the same no matter what Δx becomes. The limit of the entire expression is therefore $3ax_0^2$, showing that

$$f'(x_0) = 3ax_0^2$$

for any real number x_0. Therefore, any function of the form

$$f(x) = ax^3$$

has the derivative

$$f'(x) = 3ax^2$$

- -

Here's a challenge!

Using the calculus techniques we've learned so far, find the derivative of the function

$$f(x) = 2x^3 - 5x$$

at the general point (x_0, y_0) in rectangular xy-coordinates, where $y = f(x)$.

Solution

We can plot this function as the graph of

$$y = 2x^3 - 5x$$

We create the usual movable point (x,y) near (x_0,y_0). The x-value of this point is

$$x = x_0 + \Delta x$$

The y-value of the movable point is

$$y = 2x^3 - 5x = 2(x_0 + \Delta x)^3 - 5(x_0 + \Delta x)$$
$$= 2[x_0^3 + 3x_0^2\Delta x + 3x_0(\Delta x)^2 + (\Delta x)^3] - 5x_0 - 5\Delta x$$
$$= 2x_0^3 + 6x_0^2\Delta x + 6x_0(\Delta x)^2 + 2(\Delta x)^3 - 5x_0 - 5\Delta x$$

The slope of a line through (x_0,y_0) and (x,y) is

$$\Delta y/\Delta x = (y - y_0) / (x - x_0)$$

The original function tells us that $y_0 = 2x_0^3 - 5x_0$. In the above equation, let's substitute

- The quantity $[2x_0^3 + 6x_0^2\Delta x + 6x_0(\Delta x)^2 + 2(\Delta x)^3 - 5x_0 - 5\Delta x]$ in place of y
- The quantity $(2x_0^3 - 5x_0)$ in place of y_0
- The quantity $(x_0 + \Delta x)$ in place of x

When we make these substitutions, we get

$$\Delta y/\Delta x = [2x_0^3 + 6x_0^2\Delta x + 6x_0(\Delta x)^2 + 2(\Delta x)^3 - 5x_0 - 5\Delta x - (2x_0^3 - 5x_0)] / (x_0 + \Delta x - x_0)$$

which can be rewritten as

$$\Delta y/\Delta x = [2x_0^3 + 6x_0^2\Delta x + 6x_0(\Delta x)^2 + 2(\Delta x)^3 - 5x_0 - 5\Delta x - 2x_0^3 + 5x_0)] / (x_0 + \Delta x - x_0)$$

This can be simplified to

$$\Delta y/\Delta x = [6x_0^2\Delta x + 6x_0(\Delta x)^2 + 2(\Delta x)^3 - 5\Delta x)] / \Delta x$$

and finally to

$$\Delta y/\Delta x = 6x_0^2 + 6x_0\,\Delta x + 2(\Delta x)^2 - 5$$

The derivative $f'(x_0)$ is

$$f'(x_0) = \lim_{\Delta x \to 0} 6x_0^2 + 6x_0\Delta x + 2(\Delta x)^2 - 5$$

As Δx approaches 0, the second addend, $6x_0\Delta x$, approaches 0 because it is a multiple of Δx. The third addend, $2(\Delta x)^2$, approaches 0 because it's a constant multiple of $(\Delta x)^2$. The first addend, $6ax_0^2$, and the fourth addend, -5, are unaffected by changes in Δx. The limit of the entire expression is therefore $6x_0^2 - 5$, so

$$f'(x_0) = 6x_0^2 - 5$$

for any real number x_0. That means our original function

$$f(x) = 2x^3 - 5x$$

has the derivative function

$$f'(x) = 6x^2 - 5$$

Here's another challenge!

In the above scenario, consider the addends in the expression for f as two separate functions g and h, like this:

$$g(x) = 2x^3$$

and

$$h(x) = -5x$$

On this basis, we can say that

$$f(x) = g(x) + h(x)$$

Show that the derivative of this particular sum is equal to the sum of the derivatives:

$$f'(x) = g'(x) + h'(x)$$

Solution

First, let's figure out the derivative function g' using the general rule we've already derived. Using the rule for the derivative of a variable cubed and then multiplied by a constant, we get

$$g'(x) = 6x^2$$

Using the rule for the derivative of a variable multiplied by a constant, we get

$$h'(x) = -5$$

Adding these results gives us

$$g'(x) + h'(x) = 6x^2 - 5$$

This is what we got when we calculated $f'(x)$ in the previous "challenge."

Are you astute?

After all this repetition, you've probably noticed that some of the terms in the examples are always the same as corresponding terms in other examples. You've also seen that certain pairs of terms always cancel out, making the expressions simpler. In the numerators, you always end up subtracting $f(x_0)$ from $f(x_0 + \Delta x)$. In the denominators, you always end up subtracting x_0 from x_0. In the end, you always get a ratio of the form

$$[f(x_0 + \Delta x) - f(x_0)] / \Delta x$$

Whenever you calculate a derivative using the techniques in this chapter, you're actually generating the above expression, and then finding its limit as Δx approaches 0. In fact, the derivative $f'(x_0)$ of a function $f(x)$ at the point where $x = x_0$ is commonly defined as

$$f'(x_0) = \underset{\Delta x \to 0}{Lim} \ [f(x_0 + \Delta x) - f(x_0)] / \Delta x$$

If we can differentiate the function at every point in its domain, then the formula can be generalized to

$$f'(x) = \underset{\Delta x \to 0}{Lim} \ [f(x + \Delta x) - f(x)] / \Delta x$$

Remember that Δx can be either positive or negative. The limit, as shown above, must work from both the left and from the right, and the two results must agree. In my opinion, this definition is worth memorizing. You might put it into "mathematical verse," breaking the thoughts down line by line:

> To find the derivative of a function
> *f* of *x*,
> take *f* of "*x* plus a little something,"
> then subtract *f* of *x* from that,
> then divide by the little something,
> and finally find the limit
> as the little something
> shrinks
> until it's practically nothing.

- -

Practice Exercises

This is an open-book quiz. You may (and should) refer to the text as you solve these problems. Don't hurry! You'll find worked-out answers in App. A. The solutions in the appendix may not represent the only way a problem can be figured out. If you think you can solve a particular problem in a quicker or better way than you see there, by all means try it!

1. Find the derivative of the quadratic function:

$$f(x) = x^2$$

at the points where $x = -3$, $x = -2$, and $x = -1$.

Here's a hint for this exercise and all the rest to follow: You may use the general formulas we derived in the text to find these derivatives. Those formulas are *theorems* now—facts we can use again and again—because we've proven them true. You don't have to go through the ritual of setting up a movable point, approximating the slope, and then finding the limit of that slope as Δx approaches 0.

2. Find the derivative of the quadratic function:

$$f(x) = x^2$$

at the points where $x = 0$, $x = 1$, $x = 2$, and $x = 3$.

3. Find the derivative of the quadratic function:

$$f(x) = -2x^2$$

at the points where $x = -3$, $x = -2$, and $x = -1$.

4. Find the derivative of the quadratic function:

$$f(x) = -2x^2$$

at the points where $x = 0$, $x = 1$, $x = 2$, and $x = 3$.

5. Find the derivative of the quadratic function:

$$f(x) = -7x^2 + 2x$$

at the points where $x = -3$, $x = -2$, and $x = -1$.

6. Find the derivative of the quadratic function:

$$f(x) = -7x^2 + 2x$$

at the points where $x = 0$, $x = 1$, $x = 2$, and $x = 3$.

7. Find the derivative of the cubic function:

$$f(x) = 5x^3$$

at the points where $x = -3$, $x = -2$, and $x = -1$.

8. Find the derivative of the cubic function:

$$f(x) = 5x^3$$

at the points where $x = 0$, $x = 1$, $x = 2$, and $x = 3$.

9. Find the derivative of the cubic function:

$$f(x) = 2x^3 - 5x$$

at the points where $x = -3$, $x = -2$, and $x = -1$.

10. Find the derivative of the cubic function:

$$f(x) = 2x^3 - 5x$$

at the points where $x = 0$, $x = 1$, $x = 2$, and $x = 3$.

CHAPTER

4

Derivatives Don't Always Exist

In Chap. 3, every function we saw was *differentiable,* meaning that it had a derivative at every possible point. But there are functions for which derivatives don't exist at certain points. In this chapter, we'll learn how to tell when a function is differentiable, and when it is not.

Let's Look at the Graph

There's a quick and easy way to see if a function is differentiable at every point in its domain, or if there are some points at which it's *nondifferentiable* (impossible to differentiate). We can simply graph it! If the graph appears to have a definable slope at every point, then the function is probably differentiable. We can look for three signs that indicate differentiation problems.

Is there a gap?

Figure 4-1 is a graph of the function $f(x) = 1/x$. This graph has a gap. We can't define the function when $x = 0$. Because the function has no value when $x = 0$, the graph can't have a slope at the point where $x = 0$.

It's tempting to make an intuitive leap here. Could a line running straight up and down along the axis labeled $f(x)$ be tangent to the graph? That's a fine idea, but there's no point on the graph at which such a line could come into contact! A tangent line always touches a curve at a point.

Because we can't define the slope of a line tangent to the graph of this function at the point where $x = 0$, we can't define the derivative there, either. Even if we add the point (0,0) to define $f(x) = 1/x$ for all real numbers, we can't make the function differentiable at the point where $x = 0$, as we'll see later in this chapter.

Is there a jump?

Figure 4-2 is a graph of a function that takes a jump. This function is defined for all real numbers, although there is a discontinuity. At every point on the line except the one where the jump occurs, the slope, and therefore the derivative, appears to be 0. Let's look closely at the point where $x = 0$. At this point, we have $f(0) = 3$. What is the slope of this graph at (0,3)?

55

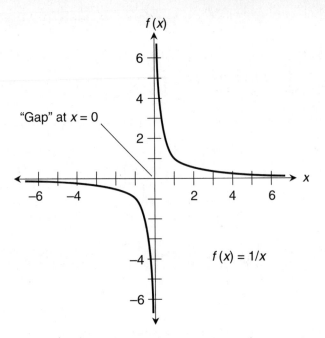

Figure 4-1   This function has a gap at $x = 0$, because it is not defined there. It is discontinuous at the point where $x = 0$.

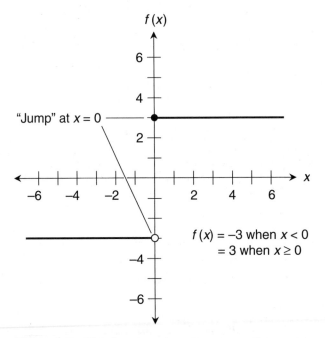

Figure 4-2   This function takes a jump at $x = 0$, even though it is defined there. It is discontinuous at the point where $x = 0$.

We might think that the slope at the point (0,3) is parallel to the *x* axis, running along the part of the graph that lies to the right of that point. After all, the whole graph looks horizontal. I might try to convince you, based on that notion, that the function has a slope of 0 at every point in its domain, including (0,3). But you could argue that a tangent line at the jump is not horizontal but vertical, running along the $f(x)$ axis and passing through (0,−3) as well as (0,3). That makes just as much (or just as little) sense.

The slope of this graph at the point where $x = 0$ can't be defined with certainty. Therefore, the derivative does not exist for this function at the point where $x = 0$. Both this function and the one with the gap (Fig. 4-1) have discontinuities. As things work out, if a function is discontinuous at a particular point, then it has no derivative there. Turning this logic "inside-out" gives us one of the most important facts in calculus:

- If a function has a derivative at a point, then the function is continuous there.

Is there a corner?

Figure 4-3 shows the graph of a function that's defined over the entire set of real numbers, and that's continuous everywhere as well. But there's something strange about this function at the point where $x = 2$. It has no gap there, and its value doesn't jump there, but the curve turns a corner.

The slope of a tangent line can't be defined at the point where $x = 2$. The trouble here is similar to the problem we encountered in Fig. 4-2. If we want to talk about the slope at the point (2,4) on this graph, should we base our idea on the half-parabola to the left, or on the half-line to the right? Neither idea is better, but they disagree.

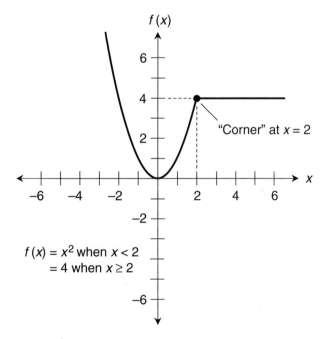

Figure 4-3   This function turns a corner at $x = 2$, although it is continuous over the entire set of real numbers.

- -

Are you confused?

You might still wonder why we can't define the slope of the graph in Fig. 4-3 at the point (2,4). You might say, "Suppose we choose two movable points, one on either side of (2,4), and draw a line through them. As we move these two points closer and closer to (2,4), one along the half-parabola and the other along the half-line, won't they approach a line having a slope of 2 and running through the point (2,4)? Won't that line be a legitimate tangent line?" No, there are two problems with that idea.

First, it's "illegal" to use two movable points in an attempt to find a line whose slope indicates a derivative. That's not the way it's done according to the definition at the end of Chap. 3. We must approach the fixed point, in this case (2,4), from one side at a time, and use it as one of the points through which we draw the line. Then we must be sure that we end up with the same line when we approach from the left as we do when we approach from the right. Only then can we assign it a slope and determine the derivative based on that slope.

Second, even if it were "legal" to use two movable points to find the line, the slope of that line would depend on the "relative rates of approach" in a situation like the one we see in Fig. 4-3. If the point on the left is always twice as far away from (2,4) as the point on the right, for example, we will end up with a different line than we will get if the two points are always at the same distance from (2,4). By adjusting the "relative rates of approach" of the two points, we could get a line with any slope between 0 and 4!

Here's a challenge!

Look again at the graph of the function $f(x) = 1/x$, shown in Fig. 4-1. Can we tell, by visual inspection, what happens to $f'(x)$ as x increases endlessly (that is, "approaches positive infinity")? Can we tell what happens to $f'(x)$ as x decreases endlessly (that is, "approaches negative infinity")?

Solution

Yes, we can get an idea of what happens in these cases, although looking at a graph doesn't constitute a mathematical proof of anything. In this particular graph, as x increases endlessly (that is, "approaches positive infinity"), the *slope* of the curve, and therefore the derivative $f'(x)$, appears to approach 0 from the *negative* direction. As x decreases endlessly ("approaches negative infinity"), the *slope* of the curve, and therefore the derivative, again appears to approach 0 from the *negative* direction. We can write these statements symbolically as

$$\text{As } x \rightarrow +\infty, f'(x) \rightarrow 0-$$

and

$$\text{As } x \rightarrow -\infty, f'(x) \rightarrow 0-$$

Don't get confused here. The function itself behaves quite differently than its derivative! As x increases endlessly in the *positive* direction, the *value* of the function approaches 0 from the *positive* direction. As x increases endlessly in the *negative* direction, the *value* of the function approaches 0 from the *negative* direction. We can write these statements symbolically as

$$\text{As } x \rightarrow +\infty, f(x) \rightarrow 0+$$

and

$$\text{As } x \rightarrow -\infty, f(x) \rightarrow 0-$$

- -

When We Can Differentiate

Now that we have a visual idea of what it means for a function to be differentiable at a point, let's get a little more formal. We can call a function differentiable at a single point, over an interval, or over its entire domain.

- A function is *differentiable at a point* if and only if the function has a derivative, as defined at the end of Chap. 3, at that point.
- A function is *differentiable over an interval* if and only if it has a derivative, as defined at the end of Chap. 3, at every point in that interval.
- A function is *differentiable in general* (or simply *differentiable*) if and only if it has a derivative, as defined at the end of Chap. 3, at every point in its domain.

Try to find the limits

When we want to know whether or not a function is differentiable at a particular point, we can try to find the derivative at that point and see if the result makes sense. If we get a meaningful and unambiguous result, then the derivative exists at the point. Otherwise, it doesn't.

For all the functions $f(x)$ we examined in Chap. 3, we were able to find the following limit at any point where the independent variable was equal to some specific value x_0:

$$f'(x_0) = \lim_{\Delta x \to 0} \ [f(x_0 + \Delta x) - f(x_0)] \ / \ \Delta x$$

If the notion of a derivative is to make any sense for a function $f(x)$ at the point where $x = x_0$, the value x_0 must be in the domain. That is to say, $f(x_0)$ must be defined. Once we know that, then we must be able to find the above limit whether Δx is positive or negative. In other words, the limit has to exist as we approach x_0 from the right, and it also has to exist as we approach x_0 from the left. That's not all! For $f(x)$ to have a derivative at the point where $x = x_0$, the right-hand and left-hand limits, as defined in Chap. 2, must be the same. Stated symbolically,

$$f'(x_0) = \lim_{\Delta x \to 0+} \ [f(x_0 + \Delta x) - f(x_0)] \ / \ \Delta x$$

must be identical to

$$f'(x_0) = \lim_{\Delta x \to 0-} \ [f(x_0 + \Delta x) - f(x_0)] \ / \ \Delta x$$

Example

Figure 4-4 is a graph of a quadratic function that is defined and continuous over the entire set of real numbers:

$$f(x) = x^2 - 3$$

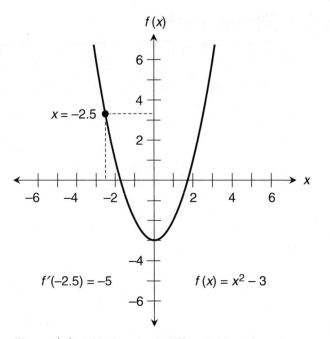

Figure 4-4   This function is differentiable at the point where $x = -2.5$.

Let's see whether or not this function is differentiable at the point where $x = -2.5$. We set up the limit

$$\underset{\Delta x \to 0}{Lim} \; [f(-2.5 + \Delta x) - f(-2.5)] \, / \, \Delta x$$

We can calculate $f(-2.5)$ easily enough:

$$f(-2.5) = (-2.5)^2 - 3$$
$$= 6.25 - 3$$
$$= 3.25$$

Now we can write the limit as

$$\underset{\Delta x \to 0}{Lim} \; [(-2.5 + \Delta x)^2 - 3 - 3.25] \, / \, \Delta x$$

which expands to

$$\underset{\Delta x \to 0}{Lim} \; [6.25 - 5\Delta x + (\Delta x)^2 - 3 - 3.25] \, / \, \Delta x$$

and then simplifies to

$$\underset{\Delta x \to 0}{Lim} \; [-5\Delta x + (\Delta x)^2] \, / \, \Delta x$$

and finally to

$$\underset{\Delta x \to 0}{Lim} \ -5 + \Delta x$$

Whether Δx is positive or negative (that is, whether Δx approaches 0 from the right or the left), it's easy to see that this limit is equal to −5. Therefore,

$$f'(-2.5) = -5$$

Now that we've found the derivative at the point where $x = -2.5$, we can conclude that the function is differentiable there.

Another example

Let's reexamine the function we saw in Fig. 4-2. Suppose we want to find out if this function is differentiable at the point where $x = 3$. In the vicinity of this point, we can imagine that the function is

$$f(x) = 3$$

as shown in Fig. 4-5. To see whether or not this function is differentiable at the point where $x = 3$, we can try to find the derivative there by setting up the limit

$$\underset{\Delta x \to 0}{Lim} \ [f(3 + \Delta x) - f(3)] \ / \ \Delta x$$

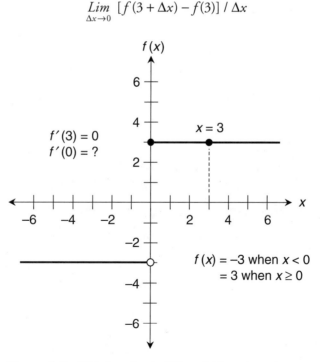

Figure 4-5 This function is differentiable at the point where $x = 3$, but not where $x = 0$.

We know that $f(3) = 3$, so we can rewrite this limit as

$$\underset{\Delta x \to 0}{Lim} \ [f(3 + \Delta x) - 3] \ / \ \Delta x$$

We must keep in mind that this limit has to be evaluated as Δx becomes very small either positively or negatively, approaching 0 from either the right or the left. In either of those scenarios, the value of $f(3 + \Delta x)$ remains constant at 3. That allows us to simplify the above limit to

$$\underset{\Delta x \to 0}{Lim} \ (3 - 3) \ / \ \Delta x$$

and then to

$$\underset{\Delta x \to 0}{Lim} \ 0/\Delta x$$

This limit is equal to 0. No matter how small we make Δx, it's still a tiny positive or negative real number (never 0), so $0/\Delta x$ is always equal to 0. Therefore

$$f'(3) = 0$$

Having found a well-defined derivative at the point where $x = 3$, we can be sure that the function is differentiable there.

- -

Are you confused?

"Wait!" you say. "In the two examples we just finished, we didn't find the left-hand and the right-hand limits separately! Doesn't the definition of the derivative, given at the end of Chap. 3, require us to do that?" Yes, it does. But in the two examples we just finished, we were able to perform both tasks together, because it was apparent that the two limits would be the same if we calculated them individually. If your mind demands absolute rigor (not a bad trait for a mathematician), you can do both limit calculations separately in every case. In the following challenge, we'll see a situation where that attitude serves us well.

Here's a challenge!

Let's try to find the derivative of the function shown in Fig. 4-5 at the point where $x = 0$, first by approaching the point from the right, and then by approaching from the left. What can we conclude from this result?

Solution

Before we start, let's state the function mathematically. To fully describe it, we must write it in two parts, like this:

$$f(x) = -3 \quad \text{when } x < 0$$
$$= 3 \quad \text{when } x \geq 0$$

First, let's approach 0 from the right, that is, from the positive side. Geometrically, we'll move from right to left along the upper half-line, or the portion to the right of the vertical axis. We set up the limit

$$\underset{\Delta x \to 0+}{Lim} \ [f(0 + \Delta x) - f(0)] \ / \ \Delta x$$

We know that $f(0) = 3$, so we can rewrite this limit as

$$\underset{\Delta x \to 0+}{Lim} \; [f(0 + \Delta x) - 3] \, / \, \Delta x$$

As Δx becomes small positively, the value of $f(0 + \Delta x)$ remains constant at 3, so we can rewrite the above limit as

$$\underset{\Delta x \to 0+}{Lim} \; (3 - 3) \, / \, \Delta x$$

and then to

$$\underset{\Delta x \to 0+}{Lim} \; 0/\Delta x$$

This limit is equal to 0. No matter how small we make Δx, it's always a positive real number, so $0/\Delta x = 0$. This result makes sense. The right-hand part of the half-line is horizontal, so we should expect the derivative, which represents the slope, to be 0.

Now let's approach 0 from the left (from the negative side). In the graph, this is the equivalent of moving from left to right along the lower half-line, which lies to the left of the vertical axis. We set up the limit

$$\underset{\Delta x \to 0-}{Lim} \; [f(0 + \Delta x) - f(0)] \, / \, \Delta x$$

Remember that Δx is negative now, not positive! We know that $f(0) = 3$, so we can rewrite this limit as

$$\underset{\Delta x \to 0-}{Lim} \; [f(0 + \Delta x) - 3] \, / \, \Delta x$$

As Δx becomes smaller and smaller negatively, $f(0 + \Delta x)$ remains constant, maintaining a value of -3 (not 3, as it does to the right of the vertical axis), so we can rewrite the above limit as

$$\underset{\Delta x \to 0-}{Lim} \; (-3 - 3) \, / \, \Delta x$$

and then to

$$\underset{\Delta x \to 0-}{Lim} \; -6/\Delta x$$

Here, the value of Δx becomes smaller and smaller negatively, becoming arbitrarily close to 0, but never quite getting there. The ratio $-6/\Delta x$ is always a positive real, because it's a negative divided by a negative. As Δx approaches 0, the ratio $-6/\Delta x$ blows up; it gets arbitrarily large. The limit is not defined. This result tells us that the function

$$f(x) = -3 \quad \text{when } x < 0$$
$$= 3 \quad \text{when } x \geq 0$$

is nondifferentiable at the point where $x = 0$.

- -

When We Can't Differentiate

The "challenge" we just finished involves a function that can be differentiated everywhere except at a single point. The problem in this particular case is caused by a discontinuity at the point where $x = 0$. There are other ways that a function can be nondifferentiable at a point.

Example

Let's consider another function with a discontinuity, but of a different sort than the one in the "challenge" we solved a moment ago. Figure 4-6 is a graph of the modified reciprocal function

$$f(x) = 1/x \quad \text{when } x \neq 0$$
$$= 0 \quad \text{when } x = 0$$

This function is defined over the entire set of real numbers, but it has a discontinuity at $(0,0)$. If we want to find out whether or not $f(x)$ is differentiable at the point where $x = 0$, we must set up the limit

$$\underset{\Delta x \to 0}{Lim} \ [f(0 + \Delta x) - f(0)] \ / \ \Delta x$$

and evaluate it both from the right and from the left. Let's do it from the right first. We write this as

$$\underset{\Delta x \to 0+}{Lim} \ [f(0 + \Delta x) - f(0)] \ / \ \Delta x$$

which simplifies to

$$\underset{\Delta x \to 0+}{Lim} \ [f(\Delta x) - f(0)] \ / \ \Delta x$$

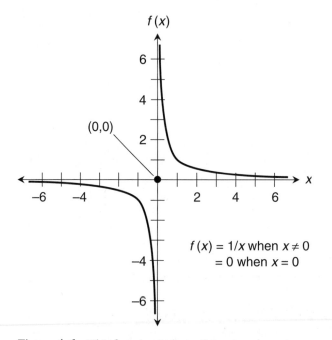

Figure 4-6   This function isn't continuous at the point where $x = 0$, and it isn't differentiable there, either.

because $0 + \Delta x = \Delta x$. We are told that $f(0) = 0$, so we can simplify further, getting

$$\underset{\Delta x \to 0+}{Lim} \ f(\Delta x) \ / \ \Delta x$$

The function tells us to take the reciprocal of whatever nonzero number we put in. That means we can rewrite the above expression as

$$\underset{\Delta x \to 0+}{Lim} \ [1/(\Delta x)] \ / \ \Delta x$$

which simplifies to

$$\underset{\Delta x \to 0+}{Lim} \ 1/(\Delta x)^2$$

As Δx approaches 0 from the right (the positive side), the ratio $1/(\Delta x)^2$ grows arbitrarily large in the positive sense. That's because we're taking the reciprocal of $(\Delta x)^2$, an endlessly shrinking positive real. The above limit blows up positively; it's not defined.

The same thing happens when we approach 0 from the left. We can rework the sequence of steps, replacing $\Delta x \to 0+$ with $\Delta x \to 0-$ in every instance, to arrive at

$$\underset{\Delta x \to 0-}{Lim} \ 1/(\Delta x)^2$$

This limit blows up positively, just as the right-hand limit does, so it's undefined as well. Actually, this second exercise is overkill. Once we showed that the right-hand limit doesn't exist at the point where $x = 0$, we gathered enough information to know that our function isn't differentiable there.

It's tempting to say that the discontinuity here is somehow "worse" than the one in the function shown by Fig. 4-5. In the earlier case, a limit could be defined from one side, at least. This time, there is no definable limit from either side. We can be "double-sure" that the function

$$f(x) = 1/x \quad \text{when } x \neq 0$$
$$= 0 \quad \text{when } x = 0$$

is nondifferentiable at the point where $x = 0$.

Another example

Figure 4-7 is a graph of the absolute-value function, which can be written in two-part fashion like this:

$$f(x) = x \quad \text{when } x \geq 0$$
$$= -x \quad \text{when } x < 0$$

This function is defined and continuous over the entire set of reals. If $x > 0$ or $x < 0$, the absolute-value function behaves as a linear function, and it is therefore differentiable when we restrict the domain to either the set of *strictly positive* reals or the set of *strictly negative* reals.

As we can see by examining Fig. 4-7, the graph turns a corner. The slope changes suddenly from -1 to 1 as we move to the right through the point where $x = 0$. When the slope of

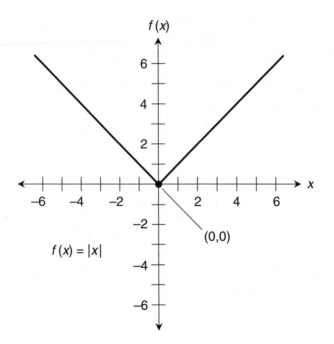

Figure 4-7   This function is continuous at the point
where $x = 0$, but it isn't differentiable there.

a graph changes abruptly at a point, we should suspect that point as a likely place where the function is nondifferentiable! Let's find out if that's the case here. First, we'll evaluate the limit of the slope as we approach the point where $x = 0$ from the right:

$$\underset{\Delta x \to 0+}{Lim} \; [f(0 + \Delta x) - f(0)] \, / \, \Delta x$$

Because $0 + \Delta x = \Delta x$, this can be simplified to

$$\underset{\Delta x \to 0+}{Lim} \; [f(\Delta x) - f(0)] \, / \, \Delta x$$

We can simplify further, knowing that $f(0) = 0$:

$$\underset{\Delta x \to 0+}{Lim} \; f(\Delta x) \, / \, \Delta x$$

When Δx is small and positive, $f(\Delta x) = \Delta x$. That means we can simplify the limit even more, writing it as

$$\underset{\Delta x \to 0+}{Lim} \; \Delta x \, / \Delta x$$

Obviously, $\Delta x \, / \Delta x = 1$. That means we have

$$\underset{\Delta x \to 0+}{Lim} \; 1$$

This limit is equal to 1. We can believe this easily enough. The right-hand half of the graph is a half-line with a slope of 1.

Now, let's take the limit of the slope as we approach the point where $x = 0$ from the left. We write

$$\underset{\Delta x \to 0-}{Lim} \ [f(0 + \Delta x) - f(0)] \ / \ \Delta x$$

Because $0 + \Delta x = \Delta x$, this can be simplified to

$$\underset{\Delta x \to 0-}{Lim} \ [f(\Delta x) - f(0)] \ / \ \Delta x$$

Knowing that $f(0) = 0$, we can substitute and rewrite this as

$$\underset{\Delta x \to 0-}{Lim} \ f(\Delta x) \ / \ \Delta x$$

When we take the absolute value of a negative number, the function simply reverses the sign of the number, so when Δx is small and negative, $f(\Delta x) = -\Delta x$. That means we can simplify the limit some more, writing it as

$$\underset{\Delta x \to 0-}{Lim} \ -\Delta x \ / \Delta x$$

The ratio $-\Delta x \ / \Delta x$ is always equal to -1, so we have

$$\underset{\Delta x \to 0-}{Lim} \ -1$$

This is equal to -1, a sensible result. The left-hand part of the graph is a half-line with a slope of -1.

We have found the limits in the prescribed form, both from the right and from the left. But they aren't the same, showing that the function

$$f(x) = |x|$$

is nondifferentiable at the point where $x = 0$. (The vertical lines in this equation indicate that the absolute value is to be taken of whatever quantity appears between them.)

- -

Are you confused?

"I see what's going on now," you say. "If the slope of a graph changes suddenly at a point, then even if the function is continuous, there's no derivative at that point. If we approach the point from one side, the slope approaches a certain value, but if we approach the point from the other side, the slope approaches a different value. A derivative can't be defined as more than one slope at a time. Is that right?" Yes, that geometrically describes the situation. It's not rigorous, but it's a good way to talk about it informally.

Then you ask, "What about a function where two different curves terminate and meet at a single point, but their slopes approach the same value at the point? Is such a two-part function differentiable at the point where the curves come together?" That is a great question! You'll learn the answer (in one case, anyway) as you work out Exercise 10 at the end of this chapter.

Here's a challenge!

Look once again at the function graphed in Fig. 4-3. Show that it's nondifferentiable at the point where $x = 2$.

Solution

There are two components to the function. When the input is smaller than 2, the output is the square of the input. When the input is 2 or larger, the output has a constant value of 4. We can express this situation by writing

$$f(x) = x^2 \quad \text{when } x < 2$$
$$= 4 \quad \text{when } x \geq 2$$

The function is defined and continuous over the set of real numbers, but the slope of its graph abruptly changes at the point where $x = 2$. We can attempt to find the derivative at the point where $x = 2$ by evaluating

$$\lim_{\Delta x \to 0} [f(2 + \Delta x) - f(2)] / \Delta x$$

from both the right and the left. Let's do it from the right first. We have

$$\lim_{\Delta x \to 0+} [f(2 + \Delta x) - f(2)] / \Delta x$$

When Δx is small and positive, we're in the part of the function with a constant value of 4. We're told that $f(2) = 4$. When we substitute 4 in place of $f(2 + \Delta x)$, and substitute 4 in place of $f(2)$ in the above expression, it becomes

$$\lim_{\Delta x \to 0+} (4 - 4) / \Delta x$$

which simplifies to

$$\lim_{\Delta x \to 0+} 0/\Delta x$$

and further to

$$\lim_{\Delta x \to 0+} 0$$

This limit is obviously equal to 0. When we examine the graph, that conclusion seems reasonable, because the part of the graph to the right of the point where $x = 2$ appears as a straight, horizontal line.

Now let's go to the left of the point where $x = 2$, into the part of the graph that has a parabolic shape. When Δx is small and negative, we're in the zone where the function squares the input. We now must work with the limit

$$\lim_{\Delta x \to 0-} [f(2 + \wedge x) - f(2)] / \Delta x$$

We must not forget that $f(2) = 4$. When we substitute $(2 + \Delta x)^2$ for $f(2 + \Delta x)$, and substitute 4 for $f(2)$ in the above expression, it becomes

$$\lim_{\Delta x \to 0-} [(2 + \Delta x)^2 - 4] / \Delta x$$

When we square the binomial, we get

$$\lim_{\Delta x \to 0-} [4 + 4\Delta x + (\Delta x)^2 - 4] / \Delta x$$

which can be simplified to

$$\lim_{\Delta x \to 0-} [4\Delta x + (\Delta x)^2] / \Delta x$$

and further to

$$\lim_{\Delta x \to 0-} 4 + \Delta x$$

This limit is equal to 4. That's a sensible outcome. The slope of the parabola, if it were complete, would be 4 at the point where $x = 2$. But that's not the same value as we got when we evaluated the right-hand limit. Having found a disagreement between the right-hand and left-hand limits of the slope, we know that this function is nondifferentiable at the point where $x = 2$.

- -

Practice Exercises

This is an open-book quiz. You may (and should) refer to the text as you solve these problems. Don't hurry! You'll find worked-out answers in App. A. The solutions in the appendix may not represent the only way a problem can be figured out. If you think you can solve a particular problem in a quicker or better way than you see there, by all means try it!

1. Sketch a graph of the following function:

$$f(x) = 1/x \quad \text{when } x < -1$$
$$= -1 \quad \text{when } x \geq -1$$

   At what point or points does it appear, based on the graph, that this function is nondifferentiable?

2. Verify the answer to Prob. 1 mathematically.

3. Sketch a graph of the following function:

$$f(x) = x \quad \text{when } x \leq 3$$
$$= 1 \quad \text{when } x > 3$$

   At what point or points does it appear, based on the graph, that this function is nondifferentiable?

4. Verify the answer to Prob. 3 mathematically.

5. Sketch a graph of the following function, noting the subtle difference between it and the function described in Prob. 3:

$$f(x) = x \quad \text{when } x < 3$$
$$= 1 \quad \text{when } x \geq 3$$

At what point or points does it appear, based on the graph, that this function is nondifferentiable?

6. Verify the answer to Prob. 5 mathematically.

7. Sketch a graph of the following function:

$$f(x) = x^2 \quad \text{when } x < 1$$
$$= x^3 \quad \text{when } x \geq 1$$

At what point or points does it appear, based on the graph, that this function is nondifferentiable?

8. Verify the answer to Prob. 7 mathematically.

9. Sketch a graph of the following function:

$$f(x) = x^2 \quad \text{when } x < 1$$
$$= 2x - 1 \quad \text{when } x \geq 1$$

At what point or points does it appear, based on the graph, that this function is nondifferentiable?

10. Verify the answer to Prob. 9 mathematically.

CHAPTER

5

Differentiating Polynomial Functions

In this chapter, we'll learn how to differentiate a function that raises a variable to a large integer power. We'll learn how to differentiate the sum of two functions. Then we'll combine these rules to differentiate functions that can be written in polynomial form.

Power Rule

In Chap. 3, we learned how to find the derivatives of *monomial* (single-term) quadratic and cubic functions. Now let's develop a rule for finding the derivative of any monomial function of the form

$$f(x) = ax^n$$

where a is a real-number constant and n is a positive integer larger than 2. We'll call this the *power rule for derivatives.*

Power of a binomial

Think of the binomial $a + b$, where a and b are nonzero real numbers. (The value of b can be either positive or negative, so we can stick with the plus sign in the notation.) When such an expression is raised to successive positive integer powers, it generates an increasingly complicated series of terms, each of which consists of a power of a, a power of b, or a product of powers of a and b. Consider

$$(a + b)^n = (a + b)(a + b)(a + b) \cdots (n \text{ times})$$

where n is an integer larger than 2. When we multiply this expression out in full (a tedious process if n is large), we always get a polynomial that starts with a^n, followed by terms that consist of successively smaller powers of a multiplied by successively larger powers of b, until we get to the last term, which is b^n. Each term in the multiplied-out polynomial, containing

71

powers of both *a* and *b*, has a numerical *coefficient*, a constant by which the product of powers of *a* and *b* is multiplied. Here are the first few cases:

$$(a + b)^3 = a^3 + 3a^2 b + 3ab^2 + b^3$$
$$(a + b)^4 = a^4 + 4a^3 b + 6a^2 b^2 + 4ab^3 + b^4$$
$$(a + b)^5 = a^5 + 5a^4 b + 10a^3 b^2 + 10a^2 b^3 + 5ab^4 + b^5$$
$$(a + b)^6 = a^6 + 6a^5 b + 15a^4 b^2 + 20a^3 b^3 + 15a^2 b^4 + 6ab^5 + b^6$$

As long as $n > 2$, we will always notice the following facts about the multiplied-out polynomial:

- The first term is a^n.
- The second term is $na^{(n-1)}b$.
- We can factor b^2 out of the third term and each term after that.

Now instead of *a* and *b*, let's call the terms in the binomial by different names. How about x_0 and Δx? That gives us

$$(x_0 + \Delta x)^n = (x_0 + \Delta x)(x_0 + \Delta x)(x_0 + \Delta x) \cdots (n \text{ times})$$

When we multiply this out, we get a polynomial that starts with $x_0{}^n$, followed by terms that consist of successively smaller powers of x_0 times successively larger powers of Δx, until we get to the last term, which is $(\Delta x)^n$. We always find that:

- The first term is $x_0{}^n$.
- The second term is $nx_0{}^{(n-1)}\Delta x$.
- We can factor $(\Delta x)^2$ out of the third term and each one after that.

Deriving the rule

Now we'll work out a formula that tells us the derivative of a function *f* in which we take the *n*th power of a real variable *x* and then multiply by a real constant *a*. It's a monomial of the form

$$f(x) = ax^n$$

where *n* is an integer larger than 2. Let's determine f' at a nonspecific point (x_0, y_0). Imagine that, in the *xy*-plane, we plot the graph of

$$y = ax^n$$

as a curve. The domain is always the entire set of reals, no matter what *n* happens to be. The range is the set of nonnegative reals if *n* is even, and the set of all reals if *n* is odd. (Try graphing a few functions of this form, and see for yourself!) The *x*-value of a movable point (x,y) in the vicinity of (x_0, y_0) is

$$x = x_0 + \Delta x$$

The *y*-value of the movable point is

$$y = ax^n = a(x_0 + \Delta x)^n = a[x_0{}^n + nx_0{}^{(n-1)}\Delta x + S]$$

where *S* is a series, each term of which has a factor of $(\Delta x)^2$, so

$$S = r(\Delta x)^2$$

where *r* is some real constant built up by adding and multiplying other constants. Substituting $r(\Delta x)^2$ for *S* in the above equation, we can say that the *y*-value of the movable point is

$$y = a[x_0{}^n + nx_0{}^{(n-1)}\Delta x + r(\Delta x)^2] = ax_0{}^n + anx_0{}^{(n-1)}\Delta x + ar(\Delta x)^2$$

The slope of a line through (x_0, y_0) and (x, y), which approximates $f'(x_0)$, is

$$\Delta y / \Delta x = (y - y_0) / (x - x_0)$$

The function tells us that $y_0 = ax_0{}^n$. In the above equation, let's substitute

- The quantity $[ax_0{}^n + anx_0{}^{(n-1)}\Delta x + ar(\Delta x)^2]$ for *y*
- The quantity $ax_0{}^n$ for y_0
- The quantity $(x_0 + \Delta x)$ for *x*

That gives us

$$\Delta y / \Delta x = [ax_0{}^n + anx_0{}^{(n-1)}\Delta x + ar(\Delta x)^2 - ax_0{}^n] / (x_0 + \Delta x - x_0)$$

which simplifies to

$$\Delta y / \Delta x = [anx_0{}^{(n-1)}\Delta x + ar(\Delta x)^2] / \Delta x$$

and further to

$$\Delta y / \Delta x = anx_0{}^{(n-1)} + ar\Delta x$$

The derivative $f'(x_0)$ is therefore

$$f'(x_0) = \lim_{\Delta x \to 0} anx_0{}^{(n-1)} + ar\Delta x$$

As Δx approaches 0, the second addend, $ar\Delta x$, approaches 0 because it is a constant multiple of Δx. The first addend, $anx_0{}^{(n-1)}$, doesn't change no matter what Δx happens to be, because Δx isn't a factor in this term at all. The limit of the entire expression is therefore $anx_0{}^{(n-1)}$, telling us that

$$f'(x_0) = anx_0{}^{(n-1)}$$

for any real number x_0. Therefore, any function of the form

$$f(x) = ax^n$$

has the derivative

$$f'(x) = anx^{(n-1)}$$

where a is a nonzero real constant and n is an integer larger than 2.

- -

Are you confused?

You might wonder, "Why do we restrict n to integers larger than 2? Won't the above formula work if $n = 1$ or $n = 2$?" Yes, it will. We covered those cases in the linear and quadratic examples that we worked out in Chap. 3, so we don't have to do them again. To make the above derivation work, we had to set things up to get a factor of $(\Delta x)^2$ out of every term in the polynomial after the second one. That can only happen if $n > 2$.

Here's a challenge!

Consider the function $f(x) = x^4$. Sketch a graph of this function for values of x between, and including, 0 and 1. Using a calculator, find the value of x (call it x_0) at which the slope of the curve is equal to 1. Round the answer off to three decimal places.

Solution

Figure 5-1 shows this situation. First, let's find the derivative using the power rule, letting $a = 1$ and $n = 4$. We have

$$f'(x) = 4x^{(4-1)} = 4x^3$$

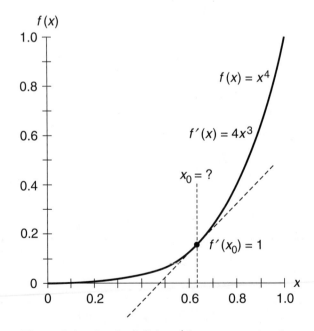

Figure 5-1   Graph of $f(x) = x^4$ for $0 \leq x \leq 1$. At what value of x is the derivative equal to 1?

We can solve the following equation to get the value of x for which the slope of the curve is equal to 1:

$$4x_0{}^3 = 1$$

Dividing through by 4, we get

$$x_0{}^3 = 1/4$$

Taking the 1/3 power of each side tells us that

$$x_0 = (1/4)^{1/3}$$

We should remember that the 1/3 power is the *positive real cube root*. We're not interested in nonreal complex-number roots in this case, because we're restricted to interval between, and including, the real numbers 0 and 1. A calculator produces the result, rounded to three decimal places, of

$$x_0 = 0.630$$

- -

Sum Rule

In Chap. 3, we found derivatives of a few functions that were sums of other functions. Perhaps you noticed that the derivative of the whole sum was the same as the sum of the derivatives of its addends. That was no accident!

Definition and notation

Let's be sure we know what's meant by the *sum of two functions*. When we want to add two functions, we simply add the expressions that define those functions. For example, if

$$f_1(x) = 6x^3$$

and

$$f_2(x) = -3x$$

then

$$f_1(x) + f_2(x) = 6x^3 - 3x$$

Here's another example. If

$$f_1(x) = -2x^2$$

and

$$f_2(x) = 5x + 3$$

then

$$f_1(x) + f_2(x) = -2x^2 + 5x + 3$$

When two functions f_1 and f_2 operate on the same variable such as x, we can denote their sum in various ways. Here are the three most common formats:

$$f_1(x) + f_2(x)$$
$$(f_1 + f_2)(x)$$
$$f_1 + f_2$$

The rule in brief

The sum rule for two derivatives tells us this:

- The derivative of the sum of two differentiable functions of a single real variable is equal to the sum of their derivatives.

This rule always works, as long we stick with the same variable in both functions, and as long as the functions are both differentiable at all the points that interest us. Stated symbolically, if f_1 and f_2 are differentiable functions of the same variable, then

$$(f_1 + f_2)' = f_1' + f_2'$$

-- -- -- -- -- -- -- -- -- -- -- -- -- -- -- -- -- -- -- --

Are you confused?

Have you noticed that sometimes we write $f(x)$ to denote a function and $f'(x)$ to denote its derivative, and then later we write f and f' to denote the same things? Don't let this bother you. As long as we know that f and f' operate on the same variable, it doesn't matter whether or not we include the variable in parentheses after the name of the function.

When we want to be sure we know which variable we're talking about, then it's a good idea to include the parentheses and the variable. This can be important in *multi-variable calculus,* which we'll study later in this book. It's also important if we're talking about something other than the independent variable itself, such as the number 4 or the expression $(x + \Delta x)$.

-- -- -- -- -- -- -- -- -- -- -- -- -- -- -- -- -- -- -- --

Example

Consider the following functions of x. Note that the third function is the sum of the first two:

$$f_1(x) = 6x^3$$
$$f_2(x) = -3x$$
$$(f_1 + f_2)(x) = 6x^3 - 3x$$

We can find the derivatives of the first two functions using rules we've already learned:

$$f_1'(x) = 18x^2$$
$$f_2'(x) = -3$$

The sum of the derivatives is

$$f_1'(x) + f_2'(x) = 18x^2 - 3$$

If we use the definition at the end of Chap. 3 to calculate the derivative of the sum function using limits, we will get

$$(f_1 + f_2)'(x) = 18x^2 - 3$$

which is the same as the result we get when we find the derivatives of component functions separately and then add them.

Another example

Let's look at an example involving functions of the variable y. Again, the third function is the sum of the first two:

$$f_1(y) = -2y^2$$
$$f_2(y) = 5y + 3$$
$$(f_1 + f_2)(y) = -2y^2 + 5y + 3$$

When we find the derivatives of these two component functions, we get

$$f_1'(y) = -4y$$
$$f_2'(y) = 5$$

The sum of these derivative functions is

$$f_1'(y) + f_2'(y) = -4y + 5$$

Again using the definition at the end of Chap. 3, we can show that

$$(f_1 + f_2)'(y) = -4y + 5$$

This is what we get when we determine the derivatives separately and add them.

Mathematical induction

If you've taken any number-theory or set-theory courses, you've learned about a technique called *mathematical induction*. If you haven't seen this technique before, or if you've forgotten how it works, here's a brief review. Mathematical induction allows you to prove infinitely many facts in a finite number of steps.

Imagine an infinite sequence of statements called S_1, S_2, S_3, and so on. Suppose that you want to prove all these statements true. You can't prove every statement one by one, because you don't have an infinite amount of time. But suppose that you can demonstrate two things:

- The first statement S_1 is true, and
- If a statement S_n (where n can be any positive integer) in the sequence is true, then the next statement S_{n+1} in the sequence is also true.

Once you have done these two things, the first statement "automatically" proves the second statement. That, in turn, "automatically" proves the third statement. The logical process can go on forever, like an endless chain reaction! It proves that all the statements in the sequence are true, even if there are infinitely many of them.

- -

Are you confused?

By now you might wonder how mathematical induction can help you prove anything that has to do with adding derivatives. The following "challenge" will answer that question!

Here's a challenge!

Extrapolate the sum rule for two derivatives to show that it works for the sum of any number of derivatives.

Solution

The sum rule for two derivatives says that for any two differentiable functions f_1 and f_2,

$$(f_1 + f_2)' = f_1' + f_2'$$

Let's call this statement S_1 in an infinite sequence of truths we intend to prove. We are given the fact that this initial statement is true. Now imagine an unlimited supply of differentiable functions called $f_1, f_2, f_3, \ldots$, and so on, forever! According to our sequencing scheme, the second statement, S_2 (which we haven't proven yet) is

$$(f_1 + f_2 + f_3)' = f_1' + f_2' + f_3'$$

The third statement, S_3 (which, again, remains to be proved) is

$$(f_1 + f_2 + f_3 + f_4)' = f_1' + f_2' + f_3' + f_4'$$

Imagine that we have been assured of the truth of statement S_n saying that, for any $n + 1$ differentiable functions $f_1, f_2, f_3, \ldots$, and so on up to f_{n+1},

$$(f_1 + f_2 + f_3 + \cdots + f_{n+1})' = f_1' + f_2' + f_3' + \cdots f_{n+1}'$$

We must show that if S_n is true, then the next statement S_{n+1} is also true, which would tell us that for any $n + 2$ differentiable functions $f_1, f_2, f_3, \ldots$, and so on up to f_{n+2},

$$(f_1 + f_2 + f_3 + \cdots + f_{n+1} + f_{n+2})' = f_1' + f_2' + f_3' + \cdots + f_{n+1}' + f_{n+2}'$$

Now let's temporarily change a couple of names, like this:

- Call the sum of functions $f_1 + f_2 + f_3 + \cdots + f_{n+1}$ by the nickname g
- Call the sum of functions $f_1' + f_2' + f_3' + \cdots + f_{n+1}'$ by the nickname h

On that basis, we can rewrite S_n as

$$g' = h$$

Let's pick a function "out of the air" and call it f_{n+2}. (We can call it anything we want, so why not that?) We add this new function to g and then take the derivative:

$$(g + f_{n+2})'$$

According to the sum rule for two derivatives, we know that

$$(g + f_{n+2})' = g' + f_{n+2}'$$

We also know that $g' = h$, so we can substitute in the above equation to get

$$(g + f_{n+2})' = h + f_{n+2}'$$

When we give g and h their original names back, the above expression expands to

$$(f_1 + f_2 + f_3 + \cdots + f_{n+1} + f_{n+2})' = f_1' + f_2' + f_3' + \cdots + f_{n+1}' + f_{n+2}'$$

That's exactly the statement S_{n+1}! Having fulfilled the requirements of the mathematical induction routine, we're entitled to claim that the derivative of a sum is equal to the sum of the derivatives, no matter how many functions are involved. We can now call this rule simply the *sum rule for derivatives*.

- -

Summing the Powers

If we give a function the name f and tell it to operate on a variable x, then f is an *nth-degree polynomial function* if and only if it can be written in this form:

$$f(x) = a_n x^n + a_{n-1} x^{n-1} + a_{n-2} x^{n-2} + \cdots + a_1 x + a_0$$

where each addend is called a *term*, the subscripted letters a_1, a_2, a_3, $\ldots$, and a_n represent real numbers called the *coefficients of the terms,* the subscripted letter a_0 represents the *stand-alone constant* or *constant term,* and n is a positive integer.

How it looks

Here are some examples of polynomial functions. They are of degree 4, 5, 7, and 11, respectively:

$$f_1(x) = 6x^4 - 3x^3 + 3x^2 + 2x + 5$$
$$f_2(x) = 3x^5 - 4x^3$$
$$f_3(x) = -x^7 - 5x^4 + 3x^3 - x^2 - 29$$
$$f_4(x) = -4x^{11}$$

In all but the first of these functions, some of the coefficients are equal to 0. The coefficient a_n by which x^n is multiplied, called the *leading coefficient,* can't be 0 in an nth-degree polynomial function. If we set $a_n = 0$ in a polynomial function $f(x)$, we end up with

$$f(x) = 0x^n + a_{n-1}x^{n-1} + a_{n-2}x^{n-2} + \cdots + a_1x + a_0$$

This expression is not technically wrong, but it contains a useless term. It's really a single-variable polynomial function of degree $n - 1$:

$$f(x) = a_{n-1}x^{n-1} + a_{n-2}x^{n-2} + \cdots + a_1x + a_0$$

We assume, of course, that $a_{n-1} \neq 0$!

Break it down

Look closely at the general form for an nth-degree polynomial function. Then look at each of the four examples above. Every one of these functions has something in common. They're all sums of *monomial power functions.* For example,

$$g(x) = 6x^4 - 3x^3 + 3x^2 + 2x + 5$$

is the sum of a monomial fourth-degree (or *quartic*) function, a monomial cubic function, a monomial quadratic function, a monomial linear function, and a constant function. Let's give each of these functions the name of g with a subscript indicating its degree, like this:

$$g_4(x) = 6x^4$$
$$g_3(x) = -3x^3$$
$$g_2(x) = 3x^2$$
$$g_1(x) = 2x$$
$$g_0(x) = 5$$

We can write $g(x)$ as the sum of these:

$$g(x) = g_4(x) + g_3(x) + g_2(x) + g_1(x) + g_0(x)$$

We've learned that the derivative of a sum is always equal to the sum of the derivatives. We've also learned how to differentiate any function that takes the variable to a nonnegative integer power and then multiplies it by a constant. Therefore, we have the tools to differentiate any polynomial function.

Differentiate each term

Once we've broken a polynomial function down into its terms, we can use the power rule to differentiate each one of them. In the above situation, we get these derivatives:

$$g_4'(x) = 24x^3$$
$$g_3'(x) = -9x^2$$
$$g_2'(x) = 6x$$
$$g_1'(x) = 2$$
$$g_0'(x) = 0$$

Put it back together

To do the final step in working out the derivative of the polynomial function, we add the derivatives of the individual terms together in order of highest-to-lowest powers:

$$g'(x) = g_4'(x) + g_3'(x) + g_2'(x) + g_1'(x) + g_0'(x)$$
$$= 24x^3 - 9x^2 + 6x + 2 + 0 = 24x^3 - 9x^2 + 6x + 2$$

- -

Are you confused?

Do you wonder what happens if we try to find the derivative of a function that raises the variable to a negative-integer power such as −5, or to a non-integer rational power such as 3/5 or −3.7, or even to an irrational power such as π or the square root of 2? What about functions that don't involve exponents, such as the sine or cosine? We'll explore the derivatives of some such functions in Chap. 7.

Here's a challenge!

The sum rule for two derivatives, stated earlier in this chapter, says that if f_1 and f_2 are differentiable functions of the same variable, then

$$(f_1 + f_2)' = f_1' + f_2'$$

We've seen some examples of this rule "in action," but we haven't proved it yet. Now is the time!

Solution

Before we begin the proof, let's make four changes to the notation:

- Include the name of the variable (we'll use x)
- Change the names of the functions to f and g
- Write δ (lowercase Greek delta) instead of Δx to represent a shrinking increment
- Leave out "$\delta \to 0$" beneath "*Lim*" (but remember that it's implied)

These changes will make the expressions clearer than they would be otherwise. They'll also help you get used to some of the alternative ways that things can be written down in calculus. We want to prove that if *f* and *g* are differentiable functions of *x*, then

$$(f + g)'\,(x) = f'\,(x) + g'\,(x)$$

According to the definition at the end of Chap. 3, the derivative of our sum function, written the "new way," is

$$(f + g)'\,(x) = Lim\ [(f + g)\,(x + \delta) - (f + g)\,(x)] / \delta$$

We can rewrite this as

$$Lim\ \{[f(x + \delta) + g\,(x + \delta)] - [f(x) + g\,(x)]\} / \delta$$

which can be rearranged to

$$Lim\ [f(x + \delta) + g\,(x + \delta) - f(x) - g\,(x)] / \delta$$

and further to

$$Lim\ \{[f(x + \delta) - f(x)] + [g\,(x + \delta) - g\,(x)]\} / \delta$$

and still further to

$$Lim\ [f(x + \delta) - f(x)] / \delta + [g\,(x + \delta) - g\,(x)] / \delta$$

In Chap. 2, we learned that the limit of a sum is equal to the sum of the limits. Using that rule, we can split the above limit into a sum of two limits, getting

$$Lim\ [f(x + \delta) - f(x)] / \delta\ +\ Lim\ [g\,(x + \delta) - g\,(x)] / \delta$$

To finish, we can apply the definition of the derivative "backward" to both of these limits and rewrite the above expression as

$$f'(x) + g'(x)$$

We started out with $(f + g)'\,(x)$ and ended up with $f'(x) + g'(x)$, showing that these two expressions are equivalent. Mission accomplished!

- -

Practice Exercises

This is an open-book quiz. You may (and should) refer to the text as you solve these problems. Don't hurry! You'll find worked-out answers in App. A. The solutions in the appendix may not represent the only way a problem can be figured out. If you think you can solve a particular problem in a quicker or better way than you see there, by all means try it!

1. Determine the derivatives of the following functions:

   (a) $f(x) = -8x^5$

   (b) $g(z) = 12z^7$

   (c) $h(t) = -21t^{21}$

2. Consider the function $f_3(x) = x^3$ in the interval $0 \leq x \leq 1$. Find x_0 in this interval, accurate to three decimal places, such that $f_3'(x_0) = 1$. (It's okay to use a calculator.)

3. Consider the function $f_5(x) = x^5$ in the interval $0 \leq x \leq 1$. Find x_0 in this interval, accurate to three decimal places, such that $f_5'(x_0) = 1$. (It's okay to use a calculator.)

4. Consider the infinite set of functions $f_n(x) = x^n$ in the interval $0 \leq x \leq 1$, where n is an integer larger than or equal to 2. Find a general expression for x_0 in this interval, such that $f_n'(x_0) = 1$. (A calculator is useless here!)

5. Sketch a multi-curve graph that shows $f_n(x) = x^n$ in the interval $0 \leq x \leq 1$ for $n = 3$, $n = 4$, and $n = 5$. On each curve, show the point x_0 at which $f_n'(x_0) = 1$ as a solid dot. To plot the points, use the values obtained in the solutions of problems we've already solved in this chapter. Draw lines tangent to the curves at these points.

6. Consider again the set of functions $f_n(x) = x^n$, where n is an integer larger than or equal to 2. Find a general expression for the value of $f_n'(1)$.

7. Consider (yet again!) the set of functions $f_n(x) = x^n$, where n is an integer larger than or equal to 2. Find a general expression for the value of $f_n'(0)$.

8. Find the derivative of the function

$$f(x) = 8x^7 + 4x^6 - 3x^5 + x^4 + x^2 - 3$$

9. Find the derivative of the function

$$f(x) = a_5 x^5 + a_4 x^4 + a_3 x^3 + a_2 x^2 + a_1 x + a_0$$

   where a_5, a_4, a_3, a_2, a_1, and a_0 are real numbers, and $a_5 \neq 0$.

10. Write an expression for the derivative of the general polynomial function

$$f(x) = a_n x^n + a_{n-1} x^{n-1} + a_{n-2} x^{n-2} + \cdots + a_2 x^2 + a_1 x + a_0$$

   where a_n, a_{n-1}, a_{n-2}, $\ldots$, a_1, and a_0 are real numbers, and $a_n \neq 0$.

6

More Rules for Differentiation

The functions we've differentiated so far have all been specialized. Often, we'll want to differentiate more general functions built up by multiplication, division, or other processes. Sometimes we'll encounter a function of another function! In this chapter, we'll learn some more useful rules for differentiation.

Multiplication-by-Constant Rule

We can take any function and multiply it by a constant, and the result is another function. There's a convenient rule that applies to derivatives when this is done.

Definition and notation

When we want to find the *product of a function and a constant,* we multiply the expression of the function by that constant. For example, if

$$f(x) = 6x^3$$

and

$$c = -3$$

then

$$c\,[f(x)] = -3 \times 6x^3 = -18x^3$$

Here's another example. If

$$f(x) = 5x^2 + 3x - 7$$

and

$$c = 2$$

then

$$c\,[f(x)] = 2 \times (5x^2 + 3x - 7) = 10x^2 + 6x - 14$$

We can write cf by itself, without including the variable in parentheses, as long as we know which variable we're dealing with. We can also write cf if we don't want to restrict f to a particular variable, or if we want to let f operate on all sorts of things. (Functions can operate on expressions more complicated than plain variables.)

The rule in brief

The multiplication-by-constant rule for differentiation tells us this:

- If we take the derivative of a differentiable function *after* it has been multiplied by a constant, we get the same result as we do if we take the derivative of the function and *then* multiply by the constant.

This rule always works, provided that the function is differentiable at every point that interests us. Stated symbolically, if f is a differentiable function and c is a real-number constant, then

$$(cf)' = c(f')$$

and

$$(cf)' = (f')c$$

Example

Let's apply this rule to the first example we saw a few moments ago:

$$f(x) = 6x^3$$

and

$$c = -3$$

When we differentiate the product of the function and the constant, we get

$$(cf)'\,(x) = d/dx\,(-3 \times 6x^3) = d/dx\,(-18x^3) = -54x^2$$

When we differentiate the function first and then multiply by the constant, we get

$$c\,[f'(x)] = -3 \times (d/dx\,6x^3) = -3 \times 18x^2 = -54x^2$$

Another example

Now let's work on the second example we saw earlier:

$$f(x) = 5x^2 + 3x - 7$$

and

$$c = 2$$

When we multiply the function by the constant before we differentiate, we get

$$(cf)'\,(x) = d\,/\,dx\,[2 \times (5x^2 + 3x - 7)] = d\,/\,dx\,(10x^2 + 6x - 14) = 20x + 6$$

When we multiply by the constant after we differentiate, we obtain

$$c\,[f'(x)] = 2 \times [d\,/\,dx\,(5x^2 + 3x - 7)] = 2 \times (10x + 3) = 20x + 6$$

- -

Are you confused?

"Wait," you might say. "What does this new notation $d\,/\,dx$ mean?" The answer is that is isn't actually new; it was introduced in Chap. 3. You can read $d\,/\,dx$ as "the derivative with respect to x." If you see

$$d\,/\,dx\,\,8x^3$$

it means "the derivative, with respect to x, of $8x^3$." The variable doesn't necessarily have to be x. If it happens to be y and you see

$$d\,/\,dy\,(y^2 + 2y + 1)$$

you would read it as "the derivative, with respect to y, of the quantity $(y^2 + 2y + 1)$."

Here's a challenge!

Prove the multiplication-by-constant rule for differentiation.

Solution

The proof of this rule is similar to the proof of the rule for the sum of two derivatives, which we carried out at the end of Chap. 5. Once again, let's make some changes to conventional notation before we begin:

- Include the name of the variable (we'll use x)
- Write δ instead of Δx to represent a shrinking increment
- Leave out "$\delta \to 0$" beneath "*Lim*" (but remember that it's implied)

We want to prove that if f is a differentiable function of x, and c is a real-number constant, then

$$(cf)'\,(x) = c\,[f'(x)]$$

According to the definition at the end of Chap. 3, the derivative of our product function, written the "new way," is

$$(cf)' (x) = Lim \ [(cf) (x + \delta) - (cf) (x)] / \delta$$

We can rewrite the right side of this equation as

$$Lim \ \{c [f(x + \delta)] - c[f(x)]\} / \delta$$

which can be rearranged to

$$Lim \ c \{[f(x + \delta) - f(x)] / \delta\}$$

In Chap. 2, we learned that a constant times the limit of an expression is equal to the limit of the expression times the constant. Using that rule, we can "pull out the constant" and rewrite the above expression as

$$c \{Lim \ [f(x + \delta) - f(x)] / \delta\}$$

Inside the curly braces, we have the derivative of the function *f* all by itself. That's because, by definition,

$$Lim \ [f(x + \delta) - f(x)] / \delta = f' (x)$$

Therefore,

$$c \{Lim \ [f(x + \delta) - f(x)] / \delta\} = c[f' (x)]$$

We began with the expression $(cf)' (x)$ and derived $c[f' (x)]$ from it, showing that these two expressions are equivalent.

- -

Product Rule

When two functions are multiplied, the result is another function. Sometimes a function can be broken down into a product of two simpler functions. But the derivative of the *product of two functions* is not the simple product of the derivatives!

Definition and notation

To determine the product of two functions, we multiply the expression of the first function by the expression of the second. For example, if

$$f(x) = 6x^3$$

and

$$g (x) = -3x$$

then

$$fg(x) = (6x^3)(-3x) = -18x^4$$

Here's another example. If we have

$$f(x) = (x^2 + 2x)$$

and

$$g(x) = (2x^2 - 3)$$

then

$$fg(x) = (x^2 + 2x)(2x^2 - 3) = 2x^4 + 4x^3 - 3x^2 - 6x$$

When we want to denote the product of functions f and g, we can write fg without including the variable in parentheses, as long as both functions operate on the same variable or expression.

The rule in brief

The two-function product rule for differentiation tells us this:

- To find the derivative of the product of two differentiable functions, we multiply the derivative of the first function by the second function, then multiply the derivative of the second function by the first function, and finally add the two products.

For this rule to apply, both functions must be differentiable at every point that interests us. Stated symbolically, if f and g are differentiable functions, then

$$(fg)' = f'g + g'f$$

Example

Let's see how this rule works, using the examples we saw a few moments ago. First, consider

$$f(x) = 6x^3$$

and

$$g(x) = -3x$$

When we differentiate the product of the functions, we get

$$(fg)'(x) = d/dx\,[(6x^3)(-3x)] = d/dx\,(-18x^4) = -72x^3$$

When we use the product rule, we get

$$[f'(x)][g(x)] + [g'(x)][f(x)] = [d/dx\,(6x^3)](-3x) + [d/dx\,(-3x)](6x^3)$$

$$= (18x^2)(-3x) + (-3)(6x^3) = -72x^3$$

Another example

In the second situation, we have

$$f(x) = (x^2 + 2x)$$

and

$$g(x) = (2x^2 - 3)$$

When we differentiate the product of the functions, we get

$$(fg)'(x) = d/dx\,[(x^2 + 2x)(2x^2 - 3)] = d/dx\,(2x^4 + 4x^3 - 3x^2 - 6x)$$

$$= 8x^3 + 12x^2 - 6x - 6$$

When we use the product rule, we get

$$[f'(x)][g(x)] + [g'(x)][f(x)] = [d/dx\,(x^2 + 2x)](2x^2 - 3) + [d/dx\,(2x^2 - 3)](x^2 + 2x)$$

$$= (2x + 2)(2x^2 - 3) + (4x)(x^2 + 2x) = 8x^3 + 12x^2 - 6x - 6$$

- -

Are you confused?

You must be careful when writing the product of two functions. It's easy to inadvertently write the expression for a function of a function instead. For example, if you put down

$$f[g(x)]$$

you tell your readers to apply the function g to the variable x, and then apply the function f to the variable $g(x)$. In other words, you indicate "f of g of x." If you want to denote the product "f of x times g of x," you should write

$$[f(x)][g(x)]$$

or

$$f(x) \times g(x)$$

If you don't want to specify the independent variable, things get trickier. If you write

$$f(g)$$

you indicate "*f* of *g*," meaning that function *f* should be applied to the output of function *g*. If you want to show "*f* times *g*," you should write

$$fg$$

or

$$f \times g$$

Here's a challenge!

Based on the rules we've learned so far, prove that if f_1 and f_2 are differentiable functions of the same variable, and if a_1 and a_2 are constants, then

$$(a_1 f_1 + a_2 f_2)' = a_1(f_1') + a_2(f_2')$$

This is called the linear combination rule for differentiation.

Solution

This proof is amazingly simple, once we realize that $a_1 f_1$ and $a_2 f_2$ are both functions, so they can both be treated that way in all respects. Therefore, according to the sum rule for two derivatives, we have

$$(a_1 f_1 + a_2 f_2)' = (a_1 f_1)' + (a_2 f_2)'$$

The multiplication-by-constant rule for differentiation tells us that

$$(a_1 f_1)' = a_1(f_1')$$

and

$$(a_2 f_2)' = a_2(f_2')$$

Substituting these expressions in the first equation, we obtain

$$(a_1 f_1 + a_2 f_2)' = a_1(f_1') + a_2(f_2')$$

- -

Reciprocal Rule

We can take the *reciprocal of a function* (as long as it's not the zero function) and get another function. But in general, the derivative of a reciprocal is not the reciprocal of the derivative.

There's another complication, too. We must be careful about specifying the domain when we work with the reciprocal of a function.

Definition and notation

To find the reciprocal of a function, we divide 1 by the expression of the function. For example, if

$$f(x) = x^2$$

then

$$1 / [f(x)] = 1 / (x^2) = x^{-2}$$

Here's another example. If

$$f(x) = 2x^2 - 8$$

then

$$1 / [f(x)] = 1 / (2x^2 - 8) = (2x^2 - 8)^{-1}$$

Caution!

We can express the reciprocal of $f(x)$ by writing $[f(x)]^{-1}$, but this brings us near dangerous territory! We can't write an exponent -1 immediately after f to indicate the reciprocal of f. If we put down f^{-1}, we denote the *inverse function,* not the reciprocal function.

With numbers and algebraic expressions, the terms "inverse" and "reciprocal" are often used interchangeably, because the multiplicative inverse and the reciprocal are the same thing. But when we deal with functions, there's a big difference between an inverse and a reciprocal. Suppose we have a function f of a variable x such that

$$f(x) = 5x$$

Then the reciprocal function is

$$[f(x)]^{-1} = 1/(5x)$$

But the inverse function is

$$f^{-1}(x) = x/5$$

The reciprocal is 1 divided by the function. The inverse is a completely different function; it "undoes" or "reverses" the work of the original function.

The rule in brief

The reciprocal rule for differentiation tells us this:

- To find the derivative of the reciprocal of a differentiable function, we must first find the derivative of the original function, then multiply by −1, and finally divide by the square of the original function.

Stated symbolically, if *f* is a differentiable function, then

$$(1/f)' = -f'/(f^2)$$

The square of a function is, as we should expect, the expression of the function multiplied by itself.

For this rule to apply, the function *f* must be differentiable at every point that interests us. Generally, that means that *f* can't have a gap, take a jump, turn a corner, or blow up. In addition, *f* can't attain the value 0 anywhere in its domain. If it does, we end up having to divide by 0 to take the reciprocal.

More caution!

If a function is differentiable throughout its domain, then the reciprocal function is not necessarily differentiable at every point in its domain. In the example above, the function

$$f(x) = 5x$$

is defined over the entire set of real numbers *x*, but the reciprocal function

$$[f(x)]^{-1} = 1/(5x)$$

is not defined for *x* = 0. There's a discontinuity, so the reciprocal function is not differentiable over the entire set of reals. We can make the reciprocal function differentiable only if we restrict the domain to keep the discontinuity out.

Example

Let's apply the reciprocal rule for differentiation to a couple of common functions. First, we'll look at

$$g(x) = 1/x^2$$

This is the reciprocal of a function we can call *f*(*x*), as follows:

$$f(x) = x^2$$

The negative of the derivative of *f*(*x*) is

$$-f'(x) = -2x$$

and the square of $f(x)$ is

$$[f(x)]^2 = (x^2)^2 = x^4$$

We can now set up the derivative of $g(x)$ using the rule:

$$g'(x) = \{1 / [f(x)]\}' = -f'(x) / [f(x)]^2 = -2x / x^4 = -2/x^3$$

This holds only as long as $x \neq 0$. If $x = 0$, then $g(x)$ is undefined, so its derivative is undefined as well.

Another example

Now let's look at another example we saw earlier in this chapter. Let

$$h(x) = 1 / (2x^2 - 8)$$

which is the reciprocal of a function we can call $f(x)$, like this:

$$f(x) = 2x^2 - 8$$

Taking the derivative of $f(x)$, we get

$$f'(x) = 4x$$

Multiplying by -1 gives us

$$-f'(x) = -4x$$

The square of $f(x)$ is

$$[f(x)]^2 = (2x^2 - 8)^2 = 4x^4 - 32x^2 + 64$$

Using the reciprocal rule for differentiation, we obtain

$$h'(x) = \{1 / [f(x)]\}' = -f'(x) / [f(x)]^2 = -4x / (4x^4 - 32x^2 + 64)$$

$$= -x / (x^4 - 8x^2 + 16)$$

If $x = 2$ or $x = -2$, then $h(x)$ is undefined, so $h'(x)$ is also undefined. Those values of x are the zeros of $f(x)$, where

$$2x^2 - 8 = 0$$

The function $h(x)$ is nondifferentiable at the points where $x = 2$ and $x = -2$.

- -

Are you confused?

Whenever you apply the reciprocal rule, a function appears in the denominator of a fraction. When that denominator function attains the value 0, a discontinuity occurs, and the main function is not differentiable at that point. You might wonder, "Does this mean that *any* reciprocal function has at least one point at which it can't be differentiated? If that's true, then reciprocal functions are *never* differentiable in the general sense." Well, as things work out, plenty of reciprocal functions are differentiable over the entire set of real numbers. You'll encounter an example as you work through the exercises at the end of this chapter.

Here's a challenge!

Find the derivative of the function

$$q(x) = 1 / (x^2 + x - 2)$$

in the set of real numbers. Indicate the real values of x, if any, for which this function is nondifferentiable.

Solution

This is the reciprocal of a function we can call $f(x)$, as follows:

$$f(x) = x^2 + x - 2$$

Taking the derivative of $f(x)$, we get

$$f'(x) = 2x + 1$$

Multiplying by -1 gives us

$$-f'(x) = -2x - 1$$

The square of $f(x)$ is

$$[f(x)]^2 = (x^2 + x - 2)^2 = x^4 + 2x^3 - 3x^2 - 4x + 4$$

Using the reciprocal rule for differentiation, we obtain

$$q'(x) = \{1 / [f(x)]\}' = -f'(x) / [f(x)]^2 = (-2x - 1) / (x^4 + 2x^3 - 3x^2 - 4x + 4)$$

If $x = 1$ or $x = -2$, then $q(x)$ is undefined, so $q'(x)$ is also undefined. Those values of x are the zeros of

$$f(x) = x^2 + x - 2$$

which forms the denominator of $q(x)$. We can find these zeros either by factoring the quadratic or by using the quadratic formula.

- -

Quotient Rule

We can take the *quotient of two functions* and get another function, but we must restrict the domain to be sure that the second function (the denominator or divisor) cannot attain the value 0. The derivative of a quotient is not, in general, merely the quotient of the derivatives. It's a little more complicated than that.

Definition and notation

When we want to divide a function by second function, we divide the expression of the first function by the expression of the second one. We find the ratio of the expressions that define the functions. Consider

$$f(x) = 8x^2$$

and

$$g(x) = 4x$$

In this case,

$$[f(x)] / [g(x)] = (8x^2) / (4x) = 2x$$

provided $x \neq 0$. Here's another example. Suppose we have

$$f(x) = x^2 - x - 20$$

and

$$g(x) = x - 5$$

The quotient in this case, as long as $x \neq 5$, is

$$[f(x)] / [g(x)] = (x^2 - x - 20) / (x - 5) = x + 4$$

The rule in brief

The quotient rule for differentiation tells us this:

- To find the derivative of the quotient of two differentiable functions, we multiply the derivative of the first function by the second function, then multiply the derivative of the second function by the first function, then subtract the second product from the first product, and finally divide by the square of the second function.

If we want to use this rule, then both functions, as well as their ratio, must be differentiable at every point that interests us. (Sometimes a derivative exists for the quotient of two functions even when this rule can't be applied. We'll see an example of this shortly.) In addition, the denominator function can't attain the value 0 anywhere in its domain. Stated symbolically, if *f* and *g* are differentiable functions, then

$$(f/g)' = (f'g - g'f) / g^2$$

for all points where *g* does not become 0.

Example

Let's see how this rule works, using the examples from above. First, consider

$$f(x) = 8x^2$$

and

$$g(x) = 4x$$

When we differentiate the quotient of the functions, we get

$$(f/g)'(x) = d/dx\,[(8x^2) / (4x)] = d/dx\,(2x) = 2$$

When we use the quotient rule, we get

$$\begin{aligned}(f/g)'(x) &= \{[f'(x)][g(x)] - [g'(x)][f(x)]\} / [g(x)]^2\\ &= \{[d/dx\,(8x^2)](4x) - [d/dx\,(4x)](8x^2)\} / (4x)^2\\ &= [(16x)(4x) - 4 \times (8x^2)] / (16x^2) = (64x^2 - 32x^2) / (16x^2)\\ &= (32x^2) / (16x^2) = 2\end{aligned}$$

This works only as long as $g(x) \neq 0$. That means the derivative of the quotient function is defined for all real numbers except 0.

Another example

Now let's look at these two functions:

$$f(x) = x^2 - x - 20$$

and

$$g(x) = x - 5$$

When we differentiate the quotient of these, we get

$$(f/g)'(x) = d/dx\,[(x^2 - x - 20) / (x - 5)] = d/dx\,(x + 4) = 1$$

When we use the quotient rule, we get

$$(f/g)'(x) = \{[f'(x)][g(x)] - [g'(x)][f(x)]\} / [g(x)]^2$$
$$= \{[d/dx\,(x^2 - x - 20)](x - 5) - [d/dx\,(x - 5)](x^2 - x - 20)\} / (x - 5)^2$$
$$= [(2x - 1)(x - 5) - 1 \times (x^2 - x - 20)] / (x^2 - 10x + 25)$$
$$= [(2x^2 - 11x + 5) - (x^2 - x - 20) / (x^2 - 10x + 25)$$
$$= (2x^2 - 11x + 5 - x^2 + x + 20) / (x^2 - 10x + 25)$$
$$= (x^2 - 10x + 25) / (x^2 - 10x + 25) = 1$$

This works provided $g(x) \neq 0$. That means the derivative of the quotient function is defined for all real numbers except 5. That value of x is the zero of

$$g(x) = x - 5$$

which forms the denominator in the quotient.

- -

Are you astute?

The quotient of two functions can be differentiable in general, even if one or both of them individually is not. At first you might think that this sort of thing can't occur. But once in awhile it happens. Here's an example.

Take a look at Fig. 6-1. The derivatives of $f(x)$ and $g(x)$ both abruptly change at the point where $x = 0$. Specifically:

$$f'(x) = -2 \quad \text{when } x < 0$$
$$f'(x) = 2 \quad \text{when } x > 0$$
$$g'(x) = -1 \quad \text{when } x < 0$$
$$g'(x) = 1 \quad \text{when } x > 0$$

Neither derivative is defined at the point where $x = 0$, so neither function is differentiable in general. You can't use the quotient rule to find $[f(x) / g(x)]'$ unless you leave $x = 0$ out of the domains.

Nevertheless, a derivative exists for $f(x) / g(x)$ over the entire set of reals, including $x = 0$. Because $g(x)$ never becomes equal to 0, you know that

$$f(x) / g(x) = 2\,(|x| + 1) / (|x| + 1) = 2$$

Therefore,

$$[f(x) / g(x)]' = d/dx\,2 = 0$$

for all real numbers x. The graph of $[f(x) / g(x)]'$ is a line with a slope of 0 at every point along its infinite length, and that passes through the point $(0,2)$. There is no discontinuity at the point $(0,2)$ or anywhere else.

Here's a challenge!

Derive the quotient rule for differentiation from the other rules we've learned so far.

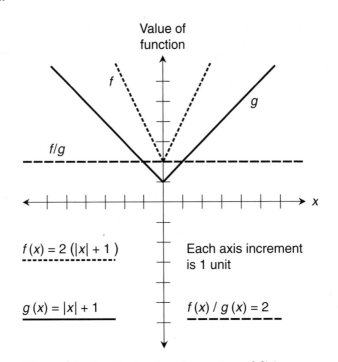

Value of
function

f

g

f/g

x

$f(x) = 2 (|x| + 1)$

Each axis increment
is 1 unit

$g(x) = |x| + 1$

$f(x) / g(x) = 2$

Figure 6-1   In this situation, the quotient of $f(x)$
and $g(x)$ is differentiable over the entire
set of reals, but neither $f(x)$ nor $g(x)$ is
differentiable at the point where $x = 0$.

Solution

Whenever we want to divide a quantity by another, we can multiply the first quantity by the reciprocal of the second, as long as the second quantity is not equal to 0. Knowing that, it seems that we should be able to derive the quotient rule for differentiation from the product and reciprocal rules for differentiation. Let's try that approach and see if it works.

Suppose that we're given two functions called f and g, and the domain is restricted so the value of g can never become 0. We can write the reciprocal of g as $(1/g)$. Our knowledge of algebra tells us that

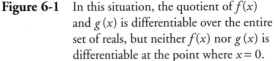

$$f/g = f \times (1/g)$$

Let's insert a "times sign" ($\times$) if there is any risk denoting a function *of* a function when we mean to denote a function *times* a function. The reciprocal rule for differentiation tells us that

$$(1/g)' = -g' / (g^2)$$

The product rule for differentiation allows us to write

$$[f \times (1/g)]' = f' \times (1/g) + (1/g)' f$$

When we substitute $(f/g)'$ for $[f \times (1/g)]'$ on the left-hand side of this equation, and we also substitute $[-g'/(g^2)]$ for $(1/g)'$ on the right-hand side, we get

$$(f/g)' = f' \times (1/g) + [-g'/(g^2)]\, f$$

We can simplify this to

$$(f/g)' = f'/g - g'\, f/g^2$$

Now let's multiply the quotient immediately after the equals sign by the quantity (g/g). We know that (g/g) is defined under our domain restriction—that is, $g \neq 0$—so it's equal to 1. When we do that, the above equation becomes

$$(f/g)' = (f'/g)(g/g) - g'\, f/g^2$$

We can multiply out the term immediately to the right of the equals sign to get

$$(f/g)' = f'\, g/g^2 - g'\, f/g^2$$

Now we have a difference of two fractions with a common denominator on the right-hand side of the equation. Consolidating this into a single fraction, we get

$$(f/g)' = (f'g - g'f)/g^2$$

That's the quotient rule for differentiation!

- -

Chain Rule

When a function operates on the output of another function, the combination is a third function known as a *composite function*. Sometimes it's called a *function of a function*. We can use a procedure called the *chain rule for differentiation* to find the derivative of a composite function.

Definition and notation

To determine a function of a function, we apply the second function to the output of the first. In other words, we apply the first (or "inner") function to its independent variable, and then we apply the second (or "outer") function to the value of the first function. For example, if

$$f(x) = 6x^3$$

and

$$g(y) = -3y$$

then

$$g[f(x)] = -3 \times (6x^3) = -18x^3$$

Here's another example. If

$$f(x) = (x^3 + 1)$$

and

$$g(y) = (4y^2 - 3y)$$

then

$$g[f(x)] = 4 \times (x^3 + 1)^2 - 3 \times (x^3 + 1)$$
$$= 4 \times (x^6 + 2x^3 + 1) - 3x^3 - 3 = 4x^6 + 5x^3 + 1$$

To make things look simpler, we can write $g(f)$ without including the first function's independent variable. But we don't want to give our readers the idea that we mean the product gf or $g \times f$. Some texts put a small letter o, which stands for "of," between the two functions. In this situation they'd write $g(f)$ as $g \circ f$.

Note that $g(f)$ is rarely the same as $f(g)$. The function-of-a-function operation is not, in general, commutative.

The rule in brief

The chain rule can be stated informally like this:

- To differentiate a function of a function, we multiply the derivative of the "outer" function by the derivative of the "inner" function.

For the chain rule to apply, both functions must be differentiable at every point that interests us. Stated symbolically, if f and g are differentiable functions of the same single variable, then

$$[g(f)]' = g'(f) \times f'$$

If we want to include an independent variable (say x) in the notation, then we can write the above statement as

$$\{g[f(x)]\}' = g'[f(x)] \times f'(x)$$

taking care not to confuse the multiplication symbol with the variable x.

- -

Are you confused?

The above notation can baffle some people, because the difference between the two expressions

$$\{g[f(x)]\}'$$

and

$$g'\,[f(x)]$$

is subtle. Let's clarify this before we go any further! The first expression, stated as an ordered list of instructions, reads like this:

- Apply the function f to the variable x.
- Apply the function g to that output.
- Differentiate the function g with respect to x to get the final result.

The second expression tells us to do these things in order:

- Apply the function f to the variable x.
- Differentiate the function g with respect to some arbitrary variable *other than x*. (We can call it anything we want, such as y.)
- Plug in $f(x)$ as that other variable.
- Apply the derivative function g' to that output to get the final result.

- -

Example

Let's see how this rule works. Consider the first of the two examples we saw a short while ago:

$$f(x) = 6x^3$$

and

$$g(y) = -3y$$

When we differentiate the composite function directly, we get

$$\{g\,[f(x)]\}' = d/dx\,(-18x^3) = -54x^2$$

The derivatives of the individual functions are

$$f'(x) = 18x^2$$

and

$$g'(y) = -3$$

When we apply the chain rule, we get

$$\{g\,[f(x)]\}' = g'\,[f(x)] \times f'(x) = -3 \times d/dx\,(6x^3) = -3 \times 18x^2 = -54x^2$$

Another example

Now let's see what happens with the composite of the two functions

$$f(x) = (x^3 + 1)$$

and

$$g(y) = (4y^2 - 3y)$$

The derivatives of the individual functions are

$$f'(x) = 3x^2$$

and

$$g'(y) = 8y - 3$$

Differentiating the composite function directly gives us

$$\{g[f(x)]\}' = d/dx \, (4x^6 + 5x^3 + 1) = 24x^5 + 15x^2$$

When we apply the chain rule, we get

$$\{g[f(x)]\}' = g'[f(x)] \times f'(x) = [8 \times (x^3 + 1) - 3] \times d/dx \, (x^3 + 1)$$
$$= (8x^3 + 5)(3x^2) = 24x^5 + 15x^2$$

- -

Are you confused?

If you aren't certain that you understand how the chain rule works, make up some more examples in the same format as the previous two. First, differentiate the composite function as a whole. Then, use the chain rule. You should always get the same final answer either way. Problems like this are "self-checking." If you make a mistake anywhere, you'll probably get answers that don't agree with each other.

Here's a challenge!

Show that the following statement is not true in general:

$$g(f) = f(g)$$

Solution

It's easy to prove that a statement isn't universally true. We simply find a situation called a *counterexample* where it's false. Consider

$$f(x) = (x^3 + 1)$$

and

$$g(y) = (4y^2 - 3y)$$

We've already determined that

$$g[f(x)] = 4x^6 + 5x^3 + 1$$

Working the other way, we get

$$f[g(y)] = (4y^2 - 3y)^3 + 1 = 64y^6 - 144y^5 + 108y^4 - 27y^3 + 1$$

The function $g(f)$ operates on a variable we call x, and the function $f(g)$ operates on a variable we call y. But the names of the variables aren't important. The composite functions $g(f)$ and $f(g)$ do different things. That's what matters!

Practice Exercises

This is an open-book quiz. You may (and should) refer to the text as you solve these problems. Don't hurry! You'll find worked-out answers in App. A. The solutions in the appendix may not represent the only way a problem can be figured out. If you think you can solve a particular problem in a quicker or better way than you see there, by all means try it!

1. Find the derivative of the polynomial function

$$f(x) = -4x^4 + 2x^3 - x^2 - x + 1$$

Then find the derivative of

$$g(x) = 2 \times (-4x^4 + 2x^3 - x^2 - x + 1) = -8x^4 + 4x^3 - 2x^2 - 2x + 2$$

Verify that $g' = 2f'$.

2. Find the derivative of the polynomial function

$$f(x) = -40x^4 + 20x^3 - 10x^2 - 10x + 10$$

Then find the derivative of

$$g(x) = (1/5) \times (-40x^4 + 20x^3 - 10x^2 - 10x + 10) = -8x^4 + 4x^3 - 2x^2 - 2x + 2$$

Verify that $g' = (1/5) f'$.

3. Show that the two-function product rule for differentiation is commutative. That is, show that if f and g are differentiable functions, then

$$(fg)' = (gf)'$$

4. Based on the two-function product rule for differentiation, derive a three-function product rule for differentiation. Assume that all three functions are differentiable.

5. Find the derivative of the function

$$p(y) = 1 / (y^2 + 1)$$

in the set of real numbers. Indicate the real values of y, if any, for which this function is nondifferentiable.

6. Find the derivative of the function

$$r(y) = 1 / (y^2 - 1)$$

in the set of real numbers. Indicate the real values of y, if any, for which this function is nondifferentiable.

7. Find the derivative of the function

$$s(z) = (z^4 - 1) / (z^2 + 1)$$

in the set of real numbers. First differentiate $s(z)$ directly after dividing out the quotient. Then differentiate $s(z)$ using the quotient rule. Indicate the real values of z, if any, for which this function is nondifferentiable.

8. Find the derivative of the function

$$t(z) = (z^4 - 1) / (z^2 - 1)$$

in the set of real numbers. First differentiate $t(z)$ directly after dividing out the quotient. Then differentiate $t(z)$ using the quotient rule. Indicate the real values of z, if any, for which this function is nondifferentiable.

9. Consider the functions

$$f(x) = x^3 - 4x^2$$

and

$$g(y) = y^2 + 5y$$

Differentiate the composite function $g(f)$ directly. Then do it using the chain rule.

10. Consider the functions

$$f(x) = x^2 - 4x$$

and

$$g(y) = 2y^2 + 7y$$

Differentiate the composite function $g(f)$ directly. Then do it using the chain rule.

7

A Few More Derivatives

In this chapter, we'll learn how to differentiate functions that raise a variable to a negative integer power, or to a non-integer real power. We'll also look at the derivatives of the sine, cosine, exponential, and logarithmic functions.

Real-Power Rule

We've seen how the power rule can be used to differentiate functions with a variable raised to a positive integer power and then multiplied by a real-number constant. Now we'll take that rule further. The power to which the variable is raised can be any real number.

The old rule extended

In Chaps. 4 and 5, we showed that any function of the form

$$f(x) = ax^n$$

has the derivative

$$f'(x) = anx^{(n-1)}$$

where a is a nonzero real constant and n is a positive integer. This rule applies not only to positive integer exponents, but for all real-number exponents. We can rewrite the rule by substituting k in place of n, where k represents a real number. If

$$f(x) = ax^k$$

then

$$f'(x) = akx^{(k-1)}$$

There's an important restriction. If $k \leq 1$, then we must be sure that $x \neq 0$. Otherwise, when we take the derivative, we'll end up dividing by 0 when we raise the variable x to the power of $k - 1$.

Three examples

Let's work out the derivatives of three simple functions that have exponents other than positive integers. First, consider

$$f(x) = 7x^{-5}$$

We multiply the coefficient 7 by the exponent -5 to get -35. That's the new coefficient. Then we reduce the exponent by 1, making it -6. The result:

$$f'(x) = -35x^{-6}$$

provided $x \neq 0$. Next, let's differentiate

$$g(y) = (5/3)y^{3/5}$$

Again, we multiply the coefficient by the exponent, getting 1. Then we reduce the exponent by 1, getting $-2/5$. The result:

$$g'(y) = y^{-2/5}$$

provided $y \neq 0$. Finally, let's consider the function

$$h(z) = -4z^{-3.7}$$

Going through the same ritual, taking care with signs and subtraction, we get a new coefficient of $-3.7 \times (-4)$ or 14.8, and a new exponent of -3.7, -1, or -4.7. The result:

$$h'(z) = 14.8z^{-4.7}$$

- -

Are you confused?

If you're not sure what a rational-number exponent is, think back to what you learned in algebra. Remember the definition of the term *rational number,* and then remember how *negative-integer powers* and *reciprocal powers* are defined. A rational number q can always be expressed as a ratio, like this:

$$q = a/b$$

where a is an integer and b is a positive integer. If you talk about the quantity x^q, you're really talking about $x^{a/b}$, or x raised to the power a/b. That's the same as the positive bth root of x^a. That is,

$$x^{a/b} = (x^a)^{1/b}$$

This definition doesn't make sense if $x = 0$ and $a \leq 0$. In that case, you end up dividing by 0 when you work out the expression.

Irrational-number exponents are a little more esoteric. You'll get a chance to "wrap your mind around them" at the end of this chapter.

Here's a challenge!

Consider the following function, in which the exponent appears as an endless, repeating decimal:

$$p(v) = 44v^{0.09090909\ldots}$$

What's the derivative of this function?

Solution

At first glance, this might seem insoluble. But let's look closely at the exponent, which is 0.09090909 . . . . In algebra, we learned that when we see an endless, repeating decimal fraction whose absolute value is less than 1, we can take the sequence of digits and divide it by an equal number of 9s, and we get a fraction equivalent to that endless decimal. In this situation we get 09/99, which reduces to 1/11. Now we have

$$p(v) = 44v^{1/11}$$

To differentiate, we multiply the coefficient 44 by the exponent 1/11 to get 4. That's the new coefficient. Then we reduce the exponent by 1, making it −10/11. The result:

$$p'(v) = 4v^{-10/11}$$

provided $v \neq 0$.

- -

Sine and Cosine Functions

Now that we've learned how to differentiate functions in which the variable is raised to a power, let's look at two functions that work in a different way: the *sine* and the *cosine*.

What's the sine?

In trigonometry, you learned that the sine function acts on angles. The *unit-circle model* (Fig. 7-1) defines the sine of x (sin x) for all possible angles x. The value of x is given in *radians* (rad). Positive angles go counterclockwise from the right-hand horizontal axis or "due east." Negative angles go clockwise from "due east." In Fig. 7-1, each axis division represents 1/4 unit, so the circle has a radius of 1 unit.

The sine of an angle is the *vertical-axis coordinate* of the point where the radial ray intersects the unit circle. Because this circle has a radius of 1 unit, the sine function can never attain values larger than 1 or smaller than −1. Figure 7-1 shows the situation for an angle of $2\pi/3$ rad ($x = 2\pi/3$), which is 1/3 of a rotation.

Remember that all angle expressions for trigonometric functions are given in radians (never in degrees) unless you're told specifically otherwise!

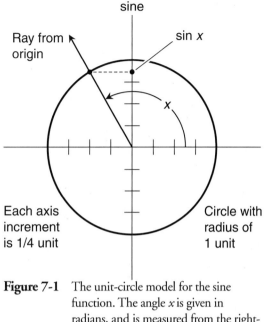

Figure 7-1   The unit-circle model for the sine
function. The angle *x* is given in
radians, and is measured from the right-
hand horizontal axis or "due east." Each
axis division represents 1/4 unit.

What's the cosine?

The cosine function can also be defined in terms of the unit circle, as shown in Fig. 7-2. As
before, positive angles are expressed as counterclockwise rotation of the radial ray, and negative angles are expressed as clockwise rotation.

   The cosine of the angle is the *horizontal-axis coordinate* of the point where the radial ray
intersects the circle. As with the sine, the cosine can never be larger than 1 or smaller than −1.
This drawing illustrates the case where $x = 2\pi/3$.

   Both the sine and the cosine functions are defined, continuous, and differentiable over
the entire set of real numbers. An angle of 2π represents a complete counterclockwise rotation
of the ray extending straight out from the origin. Angles larger than 2π represent more than
one complete counterclockwise rotation. Negative angles go clockwise instead of counterclockwise. For example, an angle of $-2\pi/3$ translates to 1/3 of a clockwise rotation. An angle
of -2π is a full clockwise rotation. Angles smaller than -2π (or larger negatively) represent
more than a full clockwise rotation.

Derivative of the sine

Suppose we graph the function $f(x) = \sin x$ for values of *x* between -3π and 3π. The result
is the solid curve in Fig. 7-3. Because of its wavelike appearance, this curve is often called a
sine wave. If we examine it closely, we can see that its slope is 0 when it reaches a *crest* or *local
maximum,* and the slope is also 0 when it reaches a *trough* or *local minimum.* A point of maximum slope occurs 1/4 of a rotation, or $\pi/2$ rad, after every trough. A point of minimum slope

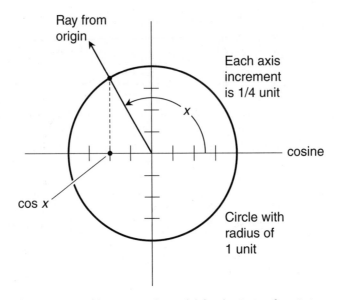

Figure 7-2 The unit-circle model for the cosine function. Each axis division represents 1/4 unit.

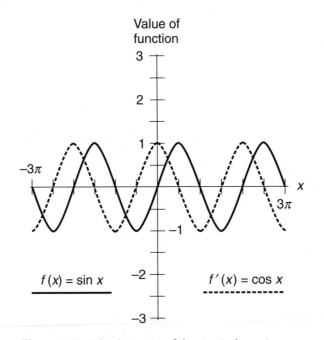

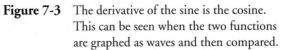

Figure 7-3 The derivative of the sine is the cosine. This can be seen when the two functions are graphed as waves and then compared.

occurs $\pi/2$ rad after every crest. The curve crosses the x axis at every point where the slope is maximum or minimum.

If we plot a graph of the slope of the sine function, which represents its derivative, for many points along the x axis, we'll see that the graph of the derivative is another wave, shown by the dashed curve in Fig. 7-3. It has the same shape and size as the sine wave. This particular wave shape occurs often in physics and engineering. The general term for it is *sinusoid*.

In Fig. 7-3, we can see that the derivative wave is displaced by $\pi/2$ rad to the left of the original wave. This amount of displacement is sometimes called a *quarter wavelength* or a *quarter cycle*. In this example, the dashed wave represents the cosine function. Therefore, if we differentiate

$$f(x) = \sin x$$

we get

$$f'(x) = \cos x$$

Derivative of the cosine

Now let's graph the function $g(x) = \cos x$ for values of x between -3π and 3π. We get the solid curve shown in Fig. 7-4. This wave, like the sine wave, has a slope of 0 when it reaches a crest or a trough. Also like the sine wave, a point of maximum slope occurs $\pi/2$ rad after

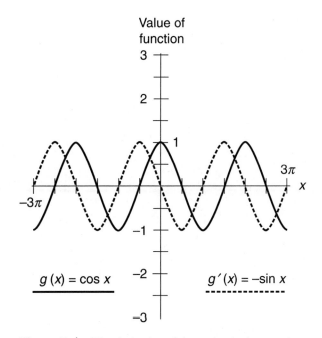

Figure 7-4 The derivative of the cosine is the negative of the sine.

every trough, and a point of minimum slope occurs $\pi/2$ rad after every crest. The *cosine wave* is a sinusoid, just like the sine wave. The only difference between the cosine wave and the sine wave is the horizontal position, also called the *phase*.

When we graph the slope of the cosine function for many points, we find that the graph of the derivative is another sinusoid, shown by the dashed curve in Fig. 7-4. The derivative wave is, as in the earlier case with the sine wave, displaced $\pi/2$ rad to the left of the original. This new wave is an "upside-down" sine wave, so it represents the negative of the sine function. If

$$g(x) = \cos x$$

then the derivative is

$$g'(x) = -\sin x$$

Example

Now that we've learned how to differentiate two new functions, we can differentiate any function that contains them in a sum, product, reciprocal, or quotient. We can also use the chain rule to differentiate composite functions containing them. Let's try an example:

$$f(x) = -3x^4 + 2 \cos x$$

Using the sum rule, we know that

$$f'(x) = d/dx\,(-3x^4) + d/dx\,(2 \cos x)$$

The power rule and the multiplication-by-constant rule allow us to rewrite this as

$$f'(x) = -12x^3 + 2\,d/dx\,(\cos x)$$

We know that the derivative of the cosine is the negative of the sine. Therefore

$$f'(x) = -12x^3 + 2 \cdot (-\sin x) = -12x^3 - 2 \sin x$$

- -

Are you confused?

The little dot in the first part of the above equation represents multiplication. It's often used instead of the traditional "times sign," which some people think looks like the letter x representing a variable. If we were to write the first part of the above equation with the conventional "times sign," we would get

$$f'(x) = -12x^3 + 2 \times (-\sin x)$$

Some people might confuse this with the completely different equation

$$f'(x) = -12x^3 + 2x\,(-\sin x)$$

Let's agree that from now on, whenever we feel tempted to use a "times sign" to represent multiplication, we'll use the dot. This dot should be elevated above the base line. Otherwise, it could be confused with a decimal point! But let's also remember that we can often indicate multiplication without writing any symbol at all, as in expressions such as $2x$, xy, or $12ab^2x^3$.

Another example

Let's try differentiating a product of trigonometric functions. Consider

$$q(x) = \sin x \cos x$$

We can name the individual functions as

$$f(x) = \sin x$$

and

$$g(x) = \cos x$$

The derivatives are

$$f'(x) = \cos x$$

and

$$g'(x) = -\sin x$$

Applying the product rule for differentiation gives us

$$q'(x) = (fg)'(x) = f'(x)\,g(x) + g'(x)\,f(x)$$
$$= \cos x \cos x + (-\sin x)\sin x = (\cos x)^2 - (\sin x)^2$$

With positive integer powers of trigonometric functions such as the sine and the cosine, the exponent is customarily written directly after the abbreviated name of the function. For example, instead of $(\sin x)^2$, we write $\sin^2 x$. We can therefore rewrite the above equation as

$$q'(x) = \cos^2 x - \sin^2 x$$

Be warned!

The above notational trick is *never* used for negative integer powers of trig functions. For example, when we write $\sin^{-1} x$, we actually mean the inverse of the sine of x, also called the *Arcsine*. This isn't even a function unless its domain is restricted.

- -

Here's a challenge!

Consider a function $p(x)$ that we get when we take sin x and then square the result:

$$p(x) = \sin^2 x$$

Differentiate this function. Indicate the values, if any, for which it is nondifferentiable.

Solution

We must use the chain rule in this situation. Let's name the component functions like this:

$$f(x) = \sin x$$

and

$$g(y) = y^2$$

The derivatives of the component functions are

$$f'(x) = \cos x$$

and

$$g'(y) = 2y$$

Now we can write the derivative of the composite function using the chain rule:

$$p'(x) = \{g[f(x)]\}' = g'[f(x)] \cdot f'(x) = 2 \sin x \cos x$$

Both of the component functions are continuous and differentiable over the entire set of reals. We never get a denominator that can become equal to 0, because we never have to divide by anything. Therefore, the composite function is differentiable over the entire set of reals.

- -

Natural Exponential Function

Imagine some function, other than the zero function, that's its own derivative. You might wonder, "Does such a function exist? If so, what does it look like?" We can say this much about it: The slope of its graph must equal the value of the function at every point in the domain. As things turn out, infinitely many such functions exist.

What's a natural exponential?

The *natural exponential* of a quantity is what you get when you raise a unique irrational-number constant, called *Euler's constant* or the *exponential constant,* to a power equal to that quantity. Euler's constant is symbolized by e.

If you have a good scientific calculator, it should have a key marked e^x. To get an idea of the approximate value of e, enter "1" and then hit the e^x key. You'll get 2.718 followed by a string of digits. Because e is irrational, this string of digits is endless and nonrepeating.

Here are some examples of natural exponentials, rounding to three decimal places except for e^0, which is exactly 1:

$$e^2 \approx 7.389$$
$$e^{1.478} \approx 4.384$$
$$e^1 \approx 2.718$$
$$e^{0.8347} \approx 2.304$$
$$e^0 = 1$$
$$e^{-0.5} \approx 0.607$$
$$e^{-1} \approx 0.368$$
$$e^{-1.7} \approx 0.183$$
$$e^{-2} \approx 0.135$$

The wavy equals sign ($\approx$) means "is approximately equal to."

What does it look like?

Figure 7-5 is a graph of the *natural exponential function*. Its domain is the entire set of real numbers, and it is continuous at every point in the domain. The function has no gap, does

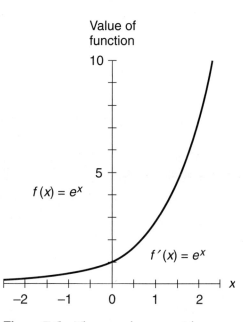

Figure 7-5   The natural exponential
function is its own derivative.

not have a *singularity* (that is, it doesn't blow up), and it doesn't turn any corners. The natural exponential function is therefore differentiable over the entire set of reals.

At any point on the graph in Fig. 7-5, the slope is equal to e^x. In more formal terms, we can write

$$d/dx\,(e^x) = e^x$$

Offspring functions

The natural exponential function is a *parent function* with infinitely many *child functions*, each of which is its own derivative. Consider the generalized function

$$g(x) = ke^x$$

where k is a constant that can be equal to any real number. Then

$$g'(x) = d/dx\,(ke^x)$$

Using the multiplication-by-constant rule for differentiation, we can rewrite this as

$$g'(x) = k\,[d/dx\,(e^x)]$$

Because $d/dx\,(e^x) = e^x$, we have

$$g'(x) = ke^x$$

which is exactly the same function as $g(x)$.

Example

Let's differentiate an exponential function that's a little more complicated than a direct off-spring of the parent. Consider

$$p(x) = e^{-x}$$

We can rewrite this as

$$p(x) = 1\,/\,(e^x)$$

Now let's invent a function f, and define it as

$$f(x) = e^x$$

Then we have

$$p(x) = 1 / [f(x)]$$

The reciprocal rule for differentiation tells us that if f is a differentiable function of a single variable, then

$$[1 / f(x)]' = -f'(x) / [f(x)]^2$$

In this situation, f is differentiable, so

$$f'(x) = e^x$$

and

$$[f(x)]^2 = (e^x)^2 = e^{2x}$$

Therefore

$$p'(x) = -e^x / (e^{2x}) = -[e^x / (e^{2x})] = -e^{(x-2x)} = -e^{-x}$$

- -

Are you confused?

A couple of steps in the above process might confuse you. Both of them involve rules for exponentials that you learned in algebra or precalculus. They're easy to forget!

You might wonder how we got from $(e^x)^2$ to e^{2x}. That comes from the rule for a power of a power. If a, b, and c are real numbers, then

$$(a^b)^c = a^{bc}$$

as long as we don't run into some undefined expression along the way, such as 0 to the 0th power, or 0 to some negative power.

The transition from $-[e^x / (e^{2x})]$ to $-e^{(x-2x)}$ is the other step that you might not understand right away. This comes from a rule in algebra that deals with quotients of expressions with exponents. If a, b, and c are real numbers and $a \neq 0$, then

$$a^b / a^c = a^{(b-c)}$$

Here's a challenge!

Differentiate the function in the example we just finished, using the chain rule instead of the reciprocal rule. Again, that function is

$$p(x) = e^{-x}$$

Solution

This time, we'll break the function p down into two component functions. Let's call them f and g, and say that

$$f(x) = -x$$

and

$$g(y) = e^y$$

Then the function p becomes

$$p = g[f(x)]$$

Both f and g are differentiable, so we have

$$f'(x) = -1$$

and

$$g'(y) = e^y$$

The chain rule tells us that if f and g are differentiable functions of the variable x, then

$$\{g[f(x)]\}' = g'[f(x)] \cdot f'(x)$$

Substituting the values into this formula, we get

$$p'(x) = e^{-x} \cdot (-1) = -e^{-x}$$

This agrees with what we got when we used the reciprocal rule. Which method do you think was easier?

Natural Logarithm Function

The natural exponential function has an *inverse* that "undoes" its work. This inverse is known as the *natural logarithm,* or simply the *natural log.*

What's a natural log?

A *logarithm* of a quantity is a power to which a positive real constant is raised to get that quantity. The constant is called the *base,* which is usually 10 or e. We'll deal only with the base-e or natural log.

In equations, the natural log is usually denoted by writing "ln" followed by the argument. Here are some equations that you can check out with your calculator. With the exceptions of ln *e*, ln 1, and ln (1/*e*), which are exact values, everything has been rounded to three decimal places.

$$\ln 100 \approx 4.605$$
$$\ln 45 \approx 3.807$$
$$\ln 10 \approx 2.303$$
$$\ln 6 \approx 1.792$$
$$\ln e = 1$$
$$\ln 1 = 0$$
$$\ln (1/e) = -1$$
$$\ln 0.5 \approx -0.693$$
$$\ln 0.1 \approx -2.303$$
$$\ln 0.07 \approx -2.659$$
$$\ln 0.01 \approx -4.605$$

While the natural log "undoes" the natural exponential, the reverse operation also works. Therefore,

$$\ln (e^x) = x$$

and

$$e^{(\ln x)} = x$$

The *natural log function* is defined only for positive real-number arguments. In the first equation above, that's not a problem, because e^x is always a positive real, no matter what real number we input for *x*. In the second equation, we must restrict the domain to the positive reals only. As long as we confine the domain to the positive reals, the natural log function is continuous and differentiable.

What does it look like?

Figure 7-6 is a graph of the natural log function. Its domain is the set of positive real numbers, and it's continuous at every point in that domain. The function is *singular* (meaning that it blows up) as *x* approaches 0 from the right, but this is not a problem as long as we keep 0 and all the negative reals out of the domain. Once we restrict the domain in that way, the natural log function doesn't have any gaps or turn any corners, so it's differentiable.

At any point on the graph in Fig. 7-6, the slope is equal to 1/*x*. We can write this as the equation

$$d/dx \, (\ln x) = 1/x$$

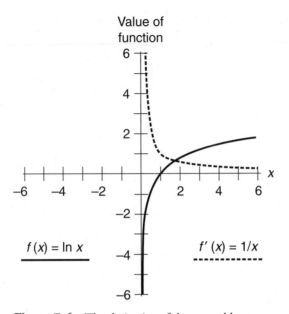

Value of
function

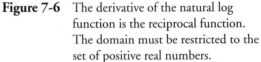

$f(x) = \ln x$

$f'(x) = 1/x$

Figure 7-6   The derivative of the natural log
function is the reciprocal function.
The domain must be restricted to the
set of positive real numbers.

Example

Let's work out a derivative involving a natural logarithm. Without getting complicated, but still requiring some thought, we'll tackle this:

$$p(x) = \ln(7x - 14)$$

The chain rule will work here. If we call the component functions f and g, and if we express the above function as $g(f)$, then

$$f(x) = 7x - 14$$

and

$$g(y) = \ln y$$

The function f is differentiable over the entire set of reals, but g is differentiable only as long as $y > 0$. That means we must restrict the domain of $g(f)$ so that

$$7x - 14 > 0$$

This works out to $x > 2$. Now we can differentiate both functions to get

$$f'(x) = 7$$

and

$$g'(y) = 1/y$$

The chain rule, stated once again, tells us that if f and g are differentiable functions of the variable x, then

$$\{g[f(x)]\}' = g'[f(x)] \cdot f'(x)$$

Substituting the values into this formula, we get

$$p'(x) = [1 / (7x - 14)] \cdot 7 = 7 / (7x - 14)$$

We can multiply both the numerator and denominator by 1/7, reducing this to

$$p'(x) = 1 / (x - 2)$$

Some people would rather write this as

$$p'(x) = (x - 2)^{-1}$$

- -

Here's a challenge!

Suppose we're confronted with a differentiation problem that involves the composite of a trig function and a log function. Let's find the derivative of

$$q(x) = \ln(\cos 2x)$$

We'll also define the values of x, if any, for which this function is nondifferentiable.

Solution

Before we get started with the differentiation, we had better find out where q is differentiable and where it is not. The natural log function is defined only for positive values of the argument. That means q is differentiable only for those values of x such that

$$\cos 2x > 0$$

This inequality is satisfied for infinitely many open intervals. It's awkward to express this symbolically, but the following arrangement gives the general idea:

From "negative infinity"

$$\downarrow$$

$$-13\pi/4 < x < -11\pi/4$$
$$-9\pi/4 < x < -7\pi/4$$
$$-5\pi/4 < x < -3\pi/4$$
$$-\pi/4 < x < \pi/4$$
$$3\pi/4 < x < 5\pi/4$$
$$7\pi/4 < x < 9\pi/4$$
$$11\pi/4 < x < 13\pi/4$$

$$\downarrow$$

Toward "positive infinity"

Our function q is defined and continuous everywhere in each of these open intervals. There are no gaps, no singularities, and no corners within any single interval. But we must take note of the fact that these are *open* intervals, meaning that the end points are *not* included. Our function q is differentiable at any point within any of these open intervals, but nowhere else. You might find it helpful to graph the function

$$y = \cos 2x$$

in the *xy*-plane to see where this function is positive, where it's zero, and where it's negative. Then you'll be able to envision the above open intervals along the x axis.

In this situation, we're dealing with a function of a function of a function! Another way to say this is to call q a triplet of *nested functions.* To differentiate it, we must employ the chain rule twice. Let's call the component functions f, g, and h, working from the inside out. Then we have

$$q(x) = h\{g[f(x)]\}$$

where

$$f(x) = 2x$$
$$g(y) = \cos y$$
$$h(z) = \ln z$$

The derivatives are

$$f'(x) = 2$$
$$g'(y) = -\sin y$$
$$h'(z) = 1/z$$

Applied to the outer two functions, the chain rule gives us

$$[h(g)]' = h'(g) \cdot g'$$

If we let g operate on f, we can rewrite this as

$$\{h[g(f)]\}' = h'[g(f)] \cdot [g(f)]'$$

The chain rule, applied to the inner pair of functions, tells us that

$$[g(f)]' = g'(f) \cdot f'$$

So, by substitution, we obtain

$$\{h[g(f)]\}' = h'[g(f)] \cdot g'(f) \cdot f'$$

The left-hand side of the above equation happens to be the same as q', which is what we seek! So let's substitute:

$$q' = h'[g(f)] \cdot g'(f) \cdot f'$$

We can include the variable x in the above expression to see how it fits in:

$$q'(x) = h'\{g[f(x)]\} \cdot g'[f(x)] \cdot f'(x)$$

Now let's plug in expressions for what each of these functions does to its argument, and also plug in the arguments themselves. When we do that, we get

$$q'(x) = [1 / (\cos 2x)] \cdot (-\sin 2x) \cdot 2 = -2 (\sin 2x) / (\cos 2x)$$

It's time for us to invoke another well-known law from trigonometry: The sine divided by the cosine is equal to the *tangent*, abbreviated *tan*. Knowing this, we can rewrite the above equation as

$$q'(x) = -2 \tan 2x$$

provided $2x$ is not an odd-integer multiple of $\pi/2$. That means x can't be an odd-integer multiple of $\pi/4$. We've already taken this constraint into account. Remember that the original function q is differentiable only within certain open intervals, none of which contains an odd-integer multiple of $\pi/4$.

Are you confused?

Way back in the 1960s, one of my math teachers asked the class at the end of an especially difficult session, "Are you confused by this?" When we all nodded, he said, "I don't blame you." The chain rule is tricky enough when applied once. When you apply it twice to a triplet of nested functions, it's worse! If you're disoriented by the process we just went through, put it aside for now, and look at it again tomorrow.

- -

Where to find more derivatives

You can find some worked-out derivatives in the back of this book. Refer to App. F. You can also find them on the Internet. Enter the phrase "table of derivatives" into your favorite search engine. A few sites will calculate derivatives for you.

Real powers in general

A general real-number exponent can be evaluated with natural logs and exponentials. From algebra and precalculus, remember that when you have an expression of the form x^k where x is a nonzero variable and k is any real number, then

$$\ln (x^k) = k \ln x$$

Because the natural exponential function "undoes" the natural log function, you can take the natural exponential of both sides of the above equation to get

$$x^k = e^{(k \ln x)}$$

If you want to raise a variable to any real power, you can take the natural log of the variable, multiply by the exponent, and finally take the natural exponential of that product. This scheme only works when $x > 0$, however, because the natural log of 0 or a negative quantity is not defined.

Practice Exercises

This is an open-book quiz. You may (and should) refer to the text as you solve these problems. Don't hurry! You'll find worked-out answers in App. A. The solutions in the appendix may not represent the only way a problem can be figured out. If you think you can solve a particular problem in a quicker or better way than you see there, by all means try it!

1. Differentiate the function

$$p (t) = 2t^2 - 4t + 5 + 4t^{-1} + 6t^{-2}$$

   Indicate the values of t, if any, for which the derivative is not defined.

2. Differentiate the function

$$q (w) = (w^{-1} + 1)(w^{-1} - 1)$$

   Indicate the values of w, if any, for which the derivative is not defined.

3. The reciprocal of the sine function is known as the *cosecant function,* abbreviated csc. Mathematically,

$$\csc x = 1 \,/\, \sin x$$

   provided x is not an integer multiple of π. Using the rules we've learned so far, differentiate the cosecant function. Indicate the values of x, if any, for which this function is nondifferentiable.

4. The reciprocal of the cosine function is known as the *secant function,* abbreviated sec. Mathematically,

$$\sec x = 1 \, / \cos x$$

provided x is not an odd-integer multiple of $\pi/2$. Using the rules we've learned so far, differentiate the secant function. Indicate the values of x, if any, for which this function is nondifferentiable.

5. As mentioned in the chapter text, the sine function divided by the cosine function is the tangent function. Mathematically,

$$\tan x = \sin x \, / \cos x$$

provided x is not an odd-integer multiple of $\pi/2$. Using the rules we've learned so far, differentiate the tangent function. Indicate the values of x, if any, for which this function is nondifferentiable.

6. The cosine function divided by the sine function is known as the *cotangent function,* abbreviated cot. Mathematically,

$$\cot x = \cos x \, / \sin x$$

provided x is not an integer multiple of π. Using the rules we've learned so far, differentiate the cotangent function. Indicate the values of x, if any, for which this function is nondifferentiable.

7. Differentiate this generalized function, where a is a real-number constant:

$$p\,(x) = e^{ax}$$

8. Differentiate this generalized function, where a and b are real-number constants:

$$q\,(x) = be^{ax}$$

9. Differentiate this generalized function, where a is a real-number constant:

$$r\,(x) = \ln ax$$

10. Differentiate this generalized function, where a and b are real-number constants:

$$s\,(x) = b \ln ax$$

8

Higher Derivatives

Functions can be differentiated more than once to get *higher derivatives*. Sometimes the derivative functions get simpler as we go, sometimes they get more complicated, and sometimes they go through repeating cycles.

Second Derivative

The *second derivative* of a function is the derivative of its derivative. Consider a function f of an independent variable x, producing a dependent variable y so that $y = f(x)$. The second derivative of f with respect to x can be denoted in various ways:

$$d^2y/dx^2$$
$$y''$$
$$d^2f(x)/dx^2$$
$$d^2/dx^2 \, f(x)$$
$$d^2f/dx^2$$
$$f''(x)$$
$$f''$$

If we want to find the second derivative of a function, we differentiate the function, then treat that derivative as a new function, and finally differentiate the new function with respect to the same variable or expression.

An example

Let's find the second derivative of a polynomial function. Consider

$$f(x) = -4x^4 + 5x^3 + 6x^2 - 7x + 8$$

Using the rules we've learned, we find the first derivative to be

$$f'(x) = -16x^3 + 15x^2 + 12x - 7$$

Differentiating again, we get

$$f''(x) = -48x^2 + 30x + 12$$

Another example

Now let's work out the second derivative of

$$f(x) = 4x^4 - 5\cos x$$

Using the sum rule, we obtain

$$f'(x) = d/dx\,(4x^4) - d/dx\,(5\cos x)$$

The power rule and the multiplication-by-constant rule can be applied to give us

$$f'(x) = 16x^3 - 5\,d/dx\,(\cos x)$$

The derivative of the cosine is the negative of the sine. Therefore

$$f'(x) = 16x^3 - 5 \cdot (-\sin x) = 16x^3 + 5\sin x$$

Now let's differentiate again. The sum rule can be used to get

$$f''(x) = d/dx\,(16x^3) + d/dx\,(5\sin x)$$

Applying the power rule and the multiplication-by-constant rule gives us

$$f''(x) = 48x^2 + 5\,d/dx\,\sin x$$

The derivative of the sine is the cosine, so we have

$$f''(x) = 48x^2 + 5\cos x$$

Still another example

Now we'll work with a function that takes the variable to various integer powers and then adds the natural logarithm. Let's find the second derivative of

$$g(t) = 2t^2 + 5t - 7 + 4t^{-1} - t^{-2} + 2t^{-3} + \ln t$$

with the constraint that $t > 0$, because the natural log function is defined only for positive real numbers. The first derivative comes out as

$$g'(t) = 4t + 5 - 4t^{-2} + 2t^{-3} - 6t^{-4} + (1/t)$$

The derivative of 7 is equal to 0, so that term has vanished. Because $1/t = t^{-1}$, we can rewrite the above equation, putting the exponents in descending order to get

$$g'(t) = 4t + 5 + t^{-1} - 4t^{-2} + 2t^{-3} - 6t^{-4}$$

Now we take the derivative again, obtaining

$$g''(t) = 4 - t^{-2} + 8t^{-3} - 6t^{-4} + 24t^{-5}$$

The derivative of 5 is equal to 0, so that term, like the constant 7 in the first derivative, goes away.

- -

Are you confused?

In the graph of a function, the derivative at a point represents the slope of the curve at that point, as you have learned. "All right," you say. "What does the second derivative represent graphically?" The second derivative at a point tells you the rate and the sense (increasing or decreasing) at which the slope changes at that point. Think about the graph of the function

$$f(x) = 2x + 4$$

This is a straight line with a slope of 2 at every point. This fact shows up in the first derivative

$$f'(x) = d/dx\,(2x + 4) = 2$$

The slope is the same everywhere along the line, so the rate at which the slope changes is 0 at any point you choose. This is apparent when you take the second derivative

$$f''(x) = d^2/dx^2\,(2x + 4) = d/dx\,2 = 0$$

Here's a challenge!

According to legend, one day several centuries ago Sir Isaac Newton sat in an orchard and saw an apple fall. At the instant it snapped off the twig, the apple's speed was zero. Then it began to move, slowly at first, then faster. Newton realized that falling objects don't merely *travel* downward; they *accelerate* downward. Returning to his study, he must have worked out the fact that, neglecting air resistance, the acceleration of a falling object is constant, and it doesn't depend on the object's mass.

The acceleration of a falling object is the rate at which its speed increases. That's the derivative of the speed with respect to time. The speed of a falling object is the rate at which its total fallen distance increases. That's the derivative of the distance with respect to time. Suppose we let h represent the vertical fallen distance in meters (m), v represent the vertical speed in meters per second (m/s), a represent the vertical acceleration in meters per second per second (m/s^2), and t represent the elapsed time in seconds (s) after the beginning of the descent. Then

$$v = dh/dt$$

and

$$a = dv/dt = d^2h/dt^2$$

Now the fiction begins (except for the mathematics, which is fact!). Imagine that Sir Isaac picked up the apple, took it to the cliffs of Dover, tossed it off from one of the highest points he could find, and watched it plunge to the beach far below (Fig. 8-1). He had already figured out that the vertical fallen distance as a function of time would be quite close to

$$h = 5t^2$$

Sir Isaac knew that the total fall time, the vertical component of the speed at any instant, and the vertical (downward) component of the acceleration at any instant didn't depend on whether he simply dropped the apple or whether he threw it as hard as he could—as long as the apple left his hand traveling *horizontally*. So Sir Isaac took aim straight at the French coast across the Channel, hurling the apple with enough horizontal force to make sure it would clear the outcroppings below. He timed the fall, and found that the apple struck the beach after 4 seconds. What was the height h_c of the cliff, as measured from the position of Sir Isaac's hand as the apple left it? What was the vertical component v_a of the apple's speed at impact? What was the vertical component a_a of the apple's acceleration at impact?

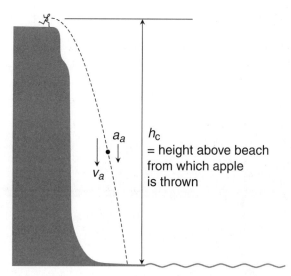

Figure 8-1   Sir Isaac hurls an apple from the cliffs of Dover. The altitude of the apple above the beach at the instant it leaves Sir Isaac's hand is h_c. The vertical speed of the apple is v_a, and the vertical acceleration is a_a.

Solution

We can calculate the height of the cliff, h_c, by plugging in 4 for t in the equation we've been given. When we do that, we get

$$h_c = 5 \cdot 4^2 = 5 \cdot 16 = 80 \text{ m}$$

The vertical speed v of a falling object at time t after its release is always

$$v = dh/dt = d/dt\,(5t^2) = 10t$$

We can calculate the vertical speed of Newton's apple at impact, v_a, by plugging in 4 for t again, getting

$$v_a = 10 \cdot 4 = 40 \text{ m/s}$$

The vertical acceleration function turns out to be a constant

$$a = d^2/dt^2\,(5t^2) = d/dt\,(10t) = 10 \text{ m/s}^2$$

The vertical acceleration a_a of Newton's apple at impact was therefore 10 m/s². In fact, the apple's vertical acceleration was 10 m/s² during its entire descent.

- -

Third Derivative

The *third derivative* of a function is the derivative of its second derivative. Suppose we have $y = f(x)$. The third derivative of f with respect to x can be written in various ways:

$$d^3y/dx^3$$
$$y'''$$
$$d^3f(x)/dx^3$$
$$d^3/dx^3\,f(x)$$
$$d^3f/dx^3$$
$$f'''(x)$$
$$f'''$$
$$f^{(3)}(x)$$
$$f^{(3)}$$

The last two expressions contain a simple numeric superscript in parentheses. The parentheses indicate that we're talking about a derivative, not a power.

If we want to find the third derivative of a function, we differentiate the function, consider that derivative as a new function, differentiate that new function again with respect to the same variable or expression, consider that as another new function, and finally differentiate it once again with respect to the same variable or expression.

An example

Earlier in this chapter, we found the second derivatives of three functions. Now let's work out the third derivatives of those same functions. Consider

$$f(x) = -4x^4 + 5x^3 + 6x^2 - 7x + 8$$

We found the second derivative to be

$$f''(x) = -48x^2 + 30x + 12$$

Differentiating again, we get

$$f'''(x) = -96x + 30$$

Another example

Let's look again at the function

$$f(x) = 4x^4 - 5 \cos x$$

We found that the second derivative is

$$f''(x) = 48x^2 + 5 \cos x$$

Now let's differentiate again. Applying the sum rule, we get

$$f'''(x) = d/dx\,(48x^2) + d/dx\,(5 \cos x)$$

The power rule and the multiplication-by-constant rule tell us that

$$f'''(x) = 96x + 5\,d/dx\,(\cos x)$$

The derivative of the cosine is the negative of the sine, so we have

$$f'''(x) = 96x + 5 \cdot (-\sin x)$$

which can be simplified to

$$f'''(x) = 96x - 5 \sin x$$

Still another example

Let's find the third derivative of

$$g(t) - 2t^2 + 5t - 7 + 4t^{-1} \quad t^{-2} + 2t^{-3} + \ln t$$

We found that the second derivative is

$$g''(t) = 4 - t^{-2} + 8t^{-3} - 6t^{-4} + 24t^{-5}$$

We can use the generalized power rule and the sum rule to differentiate term-by-term, getting

$$g'''(t) = 2t^{-3} - 24t^{-4} + 24t^{-5} - 120t^{-6}$$

The derivative of 4 is equal to 0, so that term disappears from the final result.

- -

Are you confused?

You've learned that the speed of an object falling straight down (neglecting air resistance) is the derivative of the fallen distance, and the acceleration is the second derivative of the fallen distance. "What," you ask, "is the third derivative of the fallen distance?" It's called *jerk*. It's the derivative of the acceleration, or the second derivative of the speed. You won't hear or read about jerk very often.

For a falling object, neglecting air resistance, the jerk is always 0 because the acceleration is constant, and the derivative of a constant function is the zero function. If the intensity of the earth's gravity were to increase as an object fell, the jerk would not be 0. (In fact, this does occur when a meteor "falls" from many thousands of kilometers out in space, but it's a gradual increase.) Perhaps you've been subjected to jerk when riding in a high-performance car with a bad driver. If you were pressed backward in your seat and then suddenly thrown forward against the shoulder strap, you felt it!

Here's a challenge!

Find the second derivative of the tangent function. Start with the known first derivative, which you found when you solved Practice Exercise 5 in Chap. 7:

$$d/dx \, (\tan x) = \sec^2 x$$

Indicate the values of x, if any, for which this second derivative function is nondifferentiable.

Solution

From Practice Exercise 4 in Chap. 7, we remember that the secant is the reciprocal of the cosine. We can therefore rewrite the above equation as

$$d/dx \, (\tan x) = [1 \, / \, (\cos x)]^2 = (\cos x)^{-2}$$

This function has discontinuities at all values of x where $\cos x = 0$. Those are, as we've seen, all odd-integer multiples of $\pi/2$. To find the second derivative of the tangent function, we must differentiate the cosine function raised to the -2 power:

$$d^2/dx^2 \, (\tan x) = d/dx \, (\cos x)^{-2}$$

Let's use the chain rule. We'll break the above function down into two components f and g, such that

$$f(x) = \cos x$$

and

$$g(y) = y^{-2}$$

The derivatives of the component functions are

$$f'(x) = -\sin x$$

and

$$g'(y) = -2y^{-3}$$

Now we can write the derivative of the composite function using the chain rule as follows:

$$d/dx\,(\cos x)^{-2} = \{g\,[f(x)]\}' = g'\,[f(x)] \cdot f'(x) = -2 \cdot (\cos x)^{-3} \cdot (-\sin x)$$
$$= 2 \cdot (\cos x)^{-3} \cdot \sin x$$

We've just found that

$$d^2/dx^2\,(\tan x) = 2 \cdot (\cos x)^{-3} \cdot \sin x$$

Here's an extra credit challenge!

Find the third derivative of the tangent function. Start with the known second derivative, which we just finished working out. Indicate the values, if any, for which

$$d^2/dx^2\,(\tan x) = 2 \cdot (\cos x)^{-3} \cdot \sin x$$

is nondifferentiable.

Solution

You're on your own. That's why you get extra credit for tackling this problem! Here's a hint: Use the chain rule and the product rule for differentiation.

- -

Beyond the Third Derivative

The *fourth derivative* of a function is the derivative of its third derivative, the *fifth derivative* is the derivative of the fourth derivative, and so on. The *nth derivative* of a function is what we get when we differentiate it *n* times, where *n* is a positive integer. Consider $y = f(x)$. The *n*th derivative of f with respect to x can be written in various ways:

$$d^n y / dx^n$$
$$d^n f(x) / dx^n$$
$$d^n / dx^n\, f(x)$$
$$d^n f / dx^n$$
$$f^{(n)}(x)$$
$$f^{(n)}$$

As before, the parentheses around a superscript indicate that we're talking about a derivative, not a power.

An example

Let's work out the fourth derivative of the same three functions for which we have already found the second and third derivatives. The first original function, again, is

$$f(x) = -4x^4 + 5x^3 + 6x^2 - 7x + 8$$

When we calculated the third derivative, we got

$$f'''(x) = -96x + 30$$

Differentiating again, we obtain

$$f^{(4)}(x) = -96$$

Another example

Now let's find the fourth derivative of

$$f(x) = 4x^4 - 5 \cos x$$

We found its third derivative to be

$$f'''(x) = 96x - 5 \sin x$$

We must differentiate again! The sum rule tells us that

$$f^{(4)}(x) = d/dx\,(96x) - d/dx\,(5 \sin x)$$

Using the power rule and the multiplication-by-constant rule, we get

$$f^{(4)}(x) = 96 - 5\,d/dx \sin x$$

The derivative of the sine is the cosine, so we have

$$f^{(4)}(x) = 96 - 5 \cos x$$

Still another example

Now we'll find the fourth derivative of

$$g(t) = 2t^2 + 5t - 7 + 4t^{-1} - t^{-2} + 2t^{-3} + \ln t$$

We found that the third derivative is

$$g'''(t) = 2t^{-3} - 24t^{-4} + 24t^{-5} - 120t^{-6}$$

Using the generalized power rule and working term-by-term, we get

$$g^{(4)}(t) = -6t^{-4} + 96t^{-5} - 120t^{-6} + 720t^{-7}$$

- -

Are you astute?

By now, you'll have noticed that these three functions "morph" differently as you take their derivatives repeatedly. In the first case, if you differentiate again, you'll get the zero function; after that, higher derivatives won't change anything. In the second case, the constant -96 will vanish, leaving 5 or -5 times a sine or cosine. In the last case, you'll always get a "nonstandard" polynomial. The exponents will get larger negatively, and the coefficients will alternate between positive and negative as their absolute values grow.

Can you tell, merely by looking at a function, what will occur if you differentiate it multiple times? As an extra-credit exercise, invent a few functions on your own. Then try to predict what will happen if you differentiate them over and over. Finally, do the calculations to see how accurate your predictions are.

Here's a challenge!

Find the first nine derivatives of the sine function.

Solution

To work out all these derivatives, we need to know only that

$$d/dx \, (\sin x) = \cos x$$

and

$$d/dx \, (\cos x) = -\sin x$$

Succeeding derivatives are easy to find:

$$d^2/dx^2 \, (\sin x) = d/dx \, (\cos x) = -\sin x$$
$$d^3/dx^3 \, (\sin x) = d/dx \, (-\sin x) = -\cos x$$
$$d^4/dx^4 \, (\sin x) = d/dx \, (-\cos x) = \sin x$$
$$d^5/dx^5 \, (\sin x) = d/dx \, (\sin x) = \cos x$$
$$d^6/dx^6 \, (\sin x) = d/dx \, (\cos x) = -\sin x$$
$$d^7/dx^7 \, (\sin x) = d/dx \, (-\sin x) = -\cos x$$
$$d^8/dx^8 \, (\sin x) = d/dx \, (-\cos x) = \sin x$$
$$d^9/dx^9 \, (\sin x) = d/dx \, (\sin x) = \cos x$$

As we keep going, the derivatives cycle endlessly through these four functions in order: negative sine, negative cosine, sine, and cosine.

Here's another challenge!

Find the first nine derivatives of the natural log function, given that the domain is restricted to the positive reals.

Solution

In this situation, we don't get a cycle, but we will see a pattern. To begin, remember that

$$d/dx \, (\ln x) = x^{-1}$$

We can use the generalized power rule to find higher derivatives, building our results one upon another:

$$d^2/dx^2 \, (\ln x) = d/dx \, (x^{-1}) = -x^{-2}$$
$$d^3/dx^3 \, (\ln x) = d/dx \, (-x^{-2}) = 2x^{-3}$$
$$d^4/dx^4 \, (\ln x) = d/dx \, (2x^{-3}) = -6x^{-4}$$
$$d^5/dx^5 \, (\ln x) = d/dx \, (-6x^{-4}) = 24x^{-5}$$
$$d^6/dx^6 \, (\ln x) = d/dx \, (24x^{-5}) = -120x^{-6}$$
$$d^7/dx^7 \, (\ln x) = d/dx \, (-120x^{-6}) = 720x^{-7}$$
$$d^8/dx^8 \, (\ln x) = d/dx \, (720x^{-7}) = -5{,}040x^{-8}$$
$$d^9/dx^9 \, (\ln x) = d/dx \, (-5{,}040x^{-8}) = 40{,}320x^{-9}$$

- -

Practice Exercises

This is an open-book quiz. You may (and should) refer to the text as you solve these problems. Don't hurry! You'll find worked-out answers in App. A. The solutions in the appendix may not represent the only way a problem can be figured out. If you think you can solve a particular problem in a quicker or better way than you see there, by all means try it!

1. In the chapter text, we found the second, third, and fourth derivatives of

$$f(x) = -4x^4 + 5x^3 + 6x^2 - 7x + 8$$

    Find the fifth derivative of this function.

2. Find the sixth derivative of

$$f(x) = -4x^4 + 5x^3 + 6x^2 - 7x + 8$$

3. In the chapter text, we found the second, third, and fourth derivatives of

$$f(x) = 4x^4 - 5 \cos x$$

    Find the fifth derivative of this function.

4. Find the sixth derivative of

$$f(x) = 4x^4 - 5 \cos x$$

5. Find the seventh derivative of

$$f(x) = 4x^4 - 5 \cos x$$

What happens as we keep on going to the eighth, ninth, and higher derivatives?

6. In the chapter text, we found the second, third, and fourth derivatives of

$$g(t) = 2t^2 + 5t - 7 + 4t^{-1} - t^{-2} + 2t^{-3} + \ln t$$

Find the fifth derivative of this function.

7. Find the sixth derivative of

$$g(t) = 2t^2 + 5t - 7 + 4t^{-1} - t^{-2} + 2t^{-3} + \ln t$$

8. Find the seventh derivative of

$$g(t) = 2t^2 + 5t - 7 + 4t^{-1} - t^{-2} + 2t^{-3} + \ln t$$

What happens (in general) as we keep on going to the eighth, ninth, and higher derivatives?

9. Go back to the first "challenge" in the chapter text. Suppose Sir Isaac went to another cliff only 50 m high, and repeated the experiment. How long did the apple take to fall from that cliff? What was the vertical component v_a of the apple's speed at impact? What was the vertical component a_a of the apple's acceleration at impact?

10. Look one more time at the first "challenge." Imagine that, instead of working on earth, Sir Isaac conducted his experiment on a planet where the fallen distance h (in meters) as a function of time t (in seconds) was

$$h = 3t^2$$

If the apple took 11 seconds to fall, what was the height h_c of the cliff? What was the vertical component v_a of the apple's speed at impact? What was the vertical component a_a of the apple's acceleration at impact?

Analyzing Graphs with Derivatives

Derivatives can be used to help us draw and analyze graphs of functions. Let's look at a few simple examples of how this process works.

Three Common Traits

A smooth curve, representing a continuous and differentiable function, has various characteristics. Three of the most significant and obvious traits are *concavity, extrema,* and *inflection points.*

Concavity

The concavity of a curve is the direction in which it bends. In part or all of the graph of a function, the curvature can be either *concave upward* or *concave downward.* When a curve is concave upward, the second derivative of the function is positive. When a curve is concave downward, the second derivative is negative.

Figure 9-1 shows a graph with two different regions of concavity. The portion of the curve to the left of the dashed line is concave upward, and $d^2y/d^2x > 0$. The portion of the curve to the right of the dashed line is concave downward, and $d^2y/dx^2 < 0$.

Extrema

When a curve has a peak or trough at a certain point relative to the surrounding region, that point is called an *extremum* (plural *extrema*). A peak is a *local maximum,* and a trough is a *local minimum.* If a peak represents the largest value in the entire function, it's the *absolute maximum.* If a trough is the smallest value in the function, it is the *absolute minimum.*

At any point where a graph reaches an extremum, the first derivative is 0. In Fig. 9-1, the curve has one local minimum and one local maximum. The slopes are 0 at both of these points. Some curves have multiple extrema. Once in a while, we'll see a curve that has infinitely many local extrema, and yet not a single absolute extremum! (The sine function over the domain of all reals is a good example.)

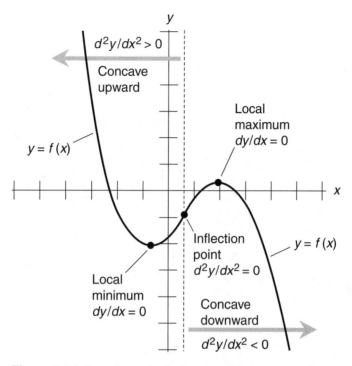

Figure 9-1   Generic graph of a function showing zones of
concavity, a local maximum, a local minimum, and
an inflection point.

Inflection point

When a graph reaches a point where it goes from concave upward to concave downward
or vice-versa, we have an inflection point. The second derivative is always 0 at an inflection
point. This means that the rate of change in the curve's slope is 0. Figure 9-2 shows six different examples of inflection points in curves representing cubic functions.

In Fig. 9-2A, C, and E, the slope of the curve is constantly decreasing (downward concavity) to the left of the inflection point, and constantly increasing (upward concavity) to
the right of that point. In Fig. 9-2B, D, and F, the slope of the curve is constantly increasing
(upward concavity) to the left of the inflection point, and constantly decreasing (downward
concavity) to the right of that point.

The rate at which the slope changes can be 0 at a point, even if the slope itself is not 0 at
that point. Figure 9-2 shows two curves (A and B) where the slope is 0 at the inflection point,
two curves (C and F) where the slope is positive at the inflection point, and two curves (D and
E) where the slope is negative at the inflection point.

- -

Are you confused?

You might ask, "What do the third and higher derivatives represent in a graph?" This question, while a
good one, is difficult to answer clearly.

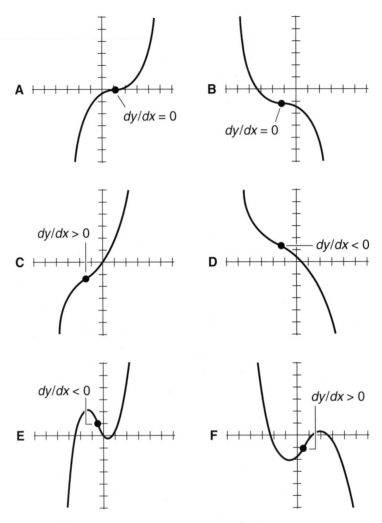

Figure 9-2   At A and B, the slope at the inflection point is 0. At C
and F, the slope at the inflection point is positive. At D
and E, the slope at the inflection point is negative.

The first derivative represents the slope of a curve. The second derivative represents how fast the slope changes. In cubic functions such as those in Fig. 9-2, the third derivative is a constant, the absolute value of which indicates how horizontally spread-out the graph appears. In Fig. 9-2A, B, C, and D, the curves are "broad," and the third derivatives have relatively small absolute values. In Fig. 9-2E and F, the curves are fairly "tight," and the third derivatives have larger absolute values.

You can make a game out of describing how the fourth and higher derivatives translate into visual characteristics. But be careful! Some curves behave in ways you might not expect. The curves representing multiple derivatives of the sine function, for example, keep the same shape. The only change is that the wave train moves $\pi/2$ units to the left every time you differentiate.

Here's a challenge!

Suppose that the function shown in Fig. 9-1 is multiplied by −1. What happens to the *x*-values of the extrema and inflection point? What happens to the range of *x*-values representing the zone where the graph is concave upward? What happens to the range of *x*-values where the graph is concave downward?

Solution

If the function is multiplied by −1, the entire curve is inverted with respect to the *x* axis. The *x*-values of the extrema and the inflection point do not change. But the local minimum becomes a local maximum, and the local maximum becomes a local minimum. The curve becomes concave downward in the region where it was concave upward before, and it becomes concave upward in the region where it was concave downward before.

Graph of a Quadratic Function

When a quadratic function has real-number coefficients and a real constant, its graph in the *xy*-plane is a *parabola*. Figure 9-3 shows several generic examples of parabolas that can represent quadratic functions of the form

$$y = ax^2 + bx + c$$

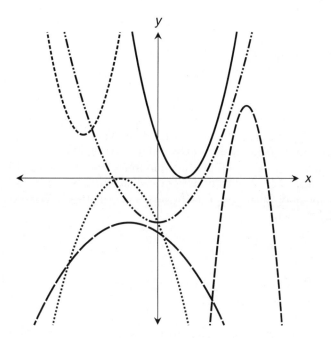

Figure 9-3   The graph of a quadratic function in the *xy*-plane is always a parabola that is either concave upward or concave downward.

where *a*, *b*, and *c* are real numbers, and $a \neq 0$. Such a parabola is either concave upward or concave downward.

An example

Consider the quadratic function

$$y = x^2 - 3x + 2$$

Let's graph this function in the *xy*-plane, and then analyze that graph. We can determine the *y*-intercept by plugging in 0 for *x* and then finding *y* by arithmetic. When we carry out the operation, we learn that the *y*-intercept is at the point $(x,y) = (0,2)$. The *x*-intercepts (or *zeros*) of the function, if any exist, can be found by solving the quadratic equation

$$x^2 - 3x + 2 = 0$$

which factors into

$$(x - 1)(x - 2) = 0$$

so its roots are $x = 1$ or $x = 2$. This tells us that the points $(1,0)$, and $(2,0)$ lie on the graph of the original quadratic function. Its first derivative is

$$dy/dx = 2x - 3$$

and its second derivative is

$$d^2y/dx^2 = 2$$

The second derivative is positive, indicating that the parabola is concave upward. We've found three points on the graph, and we know that the graph is a parabola that opens upward. Based on these facts, we can sketch the curve as shown in Fig. 9-4.

To find the *x*-value of the absolute minimum, we can average the *x*-intercepts, or we can find the *x*-value where the first derivative is equal to 0. This is a calculus course, so let's use the derivative method! When we set the first derivative of the original quadratic function equal to 0, we get the equation

$$2x - 3 = 0$$

This solves to $x = 3/2$. We can plug the *x*-value 3/2 into the original function to solve for *y*, as follows:

$$y = x^2 - 3x + 2 = (3/2)^2 - 3 \cdot (3/2) + 2 = -1/4$$

Now we know that the absolute minimum point is $(x,y) = (3/2, -1/4)$.

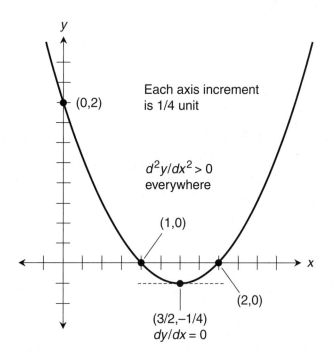

Each axis increment is 1/4 unit

$d^2y/dx^2 > 0$ everywhere

(0,2)

(1,0)

(2,0)

(3/2,−1/4)
$dy/dx = 0$

Figure 9-4   Graph of $y = x^2 - 3x + 2$. On both axes, each increment represents 1/4 unit.

Are you astute?

Have you noticed that the second derivative of a quadratic function is always a constant, equal to twice the *leading coefficient* (the coefficient of the term in which the independent variable is squared)? Geometrically, this means that the curve's slope changes at a constant rate as we move from left to right along the *x* axis at a "steady speed."

Here's a challenge!

Look again at the function graphed in Fig. 9-4. What is the slope at the *y*-intercept point, (0,2)? What are the slopes at the two *x*-intercept points, (1,0) and (2,0)?

Solution

To find the slope of the curve at the point (0,2), we can plug in the *x*-value, which is 0, to the first-derivative function and calculate

$$dy/dx = 2x - 3 = 2 \cdot 0 - 3 = -3$$

For the point (1,0), we plug in 1 for *x* to get

$$dy/dx = 2x - 3 = 2 \cdot 1 - 3 = 2 - 3 = -1$$

For the point (2,0), we plug in 2 for x to get

$$dy/dx = 2x - 3 = 2 \cdot 2 - 3 = 4 - 3 = 1$$

Graph of a Cubic Function

A cubic curve has one of the six characteristic shapes shown in Fig. 9-2. There's always one, but only one, inflection point. The *overall trend* of the curve is either upward (negative-to-positive) or downward (positive-to-negative), although the curve might reverse its direction in the vicinity of the inflection point.

Example

Let's examine the cubic function

$$y = 5x^3 + 3x^2 + 5x + 7$$

We can plug in a few x-values, calculate the y-values, and tabulate the resulting numbers. Table 9-1 is a list of coordinate values that can give us a good idea of what the curve looks like when we plot the corresponding points in the xy-plane. Figure 9-5 shows the graph. On the x axis, each division represents 1/2 unit. On the y axis, each division represents 10 units. This axis distortion allows us to graph the function within a region of reasonable dimensions.

Now let's differentiate our cubic function to see how the slope varies with the value of x. The first derivative is

$$dy/dx = 15x^2 + 6x + 5$$

The slope follows a quadratic function. This fact makes the cubic curve inherently more complicated than any quadratic curve, where the slope always follows a linear function.

The second derivative of this cubic function tells us how fast, and in what direction, the slope changes as we increase the value of x. When we differentiate the first derivative, we get

$$d^2y/dx^2 = 30x + 6$$

Table 9-1. Selected values for graphing the function
$$y = 5x^3 + 3x^2 + 5x + 7.$$

x	$5x^3 + 3x^2 + 5x + 7$
−3	−116
−2	−31
−1	0
0	7
1	20
2	69

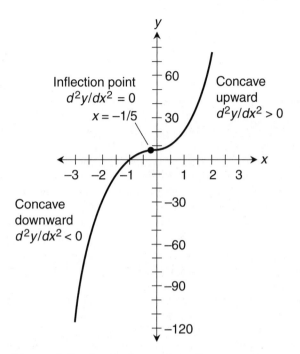

Figure 9-5   Graph of $y = 5x^3 + 3x^2 + 5x + 7$. On
the x axis, each division represents
1/2 unit. On the y axis, each division
represents 10 units.

We can find the x-value of the inflection point, where the concavity reverses, by figuring
out where the second derivative is 0. We create a linear equation by setting $d^2y/dx^2 = 0$, like
this:

$$30x + 6 = 0$$

The solution to this equation is $x = -1/5$. When we plug $-1/5$ into the original function for
x and then calculate y, we get

$$y = 5x^3 + 3x^2 + 5x + 7 = 5 \cdot (-1/5)^3 + 3 \cdot (-1/5)^2 + 5 \cdot (-1/5) + 7 = 152/25$$

The coordinates of the inflection point, written as an ordered pair, are therefore

$$(x,y) = (-1/5, 152/25)$$

This curve is concave downward to the left of the inflection point, and concave upward to the
right of the inflection point. This is a geometric indication of the facts that

$$d^2y/dx^2 < 0 \quad \text{when} \quad x < -1/5$$

and

$$d^2y/dx^2 > 0 \quad \text{when} \quad x > -1/5$$

- -

Are you confused?

It's reasonable to wonder whether there are any *local* minima or *local* maxima in this curve. There is no *absolute* minimum and no *absolute* maximum, because cubic curves never have them! When we examine Fig. 9-5, we can see that the overall trend of this curve is upward (negative-to-positive). But is this true at every point along the entire curve? Or does the curve reverse its direction near the inflection point? We can't be sure by merely looking at Fig. 9-5.

To figure this problem out by brute force, we could plot hundreds of points with the help of a computer and a good graphing program. We could also evaluate the first derivative at hundreds of *x*-values, again with the help of a computer. But there's a more elegant way to resolve this mystery. (Mathematicians use the term "elegant" to describe a proof, derivation, or process that uses finesse, rather than force, to solve a problem.)

If the slope at the inflection point is positive, then we can be sure that the curve in Fig. 9-5 trends upward all the time (like the one in Fig. 9-2C), never reverses direction, and therefore has no local minimum or maximum. If the slope at the inflection point is negative, then we can be sure that the curve in Fig. 9-5 reverses direction momentarily, trending downward (positive-to-negative) in a region near the inflection point. In that case, the curve has a local maximum slightly to the left of the inflection point, and a local minimum slightly to the right of it (like the one in Fig. 9-2E). If the slope at the inflection point is 0, then we know that the curve in Fig. 9-5 levels off at that point (like the one in Fig. 9-2A), but doesn't reverse direction and therefore has no local minimum or maximum. Let's see which of these three situations is the case here.

Here's a challenge!

Look again at Fig. 9-5 and the function that this graph represents. What is the slope of the curve at the inflection point? What does this tell us about the general nature of the curve? Which of the generic profiles in Fig. 9-2 does this curve most closely resemble?

Solution

To find the slope of the curve at the inflection point, we plug $x = -1/5$ into the formula for the first derivative, getting

$$dy/dx = 15x^2 + 6x + 5 = 15 \cdot (-1/5)^2 + 6 \cdot (-1/5) + 5 = 22/5$$

That's a positive number. We have $dy/dx > 0$ at the inflection point. The overall trend of the curve is positive, and so is the slope at the inflection point. Based on all this knowledge, we can deduce that the curve does not reverse its direction at the inflection point, and doesn't level off there, either. It therefore has no local minimum or maximum. Of the six generic profiles shown in Fig. 9-2, this curve most nearly resembles the one shown in drawing C.

- -

Graph of the Sine Function

We can use calculus to analyze the graphs of differentiable functions of many kinds, not only polynomial functions. Let's look at a continuous, differentiable function that lends itself nicely to analysis by differentiation: the sine function from trigonometry.

Analyzing the curve

Figure 9-6 is a graph of the sine function over the limited domain where x is restricted to the open interval between, but not including, 0 and 2π. Remember that the x-values are in radians, not in degrees!

We're familiar enough with trigonometry to know that this curve, with the domain restricted as shown in Fig. 9-6, has an absolute maximum at the point $(x,y) = (\pi/2,1)$. It has an absolute minimum at the point $(x,y) = (3\pi/2,-1)$. Let's verify these two facts mathematically. The first derivative is

$$dy/dx \, (\sin x) = \cos x$$

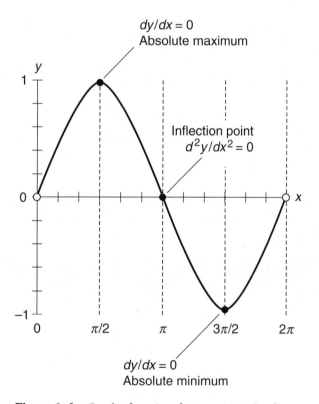

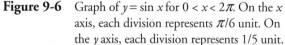

Figure 9-6 Graph of $y = \sin x$ for $0 < x < 2\pi$. On the x axis, each division represents $\pi/6$ unit. On the y axis, each division represents $1/5$ unit.

and the second derivative is

$$d^2y/dx^2 (\sin x) = d/dx (\cos x) = -\sin x$$

At the point where $x = \pi/2$, the first derivative is the cosine of $\pi/2$. That's 0. The second derivative is the negative of the sine of $\pi/2$. That's -1. The slope is 0 and the curve is concave downward, meaning that we have a maximum. At the point where $x = 3\pi/2$, the first derivative is the cosine of $3\pi/2$. That's 0. The second derivative is the negative of the sine of $3\pi/2$. That's 1. The slope is 0 and the curve is concave upward, telling us that we have a minimum. We can look at the graph and immediately see that these extrema are absolute. If we were to include multiple cycles of the wave (or even infinitely many, by letting the domain extend over the entire set of real numbers), then these extrema would be local.

It looks like the curve in Fig. 9-6 has an inflection point at $(x,y) = (\pi,0)$. If we want to be certain, we must look at the second derivative, which is the negative of the sine of π. That's 0, so we know that $(x,y) = (\pi,0)$ is indeed a point of inflection. The curve is concave downward to the left of the point, and concave upward to the right of it.

- -

Are you confused?

All of the curves we've seen in this chapter have represented differentiable functions. "But," you ask, "what happens if a function has points or intervals where it's nondifferentiable?" In cases of that sort, we can usually graph the curve, but we can't use differentiation to evaluate it at the nondifferentiable points or within the nondifferentiable intervals. Even if a function clearly has an extremum at a certain point, we can't verify this fact with differentiation if the function is nondifferentiable at that point. A good example is the absolute-value function $y = |x|$. This has an absolute minimum at the point $(x,y) = (0,0)$, but because the function isn't differentiable at that point, we can't use derivatives to verify that fact.

Here's a challenge!

Sketch a graph of the square of the sine function for $0 < x < 2\pi$. (It's all right to use a calculator, plot numerous points, and then connect them all by curve fitting to obtain this sketch.) Determine the x-values of the extrema and the inflection points. Verify these results using the calculus techniques we've learned.

Solution

Figure 9-7 is a sketch of this function. Symbolically, we can denote the function as either

$$y = (\sin x)^2$$

or

$$y = \sin^2 x$$

In Chap. 7, we found that

$$d/dx \sin^2 x = 2 \sin x \cos x$$

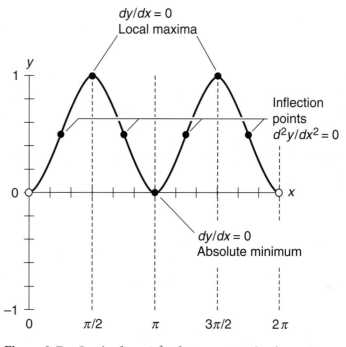

Figure 9-7   Graph of $y = \sin^2 x$ for $0 < x < 2\pi$. On the x axis, each division represents $\pi/6$ unit. On the y axis, each division represents $1/5$ unit.

We can differentiate again, obtaining the second derivative, using the multiplication-by-constant rule and the product rule:

$$d^2/dx^2 (\sin^2 x) = d/dx (2 \sin x \cos x) = 2\, [d/dx (\sin x \cos x)]$$
$$= 2\, [(d/dx \sin x) \cos x + (d/dx \cos x) \sin x] = 2\, [\cos x \cos x + (-\sin x \sin x)]$$
$$= 2\, (\cos^2 x - \sin^2 x) = 2 \cos^2 x - 2 \sin^2 x$$

To find the extrema, we must figure out all the x-values where the first derivative is 0. That means we must solve the equation

$$2 \sin x \cos x = 0$$

This is satisfied whenever $\sin x = 0$ or $\cos x = 0$. We've restricted x to positive values less than 2π, so we have

$$\sin x = 0 \quad \text{when} \quad x = \pi$$

and

$$\cos x = 0 \quad \text{when} \quad x = \pi/2 \quad \text{or} \quad x = 3\pi/2$$

When we look at Fig. 9-7, it appears that the extrema for $x = \pi/2$ and $x = 3\pi/2$ are maxima. We can verify that these points are maxima by checking the values of the second derivatives. For $x = \pi/2$, we have

$$d^2/dx^2 (\sin^2 x) = 2 \cos^2 x - 2 \sin^2 x = 2 (\cos \pi/2)^2 - 2 (\sin \pi/2)^2$$
$$= 2 \cdot 0^2 - 2 \cdot 1^2 = -2$$

The fact that the second derivative is negative indicates that the curve is concave downward, so this extremum is a maximum. In the case of $x = 3\pi/2$, we have

$$d^2/dx^2 (\sin^2 x) = 2 \cos^2 x - 2 \sin^2 x = 2 (\cos 3\pi/2)^2 - 2 (\sin 3\pi/2)^2$$
$$= 2 \cdot 0^2 - 2 \cdot (-1)^2 = -2$$

Again, the curve is concave downward, so we know that the function attains another maximum at this point.

Are you astute?

"Wait," you say. "We've found two maxima, all right. But are they local, or is one of them absolute?" If the y-values are the same for both points, then they're both local maxima. But if the y-values are different, then the point with the larger y-value is the absolute maximum. We'd better check this out! When we plug in $x = \pi/2$ to the original function, we get

$$y = \sin^2 x = (\sin \pi/2)^2 = 1^2 = 1$$

When we plug in $x = 3\pi/2$ to the original function, we get

$$y = \sin^2 x = (\sin 3\pi/2)^2 = (-1)^2 = 1$$

Now we know that these two maxima are the same, so they're both local.

Back to the challenge!

Let's return to the "challenge" we've been dealing with. There appears to be an extremum at the point where $x = \pi$. Figure 9-7 suggests that it's a minimum. To verify, let's plug in π for x and see what we get for the second derivative:

$$d^2/dx^2 (\sin^2 x) = 2 \cos^2 x - 2 \sin^2 x = 2 (\cos \pi)^2 - 2 (\sin \pi)^2$$
$$= 2 \cdot (-1)^2 - 2 \cdot 0^2 = 2$$

It's positive, all right! That means the curve is concave upward at this point, so it must represent a minimum. Within the span of x-values we've allowed here, it's the absolute minimum, because we've left $x = 0$ and $x = 2\pi$ out of the domain. (If we had left those points in the domain, they'd represent local minima, as would the point where $x = \pi$. You can verify, if you like, that all three of these minima would be the same.)

The inflection points occur where the second derivative is equal to 0. It appears that there are four such points in the curve of Fig. 9-7, but it's difficult to tell exactly where they are. We must solve the following equation to find the x-values:

$$2 \cos^2 x - 2 \sin^2 x = 0$$

Dividing through by 2, we get

$$\cos^2 x - \sin^2 x = 0$$

Adding $\sin^2 x$ to each side gives us

$$\cos^2 x = \sin^2 x$$

We can take the square root of both sides here, keeping in mind that we have to account for the positive and negative values. When we do that, we get

$$\pm\cos x = \pm\sin x$$

We know that the inflection points don't occur where either the sine or the cosine are equal to 0. Those situations represent the extrema. We've already determined that the second derivatives are nonzero at those points. Knowing that the cosine isn't 0 at any of the inflection points, we can divide each side of the above equation by $\pm\cos x$ to get

$$\pm 1 = (\pm\sin x) / (\pm\cos x)$$

From trigonometry, we remember that the sine divided by the cosine is equal to the tangent. This means we can simplify the above equation to

$$\pm\tan x = \pm 1$$

which tells us that

$$x = \text{Arctan} (\pm 1)$$

This gives us four values within the domain we've allowed, which is $0 < x < 2\pi$. Those values are

$$x = \pi/4$$
$$x = 3\pi/4$$
$$x = 5\pi/4$$
$$x = 7\pi/4$$

Here's an extra-credit challenge!

Determine the y-values of each of the inflection points on the curve shown in Fig. 9-7. Then write down the complete (x,y) coordinates of all four points.

Solution

You're on your own. That's why you get extra credit!

Practice Exercises

This is an open-book quiz. You may (and should) refer to the text as you solve these problems. Don't hurry! You'll find worked-out answers in App. A. The solutions in the appendix may not represent the only way a problem can be figured out. If you think you can solve a particular problem in a quicker or better way than you see there, by all means try it!

1. Determine the coordinates of the inflection point in the graph of the function

$$y = 3x^3 + 3x^2 - x - 7$$

2. What is the slope of the curve representing the function stated in Prob. 1 at the inflection point?

3. In what range of x-values is the graph of the function stated in Prob. 1 concave upward? In what range of x-values is it concave downward?

4. Determine the extremum in the graph of the function

$$y = 4x^2 - 7$$

   Is this an absolute maximum or an absolute minimum?

5. Look again at Fig. 9-4. What is the equation of a line tangent to the curve, and passing through the point (2,0)?

6. Consider the cubic function

$$y = x^3 + 3x^2 - 3x + 4$$

   Without drawing the graph, determine the coordinates of the inflection point. Then calculate the slope of the curve at the inflection point.

7. Evaluate the function stated in Prob. 6 for $x = 100$. We choose $x = 100$ because it's much greater than the absolute values of any of the coefficients. If we get a large positive value for y, we'll know that the curve trends upward overall, like Fig. 9-2A, C, or E. If we get a large negative value for y, we'll know that the curve trends downward overall, like Fig. 9-2B, D, or F. Note again the slope of the function at the inflection point. We found that slope in the solution to Prob. 6. From all this information, identify which of the six general contours from Fig. 9-2 applies to this curve.

8. Consider again the function we worked with in the solutions to Probs. 6 and 7. Find the y-intercept of the curve. Are there any points where the slope is 0? If so, determine the coordinates of those points. Finally, on the basis of all the information we've gathered about this function, sketch its graph.

9. Determine the slope, at the inflection point, of the curve shown in Fig. 9-6 for the function

$$y = \sin x$$

10. Determine the slope, at each of the four inflection points, of the curve shown in Fig. 9-7 for the function

$$y = \sin^2 x$$

10
Review Questions and Answers

Part One

This is not a test! It's a review of important general concepts you learned in the previous nine chapters. Read it though slowly and let it "sink in." If you're confused about anything here, or about anything in the section you've just finished, go back and study that material some more.

Chapter 1

Question 1-1

How do the domain, range, and variables relate in a mapping from one set into another?

Answer 1-1

The domain is the set of elements that the mapping "goes out from." The range is the set of elements that the mapping "goes into." The independent variable is a nonspecific element in the domain, and the dependent variable is a nonspecific element of the range. The mapping assigns specific values of the dependent variable to specific values of the independent variable.

Question 1-2

What's the difference between a relation and a function?

Answer 1-2

A relation is a mapping that can be defined as a set of ordered pairs. A function is a special sort of relation, in which a single value of the independent variable can map into *at most* one value of the dependent variable.

Question 1-3

In an ordered pair, what does the term before the comma mean? What does the term after the comma mean?

Answer 1-3

The term before the comma represents either the independent variable or some specific value of the independent variable. The term after the comma represents either the dependent variable or some value of it.

Imagine that we have a relation between two sets X and Y, where set X is the domain, and we represent the independent variable by x. Also suppose that we call set Y the range, and we represent the dependent variable by y. Then we can define our relation as a set of ordered pairs written in the form (x,y). If the relation maps, say, the value 3 in the domain to the value -5 in the range, then $(3,-5)$ is an element of the relation.

Question 1-4

How can we tell whether or not a relation is a function by looking at its graph, where the independent variable is shown on the horizontal axis and the dependent variable is shown on the vertical axis?

Answer 1-4

We can imagine a straight, vertical, infinitely long line that we can move freely to the left and right. A graph represents a function if and only if it never crosses the movable vertical line at more than one point. We can call this scheme the vertical-line test for a function.

Question 1-5

How can we tell whether or not the inverse of a function is another function by looking at a graph of the original function? Again, assume that the independent variable is represented on the horizontal axis and the dependent variable is represented on the vertical axis.

Answer 1-5

We can imagine a straight, horizontal, infinitely long line that can be freely moved up and down. The inverse of the graphed function is a function if and only if the graph of the original function never crosses the horizontal line at more than one point. We can call this scheme the horizontal-line test for an inverse function.

Question 1-6

In each of the following relations, imagine that x is the independent variable and y is the dependent variable. Which of these relations, if any, is a constant function of x? Which of these relations, if any, is not a function of x?

$$3x + 4y = 2$$
$$y = 5$$
$$x = -7$$
$$y = x + 2$$
$$x = 4y$$

Answer 1-6

The second relation is a constant function of x, because the value of the function is 5 regardless of the value of x. The third relation is not a function of x because, when x is equal to -7

(which constitutes the entire domain), the dependent variable has more than one value (in fact, it can be any real number!). We can also say that the third relation isn't a function because it doesn't even say anything about the dependent variable. The first, fourth, and fifth relations can all be expressed as functions of *x*.

Question 1-7

In each of the following functions, imagine that *t* is the independent variable and *u* is the dependent variable. Which of these functions have inverse functions? How can we tell?

$$u = 2t^2$$
$$u = 3t^3$$
$$u = 4t^4$$
$$u = 5t^5$$
$$u = 6t^6$$
$$u = 7t^7$$

Answer 1-7

All three of the functions with the odd-integer exponents have inverse functions. None of the three functions with even-integer exponents has an inverse function (although, if considered as relations, they can be said to have inverse relations). We can tell which functions have inverse functions and which do not by plotting them as graphs with *t* on the horizontal axes and *u* on the vertical axes, and then conducting horizontal-line tests on each graph.

Question 1-8

How can we recognize a discontinuity in a function by looking at its graph?

Answer 1-8

A discontinuity shows up as a jump, blow-up (singularity), or gap in the graph. The line or curve is not smooth at the discontinuity.

Question 1-9

Can the graph of a function "turn a sharp corner" at a point and still be continuous there?

Answer 1-9

Yes. A good example is the absolute-value function. It "turns a corner" at the origin point (0,0), but it's continuous there. (The absolute-value function is not *differentiable* at the origin, however.)

Question 1-10

Examine Fig. 10-1. Does this graph represent a function of *x*? At what points is the relation discontinuous? Is the relation defined at the points where it's discontinuous?

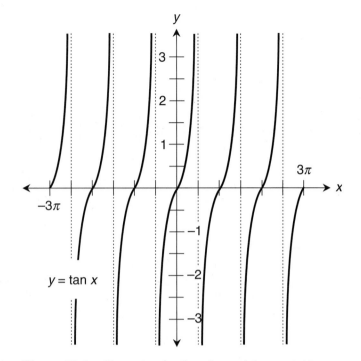

Figure 10-1   Illustration for Question and Answer 1-10.

Answer 1-10

This graph portrays a function of x, because it passes the vertical-line test. Discontinuities occur at every point where x is an odd-integer multiple of $\pi/2$. The function is not defined at the points where it is discontinuous.

Chapter 2

Question 2-1

Figure 10-2 is a graph of the inverse of the tangent function, also called the Arctangent function. For this inverse to be a function, the domain of the original tangent function is restricted to values larger than $-\pi/2$ and smaller than $\pi/2$. Here, x is the independent variable, and y is the dependent variable. We can write

$$y = \text{Arctan } x$$

or

$$y = \tan^{-1} x$$

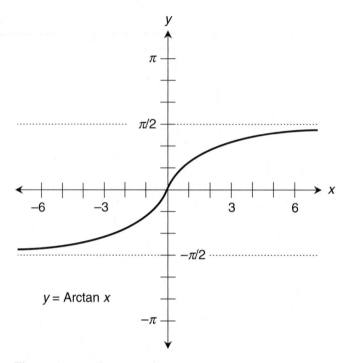

Figure 10-2   Illustration for Questions and Answers 2-1
and 2-2.

(We mustn't get confused when we write $\tan^{-1}$ in this way; it means the inverse of the tangent function, not the reciprocal!) What is the following limit? What does it mean?

$$\underset{x \to \infty}{Lim}\ \text{Arctan } x$$

Answer 2-1

This limit is equal to $\pi/2$. It means that as x increases endlessly in the positive direction, the value of the function approaches $\pi/2$.

Question 2-2

In the situation described in Question 2-1 and illustrated in Fig. 10-2, what is the following limit? What does it mean?

$$\underset{x \to -\infty}{Lim}\ \text{Arctan } x$$

Answer 2-2

This limit is equal to $-\pi/2$. It means that as x decreases endlessly (or increases endlessly in the negative direction), the value of the function approaches $-\pi/2$.

Question 2-3

What does the following symbolic expression claim? Is it true in general?

$$\lim_{x \to k} [f(x) + g(x)] = \lim_{x \to k} f(x) + \lim_{x \to k} g(x)$$

Answer 2-3

This expression claims that the limit of the sum of two functions of x, as x approaches a constant k, is equal to the sum of the limits of the individual functions as x approaches k. This is always true, assuming both functions are defined as x approaches k.

Question 2-4

What does the following symbolic expression claim? Is it true in general?

$$c \lim_{x \to k} f(x) = \lim_{x \to k} c[f(x)]$$

Answer 2-4

This expression claims that a constant c times the limit of a function of x, as x approaches another constant k, is equal to the limit of c times the function as x approaches k. This is always true, assuming the function is defined as x approaches k.

Question 2-5

What does the following expression claim? Is it true in general?

$$c \lim_{x \to k} f(x) = \lim_{x \to k} f(cx)$$

Answer 2-5

This expression claims that a constant c times the limit of a function of x, as x approaches another constant k, is equal to the limit of the same function of cx, as x approaches k. This is not always true.

Question 2-6

How does the notion of left-hand continuity at a point differ from the notion of right-hand continuity at a point?

Answer 2-6

A function is left-hand continuous at a point if and only if

- We can define the limit of the function at the point as we approach that point from the left (that is, from the negative direction).
- We can define the actual value of the function at the point.
- The limit and the actual value are the same.

A function is right-hand continuous at a point if and only if

- We can define the limit of the function at the point as we approach that point from the right (that is, from the positive direction).
- We can define the actual value of the function at the point.
- The limit and the actual value are the same.

Question 2-7

Can a function be left-hand continuous at a point, but not right-hand continuous at the same point? If so, provide an example.

Answer 2-7

Yes, this can happen. Here's an example. Let's define a function $f(x)$ like this:

$$f(x) = -1 \quad \text{when } x \leq 0$$
$$= 1 \quad \text{when } x > 0$$

This function is left-hand continuous at the point $(0,-1)$, but it's not right-hand continuous there. The conditions for left-hand continuity, as outlined in Answer 2-6, are satisfied. As we approach the point $(0,-1)$ from the left, the limit of the function is -1, and the value of the function at the point is also -1. But the conditions for right-hand continuity aren't satisfied. As we approach $(0,-1)$ from the right, the limit of the function is 1, but the value of the function at the point is -1. This situation is shown in Fig. 10-3.

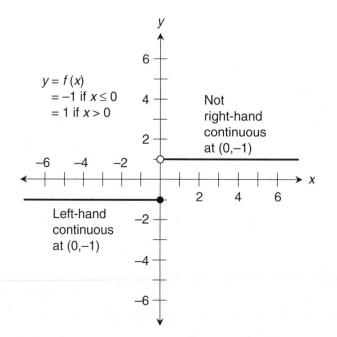

Figure 10-3   Illustration for Question and Answer 2-7.

Question 2-8

What do we mean when we say that a real function $f(x)$ is continuous in general (or simply continuous)? What does the graph of such a function look like?

Answer 2-8

A real-number function $f(x)$ is continuous in general if and only if it is both left-hand continuous and right-hand continuous at every point in its domain. When graphed, such a function looks like a line or curve that's smooth everywhere and has no gaps, blow-ups, or jumps within the domain.

Question 2-9

Which, if any, of the following functions contain at least one discontinuity, if the domain is allowed to extend over the entire set of real numbers?

$$f_1(x) = 3x^2 - 7$$
$$f_2(x) = -7x^2 + 5x - 7$$
$$f_3(x) = 4x^3 - 5x^2 + 8x - 14$$
$$f_4(x) = -6x^6 - 3x^4 + 2x^2 + 5x - 1$$
$$f_5(x) = -3 \sin x$$
$$f_6(x) = 4 \csc x$$

Answer 2-9

The first four functions are all continuous, because they're polynomial functions that can be written in the form

$$P_n(x) = a_n x^n + a_{n-1} x^{n-1} + a_{n-2} x^{n-2} + \cdots + a_1 x + b$$

where n is a positive integer, all the a's are real-number constants, and the variable x is never raised to a negative power. Such functions are always completely free of discontinuities! The fifth function, a constant multiple of the sine function, is continuous over the entire set of reals. But the sixth function, a constant multiple of the cosecant function, has discontinuities at all points where x is an integer multiple of π.

Question 2-10

What is the range of the function $f_6(x)$ stated in Question 2-9? How can this function be made continuous without restricting the range?

Answer 2-10

The range of this function is the union of two sets: the set of all reals larger than or equal to 4, and the set of all reals smaller than or equal to -4. Another way of saying this is that the range is the entire set of real numbers *except* those in the open interval $(-4,4)$.

As stated in Question 2-9, the function $f_6(x)$ has discontinuities at all integer values of π because it's undefined at those values of x. We can make $f_6(x)$ continuous if we restrict the domain by excluding all the values of x for which the function is undefined.

From trigonometry, we recall that the cosecant function is the reciprocal of the sine function. That means that in Question 2-9, we have

$$f_6(x) = 4\,[1/(\sin x)] = 4/(\sin x)$$

You can draw a graph of the sine function for the half-open interval $0 \le x < 2\pi$, and see that this span of x-values encompasses exactly one full cycle of the curve, so all possible values in the range are accounted for. But when $x = 0$ or $x = \pi$, we have $\sin x = 0$, so its reciprocal is undefined. When we work with the cosecant, we have to exclude these values of x if we want to call the function continuous. We can account for one full cycle and keep the function continuous if we define

$$f_6(x) = 4 \csc x$$
$$\text{for } 0 < x < \pi \quad \text{and} \quad \pi < x < 2\pi$$

Of course, we can make more severe restrictions than this. For example, we could say that x must be larger than 0 but less than $\pi/2$. If we do that sort of thing, however, we truncate the range. That is to say, we "cut off" part of the cycle.

Chapter 3

Question 3-1

Suppose someone says, "A line is tangent to a curve at a point (x_0, y_0) in the xy-plane." What does this statement mean?

Answer 3-1

The line touches the curve at the point (x_0, y_0), has the same slope as the curve at that point, and lies in the xy-plane along with the curve.

Question 3-2

Imagine that we want to approximate the slope M of the line tangent a curve in the xy-plane at a point (x_0, y_0). We create a movable point (x, y) on the curve near (x_0, y_0). We draw a straight line through the two points. What is the slope M^* of that line? How can this slope by symbolized?

Answer 3-2

The slope of the line through the two points is

$$M^* = (y - y_0) / (x - x_0)$$

The difference $y - y_0$ can be symbolized Δy (read "delta y"), and the difference $x - x_0$ can be symbolized Δx (read "delta x"), so we can write

$$M^* = \Delta y / \Delta x$$

Question 3-3

What happens to the slope M^* as defined in Question and Answer 3-2, as we move the point (x,y) closer to the fixed point (x_0,y_0) along the curve, without letting the points come together?

Answer 3-3

As the movable point (x,y) approaches the fixed point (x_0,y_0), the slope M^* of the line through the two points approaches the slope M of the line tangent to the curve at (x_0,y_0). This means that the slope M^* approaches the slope of the curve at (x_0,y_0).

Question 3-4

How must we be careful while defining the slope of a line tangent to a curve in the manner defined in the previous question?

Answer 3-4

For this scheme to work, we must be able to *uniquely* define a line tangent to the curve at (x_0,y_0). The curve must not have a gap, take a jump, or turn a corner at the point. If any of these things happen, then we can't uniquely define the slope of the curve at (x_0,y_0).

Question 3-5

If a curve has no gap, takes no jump, and turns no corner at a point (x_0,y_0), and if we let (x,y) be a movable point that approaches (x_0,y_0) along the curve, then Δy shrinks to almost nothing as Δx does the same. We say that the limit of $\Delta y/\Delta x$ as Δx approaches 0 is equal to M, the slope of a line tangent to the curve at the point (x_0,y_0). How do we write this symbolically?

Answer 3-5

We write the limit as

$$M = \underset{\Delta x \to 0}{Lim} \ \Delta y/\Delta x$$

Question 3-6

As the quantities Δy and Δx become arbitrarily small, we call them differentials and symbolize them dy and dx. How can we rewrite the limit in Answer 3-5 to reflect this concept? If we let $f(x)$ be the function that the curve portrays, what is this limit called?

Answer 3-6

We can rewrite the limit, which is called the derivative of f with respect to x at the point (x_0,y_0), like this:

$$dy/dx = \underset{\Delta x \to 0}{Lim} \ \Delta y/\Delta x$$

or like this:

$$f'(x_0) = \underset{\Delta x \to 0}{Lim} \ \Delta y/\Delta x$$

Other notations exist, but these are the most common.

Question 3-7

How can the derivative of a function $f(x)$ at the point where $x = x_0$ be defined without involving y in the notation?

Answer 3-7

The derivative of a function $f(x)$ at the point where $x = x_0$ can be defined as

$$f'(x_0) = \lim_{\Delta x \to 0} \; [f(x_0 + \Delta x) - f(x_0)] \, / \, \Delta x$$

This is the same as the second expression in Answer 3-6, except that Δy has been replaced by its equivalent, the quantity

$$f(x_0 + \Delta x) - f(x_0)$$

Question 3-8

How must we be careful when we interpret the definition (in Answer 3-7) of a derivative at a point?

Answer 3-8

We must remember that Δx can be either positive or negative. The process must work as we find the limit from either the left or the right, and the two results must agree.

Question 3-9

How can the derivative of a function $f(x)$ be defined in general without involving y in the notation?

Answer 3-9

The formula from Answer 3-7 can be generalized to

$$f'(x) = \lim_{\Delta x \to 0} \; [f(x + \Delta x) - f(x)] \, / \, \Delta x$$

Question 3-10

How must we be careful when we apply the definition (from Answer 3-9) of a derivative in general?

Answer 3-10

We must be sure that a derivative exists for the function at every point in its domain.

Chapter 4

Question 4-1

What are the three different characteristics a curve can have that suggest that the function it portrays is nondifferentiable?

Answer 4-1

If the curve has a gap, makes a jump, or turns a corner, it's a warning sign that the function is nondifferentiable.

Question 4-2

What do gaps, jumps, and corners in a curve look like?

Answer 4-2

Figure 10-4 shows an example of each.

Question 4-3

What other sort of gap can a function have, besides the one shown in Fig. 10-4?

Answer 4-3

When a function increases or decreases without bound (blows up) as the independent variable approaches a certain value, we have an extreme gap—an infinitely wide one! The reciprocal function, $y = 1/x$, is an example. The curve blows up when $x = 0$. The graph of the tangent function is another example. The curve for this function blows up whenever x is an odd-integer multiple of $\pi/2$.

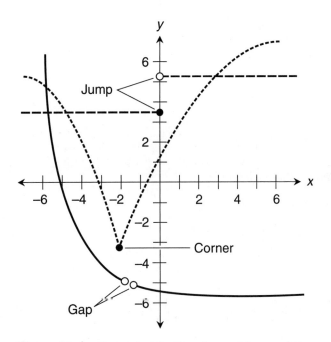

Figure 10-4   Illustration for Question and Answer 4-2.

Question 4-4

Look back again at Questions and Answers 3-2, 3-3, and 3-4. When we want to define the slope of a curve at a point (x_0, y_0), is it okay to use two movable points, one on either side of (x_0, y_0), draw a line through them, and make those two points converge on (x_0, y_0)?

Answer 4-4

No. We must approach (x_0, y_0) from one side at a time, and use that fixed point (x_0, y_0) as one of the points through which we draw the line. Then we must be sure that we get the same limit when we approach from the left as we get when we approach from the right.

Question 4-5

How can we precisely define what it means for a function to be differentiable, based on Answers 3-7 through 3-10?

Answer 4-5

We can call a function differentiable at a single point, over an interval, or over its entire domain.

- A function is differentiable at a point if and only if the function has a derivative at that point as defined in Answers 3-7 through 3-10.
- A function is differentiable over an interval if and only if it has a derivative, as defined in Answers 3-7 through 3-10, at every point in that interval.
- A function is differentiable in general (or simply differentiable) if and only if it has a derivative, as defined in Answers 3-7 through 3-10, at every point in its domain.

Question 4-6

If a function is discontinuous at a point, can it be differentiable there?

Answer 4-6

No. A function can't be differentiable at a point where a discontinuity exists. Remember that if a function is differentiable at a point, then it is continuous at that point. The *contrapositive* of this statement is logically equivalent to it: if a function is discontinuous at a point, then it is nondifferentiable at that point.

Question 4-7

If the slope of a graph changes suddenly at a point, then there's no derivative at that point, even if the function is continuous there. Why?

Answer 4-7

If we approach the point from one side, then $\Delta y / \Delta x$ (as defined in Question and Answer 3-2) approaches a certain value. But if we approach the same point from the other side, then $\Delta y / \Delta x$ approaches a different value. A derivative can't be defined as more than one limiting value for a single point!

Question 4-8

What about a situation where two different curves terminate and meet at a single point, and $\Delta y/\Delta x$ approaches the same value as we move toward the point from either direction? Is such a two-part function differentiable at the point where the curves come together?

Answer 4-8

Yes, as long as that point is on the curve. This is true even if the two curves portray completely different functions.

Question 4-9

Which of the following functions are differentiable in general, if the domain is allowed to extend over the entire set of reals?

$$f_1(x) = 3x^2 - 7$$
$$f_2(x) = -7x^2 + 5x - 7$$
$$f_3(x) = 4x^3 - 5x^2 + 8x - 14$$
$$f_4(x) = -6x^6 - 3x^4 + 2x^2 + 5x - 1$$
$$f_5(x) = -3 \sin x$$
$$f_6(x) = 4 \csc x$$

Answer 4-9

These are the same functions we saw in Question 2-9. The first four functions are all differentiable in general, because they're polynomial functions that can be written in the form

$$P_n(x) = a_n x^n + a_{n-1} x^{n-1} + a_{n-2} x^{n-2} + \cdots + a_1 x + b$$

where n is a positive integer, all the a's are real-number constants, and the variable x is never raised to a negative power. The graphs of such functions are always free of gaps, jumps, and corners. The fifth function also has a graph without any gaps, jumps, or corners, so it's differentiable in general. But the sixth function has discontinuities, so it is not differentiable in general.

Question 4-10

Is the following function differentiable over the entire set of reals? If not, how can we restrict the domain to make it differentiable in general?

$$g(v) = 8v^4 - 16v^3 + 3v^2 + 2v - 7 + 6v^{-1}$$

Answer 4-10

As stated, this function is nondifferentiable if the domain is the set of all real numbers. The term $6v^{-1}$ is not defined when $v = 0$, producing a discontinuity. If we restrict the domain to exclude the value $v = 0$, then the function becomes differentiable in general.

Chapter 5

Question 5-1

What do we get when we cube the binomial $a + b$? What do we get when we raise it to the fourth power? The fifth power? The sixth power?

Answer 5-1

Here's what happens. It's only a matter of working our way through the arithmetic:

$$(a+b)^3 = a^3 + 3a^2b + 3ab^2 + b^3$$
$$(a+b)^4 = a^4 + 4a^3b + 6a^2b^2 + 4ab^3 + b^4$$
$$(a+b)^5 = a^5 + 5a^4b + 10a^3b^2 + 10a^2b^3 + 5ab^4 + b^5$$
$$(a+b)^6 = a^6 + 6a^5b + 15a^4b^2 + 20a^3b^3 + 15a^2b^4 + 6ab^5 + b^6$$

Question 5-2

What three things can we say about the polynomial we get when we multiply out

$$(a+b)^n$$

where n is an integer larger than or equal to 3?

Answer 5-2

We can be sure that

- The first term is a^n.
- The second term is $na^{(n-1)}b$.
- We can factor b^2 out of the third term and each one after that.

Question 5-3

What is the power rule for derivatives, as it applies to positive integer exponents?

Answer 5-3

If a is a nonzero real-number constant, x is a variable, and n is a positive integer, then any function of the form

$$f(x) = ax^n$$

has the derivative

$$f'(x) = anx^{(n-1)}$$

Question 5-4

Using the power rule for derivatives, show that when we differentiate a function of the form

$$f(x) = ax$$

where a is a real-number constant and x is a nonzero variable, we always get

$$f'(x) = a$$

Explain why the power rule fails in this situation if $x = 0$.

Answer 5-4

We can rewrite the above function as

$$f(x) = ax^1$$

Using the power rule for derivatives, we get

$$f'(x) = anx^{(n-1)} = a \cdot 1 \cdot x^{(1-1)} = ax^0$$

Because any nonzero real number raised to the 0th power is equal to 1, we know that

$$f'(x) = a$$

as long as $x \neq 0$. If $x = 0$, then we get the undefined expression 0^0. (The formula stated in the question is valid even if the domain of the function includes 0, but we can't prove it from the power rule.)

Question 5-5

What is the sum rule for two derivatives?

Answer 5-5

The derivative of the sum of two functions is equal to the sum of their derivatives. Stated symbolically, if f_1 and f_2 are differentiable functions of the same variable, then

$$(f_1 + f_2)' = f_1' + f_2'$$

Question 5-6

Does the sum rule work for subtraction as well as for addition of derivatives?

Answer 5-6

Yes, the rule works for the difference of two derivatives, as long as we keep the functions and their derivatives in the same order, and as long as there are only two of them. If f_1 and f_2 are differentiable functions of the same variable, then

$$(f_1 - f_2)' = f_1' - f_2'$$

Question 5-7

What is mathematical induction? How does it work?

Answer 5-7

Mathematical induction is a trick that allows us to prove infinitely many facts. Imagine an infinite sequence of statements $S_1, S_2, S_3, \ldots$. We can prove all of them if we can demonstrate two things:

- The first statement S_1, is true, and
- If a statement S_n (where n can be any positive integer) is true, then the next statement S_{n+1} is true.

Question 5-8

Can the sum rule for two derivatives be extrapolated to work with the sum of any finite number of derivatives?

Answer 5-8

Yes. For any n differentiable functions $f_1, f_2, f_3, \ldots$, and f_n, the derivative of their sum is equal to the sum of their derivatives. Stated symbolically,

$$(f_1 + f_2 + f_3 + \ldots + f_n)' = f_1' + f_2' + f_3' + \cdots + f_n'$$

Question 5-9

Why can't the leading coefficient be 0 in an nth-degree polynomial function? Is it okay for any of the other coefficients to equal 0?

Answer 5-9

If we set the leading coefficient to 0 in an nth-degree polynomial function $f(x)$, we have $0x^n$ as the first term. That's always equal to 0, so the whole term can be removed from the polynomial. Then the "legitimate" first term contains x raised to an integer power smaller than n, and we don't really have an nth-degree function after all! It's okay for one or more of the other coefficients to be 0, however.

Question 5-10

What is the derivative of the general polynomial function

$$f(x) = a_n x^n + a_{n-1} x^{n-1} + a_{n-2} x^{n-2} + \cdots + a_2 x^2 + a_1 x + a_0$$

where $a_n, a_{n-1}, a_{n-2}, \ldots, a_1$, and a_0 are real numbers, and $a_n \neq 0$?

Answer 5-10

Using the power rule and the sum rule for derivatives, we can differentiate each term individually and then add them up to get

$$f'(x) = n a_n x^{n-1} + (n-1) a_{n-1} x^{n-2} + (n-2) a_{n-2} x^{n-3} + \cdots + 2 a_2 x + a_1$$

Chapter 6

Question 6-1

What is the multiplication-by-constant rule for differentiation?

Answer 6-1

If we take the derivative of a differentiable function after it has been multiplied by a constant, we get the same result as we do if we take the derivative of the function and then multiply by the constant. Stated symbolically, if f is a differentiable function and c is a real constant, then

$$(cf)' = c(f')$$

and

$$(cf)' = (f')c$$

Question 6-2

Does the multiplication-by-constant rule work for division by a constant?

Answer 6-2

Yes, to a limited extent. If we take the derivative of a differentiable function after it has been divided by a *nonzero* constant, we get the same result as we do if we take the derivative of the function and then divide by the constant. Stated symbolically, if f is a differentiable function and c is a nonzero real constant, then

$$(f/c)' = (f')/c$$

Question 6-3

What is the two-function product rule for differentiation?

Answer 6-3

To find the derivative of the product of two differentiable functions, we multiply the derivative of the first function by the second function, then multiply the derivative of the second function by the first function, and finally add the two products. Stated symbolically, if f and g are differentiable functions, then

$$(fg)' = f'g + g'f$$

Question 6-4

How must we be careful when expressing the product of two functions?

Answer 6-4

If we want to denote the product "f of x times g of x," we should write

$$[f(x)][g(x)]$$

or

$$f(x) \times g(x)$$

If we don't want to specify the independent variable, we should write

$$fg$$

or

$$f \times g$$

We can also use the elevated dot (·) instead of the cross to represent multiplication. The important consideration is that we don't want to mistakenly write an expression that denotes a function of another function!

Question 6-5

What is the reciprocal rule for differentiation?

Answer 6-5

To find the derivative of the reciprocal of a differentiable function, we must first find the derivative of the original function, then multiply by -1, and finally divide by the square of the original function. Stated symbolically, if f is a differentiable function, then

$$(1/f)' = -f'/(f^2)$$

Question 6-6

How must we be careful when applying the reciprocal rule for differentiation?

Answer 6-6

For this rule to apply, the original function must never become 0 anywhere in its domain. If that happens, the reciprocal function will blow up at that point, and not be differentiable there. If a function is differentiable throughout its domain, then the reciprocal function is not necessarily differentiable over the same domain. We may have to restrict the domain to keep discontinuities out.

Question 6-7

How must we be careful when expressing the reciprocal of a function?

Answer 6-7

We can express the reciprocal of $f(x)$ by writing $[f(x)]^{-1}$, but we must never write the exponent -1 immediately after the name of the function. If we write f^{-1}, we denote the inverse of f. That's a totally different notion from the reciprocal!

Question 6-8

What is the quotient rule for differentiation?

Answer 6-8

To find the derivative of the quotient (or ratio) of two differentiable functions, we multiply the derivative of the first function by the second function, then multiply the derivative of the second function by the first function, then subtract the second product from the first product, and finally divide by the square of the second function. Stated symbolically, if f and g are differentiable functions, then

$$(f/g)' = (f'g - g'f)/g^2$$

as long as g does not become 0.

Question 6-9

How must we be careful when applying the quotient rule?

Answer 6-9

If we want to use the quotient rule for differentiation, then both functions, as well as their ratio, must be differentiable at every point that interests us. In addition, the denominator function must never become 0 anywhere in its domain.

Question 6-10

What is the chain rule for differentiating a function of a function?

Answer 6-10

To differentiate a function of a function, we multiply the derivative of the "outer" function by the derivative of the "inner" function. Stated symbolically, if f and g are differentiable functions of the same single variable, then

$$[g(f)]' = g'(f) \times f'$$

Chapter 7

Question 7-1

What is the real-power rule for differentiation?

Answer 7-1

It's an extension of the power rule for integer exponents. If x is a real variable, a is a real constant, and k is a real exponent, then we can calculate the derivative of

$$f(x) = ax^k$$

by subtracting 1 from the exponent, and then multiplying a by k to get

$$f'(x) = akx^{(k-1)}$$

If $k \leq 1$, then we must be sure that $x \neq 0$. That may mean restricting the domain of the function. Otherwise, when we take the derivative, we'll end up with an undefined expression when we take x to the power of $k - 1$.

Question 7-2

How can we evaluate a variable raised to a rational-number power?

Answer 7-2

Let's remember the definition of a rational number. Any rational number q can be expressed as

$$q = a/b$$

where a is an integer and b is a positive integer. If we talk about raising a quantity to the qth power, we're talking about raising it to the power a/b. That means we take the quantity to the ath power, and then find the positive bth root of the result. If we have a function

$$f(x) = x^{a/b}$$

then we can rewrite it as

$$f(x) = (x^a)^{1/b}$$

There's a restriction here! We can't have $x = 0$ and $a = 0$ at the same time.

Question 7-3

How can we evaluate a variable raised to a real-number power (that is, rational or irrational) using natural exponential and log functions?

Answer 7-3

When we have an expression of the form x^k where x is a variable and k is any real number, then

$$\ln (x^k) = k \ln x$$

We can take the natural exponential of both sides to get

$$x^k = e^{(k \ln x)}$$

This only works if $x > 0$, because the natural log of a nonpositive quantity is not defined.

Question 7-4

Look at Fig. 10-5. Based on this information, what's the derivative of $f(x)$?

Answer 7-4

In this graph, f represents the sine function and g represents the cosine function. That is,

$$f(x) = \sin x$$

and

$$g(x) = \cos x$$

The derivative of the sine is the cosine, so

$$f'(x) = d/dx\,(\sin x) = \cos x = g(x)$$

Question 7-5

Based on the information in Fig. 10-5, what's the derivative of $g(x)$?

Answer 7-5

The derivative of the cosine is the negative of the sine, so

$$g'(x) = d/dx\,(\cos x) = -\sin x = -f(x)$$

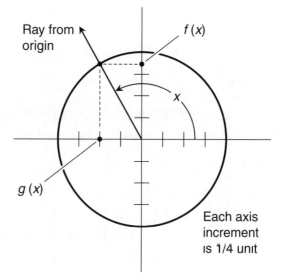

Figure 10-5   Illustration for Questions and Answers 7-4 and 7-5.

Question 7-6

How can we differentiate the following function using the product rule?

$$q(x) = \cos^2 x$$

Answer 7-6

To begin, let's rewrite the function as

$$q(x) = \cos x \cos x$$

We can name the individual functions in the product as

$$f(x) = \cos x$$

and

$$g(x) = \cos x$$

They're identical, so their derivatives are identical as well. We have

$$f'(x) = -\sin x$$

and

$$g'(x) = -\sin x$$

Applying the product rule for differentiation gives us

$$q'(x) = (fg)'(x) = f'(x)\,g(x) + g'(x)\,f(x)$$
$$= (-\sin x)(\cos x) + (-\sin x)(\cos x) = -2\sin x \cos x$$

Remember that $\sin^2 x$ and $(\sin x)^2$ mean exactly the same thing.

Question 7-7

How can we differentiate the function stated in Question 7-6 using the chain rule?

Answer 7-7

Let's break down our function $q(x)$ into two component functions. We can name the component functions

$$f(x) = \cos x$$

and

$$g(y) = y^2$$

That means our original function is

$$q(x) = g[f(x)]$$

The derivatives of the component functions are

$$f'(x) = -\sin x$$

and

$$g'(y) = 2y$$

Now we can use the chain rule to write the derivative of the composite function as

$$q'(x) = \{g[f(x)]\}' = g'[f(x)] \cdot f'(x)$$
$$= 2(\cos x)(-\sin x) = -2 \sin x \cos x$$

Question 7-8

Under what circumstances can a function be its own derivative?

Answer 7-8

If f is a function of the form

$$f(x) = ke^x$$

where x is a real variable and k is a real constant, then f is its own derivative. That is,

$$f'(x) = d/dx\,(ke^x) = ke^x$$

Question 7-9

What does the symbol e mean in the above equation?

Answer 7-9

The symbol e stands for Euler's constant, also called the exponential constant. It is an irrational number with an approximate value of 2.718.

Question 7-10

What's the derivative of the natural log function? How must the domain be restricted?

Answer 7-10

The derivative of the natural log function is the reciprocal function, but we must restrict the domain to the set of positive real numbers. We have

$$d/dx\,(\ln x) = 1/x$$

for $x > 0$. If $x \le 0$, then $\ln x$ is undefined, so its derivative is also undefined.

Chapter 8

Question 8-1

What is the second derivative of a function?

Answer 8-1

The second derivative of a function is the derivative of its derivative, taken with respect to the same variable. Suppose we have a function f of an independent variable x, producing a dependent variable y so that

$$y = f(x)$$

The second derivative of f with respect to x can be denoted in various ways. Here are the expressions most often seen:

$$d^2y/dx^2$$
$$y''$$
$$d^2f(x)/dx^2$$
$$d^2/dx^2 \, f(x)$$
$$d^2f/dx^2$$
$$f''(x)$$
$$f''$$

Question 8-2

What does the second derivative of a function represent in a graph?

Answer 8-2

The second derivative tells us the rate and the sense (increasing or decreasing) at which the slope of the graph changes.

Question 8-3

When an object falls freely after it has been dropped, what is the second derivative of the total distance it fell with respect to time? (Assume there is no air resistance.)

Answer 8-3

The second derivative of the total distance the object fell, with respect to time, is the acceleration.

Question 8-4

How does the acceleration of a freely falling object relate to its speed and the total distance it fell? (Again, assume there is no air resistance.)

Answer 8-4

The acceleration is the rate at which the speed increases. That's the derivative of the speed with respect to time. The speed of a freely falling object is the rate at which the total distance it fell increases. That's the derivative of the distance with respect to time.

Question 8-5

What's the third derivative of distance with respect to time for an object that moves in a complicated way?

Answer 8-5

The third derivative of distance relative to time is called the jerk. It's the derivative of the acceleration, and the second derivative of the speed. Jerk represents the rate (and the direction) of a change in the acceleration.

Question 8-6

What's the jerk of an object in free fall, neglecting air resistance?

Answer 8-6

For a freely falling object, neglecting air resistance, the jerk is always zero because the acceleration is constant, and the derivative of a constant function is the zero function.

Question 8-7

What's the *n*th derivative of a function?

Answer 8-7

The *nth derivative* of a function is what we get when we differentiate the function *n* times with respect to the same variable, where *n* is a positive integer. Consider

$$y = f(x)$$

The *n*th derivative of *f* with respect to *x* can be written in several ways. Here are the expressions we're most likely to see:

$$d^n y / dx^n$$
$$d^n f(x) / dx^n$$
$$d^n / dx^n \, f(x)$$
$$d^n f / dx^n$$
$$f^{(n)}(x)$$
$$f^{(n)}$$

Question 8-8

Draw graphs comparing the sine function and its first four derivatives.

Answer 8-8

Figure 10-6 is a four-part graph showing this relationship. Remember the four-way cycle that the sine function goes through as it is repeatedly differentiated:

$$d/dx \, (\sin x) = \cos x$$
$$d^2/dx^2 \, (\sin x) = d/dx \, (\cos x) = -\sin x$$
$$d^3/dx^3 \, (\sin x) = d/dx \, (-\sin x) = -\cos x$$
$$d^4/dx^4 \, (\sin x) = d/dx \, (-\cos x) = \sin x$$

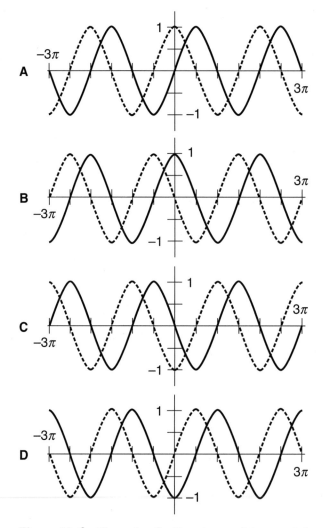

Figure 10-6   Illustration for Questions and Answers 8-8 and 9-10.

In each of the four parts of Fig. 10-6, we see the value of *x* on the horizontal axis, and the value of the function on the vertical axis.

- At A, the solid curve represents the sine function, and the dashed curve represents its derivative.
- At B, the solid curve represents the first derivative of the sine function, and the dashed curve represents its second derivative.
- At C, the solid curve represents the second derivative of the sine function, and the dashed curve represents its third derivative.
- At D, the solid curve represents the third derivative of the sine function, and the dashed curve represents its fourth derivative.

With each derivative, the whole curve moves $\pi/2$ units to the left. Therefore, when we take the fourth derivative, we end up 2π units to the left of where we started. If we let the domain extend over the entire set of real numbers, taking the fourth derivative of the sine function gives us the sine function all over again.

Question 8-9

What do we get if we differentiate the natural log function repeatedly?

Answer 8-9

In this situation, we don't get a cycle, but we get a pattern. We always get a monomial function of *x*. The power to which the argument (independent variable) is raised starts out at −1, and then decreases by 1 with each succeeding derivative. The coefficient starts out at 1, then goes to −1, and then keeps increasing in absolute value but alternating in sign. Here are the first nine derivatives:

$$d/dx \, (\ln x) = x^{-1}$$
$$d^2/dx^2 \, (\ln x) = d/dx \, (x^{-1}) = -x^{-2}$$
$$d^3/dx^3 \, (\ln x) = d/dx \, (-x^{-2}) = 2x^{-3}$$
$$d^4/dx^4 \, (\ln x) = d/dx \, (2x^{-3}) = -6x^{-4}$$
$$d^5/dx^5 \, (\ln x) = d/dx \, (-6x^{-4}) = 24x^{-5}$$
$$d^6/dx^6 \, (\ln x) = d/dx \, (24x^{-5}) = -120x^{-6}$$
$$d^7/dx^7 \, (\ln x) = d/dx \, (-120x^{-6}) = 720x^{-7}$$
$$d^8/dx^8 \, (\ln x) = d/dx \, (720x^{-7}) = -5{,}040x^{-8}$$
$$d^9/dx^9 \, (\ln x) = d/dx \, (-5{,}040x^{-8}) = 40{,}320x^{-9}$$

In every case, we must restrict the domain to *x* > 0, because the natural log function is defined only when the argument (independent variable) is positive.

Question 8-10

Imagine that we encounter a polynomial function of the form

$$f(x) = a_n x^n + a_{n-1}x^{n-1} + a_{n-2}x^{n-2} + \cdots + a_2 x^2 + a_1 x + a_0$$

where n is a positive integer, all the a's are real-number constants, and the variable x is never raised to a negative power. If we keep differentiating this function over and over, what will eventually occur?

Answer 8-10

Sooner or later, we'll get the zero function as we keep taking derivatives. This will happen no matter how high the degree of the original function might be. Each time we differentiate, the degree will decrease until, at some stage, we'll get a constant multiple of x. When we differentiate that, we'll get a constant function. When we differentiate a constant function, we always get the zero function. After that, all further derivatives will be the zero function.

Chapter 9

Question 9-1

What is concavity in the graph of a function? How does concavity relate to the first and second derivatives?

Answer 9-1

The concavity of a curve is the general direction in which it "bends," either overall or within a region. A curve can be concave upward or concave downward at a given point. When a curve is concave upward, the first derivative can be positive, negative, or 0, but the second derivative is always positive. When a curve is concave downward, the first derivative can be positive, negative, or 0, but the second derivative is always negative.

Question 9-2

What is an extremum in the graph of a function? How do extrema relate to the first and second derivatives of a function?

Answer 9-2

An extremum is a point on the curve at which the function attains a maximum or minimum value.

If an extremum represents the largest value in the entire function, it's called the absolute maximum. If it's the largest value in its immediate region, but not necessarily in the entire function, then it's called a local maximum. If an extremum represents the smallest value in the function, it's called the absolute minimum. If it's the smallest value in its immediate region, but not necessarily in the entire function, then it's called a local minimum.

At any point where a graph reaches an extremum, the function's first derivative is 0. If the extremum is a maximum, then the second derivative is negative at that point. If the extremum is a minimum, then the second derivative is positive at that point.

Question 9-3

What is an inflection point in the graph of a function? How do inflection points relate to the first and second derivatives?

Answer 9-3

An inflection point is a point where a curve goes from concave upward to concave downward or vice versa. The first derivative can be negative, positive, or 0 at an inflection point, but the second derivative is always 0.

Question 9-4

What is always true about the first derivative of a quadratic function? What does the graph of the derivative function look like?

Answer 9-4

The first derivative of a quadratic function is always a linear function. The general form of a quadratic function f in the variable x is

$$f(x) = ax^2 + bx + c$$

where a, b, and c are real numbers, and $a \neq 0$. The first derivative of this is

$$f'(x) = 2ax + b$$

That's a linear function of x. Its graph is a straight line. We might remember from our algebra courses that if we let $y = f(x)$ in this situation, then the slope of the line is $2a$, and the y-intercept is b.

Question 9-5

What's the x-value of the extremum in the general quadratic function

$$f(x) = ax^2 + bx + c$$

where a, b, and c are real numbers, and $a \neq 0$?

Answer 9-5

We can find the extremum in a quadratic function by setting the first derivative equal to 0, and then solving the linear equation

$$2ax + b = 0$$

Subtracting b from both sides, we get

$$2ax = -b$$

Dividing through by $2a$, which we know is okay because $a \neq 0$, we obtain

$$x = -b/(2a)$$

Question 9-6

What is always true about the second derivative of a quadratic function? The third derivative?

Answer 9-6

The second derivative of a quadratic function is always a constant function. When we take the derivative of f' as it appears in Answer 9-4, thereby getting the second derivative of f, we have

$$f''(x) = 2a$$

The third derivative of a quadratic function is always the zero function. When we take the derivative of f'' as shown above, we get

$$f'''(x) = 0$$

Question 9-7

How do we know whether the extremum we find in a quadratic function, according to the methods described above, is an absolute minimum or an absolute maximum?

Answer 9-7

If the second derivative is negative, we have an absolute maximum, because the curve is concave downward. If the second derivative is positive, we have an absolute minimum, because the curve is concave upward. That means we have a maximum if $a < 0$, and a minimum if $a > 0$ in the general quadratic function

$$f(x) = ax^2 + bx + c$$

where a, b, and c are real numbers, and $a \neq 0$.

Question 9-8

How can we find out whether or not the natural log function has any extrema or inflection points?

Answer 9-8

To see if there is an extremum, we can take the first derivative and see if it becomes equal to 0 anywhere in the domain. We remember from Answer 8-9 that

$$d/dx \, (\ln x) = x^{-1} = 1/x$$

There is no real number that has a reciprocal of zero. Therefore, the natural log function has no minima or maxima.

To see if there are any inflection points, we take the second derivative and see if it becomes 0 anywhere in the domain. Again consulting Answer 8-9, we have

$$d^2/dx^2 \, (\ln x) = -x^{-2} = -1/(x^2)$$

If we try to solve the equation

$$-1/(x^2) = 0$$

we find that no real number x satisfies it. That means the natural log function has no inflection points. We can see intuitively that this function has no extrema or inflection points by casually looking at its graph. Now we have mathematical proof.

Question 9-9

How can we find out whether or not the natural exponential function has any extrema or inflection points?

Answer 9-9

To see if there is an extremum, we can take the first derivative and see if it becomes 0 anywhere. The natural exponential function is its own derivative. That is,

$$d/dx \, (e^x) = e^x$$

There's no real number x such that $e^x = 0$. That means the natural exponential function has no minima or maxima.

Don't get confused here. The fact that the curve approaches the x axis as we move toward the left (that is, as x increases negatively) does not mean that the extremum is 0. The value of the function keeps getting smaller and smaller positively, approaching 0, but it never actually attains a value of 0. In fact, there is no particular real number we can choose and say, "This is the smallest value that the function attains."

To see if there are any inflection points in the graph of the natural exponential function, we take the second derivative and see if it becomes 0 anywhere in the domain. We have

$$d^2/dx^2 \, (e^x) = e^x$$

The same thing happens here as with the first derivative. There is no real number to which we can raise e and end up with 0. That means the natural exponential function has no inflection points.

Question 9-10

Look again at Fig. 10-6, which shows the sine function and its first four derivatives. Also, re-read Answer 8-8. Based on all this information, what happens to the local maxima, the local minima, and the inflection points of any of these functions when we take its derivative?

Answer 9-10

Every time we take a derivative, all the x-values of the extrema and inflection points move $\pi/2$ units to the left (that is, in the negative direction). The y-values (if we always let y equal the value of the function) do not change. The local maxima are always at $y = 1$, the local minima are always at $y = -1$, and the inflection points are always at $y = 0$.

Integration in One Variable

11

What's an Integral?

When we differentiate a function, we see its rate of change. When we *integrate* a function, we see its accumulated value. In this chapter, we'll learn how the concept of the integral translates into geometric area. Then we'll look at three ways that *integral calculus* is used.

Summation Notation

Before we start "integral-building," let's review the mathematical shorthand for writing down long sums. It's called *summation notation*.

Specify the series

Imagine a set of constants, all denoted by *a* with a subscript. Here's an example:

$$\{a_1, a_2, a_3, a_4, a_5, a_6, a_7, a_8\}$$

Suppose we add these up, getting a final sum of *b*. We can write this as

$$a_1 + a_2 + a_3 + a_4 + a_5 + a_6 + a_7 + a_8 = b$$

If we had 800 terms, getting a final sum equal to *c*, writing down the entire series would be impractical. But we could put an ellipsis (three dots) in the middle to get

$$a_1 + a_2 + a_3 + \cdots + a_{798} + a_{799} + a_{800} = c$$

Tag the terms

Let's invent a variable and call it *i* (not to be confused with the unit imaginary number). Written as a subscript, *i* can serve as a counting tag or marker in a series containing a large number of terms.

In the above situations, we can call each term by the generic name a_i. In the first series, we add up eight a_i's to get the final sum b, and the counting tag i goes from 1 to 8. In the second series, we add up 800 a_i's to get the final sum c, and the counting tag i goes from 1 to 800. Suppose that we have a series with n terms, like this:

$$a_1 + a_2 + a_3 + \cdots + a_{n-2} + a_{n-1} + a_n = d$$

In this case, we add up n a_i's to get the final sum d.

The big sigma

Let's go back to the series in which we add eight terms. It can be written in shorthand like this:

$$\sum_{i=1}^{8} a_i = b$$

We read this as, "The summation of the terms a_i, from $i = 1$ to 8, is equal to b." The large, bold symbol Σ is an uppercase Greek letter *sigma,* which stands for "summation" or "sum." Now let's look at the series in which 800 terms are added:

$$\sum_{i=1}^{800} a_i = c$$

We can read this as, "The summation of the terms a_i, from $i = 1$ to 800, is equal to c." Finally, let's express the open-ended or unbounded example:

$$\sum_{i=1}^{n} a_i = d$$

We can read this as, "The summation of the terms a_i, from $i = 1$ to n, is equal to d."

A more sophisticated example

Sometimes we'll want to use something other than a place marker as the counting variable in a summation. The variable doesn't have to start at 1. It can have an initial value of 0. It can even start out as a negative integer. And it doesn't have to end at n. It can go on forever!

Suppose we sum a series starting with 1, then adding 1/2, then adding 1/4, then adding 1/8, and going on forever, each time cutting the value in half. From precalculus, we recall that this series, even though it has infinitely many terms, adds up to 2:

$$1 + 1/2 + 1/4 + 1/8 + \cdots = 2$$

We can also write this series as

$$1/2^0 + 1/2^1 + 1/2^2 + 1/2^3 + \cdots = 2$$

In summation notation, it is

$$\sum_{i=0}^{\infty} 1/2^i = 2$$

- -

Here's a challenge!

Consider the same series as above, but only up to the reciprocal of the *n*th power of 2. Give the entire sum the name *S*. Write this fact in summation notation.

Solution

Let's use the letter *i* as the counting tag. We start at $i = 0$ and go up to $i = n$, with each term having the value $1/2^i$. Therefore, the summation notation is

$$\sum_{i=0}^{n} 1/2^i = S$$

Here's another challenge!

Think of the summation in the previous challenge, and imagine what happens as *n* increases endlessly—that is, as *n* approaches infinity. The series gets longer; as *n* grows extremely large and the sum approaches 2. Write this fact using the limit notation along with the summation notation.

Solution

We can plug the summation into a limit "template," and then state that the whole thing is equal to 2, like this:

$$\lim_{n \to \infty} \sum_{i=0}^{n} 1/2^i = 2$$

- -

Area Defined by a Curve

Now let's see how we can determine the area between the graph of a function and the independent-variable axis, as defined between two known values of the independent variable.

Defining the region

Imagine a function $f(x)$ whose graph is a curve as shown in Fig. 11-1. We want to find the area bounded by:

- The vertical line $x = a$ on the left
- The vertical line $x = b$ on the right
- The curve on the bottom when $f(x)$ is negative
- The *x* axis on the top when $f(x)$ is negative
- The *x* axis on the bottom when $f(x)$ is positive
- The curve on the top when $f(x)$ is positive

Any part of the region that appears below the *x* axis and above the curve is considered to have *negative area,* assuming we move along the *x* axis in the positive direction. Any part of the region that appears above the *x* axis and below the curve is considered to have *positive area,*

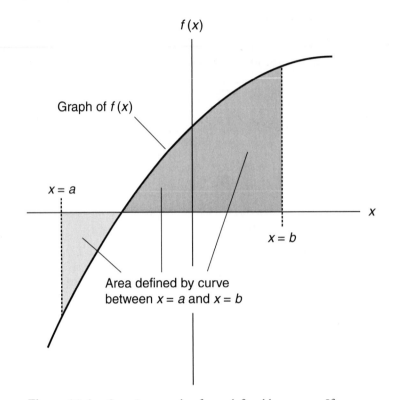

Figure 11-1   Generic example of area defined by a curve. If we
move to the right, the area below the *x* axis (light-
shaded region) is negative. The area above the *x* axis
(dark-shaded region) is positive.

again assuming we move along the *x* axis in the positive direction. (As we'll learn later, the
notions of positive area and negative area reverse if we travel along the *x* axis in the negative direction.) The total area between the two vertical lines $x = a$ and $x = b$, as defined by the
curve and the *x* axis, is the sum of the areas of both shaded regions.

   The area of a region like the one shown here is often called *area under the curve* even
though, as we can see, some of the region might lie above the curve. To keep from getting
confused or misled, let's talk about the *area defined by the curve* in situations of this kind. In
Fig. 11-1, the *net area* over the interval from $x = a$ to $x = b$ is less than the area of the dark-
shaded part, because there's some negative area involved. To get the net area, we subtract the
negative area from the positive area.

Approximating the area

Let's divide the entire region, both the light-shaded part and the dark-shaded part, into *n* rectangles, all of equal width, as shown in Fig. 11-2. For each rectangle, let's call the width Δx. Then

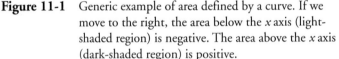

$$\Delta x = (b - a)/n$$

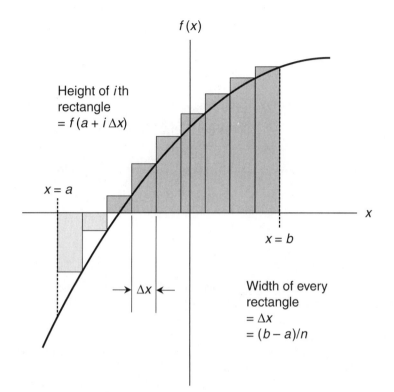

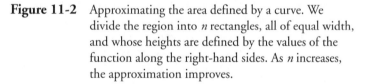

Figure 11-2   Approximating the area defined by a curve. We
divide the region into *n* rectangles, all of equal width,
and whose heights are defined by the values of the
function along the right-hand sides. As *n* increases,
the approximation improves.

As we've arranged things in Fig. 11-2, the height of any particular rectangle is equal to the
value of $f(x)$ along the rectangle's right-hand edge. Let's call the left-most rectangle number 1,
and count up as we move to the right. Suppose we make *n* large. Then:

- The height of rectangle number 1 is $f(a + \Delta x)$
- The height of rectangle number 2 is $f(a + 2\Delta x)$
- The height of rectangle number 3 is $f(a + 3\Delta x)$
- And so on . . .
- The height of rectangle number *i* is $f(a + i\Delta x)$
- And so on . . .
- The height of rectangle number *n* is $f(a + n\Delta x)$

The area of each rectangle is the width times the height. Therefore:

- The area of rectangle number 1 is $\Delta x \cdot f(a + \Delta x)$
- The area of rectangle number 2 is $\Delta x \cdot f(a + 2\Delta x)$
- The area of rectangle number 3 is $\Delta x \cdot f(a + 3\Delta x)$

- And so on . . .
- The area of rectangle number i is $\Delta x \cdot f(a + i\Delta x)$
- And so on . . .
- The area of rectangle number n is $\Delta x \cdot f(a + n\Delta x)$

We approximate the total area by adding up the areas of the rectangles:

$$\sum_{i=1}^{n} \Delta x \cdot f(a + i\Delta x)$$

Because $\Delta x = (b - a)/n$, we can substitute the quantity $(b - a)/n$ for Δx to get

$$\sum_{i=1}^{n} [(b - a)/n] \cdot f[a + i(b - a)/n]$$

The shrinking increment

In the situation shown by Fig. 11-2, we have only nine rectangles, so $n = 9$. If we were faced with an actual approximation problem, we could calculate the sum of the areas of the rectangles by working out the arithmetic. If we wanted a better approximation, an extremely large number could be chosen for n, and a computer could be used to calculate the area of every single rectangle and then add them all up. The accuracy would be limited only by how large we could make n before our computer crashed.

As n increases, the width Δx of each rectangle grows smaller, and the flat tops (or bottoms) of the rectangles get closer to the contour of the curve. The *approximation error* between the sum of the rectangle areas and the true area defined by the curve therefore decreases. If we make n extremely large, that error is tiny indeed.

The Riemann magic

In the 1800s, the mathematician Bernhard Riemann decided to carry the area-approximation process to infinite accuracy (make the error vanish). He reasoned that if he could find the limit of the sum of the rectangle areas as n approached infinity, the actual area defined by a curve could be determined. His method resembled the schemes used by Newton and Leibniz for finding the slope of a curve at a point. Riemann saw that the actual area defined by the curve, as portrayed with a scheme such as that shown in Fig. 11-2, must be equal to

$$\lim_{n \to \infty} \sum_{i=1}^{n} [(b - a)/n] \cdot f[a + i(b - a)/n]$$

We can express it in a simpler way by writing

$$\lim_{\Delta x \to 0} \sum_{i=1}^{n} \Delta x \cdot f(a + i\Delta x)$$

Sometimes the order of the product after the summation symbol is reversed, so it's written as

$$\lim_{\Delta x \to 0} \sum_{i=1}^{n} [f(a + i\Delta x)] \cdot \Delta x$$

- -

Are you confused?

"Wait!" you might say. "The method described above can work only if a limit, as we've defined it here, actually exists, isn't that so? How do we *know* that the limit exists?" These are good questions. In some cases, a limit doesn't exist, and we can't define the area in this way. But in the generic example shown by Figs. 11-1 and 11-2, our intuition ought to tell us that there is such a limit. Clearly, the area has to be definable *somehow,* because we can draw it, and it's finite. The function doesn't have a singularity, break apart, or zig-zag in some way that would make the limit impossible to define.

- -

The integral notation

Rather than scribbling out the limit notation above, the area defined by a curve is usually denoted using an *integration symbol* along with the lower and upper bounds of the interval. The integral symbol looks like a vertically stretched-out letter S. Instead of

$$\underset{\Delta x \to 0}{Lim} \sum_{i=1}^{n} [f(a + i\Delta x)] \cdot \Delta x$$

we can write

$$\int_{a}^{b} f(x)\ dx$$

This is read aloud as, "The integral from a to b of $f(x)\ dx$." The dx part of the expression comes from the shrinking of Δx to an arbitrarily small value—the differential of x—just as we saw when we worked out the definition of a derivative. It's important to write down this little dx after any *integrand* (function to be integrated).

When we express an integral with a lower bound and an upper bound so it represents a specific area defined by a curve within an interval, we call it a *definite integral.* In the foregoing example, the lower bound is $x = a$ and the upper bound is $x = b$.

- -

Here's a challenge!

Consider the function for which the output is the same as the input:

$$f(x) = x$$

Using integration, find the area defined by the curve between the bounds $a = 4$ and $a = 8$. Then calculate the area using plane geometry, recognizing that the region is a trapezoid.

Solution

Figure 11-3 shows how we can set up the integral using the Riemann method, constructing rectangles. We want to find

$$\int_{4}^{8} x\ dx$$

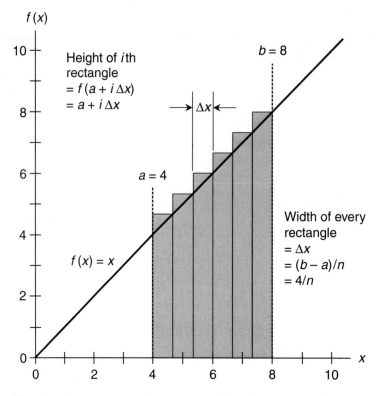

Figure 11-3   We can use integration to find the area defined by the curve for $f(x) = x$ between specific bounds.

This integral can be expressed using the long version of the limit

$$\underset{n \to \infty}{Lim} \sum_{i=1}^{n} [(b-a)/n] \cdot f[a + i(b-a)/n]$$

In this situation, $a = 4$ and $b = 8$. The output of the function is the same as the input. Knowing these things, we can rewrite the above formula as

$$\underset{n \to \infty}{Lim} \sum_{i=1}^{n} [(8-4)/n] \cdot [4 + i(8-4)/n]$$

which simplifies to

$$\underset{n \to \infty}{Lim} \sum_{i=1}^{n} (4/n) \cdot (4 + 4i/n)$$

Using algebra, we can rewrite this as

$$\underset{n \to \infty}{Lim} \sum_{i=1}^{n} 16(n + i)/n^2$$

We can factor 16 out the sum, getting

$$\underset{n \to \infty}{Lim} \; 16 \cdot \sum_{i=1}^{n} (n+i)/n^2$$

Using the multiplication-by-constant rule for limits, we can rewrite this as

$$16 \cdot \underset{n \to \infty}{Lim} \sum_{i=1}^{n} (n+i)/n^2$$

Now let's look at the summation part all by itself:

$$\sum_{i=1}^{n} (n+i)/n^2$$

We can write this series out as

$$(n+1)/n^2 + (n+2)/n^2 + (n+3)/n^2 + \cdots + (n+n)/n^2$$

which simplifies to

$$1 + (1 + 2 + 3 + \cdots + n) \, / \, n^2$$

From precalculus, we remember that

$$\underset{n \to \infty}{Lim} \; (1 + 2 + 3 + \cdots + n) \, / \, n^2 = 1/2$$

We must add 1 to this, which gives us

$$1 + (1 + 2 + 3 + \cdots + n) \, / \, n^2 = 3/2$$

Now we know that

$$\underset{n \to \infty}{Lim} \sum_{i=1}^{n} (n+i)/n^2 = 3/2$$

Therefore

$$16 \cdot \underset{n \to \infty}{Lim} \sum_{i=1}^{n} (n+i)/n^2 = 16 \cdot 3/2 = 24$$

That's the integral we seek. We've just determined that

$$\int_{4}^{8} x \, dx = 24$$

Here's a twist!

We can verify this result by looking at Fig. 11-4, which shows the actual region under the curve. It's a trapezoid. The two parallel sides have lengths of 4 units and 8 units. Averaging them gives us 6 units. We multiply that by 4 units, the lateral distance between the parallel sides, to get 24 square units for the area of the trapezoid.

The foregoing problem was easy compared to what can happen when the Riemann method is used with more complicated functions. But there's a shortcut that allows us to figure out integrals more easily. You'll learn how to apply this technique in the next few chapters.

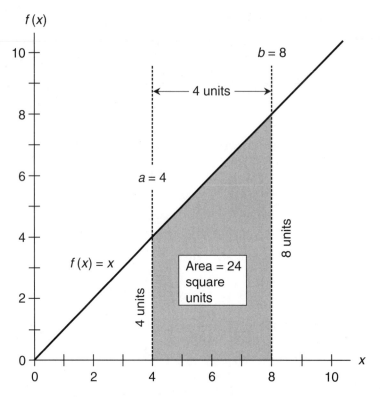

Figure 11-4   In this case, the area defined by the curve can be found using plane geometry, because it has a trapezoidal shape. This result agrees with what we got by working out the integral.

Three Applications

Let's look at how definite integrals are used in some "real-world" situations. First, we'll analyze forward motion in a straight line. Then we'll find the average value of a function over an interval. Finally, we'll examine the graph of a function that describes probability.

Displacement vs. speed

Imagine a car that starts out from a standstill and accelerates on a straight, level road. Suppose that when the driver presses her foot down on the accelerator, the car goes along with the following speed-vs.-time function s, where $s(t)$ is the speed in meters per second and t is the time in seconds:

$$s(t) = 2.4t$$

Figure 11-5 shows a graph of this function. Let's find out the total distance that the car travels between times $t = 2$ and $t = 5$. That's the distance between the car's location 2 seconds after it starts and its location 5 seconds after it starts.

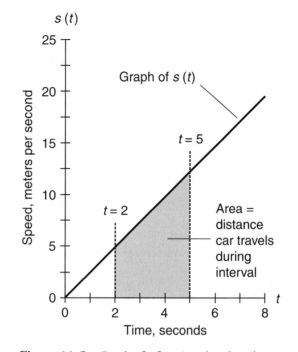

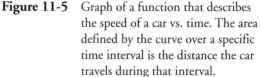

Figure 11-5 Graph of a function that describes the speed of a car vs. time. The area defined by the curve over a specific time interval is the distance the car travels during that interval.

In classical physics, the total distance traveled by an object moving in a straight line can be defined using a definite integral of the function of speed vs. time. In the situation shown by Fig. 11-5, that integral is

$$\int_{2}^{5} s(t)\, dt$$

We could go through the Riemann process to work out this integral, just as we did in the "challenge" a little while ago. But we don't have to. The graph of the function $s(t)$ is a straight line, so we can use geometry to find the area of the trapezoidal shaded region. At time $t = 2$, the speed of the car in meters per second is

$$s(2) = 2.4 \cdot 2 = 4.8$$

At time $t = 5$, the speed in meters per second is

$$s(5) = 2.4 \cdot 5 = 12$$

so the two parallel sides of the trapezoid have lengths of 4.8 units and 12 units. Averaging them gives us 8.4 units. We multiply that by 3 units, the lateral distance between the parallel sides, to get 25.2 square units for the area. That's the distance in meters that the car travels between times $t = 2$ and $t = 5$. As a definite integral, we can write this as

$$\int_{2}^{5} 2.4t \, dt = 25.2$$

Average value

Imagine a function $f(x)$ that is continuous over its domain from $x = a$ to $x = b$, as shown in Fig. 11-6. We can use integration to find the average value of the function over the interval (a,b) between $x = a$ and $x = b$. In this situation, $a < b$.

In geometric terms, we find the height (either positive or negative) of a rectangle whose area is the same as the area defined by the curve between lines $x = a$ and $x = b$. If we call the area defined by the curve A, then

$$\int_{a}^{b} f(x) \, dx = A$$

From geometry, we remember that the area of a rectangle is equal to its width times its height. Once we've found the integral, we know the area A. We also know the width of the

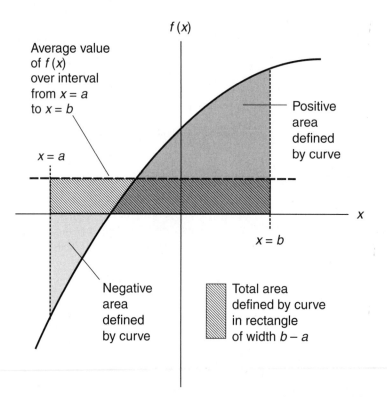

Figure 11-6   Integration can be used to find the average value of a nonlinear function over an interval.

rectangle; it's $b - a$. If we call the average value of the function over the interval by the rather fancy name $f^*_{a:b}$, then

$$f^*_{a:b} = A / (b - a)$$

We can also write

$$f^*_{a:b} = (b-a)^{-1} \int_a^b f(x) \, dx$$

Normal distribution

In the graph of a probability function called a *normal distribution*, there's a central peak, and the curve tapers off symmetrically on either side. Because of its characteristic shape, it's sometimes called a *bell-shaped curve*. Figure 11-7 shows a generic graph of a normal probability distribution. Let's call it $P(x)$. The area defined by the curve over the entire domain, which

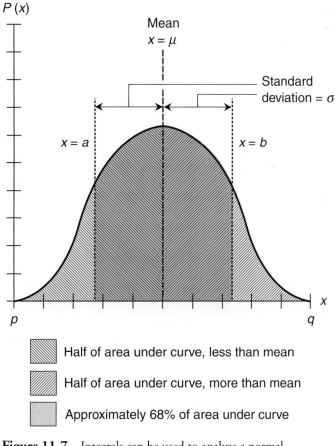

Figure 11-7   Integrals can be used to analyze a normal probability distribution. Areas under the curve define the mean and the standard deviation.

extends from $x = p$ to $x = q$, is equal to 1. All normal distributions have this property, which we can express as

$$\int_{p}^{q} P(x)\, dx = 1$$

The *mean* of normal distribution, symbolized by the lowercase Greek letter mu (μ), is the value of x that produces equal areas on either side. The mean can be found by imagining a movable, heavy, dashed vertical line that intersects the x axis. When the position of this line is such that the area defined by the curve to its left is half the total area (hatched region with down-sloping lines) and the area defined by the curve to its right is also half the total area (hatched region with up-sloping lines), then the vertical line intersects the x axis at the mean. We can describe this by writing

$$\int_{p}^{\mu} P(x)\, dx = 1/2$$

and

$$\int_{\mu}^{q} P(x)\, dx = 1/2$$

The *standard deviation* of a probability function is an expression of the extent to which the values are concentrated near the center. It's symbolized by the lowercase Greek letter sigma (σ). A small standard deviation produces a "sharp" curve with a narrow peak and steep sides. A large standard deviation produces a "broad" curve with less steep sides. Imagine two movable, heavy, dashed vertical lines, one on either side of the mean. Suppose these vertical lines, $x = a$ and $x = b$, are such that the one on the left is the same distance from μ as the one on the right, and the area defined by the curve between $x = a$ and $x = b$ is approximately 68% of the total area defined by the whole curve (dark-shaded region in Fig. 11-7). When this is the case, lines $x = a$ and $x = b$ are both exactly σ units away from μ. That is,

$$\mu - a = \sigma$$

and

$$b - \mu = \sigma$$

We can describe the area defined by the curve for the standard deviation as

$$\int_{a}^{b} P(x)\, dx \approx 0.68$$

- -

Here's a challenge!

Using the definite integral we found to describe the distance traveled by the moving car (Fig. 11-5), calculate the average value of the function

$$s(t) = 2.4t$$

over the interval from $t = 2$ to $t = 5$. What does this tell us?

Solution

To find the average value of this function, we divide the area "under the graph" in the interval by the width of the interval. Let's call the average $s^*_{2:5}$. Then

$$s^*_{2:5} = (5 - 2)^{-1} \int\limits_2^5 2.4t \, dt$$

We've already found that the area is 25.2 square units, and subtraction tells us that the width of the interval is 3 units. Therefore,

$$s^*_{2:5} = (1/3) \cdot 25.2 = 8.4$$

This is the average speed of the car during the time interval.

Here's another challenge!

Verify the solution of the above "challenge" using ordinary arithmetic. Could this alternative method be used if s were not a linear function?

Solution

The function s is linear, so its graph is a straight line. Therefore, the average of the speeds at 2 seconds and 5 seconds is equal to the speed in the middle of that time interval. Let's call that "average time" $t^*_{2:5}$. We can average the time at the beginning of the interval with the time at the end to get

$$t^*_{2:5} = (1/2) \cdot (5 + 2) = 7/2$$

The value of the function at $t = 7/2$ is the average speed of the car during the time interval:

$$t^*_{2:5} = s(t^*_{2:5}) = s(7/2) = 2.4 \cdot 7/2 = 8.4$$

This method of finding the average value of the function works because our speed-vs.-time function s is linear. If it were nonlinear, its graph wouldn't be a straight line, and we couldn't use straightforward arithmetic to find the average speed. We would have to use the integral method. Situations like this are common in physics and engineering.

Practice Exercises

This is an open-book quiz. You may (and should) refer to the text as you solve these problems. Don't hurry! You'll find worked-out answers in App. B. The solutions in the appendix may not represent the only way a problem can be figured out. If you think you can solve a particular problem in a quicker or better way than you see there, by all means try it!

1. Express the first n terms of the following series in summation notation:

$$8 + 4 + 2 + 1 + 1/2 + 1/4 + 1/8 + \cdots$$

2. Express the first n terms of the following series in summation notation:

$$1/2 + 2/3 + 3/4 + 4/5 + 5/6 + 6/7 + 7/8 + \cdots \ldots$$

3. Express the limits of the series stated in Probs. 1 and 2 as the number of terms increases endlessly. Use limit notation in conjunction with summation notation.

4. Which, if either, of the limits we obtain by working out Prob. 3 are defined?

5. Consider the function for which the output is the square of the input:

$$f(x) = x^2$$

The graph of this function is a section of a parabola passing through (0,0) and (1,1). Using the Riemann method of integration, find the area defined the curve between $x = 0$ and $x = 1$.

6. Based on the solution to Prob. 5, find the average value of the function

$$f(x) = x^2$$

over the interval from $x = 0$ to $x = 1$.

7. Consider the function for which the output is the cube of the input:

$$f(x) = x^3$$

The graph of this function is a curve, similar to a section of a parabola but somewhat "sharper," passing through (0,0) and (1,1). Using the Riemann method of integration, find the area defined by the curve between $x = 0$ and $x = 1$.

8. Based on the solution to Prob. 7, find the average value of the function

$$f(x) = x^3$$

over the interval from $x = 0$ to $x = 1$.

9. Look again at Fig. 11-5, which graphs a situation where a car's speed steadily increases from a standstill. Using the Riemann method of integration, calculate how far the car travels in the first 5 seconds (that is, from $t = 0$ to $t = 5$).

10. Verify the solution to Prob. 9 using geometry.

CHAPTER

12

Derivatives in Reverse

One of my favorite professors said, "Differentiation is like navigating downstream in a river system. *Antidifferentiation* is like navigating upstream in the same system. You're in the same scene, but the experience is completely different." That professor knew what he was talking about.

Concept of the Antiderivative

Antidifferentiation produces a new function that expresses the cumulative growth (or shrinkage) of the original function.

The notation

We usually represent functions with lowercase letters such as f, g, or h. Their antiderivatives are denoted by uppercase italic counterparts such as F, G, or H. When seeking the antiderivative of a function f, we must find a function F that, when differentiated, gives us f again.

Antiderivatives of the zero function

Let's think about a constant function f of a variable x that takes all the values of the domain and maps them into a real-number constant c. That is,

$$f(x) = c$$

When we differentiate f, we get

$$f'(x) = 0$$

Imagine another constant function g that maps all the values of x into 0. That's the zero function. It's also the derivative of f, so

$$f'(x) = g(x)$$

205

It's reasonable to suppose that we can antidifferentiate both sides here, getting

$$f(x) = G(x)$$

But $f(x) = c$, so it follows that

$$G(x) = c$$

How do we know that this constant c represents the same number that we called c at the beginning? Well, in fact, *we don't*.

- -

Are you confused?

Now you might wonder, "What's this constant c that we get when we antidifferentiate the zero function? Are we suggesting that c can be *any* real number? When we antidifferentiate the simplest imaginable function, the zero function, do we get *infinitely many* different results?" Yes. The set of all the antiderivatives of the zero function is the set of all real numbers. (We won't get into what happens with imaginary and non-real complex numbers here.)

Now perhaps you can see why my professor compared antidifferentiation with going upstream in a river system. Which tributary do we take? Going downstream, we can float with the current, and we'll always get to the ocean. But going upriver, we have many choices. If we start in New Orleans and take a steamboat up the Mississippi River, we'll end up in a different place if we take the Ohio tributary, than we will if we take the Missouri tributary.

Here's a challenge!

Draw a graph of the function

$$f(x) = 0$$

as a solid line or curve. Write down the antiderivatives for $c = -2$, $c = 3$, and $c = 5$. Draw the graphs of these antiderivatives as dashed lines or curves. What do these antiderivative graphs all have in common?

Solution

The general expression for the antiderivative is

$$F(x) = c$$

where c can be any real number. Figure 12-1 shows the graphs of the three antiderivatives where $c = -2$, $c = 3$, and $c = 5$, respectively:

$$F_{-2}(x) = -2$$
$$F_3(x) = 3$$
$$F_5(x) = 5$$

All the antiderivative graphs are straight lines with slopes of 0. The constants represent the points where the lines pass through the dependent-variable axis.

- -

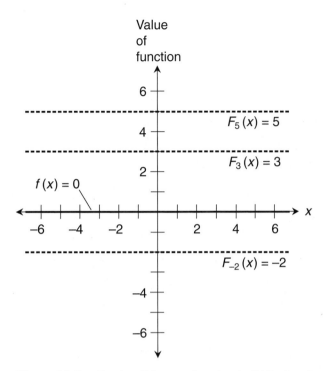

Figure 12-1 Graphs of the zero function (solid line) and three of its antiderivatives (dashed lines).

Some Simple Antiderivatives

Let's look at the antiderivatives of some basic functions. We learned how to take their derivatives in Part One, so we have a head start.

Antiderivatives of nonzero constant functions

After the zero function, the next simplest function is a constant function. If we call the function f, the variable x, and the constant a, then

$$f(x) = a$$

If we want to antidifferentiate this, we must ask ourselves, "What function or functions will give us f when differentiated?" The most immediate answer is

$$F_0(x) = ax$$

But any of the following will also work:

$$F_1(x) = ax + 1$$
$$F_3(x) = ax + 3$$
$$F_{-2}(x) = ax - 2$$

and, in general,

$$F(x) = ax + c$$

where c can be any real number.

Antiderivatives of basic linear functions

Now consider the basic linear function, which multiplies the variable by a nonzero constant. If we call the function f, the variable x, and the constant a, then

$$f(x) = ax$$

The task of antidifferentiating this is a little trickier. It might take some trial and error, but we can make educated guesses and eventually come up with

$$F_0(x) = ax^2/2$$

Let's recall the rule for differentiating a basic quadratic function, and also remember the multiplication-by-constant rule. With the help of these principles, we can differentiate the above function to get

$$F_0'(x) = 2 \cdot (ax/2) = ax = f(x)$$

But any of the following functions will also work:

$$F_4(x) = ax^2/2 + 4$$
$$F_{32}(x) = ax^2/2 + 32$$
$$F_{-298}(x) = ax^2/2 - 298$$

because the added constants always vanish when we differentiate. In general, the antiderivatives have the form

$$F(x) = ax^2/2 + c$$

where c can be any real number.

Antiderivatives of basic quadratic functions

The next level of complexity is the basic quadratic function, which squares the variable and then multiplies the result by a nonzero constant. If we call the function f, the variable x, and the constant a, then

$$f(x) = ax^2$$

We can think back to the basic rules for differentiation, and it won't take us long to see that the most obvious antiderivative is

$$F_0(x) = ax^3/3$$

Using the rule for differentiating a basic cubic function, and again remembering the multiplication-by-constant rule, we can verify that F_0 is a legitimate antiderivative by differentiating it to get

$$F_0'(x) = 3 \cdot (ax^2/3) = ax^2 = f(x)$$

As before, we can add any constant we want, and we'll still get $f(x)$ when we differentiate. For example:

$$F_{6.33}(x) = ax^3/3 + 6.33$$
$$F_{71.05}(x) = ax^3/3 + 71.05$$
$$F_{-85/13}(x) = ax^3/3 - 85/13$$

and, in general,

$$F(x) = ax^3/3 + c$$

where c can be any real number.

Antiderivatives of basic *n*th-degree functions

Can you guess what we must do when we want to find the antiderivative of a function that raises a variable to a nonnegative integer power and then multiplies by a constant? We follow the power rule for derivatives that we learned in Chap. 5, but we work it "backward." If we call the function f, the variable x, and the constant a, then

$$f(x) = ax^n$$

where n is a nonnegative integer. The most straightforward antiderivative is

$$F_0(x) = ax^{(n+1)} / (n+1)$$

Using the power rule for derivatives, we can check to be sure that F_0 is an antiderivative. When we differentiate it, we get

$$F_0'(x) = (n+1)\,[ax^{(n+1-1)}] / (n+1) = ax^n = f(x)$$

As in all the cases we've seen before, we can add any constant and still get $f(x)$ when we differentiate. For example:

$$F_{6/7}(x) = [ax^{(n+1)} / (n+1)] + 6/7$$
$$F_{-2e}(x) = [ax^{(n+1)} / (n+1)] - 2e$$
$$F_{-/8/\pi}(x) = [ax^{(n+1)} / (n+1)] - 78/\pi$$

and, in general,

$$F(x) = [ax^{(n+1)} / (n + 1)] + c$$

where c can be any real number.

Real-number exponents

The rule outlined above applies not only to nonnegative integers, but to all real-number exponents except -1. We can rewrite the rule by substituting k in place of n, where k represents any real number other than -1. If

$$f(x) = ax^k$$

then

$$F(x) = [ax^{(k+1)} / (k + 1)] + c$$

If k is negative, we must be sure that $x \neq 0$. Otherwise, ax^k is undefined. Can you see why?

We get into trouble with this rule if $k = -1$, no matter what value we plug in for the variable x. The problem is easy to see in an example. Suppose

$$f(x) = 3x^{-1}$$

When we try to apply the rule, we get

$$F(x) = [3x^{(-1+1)} / (-1 + 1)] + c = 3x^0/0 + c$$

Division by 0 is not defined, so this expression doesn't make sense! But this is an interesting situation because, as things turn out, reciprocal functions do have antiderivatives. This rule simply happens to be the wrong way to look for them. We'll learn the correct way in Chap. 17.

- -

Here's a challenge!

Imagine that you're in the gondola of a balloon hovering 1,000 meters above a deserted lake. You make sure no one is below you, and then you hold a brick over the side and let it go. Suppose the brick is so dense that air resistance does not affect it. Consider the vertical acceleration caused by gravity to be exactly 10 meters per second per second. Use antidifferentiation to derive the vertical speed as a function of time and the vertical displacement as a function of time. Assume that you start your timer at the instant you drop the brick. How long will it take for the brick to splash down after it is dropped?

Solution

The brick's initial speed is zero, and it increases downward by 10 meters per second every second until it hits the water. If you call the acceleration function f and the time variable t, then

$$f(t) = 10$$

Because you start the timer exactly when you drop the brick, and because the brick starts out at zero speed, you can set the constant c equal to 0 when you find the antiderivative. That means you can leave the constant out altogether, getting

$$F(t) = 10t$$

This function expresses the vertical speed vs. time. If you take the antiderivative again and call the new function $\Phi(t)$, then

$$\Phi(t) = 5t^2$$

The symbol Φ is the uppercase Greek letter phi. You can leave out the constant again here, because the initial vertical displacement (or "fallen distance") is 0. The function Φ tells you the vertical displacement in meters vs. the elapsed time in seconds. You can verify the relationship among Φ, F, and f by differentiating them in succession:

$$\Phi'(t) = 2 \cdot 5t^{(2-1)} = 10t = F(t)$$

and

$$F'(t) = 10 = f(t)$$

You want to know how long it will take the brick to hit the water after you release it. This is the time t at which the value of Φ is equal to the altitude of the balloon (1,000 meters). You can solve the problem by writing the equation

$$5t^2 = 1,000$$

Dividing each side by 5, you get

$$t^2 = 200$$

Therefore, $t = 200^{1/2}$, the positive square root of 200. Its value is approximately 14.14 seconds.

- -

Indefinite Integral

The *indefinite integral* of a function is the collection of all its possible antiderivatives. The indefinite integral always includes c, the so-called *constant of integration*, but does not necessarily tell us the exact value of that constant. It is quite sensible to call an indefinite integral an *ambiguous integral*, but you will probably never hear any mathematician use that slang.

The notation

Suppose that f is a continuous real-number function of a variable x. The indefinite integral of f is a function F such that

$$F'(x) = f(x)$$

added to the constant of integration c. This is written as follows:

$$\int f(x)\ dx = F(x) + c$$

where dx represents the *differential* of x. The above expression, if read out loud, is "the indefinite integral of $f(x)\ dx$," or "the indefinite integral of $f(x)$ with respect to x."

An example

Consider the function

$$f(x) = x^2$$

The derivative is

$$f'(x) = 2x$$

Now think about the following indefinite integral:

$$\int 2x\ dx$$

Let's rename the function we're integrating. We can say that

$$g(x) = 2x$$

At first thought, it's tempting to suppose that the antiderivative of g is the function G such that

$$G(x) = x^2$$

But the situation isn't quite so simple, because G isn't the only function of x that can be differentiated to get g. Any real number c can be added to G. When we differentiate the result, we always get g back again, because the derivative of the constant is always 0. We can think of it like this:

$$G(x) + c = x^2 + c$$

and therefore:

$$[G(x) + c]' = 2x + 0 = 2x = g(x)$$

That means:

$$\int 2x\ dx = x^2 + c$$

where c can be any real number.

"Pulling out" a constant

Imagine that f is a function of a variable x that can be integrated. Suppose that k is a real-number constant. The indefinite integral of k times $f(x)$ with respect to x is equal to k times the indefinite integral of $f(x)$ with respect to x. As an equation, we write this fact as

$$\int k\,[f(x)]\ dx = k \int f(x)\ dx$$

We can "pull the constant out" of an indefinite integral and get an equivalent expression, just as we can "pull the constant out" of a limit, summation, or derivative.

"Pulling out" the negative

We have a special case of the above rule when k happens to be equal to -1. In that situation, we can say that the indefinite integral of a negative is equal to the negative of the indefinite integral. We can "pull out the minus sign" like this:

$$\int -[f(x)]\ dx = - \int f(x)\ dx$$

Sum of indefinite integrals

Suppose that $f_1, f_2, f_3, \ldots$, and f_n are functions of a variable x, and all of those functions can be integrated. The indefinite integral of the sum of the functions with respect to x is equal to the sum of the indefinite integrals of each function with respect to x. As an equation, we write

$$\int [f_1(x) + f_2(x) + f_3(x) + \cdots + f_n(x)]\ dx$$
$$= \int f_1(x)\ dx + \int f_2(x)\ dx + \int f_3(x)\ dx + \cdots + \int f_n(x)\ dx$$

Again, when it comes to addition, we can work with indefinite integrals just as we can do with limits, summations, and derivatives.

- -

Are you confused?

The difference between an antiderivative and an indefinite integral is subtle. A language or grammar teacher might call it "a matter of semantics." Some texts will tell you that the antiderivative and the indefinite integral are identical. I like to think of the indefinite integral as all the antiderivatives swarming around each other in an infinitely large "mathematical cloud."

Here's a challenge!

Let's go back to the balloon, hovering 1,000 meters above the lake. You dropped a brick from the gondola and measured how long it took to splash down. The brick splashed down 14.14 seconds after you released it. You decide to conduct the experiment again. But this time, instead of dropping the brick, you lean out over the edge of the gondola and hurl the brick straight down with some force, so it starts its descent with a vertical speed of 5 meters per second. How long will this brick take to fall?

Solution

Because the brick has a "head start," the constant of integration is significant the first time you antidifferentiate! The brick's initial speed is 5 meters per second. As before, the vertical speed increases by 10 meters per second every second until the brick splashes down. (That's the acceleration caused by the earth's gravity, which never changes.) If you call the acceleration function f and the time variable t, then, as before,

$$f(t) = 10$$

You start the timer when you hurl the brick. But, although the brick starts out at zero "fallen distance," you give it an initial speed of 5 meters per second. You must account for this by adding a constant into the antiderivative the first time around. When $t = 0$ (at the moment you start the timer), the speed is 5 meters per second, so the antiderivative is

$$F(t) = 10t + 5$$

This function F expresses the downward vertical speed of the brick vs. time, assuming that you hurl the brick *straight down*. Antidifferentiating again, remembering the rule for the sum of indefinite integrals and calling the result $\Phi(t)$, you get

$$\Phi(t) = 5t^2 + 5t$$

The function Φ expresses the vertical displacement vs. the elapsed time. In this go-around, the constant of integration is 0, because the initial "fallen distance" is 0. As before, you can verify the relationship among Φ, F, and f by differentiating:

$$\Phi'(t) = 2 \cdot 5t^{(2-1)} + 5 = 10t + 5 = F(t)$$

and

$$F'(t) = 10 + 0 = 10 = f(t)$$

Now you're ready to calculate how long it will take the brick to hit the water after you throw it down. This, once again, is the time t at which the value of Φ is equal to the altitude of the balloon. You have

$$5t^2 + 5t = 1{,}000$$

This is a quadratic equation. You can get it into standard quadratic form by subtracting 1,000 from each side, obtaining

$$5t^2 + 5t - 1{,}000 = 0$$

Dividing through by 5 gives you the simpler equation

$$t^2 + t - 200 = 0$$

Using the quadratic formula to solve for t, you have

$$t = \{-1 \pm [1^2 - 4 \cdot 1 \cdot (-200)]^{1/2}\} / (2 \cdot 1) = (-1 \pm 801^{1/2}) / 2$$

The positive square root of 801 is approximately 28.30. Therefore,

$$t \approx (-1 \pm 28.30) \, / \, 2 \approx 13.65 \quad \text{or} \quad -14.65$$

What is happening here? Obviously, the brick can't fall in two different lengths of time, and it is ridiculous to suppose that it could fall in a negative time! The only solution that makes sense is $t \approx 13.65$ seconds. The other solution is a "phantom." (When you have a quadratic equation that leads to a "real-world" solution, the appearance of negative "phantom solutions" is common. Don't worry about it unless you plan to become a time traveler, enter an antimatter universe, or write science fiction.) By hurling the brick straight down with an initial speed of 5 meters per second, you got it to splash down in only a little less time than it took to fall when you merely dropped it. Does that surprise you?

- -

Definite Integral

In graphical terms, the definite integral of a function over an interval is the total area between the curve and the independent-variable axis in that interval. In Chap. 11, we added the areas of a set of rectangles, making them narrower to reduce the error, and finding the exact area by calculating the limit as the widths of the rectangles approached 0. We can also express definite integrals in terms of antiderivatives. That's easier!

The Fundamental Theorem of Calculus

Imagine that f is a continuous real-number function of a variable x. Also suppose that a and b are values in the domain of f with $a < b$, and F is a specific antiderivative with a constant of integration c. According to a law called the *Fundamental Theorem of Calculus*, the definite integral from a to b is

$$\int_{a}^{b} f(x) \, dx = F(b) - F(a)$$

Sometimes this is written as

$$\int_{a}^{b} f(x) \, dx = F(x) \, \Big]_{a}^{b}$$

where the expression on the right-hand side of the equals sign is read "$F(x)$ evaluated from a to b."

When we find a definite integral this way, the constant of integration subtracts from itself because $F(b)$ and $F(a)$ are calculated with the same antiderivative (the constant of integration is the same for both). This is convenient! It let us set the constant equal to 0 or leave it entirely out.

An example

Suppose we're given the following basic linear function, and we're told to find its definite integral from -2 to 1:

$$g(x) = 2x$$

We can easily figure out the antiderivatives from our work with the basic quadratic function. They can be written in general as

$$G(x) = x^2 + c$$

When we apply the Fundamental Theorem of Calculus to find the definite integral, the constant of integration subtracts from itself no matter which antiderivative we choose. We might as well choose the one where $c = 0$! When we do that, we get

$$\int_{-2}^{1} g(x)\, dx = G(1) - G(-2)$$

We can easily calculate

$$G(1) = 1^2 = 1$$

and

$$G(-2) = (-2)^2 = 4$$

Therefore

$$G(1) - G(-2) = 1 - 4 = -3$$

so we have the final result

$$\int_{-2}^{1} 2x\, dx = -3$$

Another example

Suppose we're given the following basic quadratic function, and we're told to find its definite integral from -2 to 1:

$$h(x) = 6x^2$$

We know the antiderivatives $H(x)$ from our work with the basic quadratic function. They can be written in general as

$$H(x) = 6x^3/3 + c = 2x^3 + c$$

As before, let's use the case where $c = 0$, because the constant of integration will cancel itself out no matter what it is. (We can always leave the constant of integration out when we take advantage of the Fundamental Theorem of Calculus.) We have

$$\int_{-2}^{1} h(x)\, dx = H(1) - H(-2)$$

Calculating the antiderivatives, we obtain

$$H(1) = 2 \cdot 1^3 = 2$$

and

$$H(-2) = 2 \cdot (-2)^3 = -16$$

Therefore

$$H(1) - H(-2) = 2 - (-16) = 18$$

so we get the definite integral

$$\int_{-2}^{1} 6x^2 \, dx = 18$$

- -

Are you confused?

If you have trouble envisioning what's taking place in the above examples, draw graphs of the functions $g(x)$ and $h(x)$, mark off the intervals of integration with the vertical lines corresponding to $x = -2$ and $x = 1$, and then fill in the areas bounded by these two lines, the x axis, and the curves.

Remember: If the complete region defined by an integral consists of more than one enclosed area, then the total area is the sum of the areas of the individual enclosed regions. Areas above the x axis are positive, and areas below the x axis are negative, assuming that we always move in the positive direction along the independent-variable axis. If the negative area is greater than the positive area, then the definite integral is negative.

Are you still confused?

You might wonder, "What if the lower and upper bounds of the interval are the same when we want to find a definite integral?" In that case, the definite integral is always 0. Suppose that you want to evaluate the definite integral of a function $f(x)$ from a to itself, with respect to x. If you apply the Fundamental Theorem of Calculus, then you get

$$\int_{a}^{a} f(x) \, dx = F(a) - F(a)$$

Remember that in any single-variable function, there can never be more than one value of the dependent variable (that is, one "output") for any single value of the independent variable ("input"). Therefore, you can be confident that

$$F(a) - F(a) = 0$$

Here's a challenge!

Once again, imagine that you're in the balloon over the lake. You're still hovering at an altitude of 1,000 meters. After expending a lot of energy (and risking your life) trying to get the brick to fall faster and gaining almost nothing for the effort, you simply drop the next brick and let it fall straight down. How much speed will the brick gain during the third second of its descent (that is, from $t = 2$ to $t = 3$)? How far will it fall in that time?

Solution

The brick's initial speed is zero, and it accelerates downward by 10 meters per second per second. As before, let's call the acceleration function f and the time variable t, so

$$f(t) = 10$$

To find out how much speed the brick gains from $t = 2$ to $t = 3$, we must find

$$\int_2^3 10 \, dt$$

We can set the constant of integration equal to 0 because the initial speed of the brick is equal to 0. Therefore the antiderivative, which tells us the instantaneous speed, is

$$F(t) = 10t$$

We evaluate the definite integral by subtracting $F(3) - F(2)$, getting

$$\int_2^3 10 \, dt = 10 \cdot 3 - 10 \cdot 2 = 10$$

We've just figured out that the brick gains 10 meters per second of speed during the third second of its fall. To find out how far the brick falls in this same time interval, we must evaluate

$$\int_2^3 10t \, dt$$

Again setting the constant of integration to 0, we get an antiderivative, which tells us the instantaneous "fallen distance," of

$$\Phi(t) = 5t^2$$

This time, we evaluate

$$\int_2^3 10t \, dt = 5 \cdot 3^2 - 5 \cdot 2^2 = 25$$

The brick falls 25 meters during the third second of its descent.

- -

Practice Exercises

This is an open-book quiz. You may (and should) refer to the text as you solve these problems. Don't hurry! You'll find worked-out answers in App. B. The solutions in the appendix may not represent the only way a problem can be figured out. If you think you can solve a particular problem in a quicker or better way than you see there, by all means try it!

1. Draw a graph of the function

$$f(x) = 2$$

as a solid line or curve. Write down the antiderivatives of this function for the constants of integration $c = 1$, $c = 3$, and $c = -2$. Draw the graphs of these antiderivatives as dashed lines or curves. What do these antiderivative graphs all have in common?

2. Draw a graph of the function

$$f(x) = 2x$$

as a solid line or curve. Write down the antiderivatives of this function for the constants of integration $c = 1$, $c = 3$, and $c = -2$. Draw the graphs of these antiderivatives as dashed lines or curves. What do these antiderivative graphs all have in common?

3. Draw a graph of the function

$$f(x) = 3x^2$$

as a solid line or curve. Write down the antiderivatives of this function for the constants of integration $c = 5$, $c = 15$, and $c = -10$. Draw the graphs of these antiderivatives as dashed lines or curves. On the graph, make each horizontal-axis division represent 1 unit, and make each vertical-axis division represent 5 units. What do these antiderivative graphs all have in common?

4. Draw a graph of the function

$$f(x) = 4x^3$$

as a solid line or curve. Write down the antiderivatives of this function for the constants of integration $c = 5$, $c = 15$, and $c = -10$. Draw the graphs of these antiderivatives as dashed lines or curves. On the graph, make each horizontal-axis division represent 1 unit, and make each vertical-axis division represent 5 units. What do these antiderivative graphs all have in common?

5. Using the rule for antidifferentiating a basic mth-degree function and the Fundamental Theorem of Calculus, evaluate the definite integral of the function

$$f(x) = 8x^3$$

with respect to x, from $x = 0$ to $x = 5$.

6. Using the rule for antidifferentiating a basic mth-degree function and the Fundamental Theorem of Calculus, evaluate the definite integral of the function

$$g(z) = -2z^4$$

with respect to z, from $z = -6$ to $z = -3$.

7. Let's go back to the balloon, hovering 1,000 meters above the lake. We conduct the experiment still another time. I lean out over the lip of the gondola (you're no longer so foolish) and toss the brick into the air with an *upward* vertical speed of 5 meters per second. It rises a short distance and then begins to fall. How long will it be from the time I toss the brick up until it hits the water?

8. How fast will the brick in Prob. 7 be traveling when it splashes down?

9. You and I are still in the balloon. You release another brick and let it fall straight down with an initial speed of 0 at time $t = 0$, just as you did in the original experiment. How fast will the brick be falling at $t = 5$? How much vertical speed will it gain between the end of the 5th second and the end of the 6th second (in the interval from $t = 5$ to $t = 6$)?

10. In the situation of Prob. 9, how far will the brick fall in the first 5 seconds? How far will it fall between $t = 5$ to $t = 6$?

13

Three Rules for Integration

Until now, the functions we've integrated have been simple and straightforward. We've always gone in the positive direction (left-to-right) over continuous intervals. In this chapter, we'll learn some rules that can help us find definite integrals in more complicated situations.

Reversal Rule

When we want to calculate a definite integral using the Fundamental Theorem of Calculus, we usually integrate from the minimum to the maximum value of the independent variable over an interval. But we can also integrate in the other direction.

The rule in brief

The reversal rule for definite integration tells us this:

- If we integrate a function over an interval from the largest to the smallest value of the independent variable, we get the negative of the integral from the smallest to the largest value. In other words, if we reverse the direction in which we integrate over a specific interval, we multiply the result by −1.

For this rule to work, the function must be *integrable* over the interval. That means the definite integral must exist, and we must be able to find it. If f is an integrable function over an interval (a,b) where $a < b$, then

$$\int_a^b f(x)\ dx = -\int_b^a f(x)\ dx$$

Example

In the solution to Practice Exercise 5 in Chap. 12, we found that

$$\int_0^5 8x^3\ dx = 1{,}250$$

Let's go the other way and find

$$\int_{5}^{0} 8x^3 \, dx$$

We can find the *basic antiderivative,* in which the constant of integration is 0 (in contrast to the *general antiderivative* where the constant of integration can be any real number). Our basic antiderivative is

$$F(x) = 8x^{(3+1)} / (3+1) = 8x^4/4 = 2x^4$$

Plugging in the values at the bounds, we obtain

$$F(0) = 2 \cdot 0^4 = 0$$

and

$$F(5) = 2 \cdot 5^4 = 1{,}250$$

Therefore

$$F(0) - F(5) = 0 - 1{,}250 = -1{,}250$$

With the help of the Fundamental Theorem of Calculus, we can conclude that

$$\int_{5}^{0} 8x^3 \, dx = -1{,}250$$

Another example

In the solution to Practice Exercise 6 in Chap. 12, we found that

$$\int_{-6}^{-3} -2z^4 \, dz = -3{,}013.2$$

We can integrate this in the other direction by finding

$$\int_{-3}^{-6} -2z^4 \, dz$$

The basic antiderivative is, as before,

$$G(z) = -2z^{(4+1)} / (4+1) = -2z^5/5$$

Calculating the antiderivatives at the bounds, we get

$$G(-6) = [-2 \cdot (-6)^5] / 5 = [-2 \cdot (-7{,}776)] / 5 = 15{,}552/5 = 3{,}110.4$$

and

$$G(-3) = [-2 \cdot (-3)^5] / 5 = [-2 \cdot (-243)] / 5 = 486/5 = 97.2$$

By arithmetic, we get

$$G(-6) - G(-3) = 3{,}110.4 - 97.2 = 3{,}013.2$$

obtaining the final answer

$$\int_{-3}^{-6} -2z^4 \, dz = 3{,}013.2$$

- -

Are you confused?

Usually, we think of areas defined by a curve above the independent-variable axis to be positive, and areas below the independent-variable axis to be negative. But, as was mentioned in Chap. 11, things work out that way only when we go from left to right, that is, in the positive direction along the independent-variable axis. If we go from right to left, then the negative and positive areas reverse, as shown in Fig. 13-1.

Here's a challenge!

Show that for any integrable function $f(x)$ over a real-number interval (a, b) where $a < b$,

$$\int_b^a f(x) \, dx = -\int_a^b f(x) \, dx$$

Solution

From the Fundamental Theorem of Calculus, we know that

$$\int_b^a f(x) \, dx = F(a) - F(b)$$

and

$$\int_a^b f(x) \, dx = F(b) - F(a)$$

where F is the basic antiderivative of f. The right-hand sides of these equations are subtractions of the same quantities done in opposite order. By algebra,

$$F(a) - F(b) = -[F(b) - F(a)]$$

We can plug in the definite integral expressions for these differences to obtain

$$\int_b^a f(x) \, dx = -\int_a^b f(x) \, dx$$

- -

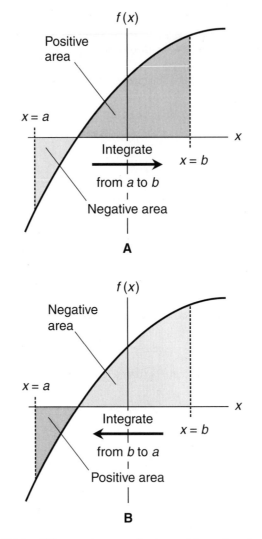

Figure 13-1 When we integrate in the positive *x* direction, areas above the *x* axis are positive, while areas below the *x* axis are negative, as shown at A. When we integrate in the negative *x* direction, areas above the *x* axis are negative, while areas below the *x* axis are positive, as shown at B.

Split-Interval Rule

Imagine that we break up an interval into two parts that don't overlap, but are *contiguous* (there is no gap between them). If we integrate the parts separately and then add the results, we'll get the same thing as we do if we integrate over the whole interval.

The rule in brief

The split-interval rule for definite integration tells us this:

- If $f(x)$ is an integrable function over a single, continuous interval containing real numbers a, b, and c, then the sum of the integral from $x = a$ to $x = b$ plus the integral from $x = b$ to $x = c$ is equal to the integral from $x = a$ to $x = c$.

When we write this symbolically, we get

$$\int_b^a f(x)\ dx + \int_b^c f(x)\ dx = \int_a^c f(x)\ dx$$

A special case exists when $a < b < c$. But this doesn't have to be true. If $a < c < b$, for example, then the second integral in the above sum goes in the negative direction. It doesn't matter. The formula works anyway!

Example

Let's integrate a function from $x = 0$ to $x = 2$, and then from $x = 2$ to $x = 5$. We'll see that the sum of these two integrals is the same as the integral from $x = 0$ to $x = 5$. First, we use the Fundamental Theorem of Calculus to evaluate

$$\int_0^2 8x^3\ dx$$

The basic antiderivative is

$$F(x) = 2x^4$$

Calculating the antiderivatives, we obtain

$$F(2) = 2 \cdot 2^4 = 32$$

and

$$F(0) = 2 \cdot 0^4 = 0$$

Therefore

$$F(2) - F(0) = 32 - 0 = 32$$

so we get the result

$$\int_0^2 8x^3\ dx = 32$$

Now, let's work out

$$\int_2^5 8x^3\ dx$$

This time, the antiderivatives are

$$F(5) = 2 \cdot 5^4 = 1{,}250$$

and

$$F(2) = 2 \cdot 2^4 = 32$$

Therefore

$$F(5) - F(2) = 1{,}250 - 32 = 1{,}218$$

so we have

$$\int_2^5 8x^3 \, dx = 1{,}218$$

Now when we add, we should get

$$\int_0^2 8x^3 \, dx + \int_2^5 8x^3 \, dx = \int_0^5 8x^3 \, dx$$

In the solution to Practice Exercise 5 in Chap. 12, we found that

$$\int_0^5 8x^3 \, dx = 1{,}250$$

Plugging in the results we've just obtained for the integrals from $x = 0$ to $x = 2$ and from $x = 2$ to $x = 5$, we get the sum

$$32 + 1{,}218 = 1{,}250$$

Another example

Let's integrate this same function from $x = 0$ to $x = 7$, and then from $x = 7$ to $x = 5$. We'll see that the sum of these two integrals is the same as the integral from $x = 0$ to $x = 5$. First, we evaluate

$$\int_0^7 8x^3 \, dx$$

Calculating the antiderivatives, we obtain

$$F(7) = 2 \cdot 7^4 = 4{,}802$$

and

$$F(0) = 2 \cdot 0^4 = 0$$

Therefore

$$F(7) - F(0) = 4{,}802 - 0 = 4{,}802$$

so we get the result

$$\int_0^7 8x^3 \, dx = 4{,}802$$

Now, let's work out

$$\int_7^5 8x^3 \, dx$$

The antiderivatives are

$$F(5) = 2 \cdot 5^4 = 1{,}250$$

and

$$F(7) = 2 \cdot 7^4 = 4{,}802$$

Therefore

$$F(5) - F(7) = 1{,}250 - 4{,}802 = -3{,}552$$

so we get the result

$$\int_7^5 8x^3 \, dx = -3{,}552$$

Now when we add, we should get

$$\int_0^7 8x^3 \, dx + \int_7^5 8x^3 \, dx = \int_0^5 8x^3 \, dx$$

We've already determined that

$$\int_0^5 8x^3 \, dx = 1{,}250$$

Let's plug in the results we just got for the integrals from $x = 0$ to $x = 7$ and from $x = 7$ to $x = 5$. When we do this, we get

$$4{,}802 + (-3{,}552) = 1{,}250$$

- -

Are you astute?

If we integrate a function from one value of the independent variable to another, it doesn't matter how we get from the starting point to the finishing point. As long as the function is integrable over the entire route, we'll get the same end result no matter how far afield we go, and no matter how many times we backtrack.

Here's a challenge!

Prove the split-interval rule in its basic form: For any function $f(x)$ that's defined and integrable over an interval containing the real numbers a, b, and c,

$$\int_b^a f(x)\,dx + \int_b^c f(x)\,dx = \int_a^c f(x)\,dx$$

Solution

From the Fundamental Theorem of Calculus, we know these three facts:

$$\int_b^a f(x)\,dx = F(b) - F(a)$$

$$\int_b^c f(x)\,dx = F(c) - F(b)$$

$$\int_a^c f(x)\,dx = F(c) - F(a)$$

where F represents the antiderivative of f with the constant of integration equal to 0. Let's rename the values of the functions at the points in the interval:

$$p = F(a)$$
$$q = F(b)$$
$$r = F(c)$$

where p, q, and r are real numbers. We know all three of these numbers exist, because we've been assured that the function $f(x)$ is defined and integrable over the interval containing the x-values a, b, and c. By substitution, we can rewrite the first two of the above integrals as

$$\int_b^a f(x)\,dx = q - p$$

and

$$\int_b^c f(x)\,dx = r - q$$

To find the sum of these, we evaluate the expression

$$(q - p) + (r - q)$$

By algebra, this simplifies to

$$r - p$$

Substituting back the original expressions for r and p, we get

$$F(c) - F(a)$$

According to the Fundamental Theorem of Calculus, that's the same as

$$\int_a^c f(x)\, dx$$

Substitution Rule

Once in awhile, we'll want to evaluate a definite integral that involves a composite function. If the integral can be written in a certain form, then it can be solved using a technique called the *substitution rule.*

The rule in brief

Imagine two functions f and g, both of them integrable, such that f operates on x while g operates on $f(x)$. Now consider

$$\int_a^b g[f(x)] \cdot f'(x)\, dx$$

We can rewrite this integral in the form

$$\int_{f(a)}^{f(b)} g(y)\, dy$$

where $y = f(x)$. Once we've made this substitution, we evaluate the definite integral of the function g with respect to y, from $y = f(a)$ to $y = f(b)$. That process gives us the value of the original definite integral.

Example

Let's evaluate this definite integral using the substitution method:

$$\int_1^2 (x+2)^4\, dx$$

We can consider this as a composite function where

$$f(x) = x + 2$$

and

$$g(y) = y^4$$

We're lucky here because

$$f'(x) = 1$$

so we can rewrite the original integral in the form

$$\int_1^2 g[f(x)] \cdot f'(x) \, dx$$

We calculate

$$f(1) = 1 + 2 = 3$$

and

$$f(2) = 2 + 2 = 4$$

Now we can write the integral as

$$\int_3^4 g(y) \, dy$$

which is

$$\int_3^4 y^4 \, dy$$

Now let's find the indefinite integral

$$\int y^4 \, dy$$

With the constant of integration set to 0, we get the antiderivative

$$G(y) = y^5/5$$

We calculate

$$G(4) - G(3) = 4^5/5 - 3^5/5 = 156.2$$

We've just found that

$$\int_1^2 (x+2)^4 \, dx = 156.2$$

- -

Are you confused?

When we work out a definite integral using substitution, the bounds of integration usually change during the intermediate steps. Don't let this baffle you! In the above situation, we must add 2 to both bounds when we integrate $g(y)$ with respect to y, because that's what the function f does to its input variable.

Here's a challenge!

Evaluate this definite integral using the substitution method:

$$\int_{-1}^{2} x \, (x^2 - 1)^4 \, dx$$

Solution

Before we start, let's multiply the entire integral by 1/2, and then multiply the quantity after the integral sign by 2. The product rule for integration allows us to do this; we're simply multiplying the whole thing by $1/2 \cdot 2$, which is 1. (You'll see why we're playing this game in a minute.) That gives us

$$(1/2) \int_{-1}^{2} 2x \, (x^2 - 1)^4 \, dx$$

We now have a composite function where

$$f(x) = x^2 - 1$$

and

$$g(y) = y^4$$

Our multiplication game has made things so that

$$f'(x) = 2x$$

Now we can rewrite the original integral in the form

$$(1/2) \int_{-1}^{2} g\,[f(x)] \cdot f'(x) \, dx$$

which gets it all ready for substitution. We calculate

$$f(-1) = (-1)^2 - 1 = 0$$

and

$$f(2) = 2^2 - 1 = 3$$

so we can write the integral as

$$(1/2) \int_{0}^{3} g(y) \, dy$$

which is

$$(1/2) \int_{0}^{3} y^4 \, dy$$

Now let's find the indefinite integral

$$\int y^4 \, dy$$

The basic antiderivative is

$$G(y) = y^5/5$$

Therefore

$$G(3) - G(0) = 3^5/5 - 0^5/5 = 48.6$$

We're not done yet! We must multiply the above result by the constant 1/2. This gives us 24.3. We can now conclude that

$$\int_{-1}^{2} x(x^2 - 1)^4 \, dx = 24.3$$

- -

Practice Exercises

This is an open-book quiz. You may (and should) refer to the text as you solve these problems. Don't hurry! You'll find worked-out answers in App. B. The solutions in the appendix may not represent the only way a problem can be figured out. If you think you can solve a particular problem in a quicker or better way than you see there, by all means try it!

1. Prove that if h is an integrable function of a variable v over an interval from $v = p$ to $v = q$ where p and q are real-number constants, then

$$\int_{p}^{q} h(v) \, dv = \int_{q}^{p} -h(v) \, dv$$

2. As an example of the rule stated in Prob. 1, show that

$$\int_{1}^{2} x^2 \, dx = \int_{2}^{1} -x^2 \, dx$$

3. Draw graphs of the two functions evaluated in the solution to Prob. 2. Show, by shading, the regions defined by the curves in the interval between $x = 1$ and $x = 2$. Explain why both areas are considered positive in this situation.

4. Consider the function $f(x) = x^3$. Evaluate each of the integrals in the following sum, demonstrating the split-interval rule:

$$\int_{-3}^{0} x^3 \, dx + \int_{0}^{3} x^3 \, dx = \int_{-3}^{3} x^3 \, dx$$

5. Draw a graph of the situation in Prob. 4. Show the negative and positive areas.

6. Evaluate each of the integrals in the following sum:

$$\int_{-3}^{-5} x^3 \, dx + \int_{-5}^{3} x^3 \, dx = \int_{-3}^{3} x^3 \, dx$$

7. Prove that for any function $f(x)$ that's defined and integrable over an interval containing the real numbers a, b, c, and d,

$$\int_{b}^{a} f(x) \, dx + \int_{b}^{c} f(x) \, dx + \int_{c}^{d} f(x) \, dx = \int_{a}^{d} f(x) \, dx$$

Here's a hint: See the "Challenge" at the end of the section "Split-Interval Rule."

8. Evaluate this definite integral using the substitution method:

$$\int_{-1}^{2} (4 - x)^{-2} \, dx$$

9. Suppose we want to evaluate

$$\int_{-1}^{4} (4 - x)^{-2} \, dx$$

There's a potential problem here. What is it?

10. Try evaluating the integral stated in Prob. 9 and see what happens.

14

Improper Integrals

In this chapter, we'll see what happens when we try to integrate a function that contains a *singularity* (blows up either positively or negatively). We'll also learn how to integrate certain functions over infinitely wide intervals. Integrals of these types are called *improper integrals*.

Variable Bounds

Let's look at definite integrals in which one of the bounds is variable rather than fixed. This can happen in either of two different ways. We can precisely adjust a bound, making it approach a certain limiting value, or we can let the bound "run away" (increase or decrease indefinitely).

Adjusting the upper bound

Consider a function f of a variable x, along with two constants a and b. Let's say that a is the lower bound of an open interval (a,b) over which the function is defined, and b is the upper bound of that same interval. That means $a < b$. As we have seen, the definite integral of $f(x)$ with respect to x, over the open interval from $x = a$ to $x = b$, can be written

$$\int_a^b f(x)\ dx$$

Suppose that the value of f increases or decreases endlessly as x approaches the upper bound b from the negative direction (from the left). The function is defined and continuous as long as $x < b$, but when $x = b$, there's a singularity. Remember that our interval is open, so *it does not include the end points*. Because of the singularity, the integral is improper. To approximate it, we can imagine some extremely small positive number ε (the lowercase Greek letter epsilon), subtract it from b, and use the difference as the upper bound instead of b. Then we get

$$\int_a^{b-\varepsilon} f(x)\ dx$$

We can make ε as small as we want, as long as we keep it positive (as shown on the left-hand side of Fig. 14-1), and we'll always be able to calculate the value of the integral. In this way, we can get an approximation of

$$\int_a^b f(x)\ dx$$

that's as close to the actual value as we want, but only if that actual value is finite.

Adjusting the lower bound

Now consider another function g of a variable x, along with two known constants c and d. Let's say that c is the lower bound of an open interval (c,d) over which g is defined, and d is

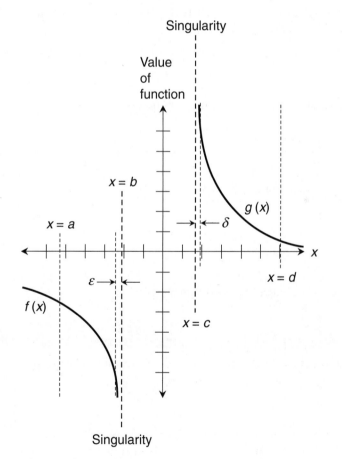

Figure 14-1 When an interval has a singularity, we
can approximate the definite integral by
choosing a bound near the singularity, but
at which the function is defined.

the upper bound of that same open interval. The definite integral of $g(x)$ with respect to x, over the open interval from $x = c$ to $x = d$, is

$$\int_c^d g(x)\, dx$$

Imagine that g has a singularity at the lower bound, c. As x approaches c from the positive direction (that is, from the right), the value of f increases or decreases endlessly. Our function g is defined and continuous as long as $x > c$. This integral, like the one in the previous section, is improper. We can take some minuscule positive number δ (the lowercase Greek letter delta), add it to c, and use the sum as the lower bound instead of c, as shown on the right-hand side of the graph in Fig. 14-1. Then we get

$$\int_{c+\delta}^d g(x)\, dx$$

We can make δ as small as we want, letting it approach 0. As long as we keep δ positive, we can get better and better approximations of

$$\int_c^d g(x)\, dx$$

as long as the actual value of the integral is finite.

"Runaway" bounds

Sometimes we'll find ourselves in a situation where we want to figure out a definite integral with an unspecified upper bound. Consider the integral

$$\int_a^q f(x)\, dx$$

where the upper bound, q, is a variable. (The lower bound, a, is a constant, as in any definite integral.) If q is allowed to increase endlessly as shown at the right-hand edge of Fig. 14-2, we approach the "one-ended" improper integral

$$\int_a^\infty f(x)\, dx$$

We can do the same thing with a lower bound. Consider the integral

$$\int_p^d g(x)\, dx$$

Here, the lower bound, p, is a variable, and the upper bound, d, is a constant. Suppose we make p shrink forever (that is, get larger and larger negatively) as shown at the left-hand edge of Fig. 14-2. In this situation, we approach the "one-ended" improper integral

$$\int_{-\infty}^d g(x)\, dx$$

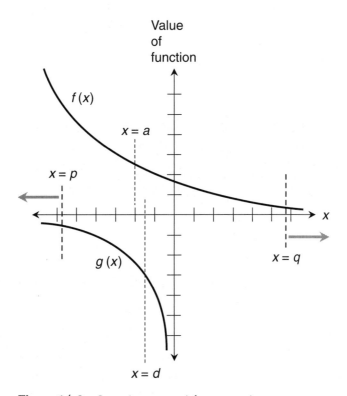

Figure 14-2 Sometimes, we might want to integrate a function over an interval where one of the bounds increases or decreases without limit ("runs away").

- -

Are you confused?

You might ask, "aren't these types of integrals always undefined? In all four of the cases shown in Figs. 14-1 and 14-2, we go off toward infinity or negative infinity." The answer, oddly enough, is that the integral is sometimes defined and finite in a situation like this. When an improper integral is infinite, it is said to *diverge*. That means the area defined by the curve over the specified open interval is infinite. But some improper integrals *converge*. In these situations, the area defined by the curve over the specified open interval is finite.

The convergence of certain improper integrals is one of the most fascinating phenomena in mathematics. We can have a geometric plane figure that's infinitely wide or infinitely tall, but nevertheless has a finite interior area. In three dimensions, the counterpart is a solid with finite volume, but with infinite surface area. In the real world, such an object would allow us to cover an infinite surface with a finite amount of paint!

Here's a challenge!

Consider the following definite integral:

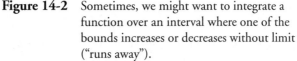

$$\int_{1}^{q} 5x^{-6} \, dx$$

where q is the upper bound. Evaluate this definite integral for

$$q = 10^2$$
$$q = 10^5$$
$$q = 10^8$$

What takes place as q grows larger endlessly? Express this integral as a limit, and also as an improper integral.

Solution

If we call our function $f(x)$, then the basic antiderivative is

$$F(x) = -x^{-5}$$

When $q = 10^2$, the definite integral is

$$F(10^2) - F(1) = -[(10^2)^{-5}] - [-(1^{-5})] = -(10^{-10}) + 1 = 1 - 10^{-10}$$

This is a little smaller than 1. When $q = 10^5$, the definite integral is

$$F(10^5) - F(1) = -[(10^5)^{-5}] - [-(1^{-5})] = -(10^{-25}) + 1 = 1 - 10^{-25}$$

This is closer to 1 than before. When $q = 10^8$, we have

$$F(10^8) - F(1) = -[(10^8)^{-5}] - [-(1^{-5})] = -(10^{-40}) + 1 = 1 - 10^{-40}$$

This is extremely close to 1, but it's still not quite there! We can now see what happens as q increases without limit: The value of the definite integral approaches 1, so

$$\lim_{q \to \infty} F(q) - F(1) = 1$$

If we want to express this fact using the integral symbology, we can write

$$\lim_{q \to \infty} \int_1^q 5x^{-6} \, dx = 1$$

As an improper integral, this can be expressed as

$$\int_1^\infty 5x^{-6} \, dx = 1$$

- -

Singularity in the Interval

When we graph a function with a singularity, the value of the function increases or decreases without bound as the independent variable approaches a certain finite value. We've seen functions like this, but we haven't really "gotten into them." Let's look at them more closely.

How to do it wrong

Suppose we want to evaluate the definite integral of a function over an interval that contains a singularity. Figure 14-3 is a graph of the function

$$f(x) = x^{-2}$$

There's a singularity at $x = 0$. Let's try to find

$$\int_{-1}^{1} x^{-2}\, dx$$

by directly applying the Fundamental Theorem of Calculus, as if this were an ordinary definite integral. Unaware of the trouble that awaits us, we begin by finding the basic antiderivative F. That turns out to be

$$F(x) = -x^{-1}$$

Next, we evaluate $F(x)$ at the bounds of integration, subtracting the value at the lower bound from the value at the upper bound. By arithmetic,

$$F(1) - F(-1) = -(1^{-1}) - [-(-1)^{-1}] = -1 - 1 = -2$$

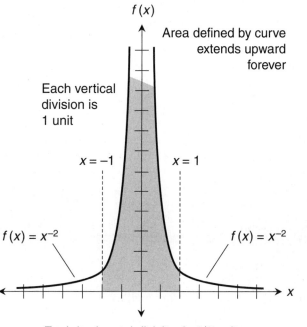

Each vertical division is 1 unit

Area defined by curve extends upward forever

$x = -1$ $x = 1$

$f(x) = x^{-2}$ $f(x) = x^{-2}$

Each horizontal division is 1/2 unit

Figure 14-3 The function $f(x) = x^{-2}$ contains a singularity at $x = 0$. In this graph, each horizontal division represents 1/2 unit, and each vertical division represents 1 unit.

This result can't be correct! The curve in the interval defines an area that's entirely above the x axis, and we integrated in the positive direction. Therefore, if the area is finite, it must be positive. It can't be -2. We must find another way to evaluate this integral.

How to do it correctly: the left-hand side

To evaluate an integral like this, we divide the interval into two parts that meet at the singularity. Then we integrate on either side of the singularity. Finally, we add the two "sub-integrals." In the situation we're looking at now, the singularity is at $x = 0$. First, let's evaluate

$$\int_{-1}^{0} x^{-2} \, dx$$

We take a tiny positive ε, subtract it from the upper bound at $x = 0$, and obtain

$$\int_{-1}^{-\varepsilon} x^{-2} \, dx$$

The basic antiderivative is

$$F(x) = -x^{-1}$$

When we evaluate this antiderivative from -1 to $-\varepsilon$, we get

$$F(-\varepsilon) - F(-1) = -(-\varepsilon)^{-1} - [-(-1)^{-1}] = \varepsilon^{-1} - 1$$

Now consider

$$\underset{\varepsilon \to 0+}{Lim} \; \varepsilon^{-1} - 1$$

This limit is infinite because, as ε approaches 0 from the positive direction, the value of the quantity $(\varepsilon^{-1} - 1)$ grows arbitrarily large. Therefore,

$$\underset{\varepsilon \to 0+}{Lim} \int_{-1}^{-\varepsilon} x^{-2} \, dx$$

is infinite, indicating that

$$\int_{-1}^{0} x^{-2} \, dx$$

is undefined. This fact is sufficient to tell us that

$$\int_{-1}^{1} x^{-2} \, dx$$

is undefined. We don't have to work out the part of the integral on the right-hand side of the singularity to reach this conclusion. We can't add anything to an undefined quantity and end up with a defined quantity! But let's work through the right-hand portion anyway.

How to do it correctly: the right-hand side

This time, we want to evaluate the improper integral in which the singularity exists at the lower bound of the open interval. That integral is

$$\int_0^1 x^{-2}\, dx$$

We'll take a tiny positive δ, add it to the lower bound at $x = 0$, and get

$$\int_\delta^1 x^{-2}\, dx$$

Again, the basic antiderivative of our function is

$$F(x) = -x^{-1}$$

When we evaluate this from δ to 1, we get

$$F(1) - F(\delta) = -(1^{-1}) - [-(\delta^{-1})] = \delta^{-1} - 1$$

Now we consider

$$\underset{\delta \to 0+}{Lim}\ \delta^{-1} - 1$$

This limit is infinite; as δ approaches 0, its reciprocal blows up. Therefore,

$$\underset{\delta \to 0+}{Lim}\ \int_\delta^1 x^{-2}\, dx$$

is infinite, telling us that

$$\int_0^1 x^{-2}\, dx$$

is undefined.

- -

Here's a challenge!

Find the definite integral of $g(x) = x^{-2/3}$ over the open interval $-1 < x < 1$:

$$\int_{-1}^1 x^{-2/3}\, dx$$

Solution

The notion of the $-2/3$ power can be confusing. Let's be sure we know what it means. Our function g takes the input x, squares it, takes the cube root of that, and finally takes the reciprocal of that.

The interval of integration includes a singularity at $x = 0$, as shown in Fig. 14-4. We must divide the interval at the singularity, integrate on either side, and add the results. First, let's work out

$$\int_{-1}^{0} x^{-2/3}\, dx$$

We take a tiny positive ε, subtract it from the upper bound at $x = 0$, and obtain

$$\int_{-1}^{-\varepsilon} x^{-2/3}\, dx$$

The antiderivative without the constant of integration is

$$G(x) = 3x^{1/3}$$

When we evaluate this antiderivative from -1 to $-\varepsilon$, we get

$$G(-\varepsilon) - G(-1) = 3 \cdot (-\varepsilon)^{1/3} - 3 \cdot (-1)^{1/3} = 3 - 3\varepsilon^{1/3}$$

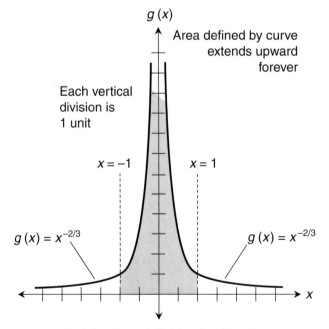

Figure 14-4   The function $g(x) = x^{-2/3}$ contains a singularity at $x = 0$, but the integral over the open interval from $x = -1$ to $x = 1$ is defined and finite. In this graph, each horizontal division represents 1/2 unit, and each vertical division represents 1 unit.

Remember that the 1/3 power is the cube root. Now we consider

$$\underset{\varepsilon \to 0+}{Lim} \; 3 - 3\varepsilon^{1/3}$$

As ε approaches 0 from the positive direction, the value of $\varepsilon^{1/3}$ also approaches 0 from the positive direction. Multiplying that vanishing quantity by 3 still gives us a term that approaches 0. That means the quantity $(3 - 3\varepsilon^{1/3})$ approaches 3. We can conclude that

$$\underset{\varepsilon \to 0+}{Lim} \int_{-1}^{-\varepsilon} x^{-2/3} \, dx = 3$$

Therefore

$$\int_{-1}^{0} x^{-2/3} \, dx = 3$$

Now let's work through the integral on the right-hand side of the singularity. This time, we want to calculate

$$\int_{0}^{1} x^{-2/3} \, dx$$

We take a tiny positive δ, add it to the lower bound at $x = 0$, and obtain

$$\int_{\delta}^{1} x^{-2/3} \, dx$$

Once again, the antiderivative of our function is

$$G(x) = 3x^{1/3}$$

When we evaluate this from δ to 1, we get

$$G(1) - G(\delta) = 3 \cdot 1^{1/3} - 3\delta^{1/3} = 3 - 3\delta^{1/3}$$

We want to determine

$$\underset{\delta \to 0+}{Lim} \; 3 - 3\delta^{1/3}$$

As δ approaches 0 from the positive direction, the value of $\delta^{1/3}$ also approaches 0 from the positive direction. Multiplying that vanishing quantity by 3 still gives us a term that approaches 0. That means the quantity $(3 - 3\delta^{1/3})$ approaches 3, so

$$\underset{\delta \to 0+}{Lim} \int_{\delta}^{1} x^{-2/3} \, dx = 3$$

Therefore

$$\int_{0}^{1} x^{-2/3} \, dx = 3$$

We've evaluated the integral from $x = -1$ up to the singularity, and found that it's equal to 3. We've also evaluated the integral from the singularity up to $x = 1$, and found that it's equal to 3. The integral from $x = -1$ to $x = 1$ is the sum of the two:

$$\int_{-1}^{1} x^{-2/3} \, dx = \int_{-1}^{0} x^{-2/3} \, dx + \int_{0}^{1} x^{-2/3} \, dx = 3 + 3 = 6$$

The shaded region in Fig. 14-4 portrays an infinitely tall geometric plane figure with a finite interior area.

- -

Infinite Intervals

We've seen what can happen when we try to integrate over an interval that's infinitely wide. Let's look further into this type of improper integral. We'll use the same two functions that we integrated from $x = -1$ to $x = 1$ in the previous section:

$$f(x) = x^{-2}$$

and

$$g(x) = x^{-2/3}$$

Example

A while ago, we figured out that the improper integral

$$\int_{-1}^{1} x^{-2} \, dx$$

is not defined. Now let's integrate the same function over the interval that includes all values of x less than -1, as shown in Fig. 14-5. We want to evaluate

$$\int_{-\infty}^{-1} x^{-2} \, dx$$

The antiderivative, as before, is

$$F(x) = -x^{-1}$$

Let's consider the definite integral

$$\int_{p}^{-1} x^{-2} \, dx$$

where p is a variable that decreases endlessly (that is, it grows larger negatively without restraint). To find the integral, we must determine

$$\mathrm{Lim}_{p \to -\infty} F(-1) - F(p)$$

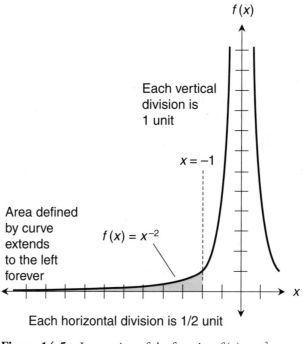

Figure 14-5   Integration of the function $f(x) = x^{-2}$ over an interval that includes all real values of x less than -1. In this graph, each horizontal division represents $1/2$ unit, and each vertical division represents 1 unit.

We calculate

$$F(-1) = -(-1)^{-1} = 1$$

As p becomes large negatively without bound, the value of $F(p)$ approaches 0. We can see this by inputting some numbers into the antiderivative and doing the arithmetic:

$$F(-10^2) = -(-10^2)^{-1} = 10^{-2}$$
$$F(-10^5) = -(-10^5)^{-1} = 10^{-5}$$
$$F(-10^8) = -(-10^8)^{-1} = 10^{-8}$$

Now we know that

$$\underset{p \to -\infty}{Lim} \; F(-1) - F(p) = 1 - 0 = 1$$

Therefore

$$\int_{-\infty}^{-1} x^{-2} \, dx = 1$$

Another example

We've already worked out the fact that

$$\int_{-1}^{1} x^{-2/3}\, dx = 6$$

Now let's integrate the same function, which we call $g(x)$, over the interval that includes all values of x larger than 1 as shown in Fig. 14-6. The improper integral is

$$\int_{1}^{\infty} x^{-2/3}\, dx$$

The antiderivative, as before, is

$$G(x) = 3x^{1/3}$$

Now let's consider the definite integral

$$\int_{1}^{q} x^{-2/3}\, dx$$

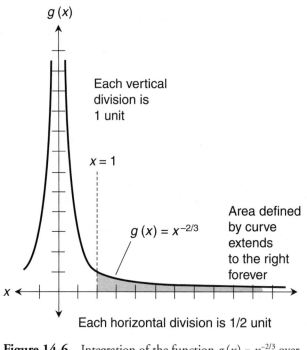

$g(x)$

Each vertical
division is
1 unit

$x = 1$

Area defined
by curve
extends
to the right
forever

$g(x) = x^{-2/3}$

x

Each horizontal division is 1/2 unit

Figure 14-6   Integration of the function $g(x) = x^{-2/3}$ over an interval that includes all real values of x larger than 1. In this graph, each horizontal division represents 1/2 unit, and each vertical division represents 1 unit.

where q is a variable that increases endlessly. To find the integral, we must determine

$$\underset{q \to \infty}{Lim}\ G(q) - G(1)$$

We calculate

$$G(1) = 3 \cdot 1^{1/3} = 3$$

As q becomes large positively without bound, $G(q)$ also increases without bound. We know this because the cube root of a growing positive real number grows, and there is no limit to how large that cube root can become. If we multiply such a number by 3, it gets bigger still. Because $G(q)$ increases without bound as q approaches infinity, we know that

$$\underset{q \to \infty}{Lim}\ G(q) - G(1)$$

is infinite. This tells us that the improper integral

$$\int_{1}^{\infty} x^{-2/3}\ dx$$

is undefined.

- -

Are you confused?

Do you wonder why we specify open intervals for definite integrals? We do this because it's the "safest" approach. Once in awhile, we'll encounter a definite integral where the function is undefined at one or both bounds. Technically, we can't integrate all the way up to a bound where a function is undefined. We can approach it, but we can't work directly with it. If we leave the endpoints out of integration intervals, we avoid this problem. (The boundary never has any "area" anyway, so we don't need to include it.)

Do you remember what the various interval notations mean? If not, here's a refresher. If we have a variable x and two constants a and b where $a < b$, then

(a,b) means the open interval where $a < x < b$

$[a,b)$ means the half-open interval where $a \le x < b$

$(a,b]$ means the half-open interval where $a < x \le b$

$[a,b]$ means the closed interval where $a \le x \le b$

Are you astute?

Have you noticed that the function f, portrayed in Figs. 14-3 and 14-5, has an undefined integral over the interval $(-1,1)$, but a defined one over the interval $(-\infty,-1)$? In the first situation, the region is infinitely tall and has infinite area. In the second situation, the region is infinitely wide but has finite area.

Have you also noticed that the function g, shown in Figs. 14-4 and 14-6, behaves in precisely the opposite manner? It has a defined integral over the interval $(-1,1)$, but an undefined one over the interval $(1,\infty)$. The infinitely tall region has finite area; the infinitely wide one has infinite area.

The curves for *f* and *g* look almost the same in the graphs, but they're a lot different mathematically. Whether a singularity gives us a finite integral or an undefined one depends on how "fast" the curve approaches the line where the singularity exists. If the curve approaches the line fast enough, then the integral is finite. Otherwise it's undefined. Usually, the only way to know what will happen when we integrate over an interval with a singularity is to try it and see. If the integral is defined, we'll come up with a limit that converges. If the integral is undefined, we'll get a limit that is infinite.

How's your imagination?

In all of the examples here, the singularities are along the coordinate axes. But there are functions with singularities that don't correspond to either axis. Can you think of any?

Practice Exercises

This is an open-book quiz. You may (and should) refer to the text as you solve these problems. Don't hurry! You'll find worked-out answers in App. B. The solutions in the appendix may not represent the only way a problem can be figured out. If you think you can solve a particular problem in a quicker or better way than you see there, by all means try it!

1. Evaluate the following integral. If it's defined, state the value.

$$\int_0^1 -7x^{-8} \, dx$$

2. Evaluate the following integral. If it's defined, state the value.

$$\int_{-\infty}^{-1} -7x^{-8} \, dx$$

3. Evaluate the following integral. If it's defined, state the value.

$$\int_0^2 x^{-3} \, dx$$

4. Evaluate the following integral. If it's defined, state the value.

$$\int_2^\infty x^{-3} \, dx$$

5. Evaluate the following integral. If it's defined, state the value.

$$\int_{-3}^0 x^{-3/5} \, dx$$

6. Evaluate the following integral. If it's defined, state the value.

$$\int_{-3}^{-\infty} x^{-3/5} \, dx$$

7. By looking at graphs of the functions, we can see *immediately* that one of the following integrals is undefined. Which one?

$$\int_{2}^{\infty} x^{-5} \, dx$$

$$\int_{2}^{\infty} -5x \, dx$$

8. Work out the integral stated in Prob. 7 that looks as if it *might* be defined. If it turns out not to be defined, indicate why. If it's defined, state the value.

9. Evaluate the following integral as if there were no singularity in the interval of integration. That's the wrong way to do it, of course! But try it anyway, and see what happens:

$$\int_{-1}^{1} x^{-3} \, dx$$

   Draw a graph that lends intuitive support to the "solution" here.

10. Evaluate the integral stated in Prob. 9 using the correct approach.

CHAPTER 15

Integrating Polynomial Functions

In this chapter, we'll integrate functions that appear in polynomial form. Working out such problems involves knowing which rules to apply, and when to apply them. The rest is, as my teachers and professors used to say, "mere busywork."

Three Rules Revisited

In Chap. 12, we learned three important rules that apply to indefinite integrals. They also apply to definite integrals under certain conditions.

The old rules

The original rules involve multiplying a function by a constant, taking the negative of a function, and adding two or more functions. Here they are again, in brief.

- If we multiply a function $f(x)$ by a constant k and then integrate with respect to x, we get the same thing as we do by integrating $f(x)$ with respect to x and then multiplying by the constant:

$$\int k\,[f(x)]\ dx = k \int f(x)\ dx$$

- If we integrate the negative of a function $f(x)$ with respect to x, we get the same thing as we do by integrating $f(x)$ with respect to x and then taking the negative:

$$\int -[f(x)]\ dx = -\int f(x)\ dx$$

- Imagine that f_1, f_2, f_3, . . . , and f_n are functions of a variable x. The integral of the sum of the functions with respect to x is equal to the sum of the integrals of the functions with respect to x:

$$\int [f_1(x) + f_2(x) + f_3(x) + \cdots + f_n(x)]\ dx$$

$$= \int f_1(x)\ dx + \int f_2(x)\ dx + \int f_3(x)\ dx + \cdots + \int f_n(x)\ dx$$

The new rules

We can modify these rules to work with definite integrals, but we must be careful. We have to stay with the same interval of integration all the time, and we must always integrate in the same direction. Suppose the bounds of the interval are $x = a$ and $x = b$, and we integrate starting at a and finishing at b. Here are the new rules.

- If we multiply a function $f(x)$ by a real constant k and then take the definite integral with respect to x, we get the same result as we do when we integrate with respect to x and then multiply by the constant:

$$\int_a^b k\,[f(x)]\ dx = k \int_a^b f(x)\ dx$$

- If we take the definite integral of the negative of a function $f(x)$ with respect to x, we get the same result as we do when we integrate the function with respect to x and then take the negative:

$$\int_a^b -[f(x)]\ dx = -\int_a^b f(x)\ dx$$

- The definite integral of the sum of two functions $f_1(x)$ and $f_2(x)$ with respect to x is equal to the sum of their definite integrals with respect to x. If f_1 and f_2 are integrable functions of the same variable x, then

$$\int_a^b [f_1(x) + f_2(x)]\ dx = \int_a^b f_1(x)\ dx + \int_a^b f_2(x)\ dx$$

This last rule, which we can call the *sum rule for definite integrals,* works when we add up any finite number of functions, as long as:

- The interval of integration is the same for each function
- All the functions are integrable over the interval
- All the integrals are done in the same direction

- -

Are you inquisitive?

You might ask, "why does the above sum-of-integrals rule work only when both, or all, of the integration intervals are the same, and only when we integrate in the same direction all the time?" The best way to answer this question is to see what happens if we violate either or both of these restrictions.

Suppose we want to consolidate a sum of two integrals into a single integral of the sum of both functions, but the intervals differ. Here's an example:

$$\int_3^5 7x^2 \, dx + \int_4^6 x^{2/3} \, dx$$

We can work out the two integrals separately (assuming they're both defined) and then add them. That's not a problem, because they're plain numbers. But if we want to consolidate the above expression into a single integral, we have no way of knowing what the interval should be. Should we use (3,5)? No; that won't let us do anything with the values of x between 5 and 6, which are included in the second integral. Should we use (4,6)? No; that leaves out the values between 3 and 4 in the first integral. How about (3,6)? No; it's too wide for either integral alone. What about (4,5)? No; that leaves out some of the values when we consider the integrals together. Now think about what happens if we integrate the functions in opposite directions:

$$\int_3^5 7x^2 \, dx + \int_5^3 x^{2/3} \, dx$$

We can rewrite this as a difference of integrals done in the same direction:

$$\int_3^5 7x^2 \, dx - \int_3^5 x^{2/3} \, dx$$

That's equal to

$$\int_3^5 (7x^2 - x^{2/3}) \, dx$$

Here's a challenge!

Consider the following two functions over the interval from $x = 2$ to $x = 4$:

$$g(x) = 8x$$

and

$$h(x) = -6x$$

Integrate these functions individually and add the results. Then add the functions and show that the integral of the sum is equal to the sum of the integrals.

Solution

First, let's figure out the definite integral

$$\int_2^4 8x \, dx$$

The basic antiderivative is

$$G(x) = 4x^2$$

When we evaluate this from $x = 2$ to $x = 4$, we get

$$G(4) - G(2) = 4 \cdot 4^2 - 4 \cdot 2^2 = 48$$

Next, we figure out

$$\int_2^4 -6x\ dx$$

The basic antiderivative is

$$H(x) = -3x^2$$

When we evaluate this from $x = 2$ to $x = 4$, we get

$$H(4) - H(2) = -3 \cdot 4^2 - (-3 \cdot 2^2) = -36$$

When we add these results, we get 48 + (−36), which is 12. Now, let's add the two original functions and then integrate their sum. If we call the sum $q(x)$, then

$$g(x) + h(x) = q(x) = 8x + (-6x) = 2x$$

We want to work out

$$\int_2^4 2x\ dx$$

The basic antiderivative is

$$Q(x) = x^2$$

When we evaluate this from $x = 2$ to $x = 4$, we get

$$Q(4) - Q(2) = 4^2 - 2^2 = 12$$

- -

Indefinite-Integral Situations

We're now able to integrate any sum of monomial terms, where each term raises the independent variable to a real power and then multiplies it by a real constant.

Example

Let's resolve the indefinite integral

$$\int (3z^2 + 2z - 6)\ dz$$

We have a sum of three monomial functions. Let's call them

$$f_1(z) = 3z^2$$
$$f_2(z) = 2z$$
$$f_3(z) = -6$$

The indefinite integrals are

$$\int f_1(z) \, dz = \int 3z^2 \, dz = z^3 + c_1$$

$$\int f_2(z) \, dz = \int 2z \, dz = z^2 + c_2$$

$$\int f_3(z) \, dz = \int -6 \, dz = -6z + c_3$$

where c_1, c_2, and c_3 are the constants of integration. These constants are not necessarily all the same. The original integral is

$$\int (3z^2 + 2z - 6) \, dz = \int f_1(z) \, dz + \int f_2(z) \, dz + \int f_3(z) \, dz$$

$$= \int 3z^2 \, dz + \int 2z \, dz + \int -6 \, dz = z^3 + c_1 + z^2 + c_2 - 6z + c_3$$

$$= z^3 + z^2 - 6z + c_1 + c_2 + c_3$$

We can consolidate $c_1 + c_2 + c_3$ into a single constant c, so

$$\int (3z^2 + 2z - 6) \, dz = z^3 + z^2 - 6z + c$$

Another example

Let's find the indefinite integral

$$\int (5y^{2/3} - y^{-1/2} - 4y^{-2}) \, dy$$

This is the integral of a sum of three monomial functions. Let's call them

$$f_1(y) = 5y^{2/3}$$
$$f_2(y) = -y^{-1/2}$$
$$f_3(y) = -4y^{-2}$$

The indefinite integrals are

$$\int f_1(y) \, dy = \int 5y^{2/3} \, dy = 3y^{5/3} + c_1$$

$$\int f_2(y) \, dy = \int -y^{-1/2} \, dy = -2y^{1/2} + c_2$$

$$\int f_3(y) \, dy = \int -4y^{-2} \, dy = 4y^{-1} + c_3$$

where c_1, c_2, and c_3 are the constants of integration. The original integral is

$$\int (5y^{2/3} - y^{-1/2} - 4y^{-2}) \, dy = \int f_1(y) \, dy + \int f_2(y) \, dy + \int f_3(y) \, dy$$

$$= \int 5y^{2/3} \, dy + \int -y^{-1/2} \, dy + \int -4y^{-2} \, dy$$

$$= 3y^{5/3} + c_1 - 2y^{1/2} + c_2 + 4y^{-1} + c_3$$

$$= 3y^{5/3} - 2y^{1/2} + 4y^{-1} + c_1 + c_2 + c_3$$

Consolidating $c_1 + c_2 + c_3$ into a single constant c, we have

$$\int (5y^{2/3} - y^{-1/2} - 4y^{-2})\ dy = 3y^{5/3} - 2y^{1/2} + 4y^{-1} + c$$

- -

Are you confused?

You might ask, "why not leave the constants of integration out until the end of the process, and then add in a single consolidated constant?" In most cases, we can do that. In scientific problems, the values of the individual constants might be significant, but in pure theoretical mathematics, they aren't.

Here's a challenge!

Determine the indefinite integral

$$\int (x + 3)(2x^2 - 4x - 5)\ dx$$

Solution

When we multiply polynomials, we get another polynomial. In this case,

$$(x + 3)(2x^2 - 4x - 5) = 2x^3 + 2x^2 - 17x - 15$$

We can therefore rewrite the above integral as

$$\int (2x^3 + 2x^2 - 17x - 15)\ dx$$

Let's take the indefinite integrals of each monomial from left to right, omit the individual constants of integration, and then add c as the last step. We get

$$\int 2x^3\ dx + \int 2x^2\ dx + \int -17x\ dx + \int -15\ dx$$
$$= (1/2)x^4 + (2/3)x^3 - (17/2)x^2 - 15x + c$$

That's it! We've determined that

$$\int (x + 3)(2x^2 - 4x - 5)\ dx = (1/2)x^4 + (2/3)x^3 - (17/2)x^2 - 15x + c$$

- -

Definite-Integral Situations

Once we know how to evaluate the indefinite integrals of polynomial functions, it's easy to work out definite integrals most of the time. But if the *integrand* (the function that we're integrating) has a singularity in the interval, then we have to use the rules we learned in Chap. 14 when we work out the definite integral.

Example

Let's find the definite integral

$$\int_2^4 (3z^2 + 2z - 6)\, dz$$

From the previous section, we know that

$$\int (3z^2 + 2z - 6)\, dz = z^3 + z^2 - 6z + c$$

We want to evaluate this from $z = 2$ to $z = 4$. Before we start, we must be sure that the integrand doesn't have any singularities in the interval for which $2 < z < 4$. To do that, we must look at each individual term as a function. Let's call the integrand f, so we have

$$f(z) = 3z^2 + 2z - 6$$

This breaks down into the three monomial functions

$$f_1(z) = 3z^2$$
$$f_2(z) = 2z$$
$$f_3(z) = 6$$

The graph of f_1 is a parabola that opens upward. The graph of f_2 is a line through the origin with a slope of 2. The graph of f_3 is a line with a slope of 0. (You might want to sketch the graphs to see these facts.) None of these three functions blows up in the interval $2 < z < 4$, so it's okay to evaluate the integral by plugging in the boundary numbers to the antiderivative $F(z)$ and then subtracting the results. That antiderivative, leaving out the constant of integration, is

$$F(z) = z^3 + z^2 - 6z$$

When we input 4 (the upper bound of the definite integral) here, we get

$$F(4) = 4^3 + 4^2 - 6 \cdot 4 = 56$$

When we input 2 (the lower bound of the definite integral), we get

$$F(2) = 2^3 + 2^2 - 6 \cdot 2 = 0$$

The difference is

$$F(4) - F(2) = 56 - 0 = 56$$

so we've determined that

$$\int_2^4 (3z^2 + 2z - 6)\, dz = 56$$

Another example

Let's find the following definite integral. The arithmetic is messy, so we'll approximate our answer with a calculator to three decimal places.

$$\int_{1}^{2} (5y^{2/3} - y^{-1/2} - 4y^{-2})\, dy$$

In the previous section, we found that

$$\int (5y^{2/3} - y^{-1/2} - 4y^{-2})\, dy = 3y^{5/3} - 2y^{1/2} + 4y^{-1} + c$$

We want to evaluate this from $y = 1$ to $y = 2$. Let's call the integrand f, so

$$f(y) = 5y^{2/3} - y^{-1/2} - 4y^{-2}$$

This breaks down into

$$f_1(y) = 5y^{2/3}$$
$$f_2(y) = -y^{-1/2}$$
$$f_3(y) = -4y^{-2}$$

The graph of f_1 is a parabola-like curve opening upward. The graph of f_2 is a curve with a singularity at $y = 0$, but we're okay over the interval where $1 < y < 2$. The graph of f_3 also has a singularity at $y = 0$, but we're okay in the interval where $1 < y < 2$. (Feel free to sketch graphs of these three monomial functions.) When we leave out the constant of integration, the antiderivative of f is

$$F(y) = 3y^{5/3} - 2y^{1/2} + 4y^{-1}$$

When we input 2 here, we get

$$F(2) = 3 \cdot 2^{5/3} - 2 \cdot 2^{1/2} + 4 \cdot 2^{-1}$$

When we work out the above expression and round off to three decimal places at the end of the calculation, we get

$$F(2) \approx 8.696$$

When we input 1 to the antiderivative F, we get

$$F(1) = 3 \cdot 1^{5/3} - 2 \cdot 1^{1/2} + 4 \cdot 1^{-1} = 5$$

The difference is

$$F(2) - F(1) \approx 8.696 - 5 \approx 3.696$$

We have determined that

$$\int_{1}^{2} (5y^{2/3} - y^{-1/2} - 4y^{-2}) \, dy \approx 3.696$$

- -

Are you confused?

"What would happen," you wonder, "if we try to evaluate the above integral from 0 to 1? Two of the three monomial terms, considered as individual functions, are singular at $y = 0$. Does that mean the entire integral is undefined?" Maybe, and maybe not! The only way to find out is to work out the definite integrals for each monomial function individually. If they're all defined, we can add them up to get the final answer. If any of them is undefined, then the entire integral is undefined.

Here's a challenge!

Evaluate the definite integral

$$\int_{0}^{1} (x^{-1/5} - x^{-1/4}) \, dx$$

Solution

The integrand is a sum of two monomial functions. Let's call them

$$f_1(x) = x^{-1/5}$$

and

$$f_2(x) = -x^{-1/4}$$

Both of these functions are singular at $x = 0$. That's the lower bound of our integration interval. One of the functions blows up positively, while the other one blows up negatively. Let's go through four steps:

- Split the original integral into two separate improper integrals.
- Evaluate those two integrals independently.
- If either of them is undefined, conclude that the entire integral is undefined.
- If they are both defined, add them to get the final result.

When we break the original integral into a sum of the integrals of the monomial functions, we get

$$\int_{0}^{1} x^{-1/5} \, dx + \int_{0}^{1} -x^{-1/4} \, dx$$

Let's evaluate the left-hand addend first. We want to find

$$\int_{0}^{1} x^{-1/5} \, dx$$

We take a tiny positive δ, add it to the lower limit at $x = 0$, and obtain

$$\int_{\delta}^{1} x^{-1/5} \, dx$$

The basic antiderivative is

$$F_1(x) = (5/4)x^{4/5}$$

When we evaluate this from δ to 1, we get

$$F_1(1) - F_1(\delta) = (5/4) \cdot 1^{4/5} - (5/4)\delta^{4/5} = 5/4 - (5/4)\delta^{4/5}$$

Now we must determine

$$\lim_{\delta \to 0+} 5/4 - (5/4)\delta^{4/5}$$

As δ approaches 0 from the positive direction, the value of $\delta^{4/5}$ also approaches 0 from the positive direction. Multiplying by 5/4 gives us a term that still approaches 0. That means the quantity $[5/4 - (5/4)\delta^{4/5}]$ approaches 5/4, so

$$\lim_{\delta \to 0+} \int_{\delta}^{1} x^{-1/5} \, dx = 5/4$$

Therefore, the left-hand addend in the "big sum" is

$$\int_{0}^{1} x^{-1/5} \, dx = 5/4$$

Now let's evaluate the right-hand addend in our sum of improper integrals. This time, our task is to figure out

$$\int_{0}^{1} -x^{-1/4} \, dx$$

We take a tiny positive δ, add it to the lower limit at $x = 0$, and get

$$\int_{\delta}^{1} -x^{-1/4} \, dx$$

The basic antiderivative is

$$F_2(x) = (-4/3)x^{3/4}$$

When we evaluate this from δ to 1, we get

$$F_2(1) - F_2(\delta) = (-4/3) \cdot 1^{3/4} - (-4/3)\delta^{3/4} = -4/3 + (4/3)\delta^{3/4} = (4/3)\delta^{3/4} - 4/3$$

Now we consider

$$\lim_{\delta \to 0+} (4/3)\delta^{3/4} - 4/3$$

As δ approaches 0 from the positive direction, the value of $\delta^{3/4}$ also approaches 0 from the positive direction. Multiplying by 4/3 gives us a term that still approaches 0. That means the quantity $[(4/3)\delta^{3/4} - 4/3]$ approaches $-4/3$, so

$$\underset{\delta \to 0+}{Lim} \int_{\delta}^{1} -x^{-1/4} \, dx = -4/3$$

This tells us that the right-hand addend in our "big sum" is

$$\int_{0}^{1} -x^{-1/4} \, dx = -4/3$$

We have now found the values of both addends in the expression

$$\int_{0}^{1} x^{-1/5} \, dx + \int_{0}^{1} -x^{-1/4} \, dx$$

The left-hand integral is 5/4, and the right-hand one is $-4/3$. They're both defined, but they have opposite sign. (That's interesting when you graph both monomial functions and see how the areas combine. Make a sketch if you're motivated!) We can apply the rule for adding two definite integrals to get

$$\int_{0}^{1} (x^{-1/5} - x^{-1/4}) \, dx = 5/4 - 4/3$$

which simplifies to

$$\int_{0}^{1} (x^{-1/5} - x^{-1/4}) \, dx = -1/12$$

- -

Practice Exercises

This is an open-book quiz. You may (and should) refer to the text as you solve these problems. Don't hurry! You'll find worked-out answers in App. B. The solutions in the appendix may not represent the only way a problem can be figured out. If you think you can solve a particular problem in a quicker or better way than you see there, by all means try it!

1. Work out the following sum of integrals:

$$\int_{5}^{6} 3x^2 \, dx + \int_{5}^{6} -12x^2 \, dx$$

Show that this is the same as

$$\int_{5}^{6} -9x^2 \, dx$$

2. Work out the following sum of integrals:

$$\int_{5}^{6} 3x^2 \, dx + \int_{6}^{5} -12x^2 \, dx$$

Note the difference in the direction of integration between the two integrals. Show that the above sum is the same as

$$\int_5^6 15x^2 \, dx$$

3. Work out the following difference of integrals:

$$\int_5^6 -12x^2 \, dx - \int_5^6 3x^2 \, dx$$

Show that this is the same as

$$\int_5^6 -15x^2 \, dx$$

4. Find the following indefinite integral:

$$\int (x^{-2} + x^{-3} + x^{-4}) \, dx$$

5. Find the following indefinite integral:

$$\int (7y)(6y - 4)(-2y + 3) \, dy$$

6. Find the following indefinite integral:

$$\int (z^8 - 1)(-3z^3 + 7z^2) \, dz$$

7. Calculate the following definite integral:

$$\int_0^1 (4x^3 - 5x^2 + 7x - 4) \, dx$$

8. Calculate the following definite integral:

$$\int_{-1}^{-2} (x^{-2} + x^{-3} + x^{-4}) \, dx$$

9. Calculate the following definite integral:

$$\int_0^1 (x - 2)^3 \, dx$$

10. Calculate the following definite integral:

$$\int_{-1}^1 (x^{1/3} + 4)^2 \, dx$$

16

Areas between Graphs

When we find the definite integral of a function over an interval, there's a geometric equivalent in rectangular coordinates: the area between the graph and the independent-variable axis. Now let's learn how to figure out the area between the graphs of two functions over an interval.

Line and Curve

Consider two functions, plotted as graphs in the xy-plane. One of the graphs is a straight line represented by

$$f(x) = x + 2$$

This line has a slope of 1 and passes through the points $(-2,0)$ and $(0,2)$. The other graph is a curve whose function is

$$g(x) = x^2 - 4$$

The curve is a parabola that opens upward, passing through $(-2,0)$, $(0,-4)$, and $(2,0)$. Both of these functions are graphed in Fig. 16-1. Let's find the total area of the region between the line and the curve. That's the shaded zone.

Solve the system

To define the enclosed region between the graphs of two functions, we must find both (or all) of the points where the graphs intersect. We know that at any intersection point,

$$f(x) = g(x)$$

If we put the value of f on the left-hand side of an equation and the value of g on the right-hand side in the situation of Fig. 16-1, we get

$$x + 2 = x^2 - 4$$

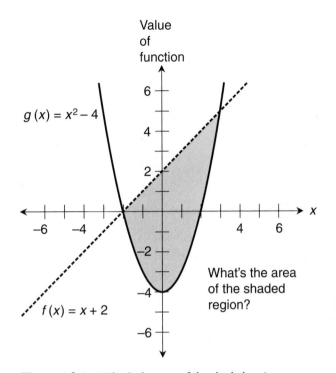

Figure 16-1 What's the area of the shaded region between the straight line and the parabola?

This can be rearranged to obtain the quadratic equation

$$x^2 - x - 6 = 0$$

which can be factored into

$$(x + 2)(x - 3) = 0$$

The roots of this equation are $x = -2$ or $x = 3$, the values of x that make one or the other factor equal to 0. When we plug $x = -2$ into the original functions, we get

$$f(-2) = -2 + 2 = 0$$

and

$$g(-2) = (-2)^2 - 4 = 0$$

This tells us that the graphs intersect at the point $(-2,0)$. When we plug $x = 3$ into the original functions, we obtain

$$f(3) = 3 + 2 = 5$$

and

$$g(3) = 3^2 - 4 = 5$$

This tells us that the graphs intersect at (3,5). Now we know the points that mark the lower and upper bounds of the region for which we want to calculate the area. These are the only two points where the curves intersect, because the equation we got from both functions is a quadratic, which can't have more than two real roots.

Work out the geometry

Now we must ask ourselves, "How do we calculate the area of an irregular region such as the shaded zone in Fig. 16-1?" In this case, we can break the region into three different plane figures, two of which contribute positively and one of which contributes negatively. Then we can add the areas that contribute positively and take away the area that contributes negatively. Let's keep an eye on Fig. 16-2 as we go through this process.

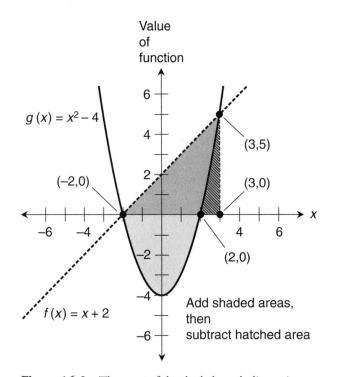

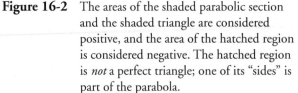

Figure 16-2 The areas of the shaded parabolic section and the shaded triangle are considered positive, and the area of the hatched region is considered negative. The hatched region is *not* a perfect triangle; one of its "sides" is part of the parabola.

As the first step, consider the lightly shaded region in Fig. 16-2 that's bounded on the top by the x axis and on the bottom by the parabola. This region lies between the points where $x = -2$ and $x = 2$. We can find its area by evaluating

$$\int_{-2}^{2} (x^2 - 4) \, dx$$

The integrand is a polynomial. The basic antiderivative is

$$G(x) = x^3/3 - 4x$$

When we evaluate this from $x = -2$ to $x = 2$, we obtain

$$G(2) - G(-2) = [(2^3/3 - 4 \cdot 2)] - [(-2)^3/3 - 4 \cdot (-2)] = -32/3$$

This area is negative, because the region is entirely below the x axis. But we want to know the *true geometric area* between the curves, so we must consider it positive! The true geometric area of the lightly shaded region between the curve and the x axis in Fig. 16-2 is 32/3 square units.

Now let's consider the area of the triangle bounded by the points $(-2,0)$, $(3,0)$, and $(3,5)$. We can find this in either of two ways. We can evaluate the definite integral

$$\int_{-2}^{3} (x + 2) \, dx$$

or we can use ordinary geometry to work out the area of the triangle. Let's use geometry; it's simpler! The base length of the triangle is 5 units, and the height is 5 units. The product of these is 25 units. Dividing by 2 gives us 25/2 square units as the area of the dark-shaded triangle in Fig. 16-2. If we add this to the area of the lightly shaded region between the x axis and the curve, we get 32/3 + 25/2. By arithmetic, we find that this total area is 139/6 square units.

We've found an area, but the region is larger than the one we're interested in. It encompasses the sum of the areas of two zones. The first zone lies above the parabola and below the x axis. The second zone lies below the line and above the x axis. The region of which we want to find the area is a little smaller. (Look again at Fig. 16-1, and compare.) To find that area, we must *subtract* the area of the *hatched* region that lies above the x axis but below the parabola in Fig. 16-2. That area is

$$\int_{2}^{3} (x^2 - 4) \, dx$$

We've figured out $G(x)$, so we can plug in the numbers to get

$$G(3) - G(2) = (3^3/3 - 4 \cdot 3) \quad (2^3/3 - 4 \cdot 2) - 7/3$$

The area of the region between the line and the curve is 139/6 square units minus 7/3 square units, or 125/6 square units.

Are you astute?

You might ask, "Isn't the above scheme unnecessarily complicated? Can we subtract the function represented by the parabola (the lower graph) from the one represented by the line (the upper graph), and then integrate the resulting *difference function* between the points where the graphs intersect?" That's a good question. The answer is yes. We can indeed!

Imagine slicing the region between the line and the curve into thin rectangles as shown in Fig. 16-3, and then working out the definite integral using the Riemann limit scheme we learned in Chap. 11. But this time, instead of using the x axis as the reference, suppose we use the graph that bounds the region on the bottom. In this case, it's the parabola. As we make the rectangles thinner and more numerous, we approximate the area of the region between the curves. We can find the limit of the sum of the rectangle areas as their widths approach 0, and we'll get exactly the area we want to find. In effect, we integrate the upper function over the interval by following the contour of lower function's graph instead of the x axis.

Here's a challenge!

Find the area between the line and the curve shown in Fig. 16-1 by subtracting the lower function g from the upper function f, and then integrating over the interval between the points where the graphs intersect.

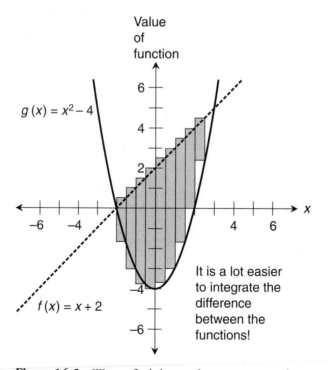

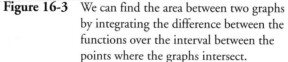

Figure 16-3   We can find the area between two graphs by integrating the difference between the functions over the interval between the points where the graphs intersect.

Solution

When we subtract the function for the parabola from the function for the straight line, we get a polynomial function. Let's call it $p(x)$. Then

$$p(x) = f(x) - g(x) = (x + 2) - (x^2 - 4) = -x^2 + x + 6$$

We want to integrate over the interval between $x = -2$ and $x = 3$, which are the x-values of the points where the line and the curve intersect. That means we must find

$$\int_{-2}^{3} (-x^2 + x + 6)\ dx$$

The basic antiderivative is

$$P(x) = -x^3/3 + x^2/2 + 6x$$

When we evaluate this from $x = -2$ to $x = 3$, we obtain

$$P(3) - P(-2) = [-(3^3/3) + 3^2/2 + 6 \cdot 3] - [-(-2)^3/3 + (-2)^2/2 + 6 \cdot (-2)] = 125/6$$

- -

Two Curves

When we want to find the area between the graphs of two polynomial functions, the hybrid geometry-and-calculus method won't work. We must use calculus exclusively, although there might be more than one way to do it.

Two "mirrored" parabolas

Let's find the area between the parabolas represented by

$$f(x) = x^2/2 - 2$$

and

$$g(x) = -x^2/2 + 2$$

From algebra, we know that the curve for f opens upward, and the curve for g opens downward. We can find the points where the parabolas intersect by setting $f(x) = g(x)$ and solving the resulting equation for x, like this:

$$x^2/2 - 2 = -x^2/2 + 2$$

Adding $x^2/2$ to both sides, we get

$$x^2 - 2 = 2$$

We can subtract 2 from each side, obtaining

$$x^2 - 4 = 0$$

This factors into

$$(x+2)(x-2) = 0$$

The roots here are $x = -2$ or $x = 2$. We can find the points where the parabolas intersect the dependent-variable axis by setting $x = 0$ for either function. When we do that, we get

$$f(0) = 0^2/2 - 2 = -2$$

and

$$g(0) = -0^2/2 + 2 = 2$$

The *vertices* of the parabolas for the functions f and g are at $(0,-2)$ and $(0,2)$, respectively. The parabolas intersect at $(-2,0)$ and $(2,0)$. Figure 16-4 shows the graphs of these two functions, along with the region they enclose.

Integrate the difference function

In this situation, the parabola for g is the upper curve, while the parabola for f is the lower curve. To find the area of the enclosed region, we must integrate the difference function

$$p(x) = g(x) - f(x) = (-x^2/2 + 2) - (x^2/2 - 2) = -x^2 + 4$$

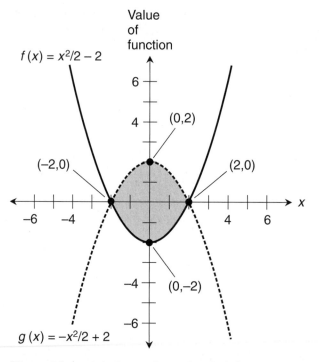

Figure 16-4   Calculus can be used to find the area between two curves. In this example, the curves are "mirror-image" parabolas.

Remember, we must subtract the function for the lower curve from the function for the upper curve to get the difference function. This ensures that we'll end up with a positive total area when we integrate over the interval in the positive direction. The interval of interest is (−2,2), so the integral we want to evaluate is

$$\int_{-2}^{2} (-x^2 + 4)\, dx$$

The basic antiderivative of the difference function is

$$P(x) = -x^3/3 + 4x$$

When we evaluate this from $x = -2$ to $x = 2$, we obtain

$$P(2) - P(-2) = [-(2^3/3) + 4 \cdot 2] - [-(-2)^3/3 + 4 \cdot (-2)] = 32/3$$

- -

Are you confused?

You might ask, "Can't we integrate the functions separately over the interval, take the absolute values of the integrals, and then add them? Won't that give us the same result as the above method?" Yes, that will work here, but only because both endpoints of the interval lie on the *x* axis.

Here's a challenge!

Find the area between the curves for $f(x)$ and $g(x)$, as above and as shown in Fig. 16-4, once again. This time, integrate the functions separately over the interval $-2 < x < 2$. Consider the areas defined by either curve (one above the *x* axis and the other below it) as positive, and then add them to get the total area of the region enclosed by the parabolas.

Solution

Let's begin by integrating the lower function, $f(x)$, with respect to *x* over the interval where $-2 < x < 2$. We want to figure out

$$\int_{-2}^{2} (x^2/2 - 2)\, dx$$

Because this region lies entirely below the *x* axis, the definite integral will be negative. But we want to consider the area positive, so we must evaluate

$$-\int_{-2}^{2} (x^2/2 - 2)\, dx$$

That's the same as

$$\int_{-2}^{2} -(x^2/2 - 2)\, dx$$

which can be simplified to

$$\int_{-2}^{2} (-x^2/2 + 2)\, dx$$

Now, notice that

$$g(x) = -x^2/2 + 2$$

That's a convenient coincidence! When we integrate $f(x)$ over the interval and then take the negative, we're actually finding

$$\int_{-2}^{2} g(x)\, dx$$

Therefore, our total area is twice what we get if we integrate $g(x)$ over the interval. That means the true geometric area between the curves is

$$2 \int_{-2}^{2} (-x^2/2 + 2)\, dx$$

We've mathematically verified a fact that appears obvious in Fig. 16-4: The upper and lower halves of the region between the curves are exact duplicates. Let's do the calculations. The antiderivative of $g(x)$, leaving out the constant of integration, is

$$G(x) = -x^3/6 + 2x$$

When we evaluate this from $x = -2$ to $x = 2$, we obtain

$$G(2) - G(-2) = [-(2^3/6) + 2 \cdot 2] - [-(-2)^3/6 + 2 \cdot (-2)] = 16/3$$

This tells us that

$$\int_{-2}^{2} (-x^2/2 + 2)\, dx = 16/3$$

The total area of the region between the curves is twice this, or

$$2 \int_{-2}^{2} (-x^2/2 + 2)\, dx = 32/3$$

- -

Singular Curves

There are infinitely many ways that two graphs can combine to create enclosed regions. Once in awhile, we'll encounter an enclosed region that has one or more infinite dimensions, but whose area is finite. Let's look at a situation of that sort.

An infinitely tall zone

Consider the following two functions. One of them has a graph that's a straight line. The other has a graph that's a curve with a singularity. Here they are:

$$f(x) = x$$

and

$$g(x) = x^{-2/3}$$

Imagine the region bounded by the graph of *f* on the bottom, the graph of *g* on the right and on top, and the dependent-variable axis on the left. Our mission is to find the area of this region, shown as the shaded zone in Fig. 16-5.

Determine the interval

Before we do any calculus, we must know the interval over which we're going to integrate. On the left, the interval is bounded by the dependent-variable axis, so the left-hand bound is $x = 0$. We can see that the line and the curve intersect somewhere. The *x*-value of the intersection point will be the right-hand end of our integration interval. To figure out that *x*-value, let's set the functions equal and then solve for *x*. Here's the equation we get:

$$x = x^{-2/3}$$

Figure 16-5 shows us that the line and the curve don't intersect at the point where $x = 0$. Therefore, $x = 0$ is not one of the roots of the above equation. That's convenient, because it means that we can multiply both sides by x^{-1}. When we do that, we obtain

$$(x^{-1})(x) = (x^{-1})(x^{-2/3})$$

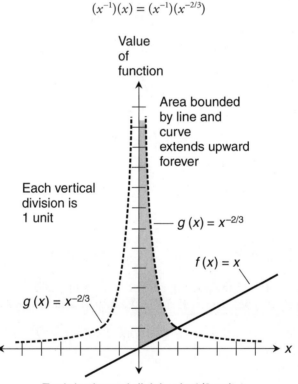

Figure 16-5   In this example, we can calculate the area of the region between a line and a curve, even though the curve blows up at one end of the interval. Each horizontal division represents 1/2 unit. Each vertical division represents 1 unit.

which simplifies to

$$1 = x^{-5/3}$$

The only real number that satisfies this equation is $x = 1$. (You might suspect that -1 would work too, but it doesn't. Plug it in and see for yourself.) Now we know that our interval of integration is $0 < x < 1$.

Integrate the difference function

In this scenario, the upper graph represents g and the lower graph represents f. That means we must subtract f from g to get the difference function. If we call our difference function p as usual, then

$$p(x) = g(x) - f(x) = x^{-2/3} - x$$

We want to integrate p over the interval $0 < x < 1$, which is bounded on the left by the singularity. That integral is

$$\int_0^1 (x^{-2/3} - x) \, dx$$

Let's take a tiny positive δ, add it to the lower limit at $x = 0$ where the singularity occurs, and obtain

$$\int_\delta^1 (x^{-2/3} - x) \, dx$$

The basic antiderivative is

$$P(x) = 3x^{1/3} - x^2/2$$

When we evaluate this from δ to 1, we get

$$P(1) - P(\delta) = (3 \cdot 1^{1/3} - 1^2/2) - (3\delta^{1/3} - \delta^2/2) = 5/2 - (3\delta^{1/3} - \delta^2/2)$$

Now we must figure out

$$\lim_{\delta \to 0+} 5/2 - (3\delta^{1/3} - \delta^2/2)$$

As δ approaches 0 from the positive direction, the values of $3\delta^{1/3}$ and $\delta^2/2$ both approach 0 as well. That means the quantity $(3\delta^{1/3} - \delta^2/2)$ approaches 0, so

$$\lim_{\delta \to 0+} 5/2 - (3\delta^{1/3} - \delta^2/2) = 5/2$$

It follows that

$$\lim_{\delta \to 0+} \int_\delta^1 (x^{-2/3} - x) \, dx = 5/2$$

Therefore

$$\int_0^1 (x^{-2/3} - x) \, dx = 5/2$$

We've found the area of the region between the curve and the line over $0 < x < 1$. This area is finite, even though the upper function blows up at the left-hand end of the interval.

- -

Here's a challenge!

Solve this problem by subtracting the area defined by f from the area defined by g over the interval $0 < x < 1$. That is, instead of integrating the difference function, find the difference between the integrals.

Solution

As part of the solution to one of our examples in Chap. 14, we found the integral of our current function g over this same interval. (You might want to look back at Chap. 14 now and review that example.) The result was

$$\int_0^1 x^{-2/3}\, dx = 3$$

This is the area of the shaded region in Fig. 16-6, bounded by the x axis on the bottom, the line $x = 1$ on the right, the function g on top, and the dependent-variable axis on the left. The hatched region in

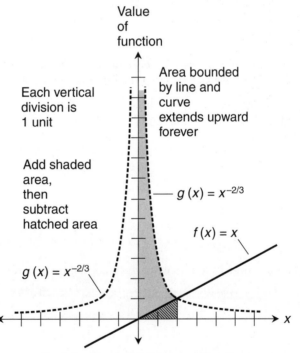

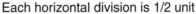

Each horizontal division is 1/2 unit

Figure 16-6 Alternative method of calculating the area of the region between the line and the curve. The shaded area is considered positive, and the hatched area is considered negative. Each horizontal division represents 1/2 unit. Each vertical division represents 1 unit.

Fig. 16-6 represents the area defined by f over $0 < x < 1$. That region is a triangle with a base length of 1 unit and a height of 1 unit. Simple geometry tells us that its area is 1/2 square unit. When we subtract this area from the area defined by the curve for g, we obtain the net area between the curve for g and the line for f. If we write this as a difference of integrals, we get

$$\int_0^1 x^{-2/3}\, dx - \int_0^1 x\, dx = 3 - 1/2 = 5/2$$

- -

Practice Exercises

This is an open-book quiz. You may (and should) refer to the text as you solve these problems. Don't hurry! You'll find worked-out answers in App. B. The solutions in the appendix may not represent the only way a problem can be figured out. If you think you can solve a particular problem in a quicker or better way than you see there, by all means try it!

1. Suppose the equation of the line shown in Fig. 16-1 is changed to

$$f(x) = 2x - 1$$

    but the equation of the parabola stays the same. Outline, step-by-step, the strategy we should use to find the area between the line and the curve in this situation.

2. Using the procedure outlined in the solution to Prob. 1, calculate the area of the region between the graphs of the functions

$$f(x) = 2x - 1$$

    and

$$g(x) = x^2 - 4$$

3. Find the area of the region between the graphs of the functions

$$f(x) = x$$

    and

$$g(x) = x^2$$

4. Using a combination of geometry and calculus, figure out the area of the region between the graphs of the functions

$$f(x) = x$$

and

$$g(x) = x^2$$

as the difference between the area under the straight line and the area under the parabola over the interval between the intersection points.

5. Find the area of the region between the graphs of the functions

$$f(x) = x^2$$

and

$$g(x) = x^3$$

6. Find the total area of the region between the graphs of the functions

$$f(x) = x$$

and

$$g(x) = x^3$$

Here's a hint: We must evaluate the areas over the intervals $-1 < x < 0$ and $0 < x < 1$ separately. Explain why.

7. Find the area of the region between the graphs of the functions

$$f(x) = 1$$

and

$$g(x) = x^{-2/3}$$

over the interval $0 < x < 1$.

8. Find the area of the region between the graphs of the functions

$$f(x) = 1$$

and

$$g(x) = x^{-2/3}$$

over the interval $1 < x < 2$.

9. Find the total area of the region between the graphs of the functions

$$f(x) = 1$$

and

$$g(x) = x^{-2/3}$$

over the interval $0 < x < 2$.

10. Find the area of the region between the graphs of the functions

$$f(x) = x^{-2}$$

and

$$g(x) = -x^{-3}$$

over the interval $1 < x$, which can also be written as $(1, \infty)$.

A Few More Integrals

Now that we're experts at integrating polynomial functions, let's work with the sine, cosine, and natural exponential. We'll also figure out how to integrate the reciprocal function, which doesn't follow the power rule for antiderivatives.

Sine and Cosine Functions

When we integrate the sine or cosine function, we obtain a function whose derivative is that sine or cosine function. Table 17-1 is a quick reference. Remember that general antiderivatives include constants of integration, while basic antiderivatives don't.

Indefinite integrals of cosine and sine

The derivative of the sine function is the cosine function, and the derivative of any real-number constant c is the zero function. Therefore,

$$d/dx \, (\sin x + c) = \cos x + 0 = \cos x$$

When we go through this process the other way, we get

$$\int \cos x \, dx = \sin x + c$$

The derivative of the negative of the cosine function is the sine function, so

$$d/dx \, (-\cos x + c) = \sin x + 0 = \sin x$$

Going backward again,

$$\int \sin x \, dx = -\cos x + c$$

Table 17-1.   Derivatives and antiderivatives for positive and negative sine and cosine. The variable is *x*, and the constant of integration is *c*.

Function	Derivative	General antiderivative	Basic antiderivative
$\sin x$	$\cos x$	$-\cos x + c$	$-\cos x$
$\cos x$	$-\sin x$	$\sin x + c$	$\sin x$
$-\sin x$	$-\cos x$	$\cos x + c$	$\cos x$
$-\cos x$	$\sin x$	$-\sin x + c$	$-\sin x$

Example

Let's figure out the area defined by the graph of the sine function over its first 1/2 cycle, as shown in Fig. 17-1. To do this, we must evaluate

$$\int_0^\pi \sin x \, dx$$

Note that 1/2 cycle is π radians. The basic antiderivative of the sine function, which we can call $f(x)$, is

$$F(x) = -\cos x$$

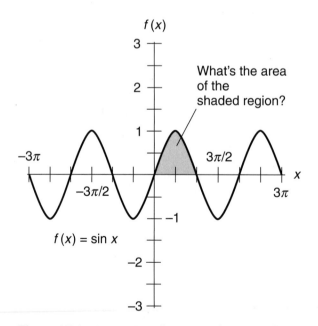

Figure 17-1   Integration allows us to determine the area defined by the curve for the sine function over its first 1/2 cycle.

When we evaluate this from $x = 0$ to $x = \pi$, we get

$$F(\pi) - F(0) = -\cos \pi - (-\cos 0) = -\cos \pi + \cos 0 = -(-1) + 1 = 2$$

We've discovered that

$$\int_0^\pi \sin x \, dx = 2$$

The area of the shaded region in Fig. 17-1 is 2 square units.

Another example

Now let's find the area defined by the graph of

$$f(x) = \sin x + 1$$

over the first 3/4 cycle, as shown in Fig. 17-2. We go from $x = 0$ to $x = 3\pi/2$. The entire wave is elevated by 1 unit compared with the wave in Fig. 17-1. We must work out

$$\int_0^{3\pi/2} (\sin x + 1) \, dx$$

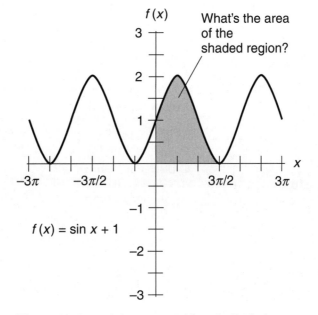

Figure 17-2   Here's an area problem that's a little more complicated, because a constant has been added to the sine function.

The basic antiderivative is

$$F(x) = -\cos x + x = x - \cos x$$

When we evaluate this from $x = 0$ to $x = 3\pi/2$, we get

$$F(3\pi/2) - F(0) = (3\pi/2 - \cos 3\pi/2) - (0 - \cos 0) = (3\pi/2 - 0) - (0 - 1)$$
$$= 3\pi/2 + 1$$

We've just determined that

$$\int_0^{3\pi/2} (\sin x + 1)\ dx = 3\pi/2 + 1$$

which tells us that the area of the shaded region in Fig. 17-2 is $3\pi/2 + 1$ square units. This is an irrational number approximately equal to 5.712.

- -

Are you confused?

Do you wonder where we get the values of the trigonometric functions for various inputs? If you've forgotten the "landmarks" for the sine and the cosine, Table 17-2 can refresh your memory. These values are exact. The positive square root of 2, represented by $2^{1/2}$, is approximately 1.414 (rounded to three decimal places).

Table 17-2.   Selected values for the sine and cosine.
Inputs are in radians.

x	$\sin x$	$\cos x$
	$\uparrow$	
	Cycle continues for $x < 0$	
0	0	1
$\pi/4$	$2^{1/2}/2$	$2^{1/2}/2$
$\pi/2$	1	0
$3\pi/4$	$2^{1/2}/2$	$-2^{1/2}/2$
π	0	-1
$5\pi/4$	$-2^{1/2}/2$	$-2^{1/2}/2$
$3\pi/2$	-1	0
$7\pi/4$	$-2^{1/2}/2$	$2^{1/2}/2$
2π	0	1
	Cycle continues for $x > 2\pi$	
	$\downarrow$	

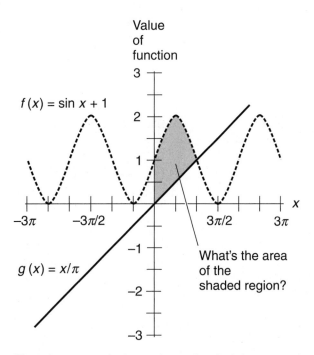

Figure 17-3   Calculus can be used to find the area between this curve and line.

Here's a challenge!

Figure 17-3 shows the graphs of two functions and a region they define:

$$f(x) = \sin x + 1$$

and

$$g(x) = x/\pi$$

The region is bounded on the bottom by the line, on the top by the curve, and on the left by the dependent-variable axis. What's the area of this zone?

Solution

Let's begin by defining the function we're going to integrate: the difference between *f* and *g*. Because *f* has the upper graph and *g* has the lower graph, we subtract *g* from *f*. If we call our difference function *q*, then

$$q(x) = f(x) - g(x) = \sin x + 1 - x/\pi$$

The next step is to figure out the interval of integration. The lower (or left-hand) bound of the interval is the dependent-variable axis. The upper (or right-hand) bound is the *x*-value of the point at which the line and the curve intersect. That's the point where the two functions have equal values. Setting $f(x) = g(x)$, we obtain the following equation to solve:

$$\sin x + 1 = x/\pi$$

Let's plug in π for the variable x and see what we get:

$$\sin \pi + 1 = \pi / \pi$$
$$0 + 1 = 1$$
$$1 = 1$$

Sometimes a good guess can save us a lot of algebra! Now we know that the interval of integration is $0 < x < \pi$. To find the area of the shaded region, we evaluate

$$\int_0^\pi (\sin x + 1 - x / \pi) \, dx$$

Remember that we defined

$$q(x) = \sin x + 1 - x / \pi$$

The basic antiderivative is

$$Q(x) = -\cos x + x - x^2 / (2\pi)$$

When we evaluate Q at the upper bound where $x = \pi$, we get

$$Q(\pi) = -\cos \pi + \pi - \pi^2 / 2\pi = -(-1) + \pi - \pi/2 = 1 + \pi/2$$

When we evaluate Q at the lower bound where $x = 0$, we get

$$Q(0) = -\cos 0 + 0 - 0^2 / 2\pi = -1 + 0 - 0 = -1$$

The definite integral is equal to

$$Q(\pi) - Q(0) = 1 + \pi/2 - (-1) = 1 + \pi/2 + 1 = \pi/2 + 2$$

We've determined that

$$\int_0^\pi (\sin x + 1 - x / \pi) \, dx = \pi/2 + 2$$

so the area of the shaded region in Fig. 17-3 is $\pi/2 + 2$ square units. This is an irrational number approximately equal to 3.571.

- -

Natural Exponential Function

In Chap. 7, we learned how to differentiate the natural exponential function and its multiples. These functions don't change when we differentiate them. Antidifferentiation doesn't change them either, except for adding a constant of integration if we want to work with the general case.

Indefinite integrals of basic exponential functions

Because any constant multiple of the exponential function is its own derivative, it follows that

$$d/dx \, (ke^x + c) = ke^x + 0 = ke^x$$

where e is the exponential constant, and k and c are real-number constants. When we "turn things around," we obtain

$$\int ke^x \, dx = ke^x + c$$

where c is the constant of integration.

Example

Figure 17-4 shows a graph of the function

$$f(x) = e^x$$

along with a shaded region defined by the curve over the interval $0 < x < 1$. To find the area of this region, we must figure out the value of

$$\int_0^1 e^x \, dx$$

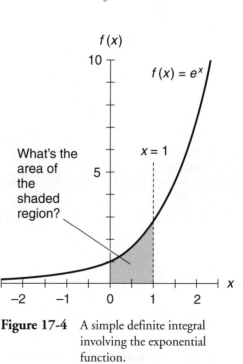

Figure 17-4   A simple definite integral involving the exponential function.

The basic antiderivative is

$$F(x) = e^x$$

When we evaluate this from $x = 0$ to $x = 1$, we get

$$F(1) - F(0) = e^1 - e^0 = e - 1$$

which tells us that

$$\int_0^1 e^x \, dx = e - 1$$

That's the area of the shaded region in Fig. 17-4, expressed in square units. A calculator can approximate this as 1.718.

Another example

Here's a situation involving the area between two graphs as shown in Fig. 17-5:

$$f(x) = e^x$$

and

$$g(x) = 2x$$

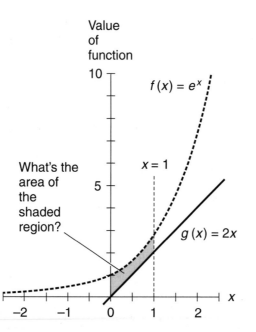

Figure 17-5   Determination of the area of a region between the graphs of the exponential function and a straight line.

We can approach this problem in two different ways. The first strategy involves taking away the area A of the triangle connecting the points $(0,0)$, $(1,0)$, and $(1,2)$ from the area we derived in the previous example. That's a triangle with a base length of 1 unit and a height of 2 units, so

$$A = (1 \cdot 2)/2 = 1$$

When we subtract this from the area defined by the exponential curve alone over the interval for which $0 < x < 1$, we obtain

$$\int_0^1 e^x \, dx - 1$$

which is $e - 1 - 1$, or $e - 2$ square units. The second method involves integrating the difference between the function for the line and the curve over the interval $0 < x < 1$. In Fig. 17-5, the function f is associated with the upper graph, and the function g is associated with the lower graph. If we call our difference function q, then

$$q(x) = f(x) - g(x) = e^x - 2x$$

We must work out the integral

$$\int_0^1 (e^x - 2x) \, dx$$

The basic antiderivative is

$$Q(x) = e^x - x^2$$

When we evaluate this from $x = 0$ to $x = 1$, we get

$$Q(1) - Q(0) = (e^1 - 1^2) - (e^0 - 0^2) = (e - 1) - (1 - 0) = e - 2$$

This agrees with the "subtract-the-triangle" scheme. We've shown that

$$\int_0^1 (e^x - 2x) \, dx = e - 2$$

A calculator can approximate this as 0.718.

- -

Here's a challenge!

Find the area of the shaded region in Fig. 17-6, defined by the graph of the exponential function over the interval for $x < 1$. We can also write this interval as $(-\infty, 1)$.

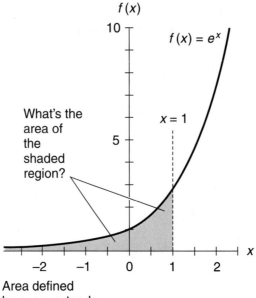

Figure 17-6   Determination of the area defined by the graph of $f(x) = e^x$ over the interval for which $x < 1$.

Solution

We want to evaluate the improper integral

$$\int_{-\infty}^{1} e^x \, dx$$

Consider this definite integral:

$$\int_{p}^{1} e^x \, dx$$

where p is an arbitrary real number less than 1. If we call the exponential function f, then the basic anti-derivative is

$$F(x) = e^x$$

To find the value of the integral, we must determine

$$\lim_{p \to -\infty} F(1) - F(p)$$

We can quickly calculate

$$F(1) = e^1 = e$$

As p becomes large negatively without bound, the value of $F(p)$ approaches 0. We can see this by inputting some increasingly negative numbers into the antiderivative and watching what happens:

$$F(-10) = e^{-10} = 1/e^{10}$$
$$F(-100) = e^{-100} = 1/e^{100}$$
$$F(-1,000) = e^{-1,000} = 1/e^{1,000}$$
$$F(-10,000) = e^{-10,000} = 1/e^{10,000}$$

These are quotients with numerators always equal to 1, but with increasingly large denominators. There's no limit to how large the denominators can get as we let x "run away" in the negative direction, so the quotients approach 0. Now we know that

$$\lim_{p \to -\infty} F(1) - F(p) = e - 0 = e$$

Therefore

$$\int_{-\infty}^{1} e^x \, dx = e$$

The area defined by the curve in Fig. 17-6 is equal to e square units.

Are you astute?

Suppose that we allow the upper bound of the interval in the previous "challenge" to vary. Imagine that, instead of being fixed with a value of 1, it can be any real number r we want, as shown in Fig. 17-7. We can show that the area defined by the curve over the interval $(-\infty, r)$ is always equal to e^r.

Let's generalize the solution, substituting r for 1 and watching what happens as we go through the process. This time, we want to determine

$$\int_{-\infty}^{r} e^x \, dx$$

where r can be any real-number constant. Now consider

$$\int_{p}^{r} e^x \, dx$$

where p is an arbitrary real number smaller than r. If we again call the exponential function f, the basic antiderivative is

$$F(x) = e^x$$

To find the value of the integral, we must determine

$$\lim_{p \to -\infty} F(r) - F(p)$$

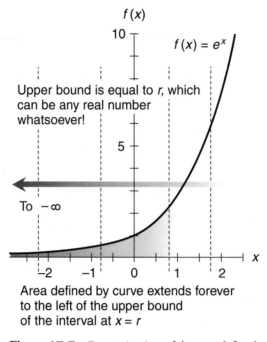

Upper bound is equal to *r*, which can be any real number whatsoever!

To −∞

Area defined by curve extends forever to the left of the upper bound of the interval at *x = r*

Figure 17-7   Determination of the area defined by the graph of $f(x) = e^x$ over the interval for which $x < r$, where *r* can be any real number.

We know straightaway that

$$F(r) = e^r$$

As *p* becomes large negatively without bound, the value of $F(p)$ approaches 0, just as it did in the solution to the "challenge." The same thing happens when we plug in some test numbers:

$$F(-10) = e^{-10} = 1/e^{10}$$
$$F(-100) = e^{-100} = 1/e^{100}$$
$$F(-1,000) = e^{-1,000} = 1/e^{1,000}$$
$$F(-10,000) = e^{-10,000} = 1/e^{10,000}$$

The quotients approach 0, so

$$\lim_{p \to -\infty} F(r) - F(p) = e^r - 0 = e^r$$

Therefore

$$\int_{-\infty}^{r} e^x \, dx = e^r$$

We've shown that the area defined by the graph of the exponential function $f(x) = e^x$ over the interval $(-\infty, r)$ is always equal to e^r.

Reciprocal Function

In Chap. 12, we saw that the general power rule can't be used to find antiderivatives or integrals of the reciprocal function. When we try to apply that rule to a function that raises the independent variable to the −1 power, we end up dividing by 0. But the reciprocal function *can* be integrated!

Indefinite integral of basic reciprocal function

We've learned that the derivative of the natural logarithm function is the reciprocal function. That is,

$$d/dx\,(\ln\,x) = x^{-1}$$

If we add a constant c to the logarithm function and then differentiate, we get

$$d/dx\,(\ln\,x + c) = x^{-1} + 0 = x^{-1}$$

When we antidifferentiate through the above equation, we obtain

$$\int x^{-1}\,dx = \ln\,x + c$$

where c is the constant of integration.

- -

Are you confused?

"Wait," you say. "The above formula suggests that if $x < 0$, we can take the natural log of a negative number to get the integral. But the natural logs of negative numbers aren't defined. Do we have to leave all the negative numbers out?" The answer is no. We can integrate this function even when $x < 0$.

Look at the graph of the reciprocal function as shown in Fig. 17-8. The pair of curves is symmetrical with respect to the origin. We can integrate this function over any interval where x is negative, and treat that interval like a "mirror image" of an interval equally far from the vertical axis on the other side. If we take the absolute value of x before we take the logarithm, then we can define the antiderivative of the reciprocal function for all real-number values of x except 0. The formula is

$$\int x^{-1}\,dx = \ln\,|x| + c$$

We still have trouble when $x = 0$, because the reciprocal function is singular there. We can't get around that.

- -

Example

Let's evaluate a simple definite integral for the reciprocal function. Figure 17-8 shows the graphical representation of

$$\int_{1}^{e} x^{-1}\,dx$$

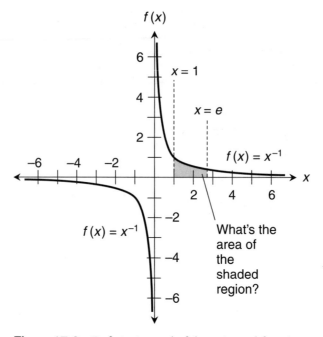

Figure 17-8   Definite integral of the reciprocal function over the interval for which $1 < x < e$.

If we call this function f, then the basic antiderivative is

$$F(x) = \ln |x|$$

Because x is positive throughout our interval, we can simplify this to

$$F(x) = \ln x$$

When we evaluate from $x = 1$ to $x = e$, we get

$$F(e) - F(1) = \ln e - \ln 1 = 1 - 0 = 1$$

We've just found that

$$\int_{1}^{e} x^{-1} \, dx = 1$$

Another example

Here's a more exotic problem. Let's find the area between the curves as shown in Fig. 17-9, over the interval for which $-2 < x < -1$. The function corresponding to the upper curve is

$$f(x) = x^2$$

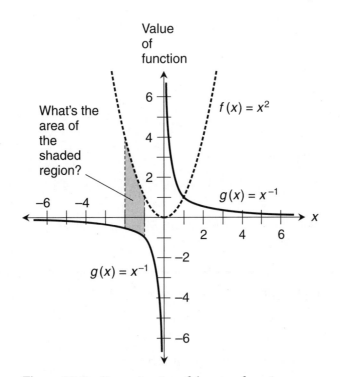

Figure 17-9   Determination of the area of a region between the graphs of the reciprocal function and a parabola over the interval for which $-2 < x < -1$.

and the function corresponding to the lower curve is

$$g(x) = x^{-1}$$

We can find the difference function p by subtracting g from f, getting

$$p(x) = f(x) - g(x) = x^2 - x^{-1}$$

The area of the shaded region can be found by evaluating

$$\int_{-2}^{-1} (x^2 - x^{-1})\, dx$$

The basic antiderivative is

$$P(x) = x^3/3 - \ln |x|$$

First, we look at the situation for $x = -1$. We see that

$$P(-1) = (-1)^3/3 - \ln |-1| = -1/3 - \ln 1 = -1/3 - 0 = -1/3$$

Next, we look at the situation for $x = -2$. We have

$$P(-2) = (-2)^3/3 - \ln |-2| = -8/3 - \ln 2$$

That's as simple as we can get this expression, because $\ln 2$ is irrational. Now we must subtract $P(-2)$ from $P(-1)$ to find the value of the integral. We get

$$P(-1) - P(-2) = -1/3 - (-8/3 - \ln 2) = -1/3 + 8/3 + \ln 2 = 7/3 + \ln 2$$

We've figured out that

$$\int_{-2}^{-1} (x^2 - x^{-1})\, dx = 7/3 + \ln 2$$

That's the area of the shaded zone in Fig. 17-9. We can use a calculator to approximate this as 3.026 square units.

- -

Here's a challenge!

Figure 17-10 shows the graphs of three functions:

$$f(x) = x^2$$
$$g(x) = x^{-1}$$
$$h(x) = 4$$

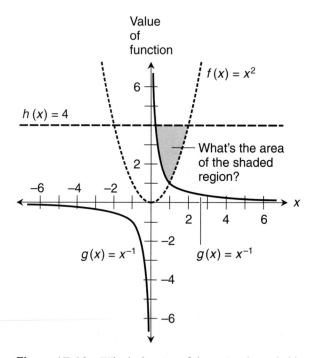

Figure 17-10   What's the area of the region bounded by the two curves and the straight line?

These three graphs enclose a region, which is shaded. The region is bounded on the right and the bottom by the curve for *f*, on the left and the bottom by the curve for *g*, and on the top by the line for *h*. What's the area of the shaded zone?

Solution

Let's divide the enclosed region into two parts, each of which is bounded by the graphs of two of the functions. Figure 17-11 shows how this is done. Only the first quadrant of the coordinate plane is shown, because our entire region lies within that quadrant.

To begin, we find the three *x*-values of the points where the function graphs intersect. The *x*-value of the intersection point between the curves for *f* and *g* can be found by setting them equal, producing the equation

$$x^2 = x^{-1}$$

If we multiply through by *x*, we obtain

$$x^3 = 1$$

which solves to $x = 1$. We draw a dashed, vertical line representing this *x*-value to mark the boundary between the two "sub-zones" in the enclosed region. We can find the *x*-value of the intersection point between the curves for *g* and *h* by setting those functions equal, getting

$$x^{-1} = 4$$

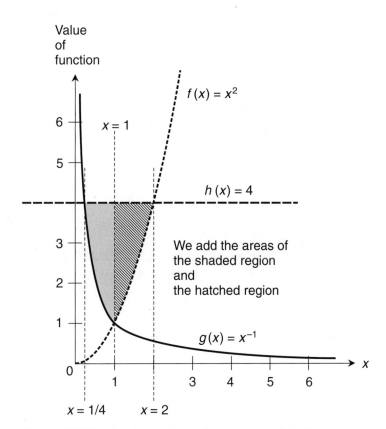

Figure 17-11   Solution to the challenge presented in Fig. 17-10.

Multiplying through by x gives us

$$1 = 4x$$

When we divide through by 4 and transpose the sides of the equation, we get the solution

$$x = 1/4$$

We draw another dashed, vertical line representing this x-value to mark the extreme left-hand boundary of the enclosed region. In a similar way, the x-value of the intersection point between the curves for f and h can be found by solving

$$x^2 = 4$$

When we take the 1/2 power of each side, we get

$$x = \pm 2$$

We aren't interested in the root $x = -2$, but only in the root $x = 2$. We draw a third dashed, vertical line for this x-value to mark the extreme right-hand boundary of the enclosed region. We now have three bounds that define the intervals for two different definite integrals.

Let's define the integrals we have to work out. In the shaded zone, which lies to the left of the vertical dashed line $x = 1$, we have the difference function

$$p(x) = h(x) - g(x) = 4 - x^{-1}$$

On the right-hand side of the vertical dashed line $x = 1$ (hatched zone), we have the difference function

$$q(x) = h(x) - f(x) = 4 - x^2$$

The area of the shaded zone (call it A) is

$$\int_{1/4}^{1} (4 - x^{-1})\, dx$$

The basic antiderivative is

$$P(x) = 4x - \ln |x|$$

When we evaluate this from $x = 1/4$ to $x = 1$, we get

$$P(1) - P(1/4) = (4 \cdot 1 - \ln |1|) - (4 \cdot 1/4 - \ln |1/4|) = 3 + \ln(1/4)$$

This is an irrational number and we can't simplify it any further, so

$$A = \int_{1/4}^{1} (4 - x^{-1})\, dx = 3 + \ln(1/4)$$

The area of the hatched zone (call it *B*) is

$$\int_1^2 (4 - x^2)\, dx$$

The basic antiderivative is

$$Q(x) = 4x - x^3/3$$

When we evaluate this from $x = 1$ to $x = 2$, we obtain

$$Q(2) - Q(1) = (4 \cdot 2 - 2^3/3) - (4 \cdot 1 - 1^3/3) = 5/3$$

We have determined that

$$B = \int_1^2 (4 - x^2)\, dx = 5/3$$

The total area enclosed by the three graphs, in square units is therefore

$$A + B = 3 + \ln(1/4) + 5/3 = 14/3 + \ln(1/4)$$

This is an irrational number, but a calculator can approximate it as 3.280 square units.

- -

Where to find more integrals

You can find a table of indefinite integrals in the back of this book. Refer to App. G. You can also find them on the Internet. Enter the phrase "table of integrals" or "table of indefinite integrals" into your favorite search engine. A few sites will calculate definite integrals if you input the bounds.

Practice Exercises

This is an open-book quiz. You may (and should) refer to the text as you solve these problems. Don't hurry! You'll find worked-out answers in App. B. The solutions in the appendix may not represent the only way a problem can be figured out. If you think you can solve a particular problem in a quicker or better way than you see there, by all means try it!

1. Evaluate the following definite integral:

$$\int_0^{2\pi} -5 \sin x\, dx$$

2. What happens to the definite integral of the cosine function over intervals whose lower bounds are always 0 and whose upper bounds keep increasing positively? Here's a hint: Evaluate

$$\underset{s \to \infty}{Lim} \int_0^s \cos x\, dx$$

3. Evaluate the following definite integral:

$$\int_0^1 (\cos x^2)(2x)\ dx$$

Here's a hint: Use the substitution method for integrating composite functions.

4. Evaluate the following definite integral:

$$\int_0^\pi (4e^x + 4 \sin x)\ dx$$

5. Evaluate the following definite integral:

$$\int_0^1 2e^{2x}\ dx$$

Here's a hint: Use the substitution method for integrating composite functions.

6. Evaluate the following definite integral:

$$\int_0^1 (\cos e^x)(e^x)\ dx$$

Here's a hint: Use the substitution method for integrating composite functions.

7. Evaluate the following integral. If it's defined, state the value.

$$\int_0^1 x^{-1}\ dx$$

8. Evaluate the following integral. If it's defined, state the value.

$$\int_{-\infty}^{-1} x^{-1}\ dx$$

9. Evaluate the following integral as though there were no singularity in the interval. That's the wrong way to do it! But try it anyway, and see what happens:

$$\int_{-e}^{e} x^{-1}\ dx$$

Note that the limits of integration are $-e$ and e, where e is the exponential constant. Draw a graph that lends intuitive support to this "solution."

10. Evaluate the integral stated in Prob. 9 using the correct approach.

18

How Long Is the Arc?

We've seen how integration can help us find the areas defined by, and between, graphs. In this chapter, we'll learn how to determine *arc length,* or the distance along a curve between two points.

A Chorus of Chords

Let's derive a formula for arc length along the graph of a function, as defined between two values of the independent variable. We'll start by approximating, and then we'll refine the estimate until we get a formula for the exact value.

Breaking up the arc

Imagine a function $f(x)$ whose graph is a curve (Fig. 18-1). We want to find the length L of the arc between the points where $x = a$ and $x = b$, assuming that $a < b$. If we move along the arc in the positive-x direction, let's call the arc length positive. If we go in the negative-x direction, let's call the length negative. No matter which interval we choose, we know that we will always travel either to the right or to the left over the entire interval. (Otherwise, there would be more than one value of the dependent variable for some values of the independent variable, and the graph would not represent a function.)

If the graph were a straight line, the length of the segment between the points where $x = a$ and $x = b$ could be found easily. We could take $b - a$ as the base length of a right triangle, take $f(b) - f(a)$ and call it the height of the triangle, and then calculate the length L of the segment connecting the points using the distance formula

$$L = \{(b - a)^2 + [f(b) - f(a)]^2\}^{1/2}$$

Obviously, the graph in Fig. 18-1 isn't a line segment over the interval shown. But we can split it up into a set of line segments. Suppose we select several points on the arc with equally spaced x-values (Fig. 18-2). These segments are called *chords.* If we break the arc up into

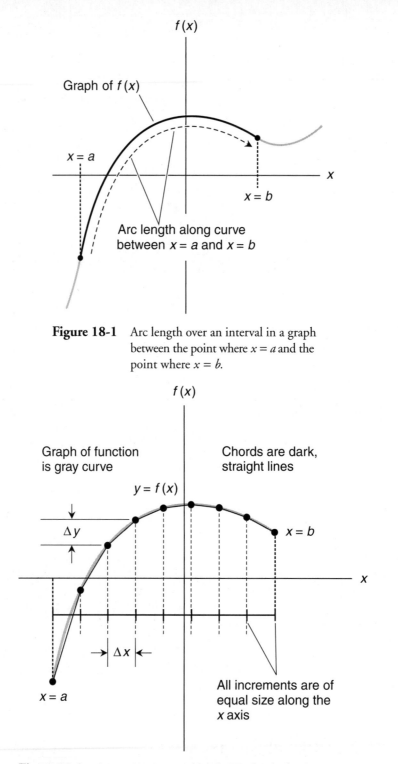

Figure 18-1   Arc length over an interval in a graph between the point where $x = a$ and the point where $x = b$.

Figure 18-2   Approximating arc length. We divide the arc into n chords. The x-values of the endpoints are equally spaced. As n increases, we get better approximations.

n chords, then the difference Δx in the *x*-values between adjacent pairs of points is

$$\Delta x = (b - a) / n$$

Each chord forms the *hypotenuse* (longest side) of a small right triangle. The base length of each triangle is Δx. The heights vary.

Let's say that $y = f(x)$. Then we can call the height of a particular triangle, say the *i*th triangle, $\Delta_i y$, as shown in Fig. 18-3. Let's call the left-most triangle "number 1," and count up as we move toward the right. If we make *n* large, then:

- The height of triangle number 1 is $\Delta_1 y$
- The height of triangle number 2 is $\Delta_2 y$
- The height of triangle number 3 is $\Delta_3 y$
- And so on . . .
- The height of triangle number *i* is $\Delta_i y$
- And so on . . .
- The height of triangle number *n* is $\Delta_n y$

The length of each chord can be found with the Pythagorean formula:

- The length of chord number 1 is $[(\Delta x)^2 + (\Delta_1 y)^2]^{1/2}$
- The length of chord number 2 is $[(\Delta x)^2 + (\Delta_2 y)^2]^{1/2}$

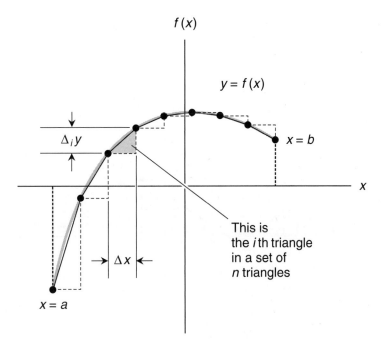

Figure 18-3 Every chord is the hypotenuse of a right triangle. Each triangle has the same base length Δx, but the heights vary. Here, we're looking at the *i*th triangle, whose height we call $\Delta_i y$.

- The length of chord number 3 is $[(\Delta x)^2 + (\Delta_3 y)^2]^{1/2}$
- And so on . . .
- The length of chord number i is $[(\Delta x)^2 + (\Delta_i y)^2]^{1/2}$
- And so on . . .
- The length of chord number n is $[(\Delta x)^2 + (\Delta_n y)^2]^{1/2}$

We can get an approximation of the arc length (call it L_{app}) by adding up the lengths of all the chords from $i = 1$ to $i = n$. Stated symbolically,

$$L_{\text{app}} = \sum_{i=1}^{n} [(\Delta x)^2 + (\Delta_i y)^2]^{1/2}$$

The law of the mean

We've described how to approximate the arc length L between the two points on the graph of $y = f(x)$ for which $x = a$ and $x = b$. Before we can work out a way to determine L exactly, we must know an important principle called the *law of the mean*.

Suppose we choose two points on the graph of a function $y = f(x)$, and these points describe one of the chords in an approximation of the arc length between the points. For the ith chord, the endpoints are where $x = x_{i-1}$ and $x = x_i$, as shown in Fig. 18-4. That chord has slope

$$\Delta_i y / \Delta x = [f(x_i) - f(x_{i-1})] / (x_i - x_{i-1})$$

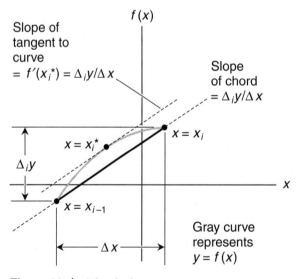

Figure 18-4   There's always a point on a curve, somewhere between the endpoints of a chord, for which the derivative of the function is equal to the slope of the chord.

If $f(x)$ is defined and continuous between the chord endpoints, then there is at least one point on the curve between the chord endpoints for which the derivative of f is equal to the slope of the chord. If we call the x-value of such a point x_i^*, then

$$x_{i-1} \leq x_i^* \leq x_i$$

and

$$f'(x_i^*) = \Delta_i y / \Delta x$$

We *don't* necessarily know the exact location of any point for which $x = x_i^*$. But we *do* know that at least one such point exists on the curve between the endpoints of the chord.

- -

Are you confused?

We're using a lot of symbols and subscripts here. Don't let them overwhelm you! Here's a summary of what they mean.

- The x-values of the endpoints of the arc are $x = a$ and $x = b$.
- Δx is the difference in the x-values between any two adjacent points on the arc.
- Δx also refers to the width of every small right triangle along the arc.
- $\Delta_i y$ refers to the height of the ith small right triangle in the set of n triangles along the arc.
- The value x_i^* is the x-coordinate of a point on the curve where the derivative of the function is equal to the slope of the ith chord.
- L_{app} is the approximate arc length, expressed as the sum of the lengths of a set of chords.
- L is the actual arc length, as we would measure it if we could move exactly along the curve.

- -

Finding the true arc length

Now we're ready to derive a formula that will tell us precisely how long the arc is. Look again at the sum that expresses the approximate length:

$$L_{app} = \sum_{i=1}^{n} [(\Delta x)^2 + (\Delta_i y)^2]^{1/2}$$

We can use algebra to rewrite this as

$$L_{app} = \sum_{i=1}^{n} [1 + (\Delta_i y / \Delta x)^2]^{1/2} \, \Delta x$$

By the law of the mean, there's a point on the arc such that $x = x_i^*$, somewhere between the point where $x = x_{i-1}$ and $x = x_i$. (Look again at Fig. 18-4.) That point is in the ith interval along the arc, corresponding to the ith chord. Now that we know this, we can substitute $f'(x_i^*)$ for $\Delta_i y / \Delta x$ in the above equation to get

$$L_{app} = \sum_{i=1}^{n} \{1 + [f'(x_i^*)]^2\}^{1/2} \, \Delta x$$

Imagine that we make n larger without end, so the chords approximate the actual arc length L ever more closely. As n approaches infinity, L_{app} approaches L, so

$$L = \lim_{n \to \infty} \sum_{i=1}^{n} \{1 + [f'(x_i^*)]^2\}^{1/2} \, \Delta x$$

This scheme is similar to the theory Riemann used to find the area defined by a curve. When written in integral notation, the above statement becomes

$$L = \int_a^b \{1 + [f'(x)]^2\}^{1/2} \, dx$$

A Monomial Curve

Let's try out our newly discovered formula with a monomial function f that cubes the variable x and then takes the positive square root of the result:

$$f(x) = x^{3/2}$$

Suppose we want to find the arc length along this curve from the point where $x = 0$ to the point where $x = 1$, as shown in Fig. 18-5.

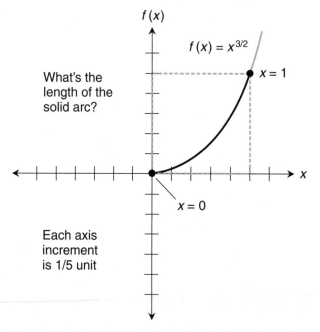

Figure 18-5   Calculating the arc length along the graph of $f(x) = x^{3/2}$, from $x = 0$ to $x = 1$.

Arc-in-a-box method

We can find a range of possible absolute values for the arc length between two points, getting a way to check arc-length calculations for major errors. I call this scheme the *arc-in-a-box method*. Here's how it works.

- Construct a box (rectangle or square) with vertical and horizontal edges, such that the endpoints of the arc are at opposite corners of the box.
- Be sure that the arc is entirely contained within the box.
- Be sure that the arc bends in the same sense (either clockwise or counterclockwise) everywhere within the box.

Once we've done these things, we know that the absolute value of the arc length is greater than or equal to the diagonal measure of the box, but less than or equal to half the perimeter of the box. In the situation of Fig. 18-5,

$$f(0) = 0^{3/2} = 0$$

and

$$f(1) = 1^{3/2} = 1$$

The arc length must be greater than or equal to the diagonal of the dashed gray square whose edges measure 1 unit. That diagonal is $2^{1/2}$ (about 1.414) units long. Also, the arc length must be less than or equal to half the perimeter of the square. That's 2 units. If we call the arc length L, then we know that

$$2^{1/2} \leq L \leq 2$$

If we get an answer that lies outside these extremes, we'll know that we've made a mistake in our calculations somewhere along the way!

Setting up the integral

Before we use the arc-length formula, we must differentiate the function. Again, our function is

$$f(x) = x^{3/2}$$

The derivative is

$$f'(x) = (3/2)x^{1/2}$$

We plug $(3/2)x^{1/2}$ into the formula for the arc length in place of $f'(x)$, and we also include the bounds of the integration interval, getting

$$L = \int_0^1 \{1 + [(3/2)\,x^{1/2}]^2\}^{1/2}\ dx$$

This can be rewritten as

$$L = \int_0^1 [1 + (9/4)x]^{1/2} \, dx$$

Let's refer to a table of integrals to resolve this. (The creators of such tables compiled them especially to help people like us in situations like this!) Looking through App. G in the back of this book, we find the form

$$\int (ax + b)^{1/2} \, dx = (2/3)(ax + b)^{3/2} \, a^{-1} + c$$

where c is the constant of integration. We can reverse the addends in the binomials to get this into a form more convenient to use directly in the problem at hand:

$$\int (b + ax)^{1/2} \, dx = (2/3)(b + ax)^{3/2} \, a^{-1} + c$$

If we let $a = 9/4$ and $b = 1$, we obtain

$$\int [1 + (9/4)x]^{1/2} \, dx = (2/3) \cdot [1 + (9/4)x]^{3/2} \cdot (4/9) + c$$

which simplifies to

$$\int [1 + (9/4)x]^{1/2} \, dx = (8/27) \cdot [1 + (9/4)x]^{3/2} + c$$

Working out the integral

To determine the arc length, we must evaluate the following expression over the interval from $x = 0$ to $x = 1$:

$$(8/27) \cdot [1 + (9/4)x]^{3/2}$$

Plugging in $x = 1$, we get

$$(8/27) \cdot [(1 + (9/4) \cdot 1]^{3/2} = 2{,}197^{1/2} \, / \, 27$$

Plugging in $x = 0$, we get

$$(8/27) \cdot [(1 + (9/4) \cdot 0]^{3/2} = 8/27$$

The arc length is the difference between these results, or

$$L = (2{,}197^{1/2} - 8) \, / \, 27$$

A calculator can approximate this to 1.440 units. That's between the constraints we derived: a minimum of $2^{1/2}$ (approximately 1.414) and a maximum of 2.

- -

Here's a challenge!

Using the techniques developed in this chapter, derive the formula for the distance of a point $(x,y) = (a,b)$ from the origin in Cartesian coordinates.

Solution

To begin, we must be sure we know the function we want to evaluate! Its graph is a straight line connecting the points $(0,0)$ and (a,b), as shown in Fig. 18-6. The slope of this line is b/a. The y-intercept is 0, because the line passes through the origin. Therefore, if we call our function f, we have

$$f(x) = (b/a)x$$

The derivative of this is

$$f'(x) = b/a$$

Keeping in mind that a and b are constants, we can plug b/a into the formula for arc length in place of $f'(x)$. The bounds of our interval are $x = 0$ and $x = a$, so

$$L = \int_0^a [1 + (b/a)^2]^{1/2}\, dx$$

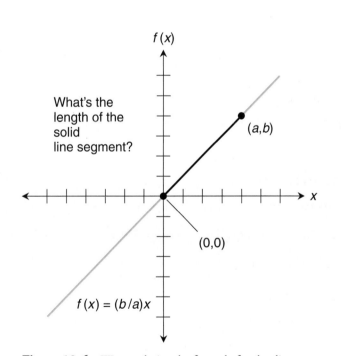

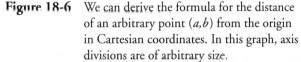

Figure 18-6 We can derive the formula for the distance of an arbitrary point (a,b) from the origin in Cartesian coordinates. In this graph, axis divisions are of arbitrary size.

Using algebra, we can rewrite this as

$$L = \int_0^a \left[(a^2 + b^2)^{1/2} \, / \, a \right] \, dx$$

If this looks complicated, note that the quantity $\left[(a^2 + b^2)^{1/2} \, / \, a \right]$ is a constant, because it's built up from other constants. For a moment, let's rename it k, so

$$k = \left[(a^2 + b^2)^{1/2} \, / \, a \right]$$

That gives us

$$L = \int_0^a k \, dx$$

The indefinite integral works out easily as

$$\int k \, dx = kx + c$$

To obtain L, we evaluate the expression kx from $x = 0$ to $x = a$, getting

$$L = k \cdot a - k \cdot 0 = k \cdot a$$

Let's give k its original name back. Now we have

$$L = \left[(a^2 + b^2)^{1/2} \, / \, a \right] \cdot a = (a^2 + b^2)^{1/2}$$

This is the formula we learned in precalculus. It defines the distance in Cartesian coordinates from the origin to an arbitrary point $(x, y) = (a, b)$.

A More Exotic Curve

Now let's try the formula with a more complicated curve. Suppose f is a function of x such that

$$f(x) = x^3/24 + 2x^{-1}$$

Let's find the arc length from the point where $x = 2$ to the point where $x = 3$.

Setting up the integral

The first thing to do is differentiate our function. That gives us

$$f'(x) = x^2/8 - 2x^{-2}$$

When we square this, we get

$$[f'(x)]^2 = x^4/64 - 1/2 + 4x^{-4}$$

Adding 1 gives us

$$1 + [f'(x)]^2 = x^4/64 + 1/2 + 4x^{-4}$$

We can morph the right-hand side of the above equation as follows:

$$x^4/64 + 1/2 + 4x^{-4} = x^{-4}(x^8/64 + x^4/2 + 4) = x^{-4}(x^4/8 + 2)^2$$
$$= (x^{-2})^2(x^4/8 + 2)^2 = (x^2/8 + 2x^{-2})^2$$

When we plug $(x^2/8 + 2x^{-2})^2$ into the arc-length formula for $\{1 + [f'(x)]^2\}$ and include the bounds of integration, we obtain

$$L = \int_2^3 [(x^2/8 + 2x^{-2})^2]^{1/2} \, dx$$

which simplifies to

$$L = \int_2^3 (x^2/8 + 2x^{-2}) \, dx$$

The indefinite integral, leaving out the constant of integration, is

$$\int (x^2/8 + 2x^{-2}) \, dx = x^3/24 - 2x^{-1}$$

- -

Are you astute?

Have you noticed that this is almost identical to our original function? The only difference is that we subtract the two monomials, rather than adding them. This is sheer coincidence.

- -

Working out the value

To determine the arc length, we must evaluate the following expression over the interval from $x = 2$ to $x = 3$:

$$x^3/24 - 2x^{-1}$$

Plugging in $x = 3$, we get

$$3^3/24 - 2 \cdot 3^{-1} = 11/24$$

Plugging in $x = 2$, we get

$$2^3/24 - 2 \cdot 2^{-1} = -2/3$$

The arc length is the difference between these results, or

$$L = 11/24 - (-2/3) = 9/8$$

- -

Here's a challenge!

Plot a graph of the function we just finished working with. Indicate the location of the arc whose length we found. Determine a range of lengths within which the arc length must lie, so we can be fairly sure that we didn't make any errors in our calculation.

Solution

Figure 18-7 is a graph of the function in the first quadrant. The black portion of the curve is the arc. The endpoints are labeled. Function values have been worked out using arithmetic. The dashed gray box is constructed so the endpoints of the arc are at opposite corners. The arc length must be greater than or equal to diagonal measure of the box, but less than or equal to half the perimeter of the box. The diagonal measure is

$$[(43/24 - 4/3)^2 + (3 - 2)^2]^{1/2} = (697/576)^{1/2}$$

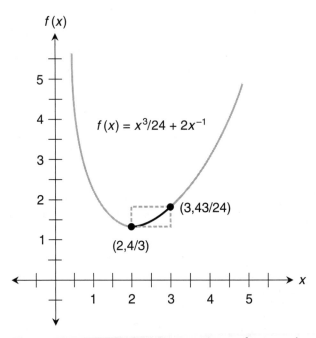

Figure 18-7   Graph of the function $f(x) = x^3/24 + 2x^{-1}$, showing the arc for which we found the length. The dashed gray box defines the range of possible values for the arc length.

Using a calculator, we get approximately 1.100 units. The arc must be at least as long as this. Checking it; we found $L = 9/8$, which is 1.125 units. Now let's determine half the perimeter of the box. That's the sum of the lengths of the long side and the short side. Calculating, we get

$$(43/24 - 4/3) + (3 - 2) = 35/24$$

A calculator approximates this as 1.458 units. The arc is shorter, as we should expect.

- -

Practice Exercises

This is an open-book quiz. You may (and should) refer to the text as you solve these problems. Don't hurry! You'll find worked-out answers in App. B. The solutions in the appendix may not represent the only way a problem can be figured out. If you think you can solve a particular problem in a quicker or better way than you see there, by all means try it! If you need to refer to integral tables, feel free to use them.

1. Using the techniques we've learned in this chapter, find the length of the arc that goes 1/4 of the way around the unit circle *in the first quadrant* of the Cartesian coordinate plane, from the point where $x = 0$ to the point where $x = 1$. Remember that the unit circle is centered at the origin and has a radius of 1 unit. Its equation is

$$x^2 + y^2 = 1$$

2. Draw a graph of the function and the arc we evaluated in the solution to Prob. 1. Then show that when we multiply this arc length by 4, we get the circumference of the circle, just as we can find it using the rules of geometry. Remember that the formula for the circumference of a circle is

$$C = \pi r^2$$

where C is the circumference and r is the radius, both expressed in the same units.

3. Using the arc-in-a-box method, find the minimum and maximum possible lengths for the arc we evaluated in the solution to Prob. 1. Then verify that the arc length is indeed within that range.

4. Using the techniques we've learned in this chapter, find the length of the arc that goes part of the way around the circle

$$x^2 + y^2 = 4$$

in the second quadrant of the Cartesian coordinate plane, from the point where $x = -2$ to the point where $x = -1$.

5. Draw a graph of the function and the arc we evaluated in the solution to Prob. 4.

6. Using the arc-in-a-box method, find a span of values between which the arc length we found in the solution to Prob. 4 must be. Verify that the arc length is within that range.

7. Verify that the arc length we found in the solution to Prob. 4 is exact, using ordinary geometry and trigonometry.

8. Using the techniques we've learned in this chapter, find the length of the arc in the graph of the function

$$f(x) = x^2/2$$

from the point where $x = 0$ to the point where $x = 1$.

9. Draw a graph of the function and the arc we evaluated in the solution to Prob. 8.

10. Using the arc-in-a-box method, find a span of values between which the arc length we found in the solution to Prob. 8 must be. Verify that the arc length is within that range.

19

Special Integration Tricks

It's okay to use a reference table to resolve a difficult or unfamiliar integral. But it's also a good idea to get some practice working out integrals. In this chapter, we'll look at three techniques for integrating certain functions or combinations of functions.

Principle of Linearity

In Chap. 15, we learned some rules that apply to definite integrals. We can expand on them to get the *principle of linearity*, also known as the *linear-combination rule for definite integration*.

The old rules

Let's review the two rules that we're about to expand upon. They involve multiplication of a function by a constant, and the addition of two different functions. If we have a variable x, a continuous interval (a,b), a constant k, and functions $f_1(x)$ and $f_2(x)$ that are integrable from $x = a$ to $x = b$, then

$$\int_a^b k\,[f_1(x)]\,dx = k \int_a^b [f_1(x)]\,dx$$

and

$$\int_a^b [f_1(x) + f_2(x)]\,dx = \int_a^b f_1(x)\,dx + \int_a^b f_2(x)\,dx$$

The new rule in brief

The principle of linearity combines the best features of these two older rules. We can write it out in formal mathematical language or as an equation. Let's do both! First, the words:

- Let f_1 and f_2 be integrable functions of x over a continuous interval (a,b). Let k_1 and k_2 be constants. The integral of k_1 times f_1 plus k_2 times f_2 over (a,b) is equal to k_1 times the integral of f_1 over (a,b) plus k_2 times the integral of f_2 over (a,b).

And now, the equation:

$$\int_a^b \{k_1 [f_1 (x)] + k_2 [f_2 (x)]\} \, dx = k_1 \int_a^b f_1 (x) \, dx + k_2 \int_a^b f_2 (x) \, dx$$

The principle of linearity isn't limited to two functions and constants. It works with any finite linear combination if we stay with the same interval, stay with the same variable, and make sure that each function can be integrated over the entire interval (a,b).

Example

To demonstrate how this principle works, suppose we come across

$$\int_0^\pi (3 \sin x + 5e^x) \, dx$$

The principle of linearity allows us to rewrite this as

$$3 \int_0^\pi \sin x \, dx + 5 \int_0^\pi e^x \, dx$$

The basic antiderivative of the sine function is the negative of the cosine function. The exponential function is its own basic antiderivative. We can therefore resolve the above sum of integrals to

$$3 \cdot [-\cos \pi - (-\cos 0)] + 5 \cdot (e^\pi - e^0)$$

which can be simplified to

$$3 \cdot [\cos 0 - \cos \pi] + 5 \cdot (e^\pi - e^0)$$

We know that $\cos 0 = 1$, $\cos \pi = -1$, and $e^0 = 1$, so the above expression works out to

$$3 \cdot [1 - (-1)] + 5 \cdot (e^\pi - 1) = 1 + 5e^\pi$$

The complete integral and its value can now be stated as

$$\int_0^\pi (3 \sin x + 5e^x) \, dx = 1 + 5e^\pi$$

Don't be fooled!

It's easy to look at the above result and suspect that

$$\int_0^\pi 3 \sin x \, dx = 1$$

and

$$\int_0^\pi 5e^x \, dx = 5e^\pi$$

These conclusions are wrong. The principle of linearity doesn't imply anything like this. If you like, work out these integrals and see what their values actually are.

Are you confused?

Do you suspect that the principle of linearity can apply to a difference, just as it does for a sum? If so, you're right, as long we keep the order of subtraction the same. We can express this in formal jargon:

- Let f_1 and f_2 be integrable functions of x over a continuous interval (a,b). Let k_1 and k_2 be constants. The integral of k_1 times f_1 minus k_2 times f_2 over (a,b) is equal to k_1 times the integral of f_1 over (a,b) minus k_2 times the integral of f_2 over (a,b).

Written as an equation, it is

$$\int_a^b \{k_1\,[f_1\,(x)] - k_2\,[f_2\,(x)]\}\,dx = k_1 \int_a^b f_1\,(x)\,dx \ - \ k_2 \int_a^b f_2\,(x)\,dx$$

Here's a challenge!

Prove the principle of linearity for the definite integrals of two functions, based on the sum rule and the multiplication-by-constant rule stated at the beginning of this chapter.

Solution

Suppose that we want to integrate a constant k_1 times a function f_1, added to another constant k_2 times another function f_2, like this:

$$\int_a^b \{k_1\,[f_1\,(x)] + k_2\,[f_2\,(x)]\}\,dx$$

The variable is x, and the integration interval is (a,b). The sum rule tells us that we can split this integral so each addend becomes a new integrand, and both of the new integrals go over the same original interval (a,b). When we do that, we get

$$\int_a^b k_1\,[f_1\,(x)]\,dx + \int_a^b k_2\,[f_2\,(x)]\,dx$$

We can pull the constants out of both integrals according to the multiplication-by-constant rule, which gives us

$$k_1 \int_a^b f_1\,(x)\,dx + k_2 \int_a^b f_2\,(x)\,dx$$

Integration by Parts

Once in awhile, you'll encounter an integral that consists of a product of two functions, but can be rearranged so it becomes easier to resolve using a scheme called *integration by parts*. Here's how it works.

An old idea revisited

Do you remember the two-function product rule for differentiation from Chap. 6? That rule says that if f and g are differentiable functions of the same variable, then we can find the derivative of their product as a sum of two other products:

$$(fg)' = f'g + g'f$$

Suppose the independent variable is x for both functions. We can integrate both sides of the above equation with respect to x, getting

$$\int (fg)' \, dx = \int (f'g + g'f) \, dx$$

Because the integral of a sum is equal to the sum of the integrals, we can rewrite the right-hand side to get

$$\int (fg)' \, dx = \int f'g \, dx + \int g'f \, dx$$

Now let's take a close look at the left-hand side of this equation. In effect, it's telling us to take the antiderivative of a derivative! These two operations "undo" each other, so we can simplify the preceding equation to

$$fg = \int f'g \, dx + \int g'f \, dx$$

When we subtract $\int f'g \, dx$ from both sides, we get

$$fg - \int f'g \, dx = \int g'f \, dx$$

Let's transpose the right-hand and left-hand sides. That gives us

$$\int g'f \, dx = fg - \int f'g \, dx$$

This formula is useful in certain situations where we must integrate a function that's a product of two others. It leaves us with another integral, as you can see. But if we're clever (and lucky), we can arrange things so the second integral is easier to figure out than the original integral would be if we attempted to resolve it directly.

- -

Are you confused?

You might ask, "Aren't the constants of integration important in the process we just finished? In particular, taking the antiderivative of a derivative gives us the same function again, but only if the constant of integration is 0, isn't that right?" Yes, that's true. However, we always have at least one indefinite integral somewhere in the equation in each step of the process. Those constants are always there, and because we never have to specify their values, we don't have to worry about making a mistake when we manipulate the equations. But we must never forget to include a constant of integration as the last step in the process of resolving an indefinite integral.

- -

Variations on a theme

The formula we've just derived can be presented in several different forms. We can write the formula out in full, showing the variable x and including multiplication symbols for clarity. It's messier but more revealing than the shorter version:

$$\int g'(x) \cdot f(x) \, dx = f(x) \cdot g(x) - \int f'(x) \cdot g(x) \, dx$$

The commutative law for multiplication (from basic algebra) allows us to reverse the orders of products. Let's use that rule with two of the three products here, and rewrite the above equation as

$$\int f(x) \cdot g'(x) \, dx = f(x) \cdot g(x) - \int g(x) \cdot f'(x) \, dx$$

Now let's use an alternative notation for the derivatives. In Chap. 3, we saw several different ways that derivatives can be denoted. Let's substitute like this:

$$f'(x) = df(x)/dx$$

and

$$g'(x) = dg(x)/dx$$

Now we get the sloppy formula

$$\int f(x) \cdot [dg(x)/dx] \, dx = f(x) \cdot g(x) - \int g(x) \cdot [df(x)/dx] \, dx$$

In both of the integrals here, we have differentials in numerators and denominators right next to each other. They're the little dx quantities that we always write with integrals. (Sometimes writing them gets to be such a habit that we forget what they really mean! If you don't remember their significance, look back at Chap. 11 under the section "The Integral Notation.") These differentials divide out to give us

$$\int f(x) \cdot dg(x) = f(x) \cdot g(x) - \int g(x) \cdot df(x)$$

When we "hide" the variable x and the multiplication symbols, we get back to an abbreviated version of this same formula:

$$\int f \, dg = fg - \int g \, df$$

Some texts use the letters u and v instead of f and g to represents the functions. In those books, you'll see the rule for integration by parts written as

$$\int u \, dv = uv - \int v \, du$$

- -

Are you confused?

If you're baffled by all these different ways of saying the same thing, I understand. It's as if I've told you "This is a calculus book" in half a dozen different languages. You may ask, "What's the point?" I've gone through them to give you a chance to see alternative notations. That way, you won't be surprised when you're reading a textbook or thesis and you come across a nonstandard form. My favorite version is

$$\int f(x) \cdot g'(x) \, dx = f(x) \cdot g(x) - \int g(x) \cdot f'(x) \, dx$$

- -

Example

Consider the following indefinite integral, which is the product of a simple linear function and an exponential function:

$$\int x \, e^x \, dx$$

To set up the parts, let's say that

$$f(x) = x$$

and

$$g'(x) = e^x$$

The basic antiderivative of g' is

$$g(x) = e^x$$

The derivative of f is

$$f'(x) = 1$$

The formula for integration by parts tells us that

$$\int x \, e^x \, dx = x \, e^x - \int e^x \cdot 1 \, dx = x \, e^x - \int e^x \, dx$$

The last term can be considered as the basic antiderivative of e^x, which is, once again, simply the exponential function e^x. Substituting back, we get

$$\int x \, e^x \, dx = x \, e^x - e^x + c = e^x (x - 1) + c$$

where c is the constant of integration.

- -

Are you confused?

You might wonder why we don't include constants of integration in either of the two cases above where we take the antiderivative of e^x. The reason is the same as when we derived the formula for integration by parts. We *can* include all those constants, and it won't do any harm, except to make things sloppier than necessary. But when we have a combination of antiderivatives in an ongoing process that eventually produces a single indefinite integral, we can wait until the last step before we add in the constant.

Here's a challenge!

Use integration by parts to find

$$\int -3x \, e^{-x} \, dx$$

Solution

Solving this requires some intuition. To begin, we can rewrite this integral as

$$\int 3x \left(-e^{-x}\right) dx$$

Does this maneuver seem strange? We move the minus sign "legally," but instead of pulling it out in front of the integral sign, we move it further into the integrand! As things turn out, that will make the problem easier to solve. To set up the parts, let's say that

$$f(x) = 3x$$

and

$$g'(x) = -e^{-x}$$

The basic antiderivative of g' is

$$g(x) = e^{-x}$$

If you wonder how we come to this conclusion, you can differentiate g as we've stated it above, using the reciprocal rule for differentiation (from Chap. 6) and the fact that the exponential function is its own derivative. You'll end up with g'. The derivative of f in this situation is

$$f'(x) = 3$$

The formula for integration by parts tells us that

$$\int 3x \left(-e^{-x}\right) dx = 3x \, e^{-x} - \int e^{-x} \cdot 3 \, dx = 3x \, e^{-x} - 3 \int e^{-x} \, dx$$

The last integral in this equation resolves to $-e^{-x}$ plus a constant. (That's another "backward application" of the reciprocal rule for differentiation, along with multiplication by -1.) Substituting back into the preceding equation, we get

$$\int 3x \left(-e^{-x}\right) dx = 3x \, e^{-x} - 3 \left(-e^{-x}\right) + c = 3e^{-x} \left(x + 1\right) + c$$

where c is the constant of integration.

Partial Fractions

Occasionally you'll see an integral of a fraction that contains a constant in the numerator and a polynomial in the denominator. Some such integrals can be split up into sums of simpler integrals using the *method of partial fractions.*

A helpful formula

Let's look at an indefinite integral with a constant in the numerator and linear function in the denominator. The general form is

$$\int k/(ax+b)\ dx$$

where x is the variable, and a, b, and k are constants. We can rewrite this using exponent notation instead of fraction notation. That gives us

$$\int k(ax+b)^{-1}\ dx$$

When we pull out the constant, we get

$$k\int (ax+b)^{-1}\ dx$$

Referring to the table of integrals in App. G, we find

$$\int (ax+b)^{-1}\ dx = a^{-1} \ln |ax+b| + c$$

When we include the constant k, we get

$$k\int (ax+b)^{-1}\ dx = k\,(a^{-1} \ln |ax+b| + c)$$

where c is the constant of integration. We can also write this as

$$k\int (ax+b)^{-1}\ dx = ka^{-1} \ln | ax+b| + kc$$

which can be simplified to

$$k\int (ax+b)^{-1}\ dx = ka^{-1} \ln |ax+b| + c$$

We can "recycle" the constant c because it doesn't have any value in particular.

A preliminary example

Let's resolve the indefinite integral

$$\int 2\,(3x+5)^{-1}\ dx$$

To begin, we move the constant in front of the integral symbol, getting

$$2 \int (3x + 5)^{-1} \, dx$$

Then we apply the formula we derived above to get

$$2 \cdot 3^{-1} \ln |3x + 5| + c$$

which can be simplified to

$$(2/3) \ln |3x + 5| + c$$

We have therefore determined that

$$\int 2 \, (3x + 5)^{-1} \, dx = (2/3) \ln |3x + 5| + c$$

"Reverse engineering"

Suppose we come across an integral of a ratio that has a linear function of x in the numerator and a quadratic function of x in the denominator. Here's the general form:

$$\int [(ax + b)(cx^2 + dx + e)^{-1}] \, dx$$

where a, b, c, d, and e are constants. (In this case, e has nothing to do with the exponential constant. It's an ordinary constant like the others.) If we try to resolve this integral straight-away, we're bound to be frustrated. But we can use algebra to rewrite many integrals of this sort in the form

$$k_1 \int (a_1 x + b_1)^{-1} \, dx \ + \ k_2 \int (a_2 x + b_2)^{-1} \, dx$$

where x is the variable, and a_1, a_2, b_1, b_2, k_1, and k_2 are constants. Both of the integrands here are reciprocals of linear functions. We know how to work them out!

The transformation from the first integral above to the second one involves a big intuitive leap. It's unlikely to be obvious to anyone but a high-caliber mathematician. But we can get an idea of how it happens by doing a little bit of "reverse engineering." We can work backward from the above expression until we get the one before it. That way, we can see the logic behind each step. Once we've finished working through the derivation backward, we can turn around and "reverse engineer" the backward derivation to get the real one.

Let's start with the above expression, in which both addends are integrals that we already know how to do. We can move the constants into the integrands, getting

$$\int k_1 (a_1 x + b_1)^{-1} \, dx \ + \ \int k_2 (a_2 x + b_2)^{-1} \, dx$$

Remembering that the sum of two integrals is the same as the integral of the sum when both functions involve the same variable, we can rewrite this as

$$\int \left[k_1(a_1x + b_1)^{-1} + k_2(a_2x + b_2)^{-1} \right] dx$$

From algebra, we can use the general rule for adding two fractions, getting

$$\int (k_1a_2x + k_1b_2 + k_2a_1x + k_2b_1) \left[(a_1x + b_1)(a_2x + b_2) \right]^{-1} dx$$

The numerator can be rearranged, and the denominator multiplied out, to obtain

$$\int \left[(k_1a_2 + k_2a_1)x + (k_1b_2 + k_2b_1) \right] \left[a_1a_2x^2 + (a_1b_2 + b_1a_2)x + b_1b_2 \right]^{-1} dx$$

This looks a lot more complicated than it actually is. All combinations of *a*'s, *b*'s, and *k*'s are constants, because they're sums and products of constants. When we look closely, we can see that this integrand is the ratio of a linear function to a quadratic function. That's obvious if we make the following substitutions:

- Call $(k_1a_2 + k_2a_1)$ by the nickname *a*
- Call $(k_1b_2 + k_2b_1)$ by the nickname *b*
- Call a_1a_2 by the nickname *c*
- Call $(a_1b_2 + b_1a_2)$ by the nickname *d*
- Call b_1b_2 by the nickname *e*

When we assign these nicknames to our original constants, the above integral becomes

$$\int \left[(ax + b)(cx^2 + dx + e)^{-1} \right] dx$$

which is the original form we started with at the beginning of this section!

A working example

Now let's work out an integral of the above form. Consider this:

$$\int (2x + 8)(x^2 + 8x + 15)^{-1} dx$$

We want to get the integrand into a sum of fractions, each of which has a constant in the numerator and a linear function in the denominator. This is the difficult part of the process. From intermediate algebra, we can figure out that the integrand breaks down into the sum of $(x + 3)^{-1}$ and $(x + 5)^{-1}$. We're lucky in this case because the numerators are both 1, and the coefficients of *x* are also both 1. That means the above integral can be rewritten as

$$\int \left[(x + 3)^{-1} + (x + 5)^{-1} \right] dx$$

Because the integral of a sum is equal to the sum of the integrals, we have

$$\int (x + 3)^{-1} dx + \int (x + 5)^{-1} dx$$

Using the formula we took earlier from App. G, we get

$$\ln |x+3| + \ln |x+5| + c$$

We have therefore found that

$$\int (2x+8)(x^2 + 8x + 15)^{-1}\, dx = \ln |x+3| + \ln |x+5| + c$$

- -

Are you confused?

Do you have trouble with the algebra here? It's not easy. The goal always the same: Get the integrand into a sum of fractions in which the numerators are all constants, and the denominators are linear functions. Let's go through the algebra in the example we just finished, so you can get the general idea for future problems of this type. Take a look at the integral we resolved:

$$\int (2x+8)(x^2 + 8x + 15)^{-1}\, dx$$

The denominator, a quadratic function, can (fortunately) be factored like this:

$$x^2 + 8x + 15 = (x+3)(x+5)$$

In a sum of fractions, the denominator of the sum can always be expressed as the product of the individual denominators. In this situation, therefore, we will have two fractions, one with a denominator of $(x+3)$ and the other with a denominator of $(x+5)$. To find the numerators, we must look for constants p and q such that

$$p\,(x+3)^{-1} + q\,(x+5)^{-1} = (2x+8)(x^2 + 8x + 15)^{-1}$$

Do you remember the rule for adding two fractions or ratios in algebra? If not, here's a reminder. If we have numbers a, b, c, and d, then

$$ab^{-1} + cd^{-1} = (ad + bc)(bd)^{-1}$$

provided, of course, that neither b nor d are equal to 0. If we apply this rule to the left-hand side of the previous equation, the expression becomes

$$[p(x+5) + q\,(x+3)]\,[(x+3)(x+5)]^{-1}$$

which can be rewritten as

$$[p(x+5) + q(x+3)]\,(x^2 + 8x + 15)^{-1}$$

Substituting this on the left-hand side of the long equation we got a minute ago (count up to the fourth equation before this one), we obtain

$$[p(x+5) + q(x+3)]\,(x^2 + 8x + 15)^{-1} = (2x+8)(x^2 + 8x + 15)^{-1}$$

Notice that the denominators are the same on both sides of this equation. (We must assume that they can never become 0 if the equation is to make sense.) When we multiply through by the quantity $(x^2 + 8x + 15)$, we can get rid of the denominators, leaving ourselves with

$$p(x+5) + q(x+3) = (2x+8)$$

Using the distributive law of multiplication over addition, we can rewrite this as

$$px + 5p + qx + 3q = 2x + 8$$

which can be rearranged to

$$(p+q)x + (5p+3q) = 2x + 8$$

We can split this into a two-by-two linear system:

$$p + q = 2 \text{ (the coefficient of } x)$$

and

$$5p + 3q = 8 \text{ (the stand-alone constant)}$$

This pair of *simultaneous linear equations* solves to $p = 1$ and $q = 1$. Those are the numerators of the two fractions we've been looking for. That means the sum of partial fractions is

$$(x+3)^{-1} + (x+5)^{-1}$$

giving us the integral

$$\int [(x+3)^{-1} + (x+5)^{-1}]\, dx$$

From here, we have only some calculus to go through, and we'll finish the problem with ease.

Here's a challenge!

Use the technique of partial fractions to resolve the indefinite integral

$$\int (8x+1)(2x^2 - x - 1)^{-1}\, dx$$

Solution

Employing our algebra skills, we can figure out that the integrand breaks down into a sum of two constant multiples of reciprocal linear functions, like this:

$$\int [2\,(2x+1)^{-1} + 3\,(x-1)^{-1}]\, dx$$

When we split this into a sum of two integrals and pull out the constants, we get

$$2\int (2x+1)^{-1}\, dx + 3\int (x-1)^{-1}\, dx$$

Applying the "old reliable" formula from App. G to each addend gives us

$$2 \cdot 2^{-1} \ln |2x + 1| + 3 \ln |x - 1| + c$$

which simplifies to

$$\ln |2x + 1| + 3 \ln |x - 1| + c$$

The complete statement of the integral and its resolution is therefore

$$\int (8x + 1)(2x^2 - x - 1)^{-1} \, dx = \ln |2x + 1| + 3 \ln |x - 1| + c$$

- -

Practice Exercises

This is an open-book quiz. You may (and should) refer to the text as you solve these problems. Don't hurry! You'll find worked-out answers in App. B. The solutions in the appendix may not represent the only way a problem can be figured out. If you think you can solve a particular problem in a quicker or better way than you see there, by all means try it!

1. Evaluate the following definite integral using the principle of linearity:

$$\int_{0}^{\pi} (3 \sin x - 5e^x) \, dx$$

2. Evaluate the following definite integral using the principle of linearity:

$$\int_{1}^{2} (5e^x + 7x^{-1}) \, dx$$

3. Evaluate the following definite integral using the principle of linearity:

$$\int_{-\pi/2}^{\pi/2} (3 \sin x + 2 \cos x) \, dx$$

4. Based on the principles we've learned so far, prove that if f_1 and f_2 are integrable functions of x over a continuous interval (a, b), and if k_1 and k_2 are constants, then

$$\int_{a}^{b} \{k_1 [f_1 (x)] - k_2 [f_2 (x)]\} \, dx = k_1 \int_{a}^{b} f_1 (x) \, dx + k_2 \int_{b}^{a} f_2 (x) \, dx$$

Note the plus and minus signs, and pay attention to the direction we go over the interval in each case!

5. Use integration by parts to find

$$\int (3 \cos x)(2x - 4) \, dx$$

6. Use integration by parts to find

$$\int 7x^2 \cos x \, dx$$

7. Use integration by parts to evaluate

$$\int_0^\pi (\pi/2 - 2x)(5 \cos x) \, dx$$

Here's a hint: When we integrate a definite integral by parts, we can work through the calculations with indefinite integrals, and restore the original bounds at the end of the process.

8. Resolve the indefinite integral

$$\int -5 \, (2x - 7)^{-1} \, dx$$

9. Use the technique of partial fractions to resolve the indefinite integral

$$\int 2x \, (x^2 - 1)^{-1} \, dx$$

10. Use the technique of partial fractions to resolve the indefinite integral

$$\int (5x - 1)(x^2 - x - 2)^{-1} \, dx$$

Review Questions and Answers

Part Two

This is not a test! It's a review of important general concepts you learned in the previous nine chapters. Read it though slowly and let it "sink in." If you're confused about anything here, or about anything in the section you've just finished, go back and study that material some more.

Chapter 11

Question 11-1

How can we write the series

$$a_1 + a_2 + a_3 + a_4 + a_5 + a_6 = 200$$

using the summation notation?

Answer 11-1

It can be written like this:

$$\sum_{i=1}^{6} a_i = 200$$

Question 11-2

Consider an open-ended series, in which n terms are added up to obtain a final sum of x. How can this be written using the summation notation?

Answer 11-2

It can be written like this:

$$\sum_{i=1}^{n} a_i = x$$

Question 11-3

Suppose we sum a series starting with 1/3, then add 1/9, then add 1/27, then add 1/81, and go on forever, each time cutting the term to 1/3 of its previous value. How do we express this series in summation notation?

Answer 11-3

We write this series in summation notation as

$$\sum_{i=1}^{\infty} 1/3^i$$

Question 11-4

Suppose we sum a series starting with 1, then add 1/3, then add 1/9, then add 1/27, then add 1/81, and continue forever, each time cutting the term to 1/3 of its previous value. How do we express this series in summation notation? How does this differ from the series in Answer 11-3?

Answer 11-4

We write this series in summation notation as

$$\sum_{i=0}^{\infty} 1/3^i$$

We begin at $i = 0$, raising 1/3 to the zeroth power (which is 1). In Answer 11-3, we started out at $i = 1$, raising 1/3 to the first power (which is 1/3).

Question 11-5

Imagine a function f graphed in the xy-plane, such that $y = f(x)$. Part of the graph lies above the x axis, and another part lies below the x axis, as shown in Fig. 20-1. If we integrate from $x = a$ to $x = b$ while moving from left to right (so $a < b$), how are areas defined by the curve with respect to the x axis?

Answer 11-5

If we integrate in the positive-x direction (from left to right), then any part of the region underneath the x axis but above the curve has negative area, and any part of the region above the x axis but below the curve has positive area.

Question 11-6

Bernhard Riemann found that the area defined by a curve can be approximated by adding up the areas of rectangles as shown in Fig. 20-2. The width of each rectangle is Δx. How can this width be expressed in terms of the bounds $x = a$ and $x = b$, and the number of rectangles n?

Answer 11-6

The widths of all the rectangles are the same. Therefore, the width of each one is equal to the width of the interval from $x = a$ to $x = b$, divided by the number of rectangles n:

$$\Delta x = (b - a)/n$$

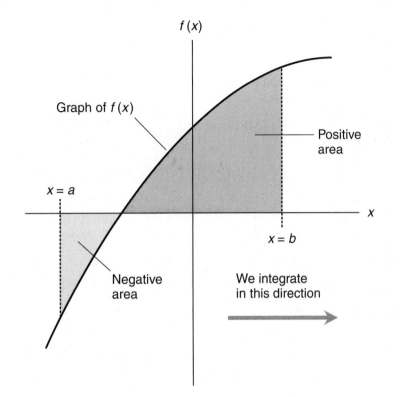

Figure 20-1   Illustration for Question and Answer 11-5.

Question 11-7

Consider the *i*th rectangle in Fig. 20-2. Assume that *i* is an integer somewhere between, and including, 1 (representing the leftmost rectangle) and *n* (representing the rightmost rectangle). If the curve represents our function *f,* what is the height of the *i*th rectangle?

Answer 11-7

The height of the *i*th rectangle is equal to

$$f(a + i\Delta x)$$

which is the same as

$$f[a + i(b - a)/n]$$

Question 11-8

Again, consider the *i*th rectangle in Fig. 20-2. What is the area of the *i*th rectangle?

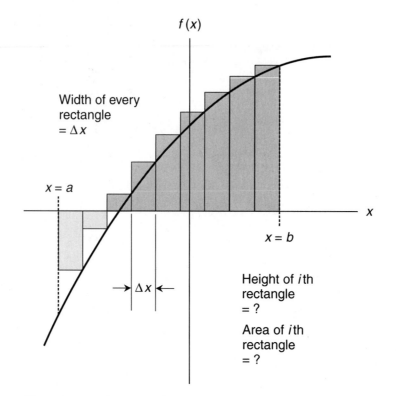

Figure 20-2   Illustration for Questions and Answers 11-6 through 11-9.

Answer 11-8

The area of the *i*th rectangle is equal to its width times its height, or

$$\Delta x \cdot f(a + i\Delta x)$$

which is the same as

$$[(b-a)/n] \cdot f[a + i(b-a)/n]$$

Question 11-9

How can we express the *exact* area defined by the curve in Fig. 20-2 as a limit in terms of Δx, *i*, *n*, *a*, and the function *f*? How can we express that area as a limit in terms of *i*, *n*, *a*, *b*, and the function *f*?

Answer 11-9

The exact area is the limit from $i = 1$ to n, as Δx approaches 0, of the sum of the rectangle areas. That's written formally as

$$\underset{\Delta x \to \infty}{Lim} \sum_{i=1}^{n} \Delta x \cdot f(a + i\Delta x)$$

or as

$$\underset{n \to \infty}{Lim} \sum_{i=1}^{n} [(b - a)/n] \cdot f[a + i(b - a)/n]$$

Question 11-10

How can we find the average value of a function over a specific interval, assuming the function is continuous and can be integrated over that interval?

Answer 11-10

Consider a function $f(x)$ that is continuous from $x = a$ to $x = b$, where $a < b$. We can determine the average value $f^*_{a:b}$ over the interval (a,b) by integrating f from a to b, and then dividing by the interval width $(b - a)$. Mathematically,

$$f^*_{a:b} = (b - a)^{-1} \int_{a}^{b} f(x)\ dx$$

Chapter 12

Question 12-1

What is the antiderivative of the zero function?

Answer 12-1

The antiderivative of the zero function can be any imaginable constant function.

Question 12-2

Why are all antiderivatives ambiguous by nature?

Answer 12-2

Whenever we antidifferentiate a function, we must add a constant, but we don't necessarily know its value.

Question 12-3

Imagine a function f of a variable x, such that

$$f(x) = 4$$

What's the general antiderivative?

Answer 12-3

The general antiderivative is

$$F(x) = 4x + c$$

where c is an unspecified real-number constant.

Question 12-4

Imagine a function g of a variable y, such that

$$g(y) = a$$

where a is some specific real number. What's the general antiderivative?

Answer 12-4

The general antiderivative is

$$G(y) = ay + c$$

where c is an unspecified real-number constant.

Question 12-5

Imagine a function h of a variable z, such that

$$h(z) = bz$$

where b is some specific real number. What's the general antiderivative?

Answer 12-5

The general antiderivative is

$$H(z) = bz^2/2 + c$$

where c is an unspecified real-number constant.

Question 12-6

How do we find the general antiderivative of a function that raises a variable to a nonnegative integer power and then multiplies by a constant? Call the function f, the variable x, and the constant a, so that

$$f(x) = ax^n$$

where n is a nonnegative integer.

Answer 12-6

The general antiderivative is

$$F(x) = ax^{(n+1)}/(n+1) + c$$

where c is an unspecified real-number constant.

Question 12-7

How can we expand the above rule so that it applies to any real-number power of the variable except the -1 power (reciprocal)?

Answer 12-7

Let k represent any real number other than -1. If

$$f(x) = ax^k$$

then

$$F(x) = ax^{(k+1)}/(k+1) + c$$

where c is an unspecified real-number constant. If k is negative, we must be sure that $x \neq 0$. Otherwise, $F(x)$ is undefined.

Question 12-8

Why doesn't the above rule work if $k = -1$?

Answer 12-8

We get into trouble with this rule if $k = -1$, because it results in our having to divide by 0. Consider the function

$$f(x) = ax^{-1}$$

where a is a specific constant. When we try to apply the above described rule, we get

$$F(x) = [ax^{(-1+1)}/(-1 + 1)] + c = ax^0/0 + c$$

Question 12-9

What are the two general rules for indefinite integration that allow us to "pull out" constants or negatives? What's the rule concerning the sum of indefinite integrals?

Answer 12-9

If f is an integrable function of a variable x, and if k is a real-number constant, then

$$\int k\,[f(x)]\ dx = k \int f(x)\ dx$$

In the specific case where $k = -1$, we have

$$\int -[f(x)]\, dx = -\int f(x)\, dx$$

If $f_1, f_2, f_3, \ldots$, and f_n are integrable functions of a variable x, then

$$\int [f_1(x) + f_2(x) + f_3(x) + \cdots + f_n(x)]\, dx$$

$$= \int f_1(x)\, dx + \int f_2(x)\, dx + \int f_3(x)\, dx + \cdots + \int f_n(x)\, dx$$

Question 12-10

What is the Fundamental Theorem of Calculus for integrals? What happens to the constant of integration when we find a definite integral using this theorem?

Answer 12-10

Imagine that f is a continuous real-number function of a variable x. Let a and b be values in the domain of f with $a < b$, and let F be a specific antiderivative with a constant of integration c. Then the definite integral from a to b is

$$\int_a^b f(x)\, dx = F(b) - F(a)$$

Sometimes this is written as

$$\int_a^b f(x)\, dx = F(x) \Big]_a^b$$

where the expression on the right-hand side of the equals sign is read "$F(x)$ evaluated from a to b." When we find a definite integral this way, the constant of integration subtracts from itself and disappears.

Chapter 13

Question 13-1

What happens when we integrate a function over a certain interval in one direction, and then integrate the same function over the same interval in the opposite direction?

Answer 13-1

The reversal rule for definite integration tells that if f is an integrable function of x over an interval between two limits a and b, then

$$\int_a^b f(x)\, dx = -\int_b^a f(x)\, dx$$

Question 13-2

How can we illustrate the reversal rule as a graph, showing the areas defined by a curve that crosses the independent-variable axis within the interval of integration?

Answer 13-2

When we integrate a function $f(x)$ over an interval (a,b) in the positive x direction, areas above the x axis are positive while areas below the x axis are negative, as shown in Fig. 20-3A. When we integrate in the negative x direction, areas above the x axis are negative while areas below the x axis are positive, as shown in Fig. 20-3B.

Question 13-3

Imagine an interval split into two contiguous subintervals that don't overlap. Suppose that we integrate the subintervals from left to right separately and then add the results. Then, we integrate over the whole interval from left to right. How do the integrals compare?

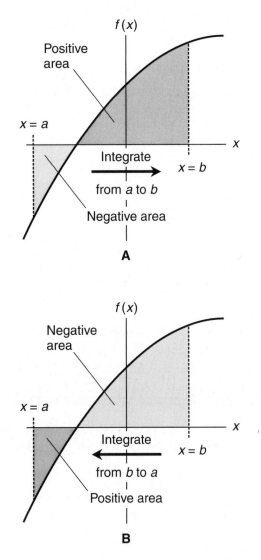

Figure 20-3 Illustration for Question and Answer 13-2.

Answer 13-3

The split-interval rule for definite integration says that if $f(x)$ is a function that's integrable over a continuous interval containing real numbers a, b, and c, then

$$\int_b^a f(x)\ dx + \int_b^c f(x)\ dx = \int_a^c f(x)\ dx$$

Question 13-4

Is it necessary to have $a < b < c$ for the above formula to work?

Answer 13-4

No, but the two subintervals must not overlap, and there must be no gap between them.

Question 13-5

Consider the function

$$f(x) = x^{\pi}$$

If the domain is the set of all nonnegative real numbers, then this function is continuous and integrable over any interval in that domain. Now suppose we find

$$\int_0^4 x^{\pi}\ dx$$

Then we find

$$\int_4^3 x^{\pi}\ dx$$

Finally, we add the two definite integrals. What's the result as a single definite integral? Here's a hint: Note the directions of integration in each case.

Answer 13-5

We can apply the split-interval rule, letting $a = 0$, $b = 4$, and $c = 3$, obtaining

$$\int_0^4 x^{\pi}\ dx + \int_4^3 x^{\pi}\ dx = \int_0^3 x^{\pi}\ dx$$

Question 13-6

If we integrate a function from one value of the independent variable to another, does it matter how we get from the starting point to the finishing point?

Answer 13-6

No. As long as the function is integrable everywhere we go, and provided we don't change the starting or finishing points, we'll always get the same final answer.

Question 13-7

What is the substitution rule for definite integration?

Answer 13-7

Suppose that we have two integrable functions f and g, such that f operates on x while g operates on $f(x)$. If a and b are the bounds of integration, we can rewrite

$$\int_a^b g[f(x)] \cdot f'(x)\,dx$$

in the form

$$\int_{f(a)}^{f(b)} g(y)\,dy$$

where $y = f(x)$. Then we can evaluate this new integral to obtain the value of the original definite integral.

Question 13-8

How can we evaluate this definite integral using the substitution method?

$$\int_0^1 3\,(x+5)^2\,dx$$

Answer 13-8

We can consider the integrand as a composite function where

$$f(x) = x+5$$

and

$$g(y) = 3y^2$$

In this case,

$$f'(x) = 1$$

so the original integral is in the form

$$\int_0^1 g[f(x)] \cdot f'(x)\,dx$$

We calculate

$$f(0) = 0 + 5 = 5$$

and

$$f(1) = 1 + 5 = 6$$

Now we can write the integral as

$$\int_5^6 g(y)\,dy = \int_5^6 3y^2\,dy$$

where $y = f(x)$. With the constant of integration set to 0, we get the basic antiderivative

$$G(y) = y^3$$

We calculate

$$G(6) - G(5) = 6^3 - 5^3 = 91$$

We've found that

$$\int_0^1 3(x+5)^2\,dx = 91$$

Question 13-9

When we evaluate a definite integral using the substitution method, is it normal for the bounds of integration to change during part of the process?

Answer 13-9

Yes. In the above situation, we must add 5 to both bounds when we integrate g with respect to y. In the problem we just solved, g operates on $(x+5)$, not on x, so we must integrate g from 5 to 6, not from 0 to 1.

Question 13-10

How can we evaluate the following integral using the substitution technique?

$$\int_{-5}^{-4} e^{(x+5)}\,dx$$

Answer 13-10

We can consider the integrand as a composite function in which

$$f(x) = x + 5$$

and

$$g(y) = e^y$$

In this situation, we have

$$f'(x) = 1$$

so the original integral takes the form

$$\int_{-5}^{-4} g[f(x)] \cdot f'(x)\, dx$$

We calculate the new bounds as

$$f(-5) = -5 + 5 = 0$$

and

$$f(-4) = -4 + 5 = 1$$

We rewrite the integral as

$$\int_{0}^{1} g(y)\, dy = \int_{0}^{1} e^y\, dy$$

The exponential function is its own basic antiderivative, so

$$G(y) = e^y$$

When we evaluate from 0 to 1, we get

$$G(1) - G(0) = e^1 - e^0 = e - 1$$

We've determined that

$$\int_{-5}^{-4} e^{(x+5)}\, dx = e - 1$$

Chapter 14

Question 14-1
What's an improper integral?

Answer 14-1
There are two major types of improper integrals. One type is a definite integral of a function that contains a singularity in, or at either bound of, the interval over which we want to integrate. The other type is a definite integral evaluated over an infinitely wide interval.

Question 14-2
An improper integral involves a region that's "infinitely stretched out." Aren't such integrals always undefined?

Answer 14-2
Some improper integrals are undefined, but many are defined and finite.

Question 14-3

Consider the following definite integral of $h(x)$ over an open interval (a,b) with $a < b$:

$$\int_a^b h(x)\,dx$$

Suppose that the value of h blows up (or down) as x approaches the upper bound b from the negative direction. How can we approximate the integral?

Answer 14-3

We can invent a tiny, "adjustable" positive number ε, and then evaluate

$$\int_a^{b-\varepsilon} h(x)\,dx$$

As we bring ε closer and closer to 0, this integral approaches the actual value of

$$\int_a^b h(x)\,dx$$

assuming that this improper integral is defined and finite.

Question 14-4

Consider the following integral of $h(t)$ over an open interval (m,n) with $m < n$:

$$\int_m^n h(t)\,dt$$

Suppose that the value of h blows up (or down) as t approaches the lower bound m from the positive direction. How can we approximate the integral?

Answer 14-4

We can invent a tiny, "adjustable" positive number δ, and then evaluate

$$\int_{m+\delta}^n h(t)\,dt$$

As we make δ approach 0, the above integral approaches

$$\int_m^n h(t)\,dt$$

assuming that the improper integral is defined and finite.

Question 14-5

Consider the following integral of $g(z)$ over an open interval (c,q) with $c < q$:

$$\int_c^q g(z)\,dz$$

where the lower bound, c, is a constant but the upper bound, q, is a variable. What happens if we make q increase endlessly?

Answer 14-5

If q increases indefinitely, then the definite integral of $g(z)$ with respect to z, evaluated from c to q, approaches

$$\int_{c}^{\infty} g(z)\, dz$$

Question 14-6

Consider the following integral of $f(s)$ over an open interval (p,a) with $p < a$:

$$\int_{p}^{a} f(s)\, ds$$

where the upper bound, a, is a constant but the lower bound, p, is a variable. What happens if we make p to decrease (grow larger negatively) without end?

Answer 14-6

If p becomes larger negatively without end, then the definite integral of $f(s)$ with respect to s, evaluated from p to a, approaches

$$\int_{-\infty}^{a} f(s)\, ds$$

Question 14-7

Figure 20-4 is a graph of a function with a singularity. Suppose we want to find

$$\int_{a}^{b} g(x)\, dx$$

What's the correct way to work out this problem?

Answer 14-7

We must split the interval into two parts, one on either side of the singularity. Because the singularity is at $x = 0$, we should use the intervals $(a,0)$ and $(0,b)$. First, we try to evaluate

$$\int_{a}^{0} g(x)\, dx$$

If this is undefined, then the original integral is undefined. If the above integral is defined and finite, then we try to evaluate

$$\int_{0}^{b} g(x)\, dx$$

If this is undefined, then the original integral is undefined. But if it's defined and finite, then

$$\int_{a}^{0} g(x)\, dx + \int_{0}^{b} g(x)\, dx = \int_{a}^{b} g(x)\, dx$$

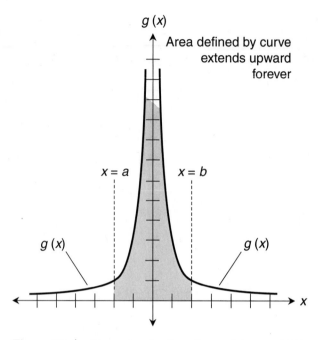

Figure 20-4   Illustration for Question and Answer 14-7.

Question 14-8

How can we find the following integral?

$$\int_{-1}^{0} x^{-4/5}/5 \; dx$$

Answer 14-8

There's a singularity at $x = 0$. We can verify this fact by plugging in 0 for x in the function and doing the arithmetic:

$$g(0) = 0^{-4/5}/5 = [1/(0^{4/5})] \,/\, 5 = (1/0) \,/\, 5$$

That's undefined because there's a denominator of 0. To evaluate the integral, we take a tiny positive number ε, subtract it from the upper bound at $x = 0$, and obtain

$$\int_{-1}^{-\varepsilon} x^{-4/5}/5 \; dx$$

The antiderivative without the constant of integration is

$$G(x) = x^{1/5}$$

When we evaluate this antiderivative from -1 to $-\varepsilon$, we get

$$G(-\varepsilon) - G(-1) = (-\varepsilon)^{1/5} - (-1)^{1/5} = 1 - \varepsilon^{1/5}$$

Remember that the 1/5 power is the fifth root. Now we consider

$$\underset{\varepsilon \to 0+}{Lim} \ 1 - \varepsilon^{1/5}$$

As ε approaches 0 from the positive direction, the value of $\varepsilon^{1/5}$ also approaches 0 from the positive direction. That means the quantity $(1 - \varepsilon^{1/5})$ approaches 1, so

$$\underset{\varepsilon \to 0+}{Lim} \int_{-1}^{-\varepsilon} x^{-4/5}/5 \ dx = 1$$

Therefore

$$\int_{-1}^{0} x^{-4/5}/5 \ dx = 1$$

Question 14-9

How can we find the following integral?

$$\int_{0}^{1} x^{-4/5}/5 \ dx$$

Answer 14-9

We take a tiny positive number δ, add it to the lower bound at $x = 0$, and obtain

$$\int_{\delta}^{1} x^{-4/5}/5 \ dx$$

Once again, the antiderivative of our function is

$$G(x) = x^{1/5}$$

When we evaluate this from δ to 1, we get

$$G(1) - G(\delta) = 1^{1/5} - \delta^{1/5} = 1 - \delta^{1/5}$$

Now we consider

$$\underset{\delta \to 0+}{Lim} \ 1 - \delta^{1/5}$$

As δ approaches 0 from the positive direction, the value of $\delta^{1/5}$ also approaches 0 from the positive direction. That means the quantity $(1 - \delta^{1/5})$ approaches 1, so

$$\underset{\delta \to 0+}{Lim} \int_{\delta}^{1} x^{-4/5}/5 \ dx = 1$$

Therefore

$$\int_{0}^{1} x^{-4/5}/5 \ dx = 1$$

Question 14-10

How can we find the following integral?

$$\int_{-\infty}^{-2} 384x^{-7}\, dx$$

Answer 14-10

If we call our integrand function f, then the basic antiderivative is

$$F(x) = -64x^{-6}$$

Now let's consider the definite integral

$$\int_{p}^{-2} 384x^{-7}\, dx$$

where p is a variable that decreases endlessly (grows larger in the negative direction forever). To find this integral, we must determine

$$\underset{p \to -\infty}{Lim}\ F(-2) - F(p)$$

We can calculate

$$F(-2) = -64 \cdot (-2)^{-6} = -1$$

As p becomes large negatively without bound, $F(p)$ approaches 0. We can see this by inputting some numbers into the antiderivative and doing the arithmetic:

$$F(-10^2) = -64 \cdot (-10^2)^{-6} = -64 \cdot 10^{-12}$$
$$F(-10^5) = -64 \cdot (-10^5)^{-6} = -64 \cdot 10^{-30}$$
$$F(-10^2) = -64 \cdot (-10^8)^{-6} = -64 \cdot 10^{-48}$$

Now we know that

$$\underset{p \to -\infty}{Lim}\ F(-2) - F(p) = -1 - 0 = -1$$

Therefore

$$\int_{-\infty}^{-2} 384x^{-7}\, dx = -1$$

Chapter 15

Question 15-1

Look back and reexamine Question and Answer 12-9 on pages 331 and 332. How can we modify these rules if we want to use them with definite integrals?

Answer 15-1

Consider the functions $f(x)$, $f_1(x)$, and $f_2(x)$. Suppose the bounds of integration are $x = a$ and $x = b$. Then these three rules apply:

$$\int_a^b k\,[f(x)]\,dx = k\int_a^b f(x)\,dx$$

$$\int_a^b -[f(x)]\,dx = -\int_a^b f(x)\,dx$$

$$\int_a^b [f_1(x) + f_2(x)]\,dx = \int_a^b f_1(x)\,dx + \int_a^b f_2(x)\,dx$$

The last rule can be extrapolated to any finite number of functions of x.

Question 15-2

What precautions must we take when applying the rules in Answer 15-1?

Answer 15-2

We must be sure that the interval is the same for each function, all the functions are integrable over the interval, and all the integrals are done in the same direction.

Question 15-3

Consider the following functions over the interval from $x = 2$ to $x = 4$:

$$g(x) = 4x$$

and

$$h(x) = 10x$$

Integrate these functions individually. What happens when we add the results?

Answer 15-3

First, we find the definite integral of $g(x)$ from $x = 2$ to $x = 4$:

$$\int_2^4 4x\,dx$$

The basic antiderivative is

$$G(x) = 2x^2$$

When we evaluate this from $x = 2$ to $x = 4$, we get

$$G(4) - G(2) = 2 \cdot 4^2 - 2 \cdot 2^2 = 24$$

Next, we find the definite integral of $h(x)$ from $x = 2$ to $x = 4$:

$$\int_2^4 10x \, dx$$

The basic antiderivative is

$$H(x) = 5x^2$$

When we evaluate this from $x = 2$ to $x = 4$, we get

$$H(4) - H(2) = 5 \cdot 4^2 - 5 \cdot 2^2 = 60$$

When we add these results, we get $24 + 60$, which is 84.

Question 15-4

Consider again the following two functions over the interval from $x = 2$ to $x = 4$:

$$g(x) = 4x$$

and

$$h(x) = 10x$$

Add the functions and then integrate. How does this compare with Answer 15-3?

Answer 15-4

If we call the sum $p(x)$, we have

$$g(x) + h(x) = p(x) = 4x + 10x = 14x$$

We want to work out the integral

$$\int_2^4 14x \, dx$$

The basic antiderivative is

$$P(x) = 7x^2$$

When we evaluate this from $x = 2$ to $x = 4$, we get

$$P(4) - P(2) = 7 \cdot 4^2 - 7 \cdot 2^2 = 84$$

That's the same result as we got in Answer 15-3.

Question 15-5

Consider the following two functions over the interval from $x = -1$ to $x = 1$:

$$g(x) = -2x$$

and

$$h(x) = 3x^2$$

Integrate these functions individually. What happens when we add the results?

Answer 15-5

First, we find the definite integral of $g(x)$ from $x = -1$ to $x = 1$:

$$\int_{-1}^{1} -2x \, dx$$

The basic antiderivative is

$$G(x) = -x^2$$

When we evaluate this from $x = -1$ to $x = 1$, we get

$$G(1) - G(-1) = -1^2 - [-(-1)^2] = 0$$

Next, we find the definite integral of $h(x)$ with respect to x from $x = -1$ to $x = 1$:

$$\int_{-1}^{1} 3x^2 \, dx$$

The basic antiderivative is

$$H(x) = x^3$$

When we evaluate this from $x = -1$ to $x = 1$, we get

$$H(1) - H(-1) = 1^3 - (-1)^3 = 2$$

When we add these results, we get $0 + 2$, which is 2.

Question 15-6

Consider again the following two functions over the interval from $x = -1$ to $x = 1$:

$$g(x) = -2x$$

and

$$h(x) = 3x^2$$

Add the functions and integrate. How does this result compare with Answer 15-5?

Answer 15-6

If we call the sum $q(x)$, we can write it in order of descending exponents as

$$q(x) = 3x^2 - 2x$$

We want to work out

$$\int_{-1}^{1} (3x^2 - 2x)\ dx$$

The basic antiderivative is

$$Q(x) = x^3 - x^2$$

When we evaluate this from $x = -1$ to $x = 1$, we get

$$Q(1) - Q(-1) = (1^3 - 1^2) - [(-1)^3 - (-1)^2] = 2$$

That's the same result as we got in Answer 15-5.

Question 15-7

How can we find the following integral by breaking the integrand into monomials?

$$\int (9z^2 + 8z - 7)\ dz$$

Answer 15-7

This is the integral of a sum of three monomial functions. Let's call them

$$f_1(z) = 9z^2$$
$$f_2(z) = 8z$$
$$f_3(z) = -7$$

The indefinite integrals of the individual monomials are

$$\int f_1(z)\ dz = \int 9z^2\ dz = 3z^3 + c_1$$
$$\int f_2(z)\ dz = \int 8z\ dz = 4z^2 + c_2$$
$$\int f_3(z)\ dz = \int -7\ dz = -7z + c_3$$

where c_1, c_2, and c_3 are the constants of integration. The original integral is

$$\int (9z^2 + 8z - 7)\ dz = \int f_1(z)\ dz + \int f_2(z)\ dz + \int f_3(z)\ dz$$
$$= \int 9z^2\ dz + \int 8z\ dz + \int -7\ dz = 3z^3 + c_1 + 4z^2 + c_2 - 7z + c_3$$
$$= 3z^3 + 4z^2 - 7z + c_1 + c_2 + c_3$$

If we consolidate $c_1 + c_1 + c_3$ into a single constant of integration c, then

$$\int (9z^2 + 8z - 7)\ dz = 3z^3 + 4z^2 - 7z + c$$

Question 15-8

How can we find the following definite integral with respect to z?

$$\int_0^1 (9z^2 + 8z - 7) \, dz$$

Answer 15-8

From Answer 15-7, we know that

$$\int (9z^2 + 8z - 7) \, dz = 3z^3 + 4z^2 - 7z + c$$

We want to evaluate this from $z = 0$ to $z = 1$. We must be sure that the integrand doesn't blow up in the interval for which $0 < z < 1$. To do that, we must look at each individual term as a function. Let's call the integrand f, so we have

$$f(z) = 9z^2 + 8z - 7$$

This breaks down into

$$f_1(z) = 9z^2$$
$$f_2(z) = 8z$$
$$f_3(z) = 7$$

The graph of f_1 is a parabola that opens upward. The graph of f_2 is a line through the origin with a slope of 8. The graph of f_3 is a line with a slope of 0. (Feel free to sketch the graphs to visualize these facts.) None of these functions is singular for $0 < z < 1$, so we can input the boundary values to the antiderivative $F(z)$ and then subtract the results. Without the constant of integration, that antiderivative is

$$F(z) = 3z^3 + 4z^2 - 7z$$

When we input 1, we get

$$F(1) = 3 \cdot 1^3 + 4 \cdot 1^2 - 7 \cdot 1 = 0$$

When we input 0, we get

$$F(0) = 3 \cdot 0^3 + 4 \cdot 0^2 - 7 \cdot 0 = 0$$

The definite integral is

$$F(1) - F(0) = 0 - 0 = 0$$

so we've determined that

$$\int_0^1 (9z^2 + 8z - 7) \, dz = 0$$

Question 15-9

How can we find the following integral by breaking the integrand into monomials?

$$\int (y^{1/3} - y^{-1/3} - y^{-2/3})\, dy$$

Answer 15-9

This is the integral of a sum of three monomial functions. Let's call them

$$f_1(y) = y^{1/3}$$
$$f_2(y) = -y^{-1/3}$$
$$f_3(y) = -y^{-2/3}$$

The indefinite integrals of the individual monomials are

$$\int f_1(y)\, dy = \int y^{1/3}\, dy = (3/4)y^{4/3} + c_1$$

$$\int f_2(y)\, dy = \int -y^{-1/3}\, dy = -(3/2)y^{2/3} + c_2$$

$$\int f_3(y)\, dy = \int -y^{-2/3}\, dy = -3y^{1/3} + c_3$$

where c_1, c_2, and c_3 are the constants of integration. The original integral is

$$\int (y^{1/3} - y^{-1/3} - y^{-2/3})\, dy = \int f_1(y)\, dy + \int f_2(y)\, dy + \int f_3(y)\, dy$$

$$= \int y^{1/3}\, dy + \int -y^{-1/3}\, dy + \int -y^{-2/3}\, dy$$

$$= (3/4)y^{4/3} + c_1 - (3/2)y^{2/3} + c_2 - 3y^{1/3} + c_3$$

$$= (3/4)y^{4/3} - (3/2)y^{2/3} - 3y^{1/3} + c_1 + c_2 + c_3$$

Consolidating $c_1 + c_2 + c_3$ into a single constant c, we obtain

$$\int (y^{1/3} - y^{-1/3} - y^{-2/3})\, dy = (3/4)y^{4/3} - (3/2)y^{2/3} - 3y^{1/3} + c$$

Question 15-10

How can we calculate the following definite integral?

$$\int_0^1 [(6/7)x^{-1/7} - (5/6)x^{-1/6}]\, dx$$

Answer 15-10

The integrand is a sum of two monomial functions. Let's call them

$$f_1(x) = (6/7)x^{-1/7}$$

and

$$f_2(x) = -(5/6)x^{-1/6}$$

Both of these functions are singular at $x = 0$, the lower bound of our interval. That means we must use the techniques for evaluating improper integrals, along with the sum rule. Let's go through four steps:

- Split the original integral into two separate improper integrals
- Evaluate those two integrals independently
- If either of them is undefined, conclude that the entire integral is undefined
- If they are both defined, add them to get the final result

When we break the original integral into a sum of two integrals, we get

$$\int_0^1 (6/7)x^{-1/7}\, dx + \int_0^1 -(5/6)x^{-1/6}\, dx$$

Let's evaluate the left-hand integral first. We take a tiny positive δ, add it to the lower limit at $x = 0$, and obtain

$$\int_\delta^1 (6/7)x^{-1/7}\, dx$$

The basic antiderivative is

$$F_1(x) = x^{6/7}$$

When we evaluate this from δ to 1, we get

$$F_1(1) - F_1(\delta) = 1^{6/7} - \delta^{6/7} = 1 - \delta^{6/7}$$

Now we must determine

$$\underset{\delta \to 0+}{Lim}\ 1 - \delta^{6/7}$$

As δ approaches 0 from the positive direction, the value of $\delta^{6/7}$ approaches 0. That means the quantity $(1 - \delta^{6/7})$ approaches 1, so

$$\underset{\delta \to 0+}{Lim} \int_\delta^1 (6/7)x^{-1/7}\, dx = 1$$

Therefore, the left-hand addend in the "big sum" is

$$\int_0^1 (6/7)x^{-1/7}\, dx = 1$$

Now let's evaluate the right-hand integral in our "big sum." We take a tiny positive δ, add it to the lower limit at $x - 0$, and get

$$\int_\delta^1 -(5/6)x^{-1/6}\, dx$$

The basic antiderivative is

$$F_2(x) = -x^{5/6}$$

When we evaluate this from δ to 1, we get

$$F_2(1) - F_2(\delta) = -1^{5/6} - (-\delta^{5/6}) = \delta^{5/6} - 1$$

Now we consider

$$\underset{\delta \to 0+}{Lim}\ \delta^{5/6} - 1$$

As δ approaches 0 from the positive direction, the value of $\delta^{5/6}$ approaches 0. Therefore, the quantity $(\delta^{5/6} - 1)$ approaches -1, so

$$\underset{\delta \to 0+}{Lim} \int_{\delta}^{1} -(5/6)x^{-1/6}\ dx = -1$$

This tells us that the right-hand addend in our "big sum" is

$$\int_{0}^{1} -(5/6)x^{-1/6}\ dx = -1$$

We have now found the values of both addends. The left-hand integral is 1, and the right-hand integral is -1. We can apply the sum rule to conclude that

$$\int_{0}^{1} [(6/7)x^{-1/7} - (5/6)x^{-1/6}]\ dx = 1 + (-1) = 0$$

Chapter 16

Question 16-1

How can we find the true geometric area of the region between the line and the parabola in Fig. 20-5?

Answer 16-1

We can proceed by following these steps, in order:

- Find the function $f(x)$ represented by the parabola.
- Find the function $g(x)$ represented by the line.
- Find the difference function $f(x) - g(x)$, remembering that must subtract the "bottom function" from the "top function" over the interval between the points where the graphs intersect.
- Integrate the difference function over the interval between the points where the graphs intersect.

Question 16-2

What is the function $f(x)$ represented by the parabola in Fig. 20-5?

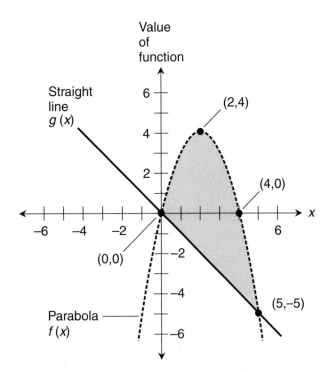

Figure 20-5   Illustration for Questions and Answers 16-1 through 16-6.

Answer 16-2

The parabola crosses the x axis at (0,0) and (4,0), so it represents a quadratic function whose zeros are $x = 0$ and $x = 4$. From algebra, we can deduce that the function is of the form

$$f(x) = k(x - 0)(x - 4) = kx(x - 4) = kx^2 - 4kx$$

where *k* is a constant. To find *k*, we note the point (2,4) at the vertex of the parabola. These coordinates tell us that

$$f(2) = 4 = k \cdot (2^2) - 4k \cdot 2 = -4k$$

This simplifies to

$$4 = -4k$$

which solves to $k = -1$. By substitution,

$$f(x) = -x^2 + 4x$$

Question 16-3

What is the function $g(x)$ represented by the straight line in Fig. 20-5?

Answer 16-3

The slope of the line is -1 and it passes through $(0,0)$, so we can see straightaway that

$$g(x) = -x$$

Question 16-4

What's the difference function $q(x) = f(x) - g(x)$ in the situation of Fig. 20-5?

Answer 16-4

We have found that

$$f(x) = -x^2 + 4x$$

and

$$g(x) = -x$$

Therefore

$$q(x) = f(x) - g(x) = (-x^2 + 4x) - (-x) = -x^2 + 5x$$

Question 16-5

What's the definite integral that represents the true geometric area between the line and the curve in Fig. 20-5?

Answer 16-5

The bounds of integration are 0 and 5, the x-values of the points where the graphs intersect, so the true geometric area between the graphs is

$$\int_0^5 (-x^2 + 5x)\, dx$$

Question 16-6

What's the true geometric area between the line and the curve in Fig. 20-5?

Answer 16-6

First, we find the basic antiderivative function $Q(x)$. It is

$$Q(x) = -x^3/3 + (5/2)x^2$$

When we plug in $x = 5$ to this, we get

$$Q(5) = -(5^3)/3 + (5/2) \cdot 5^2 = 125/6$$

When we plug in $x = 0$, we get

$$Q(0) = -(0^3)/3 + (5/2) \cdot 0^2 = 0$$

The definite integral is the difference between these, which is

$$Q(5) - Q(0) = 125/6 - 0 = 125/6$$

Therefore, the true geometric area between the line and the curve is

$$\int_0^5 (-x^2 + 5x)\ dx = 125/6$$

Question 16-7

How can we find the true geometric area of the shaded region between the two parabolas shown in Fig. 20-6?

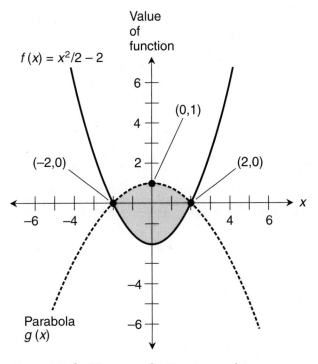

Figure 20-6 Illustration for Questions and Answers 16-7 through 16-10.

Answer 16-7

We can proceed by following these steps, in order:

- Note the function $f(x)$ represented by the parabola that opens upward.
- Find the function $g(x)$ represented by the parabola that opens downward.
- Find the difference function $g(x) - f(x)$, remembering that must subtract the "bottom function" from the "top function" over the interval between the points where the graphs intersect.
- Integrate the difference function over the interval between the points where the graphs intersect.

Question 16-8

What's the function $g(x)$ represented by the downward-opening parabola in Fig. 20-6?

Answer 16-8

The parabola crosses the x axis at $(-2,0)$ and $(2,0)$. Therefore, it represents a quadratic function $g(x)$ whose zeros are $x = -2$ and $x = 2$. With algebra, we can work out the fact that the function must be of the form

$$g(x) = k(x - 2)(x + 2) = kx^2 - 4k$$

where k is a constant. To find k, we note the point labeled at the vertex of the parabola. This point is $(0,1)$, so

$$f(0) = 1 = k \cdot (0^2) - 4k = -4k$$

which simplifies to

$$1 = -4k$$

This equation solves to $k = -1/4$. By substitution, we get

$$g(x) = (-1/4)x^2 + 1 = -x^2/4 + 1$$

Question 16-9

What's the difference function $p(x) = g(x) - f(x)$ in the situation shown by Fig. 20-6? What's the integral that represents the true geometric area between the curves?

Answer 16-9

The "top function" relative to the shaded region is

$$g(x) = -x^2/4 + 1$$

and the "bottom function" relative to the shaded region, as we are given in Fig. 20-6, is

$$f(x) = x^2/2 - 2$$

The difference function is therefore

$$p(x) = g(x) - f(x) = (-x^2/4 + 1) - (x^2/2 - 2) = -3x^2/4 + 3$$

The bounds of integration are −2 and 2, the x-values of the points where the graphs intersect. That means the definite integral representing the true geometric area between the graphs is

$$\int_{-2}^{2} (-3x^2/4 + 3)\ dx$$

Question 16-10

What's the true geometric area between the curves in Fig. 20-6?

Answer 16-10

We must find the basic antiderivative function $P(x)$ and evaluate it from −2 to 2. That antiderivative is

$$P(x) = -x^3/4 + 3x$$

When we plug in $x = 2$ to this, we get

$$P(2) = -(2^3) / 4 + 3 \cdot 2 = 4$$

When we plug in $x = -2$, we get

$$P(-2) = -(-2)^3 / 4 + 3 \cdot (-2) = -4$$

The definite integral is the difference between these, which is

$$P(2) - P(-2) = 4 - (-4) = 8$$

Therefore, the true geometric area between the line and the curve is

$$\int_{-2}^{2} (-3x^2/4 + 3)\ dx = 8$$

Chapter 17

Question 17-1

What are the indefinite integrals of the sine and cosine functions and their negatives?

Answer 17-1

We find the functions whose derivatives are the functions we want to antidifferentiate, and then add constants of integration for each. Here they are:

$$\int \sin x \, dx = -\cos x + c$$

$$\int -\sin x \, dx = \cos x + c$$

$$\int \cos x \, dx = \sin x + c$$

$$\int -\cos x \, dx = -\sin x + c$$

Question 17-2

What's the area defined by the graph of the sine function over its first 1/4 cycle?

Answer 17-2

To do this, we must evaluate

$$\int_0^{\pi/2} \sin x \, dx$$

If we call the sine function $f(x)$, then its basic antiderivative is

$$F(x) = -\cos x$$

When we evaluate this from $x = 0$ to $x = \pi/2$, we get

$$F(\pi/2) - F(0) = -\cos(\pi/2) - (-\cos 0) = -\cos(\pi/2) + \cos 0 = 0 + 1 = 1$$

We've discovered that

$$\int_0^{\pi/2} \sin x \, dx = 1$$

The area defined by the graph of the sine function over its first 1/4 cycle is 1 square unit.

Question 17-3

What's *true geometric area* between the x axis and the graph of

$$f(x) = \cos x - 1$$

over the interval from $x = 0$ to $x = \pi$?

Answer 17-3

We must work out the value of

$$\int_0^{\pi} (\cos x - 1) \, dx$$

If we call our function $g(x)$, then the basic antiderivative is

$$G(x) = \sin x - x$$

When we evaluate this from $x = 0$ to $x = \pi$, we get

$$G(\pi) - G(0) = (\sin \pi - \pi) - (\sin 0 - 0) = (0 - \pi) - (0 - 0) = -\pi$$

We've determined that

$$\int_0^\pi (\cos x - 1)\, dx = -\pi$$

which tells us that the *theoretical area* (in terms of definite integration done directly) defined by the graph of

$$f(x) = \cos x - 1$$

over the interval $(0,\pi)$ is equal to $-\pi$ square units. The *true geometric area* between the x axis and the graph, in this particular case, is the absolute value of the theoretical area, or π square units, because the entire region lies below the x axis. (Feel free to sketch a graph if you need help visualizing this.)

Question 17-4

What's the value of

$$\int_0^\pi (\cos x + \sin x)\, dx$$

representing the sum of the cosine and sine functions over the interval $(0,\pi)$?

Answer 17-4

If we call our function $g(x)$, then the basic antiderivative is

$$G(x) = \sin x - \cos x$$

When we evaluate this from $x = 0$ to $x = \pi$, we get

$$G(\pi) - G(0) = (\sin \pi - \cos \pi) - (\sin 0 - \cos 0) = [0 - (-1)] - (0 - 1) = 2$$

We've just determined that

$$\int_0^\pi (\cos x + \sin x)\, dx = 2$$

Question 17-5

The result in Answer 17-4 is the same as the result we got in one of the examples in Chap. 17, where we found that

$$\int_0^\pi \sin x\, dx = 2$$

What's the reason for this?

Answer 17-5

The definite integral of the cosine function over the interval $(0,\pi)$ is equal to 0. Half of the curve lies above the *x* axis and half lies below, and the two halves are mirror images over the interval. (If it will help you see this more clearly, sketch a graph.) That makes the cosine part of the integrand become 0 when we integrate it from 0 to π.

Question 17-6

How can we verify, in mathematical terms, the claim made in Answer 17-5?

Answer 17-6

Let's see precisely what the cosine part of the integrand in Question 17-4 contributes to the value of the entire integral. We can do this by working out

$$\int_0^\pi \cos x \, dx$$

If we call the cosine function $f(x)$, then the basic antiderivative is

$$F(x) = \sin x$$

When we evaluate this from $x = 0$ to $x = \pi$, we get

$$F(\pi) - F(0) = \sin \pi - \sin 0 = 0 - 0 = 0$$

which shows that

$$\int_0^\pi \cos x \, dx = 0$$

Using the sum rule, the integral stated in Question 17-4 can be written as

$$\int_0^\pi \cos x \, dx + \int_0^\pi \sin x \, dx$$

which, based on what we've just discovered, can be simplified to

$$0 + \int_0^\pi \sin x \, dx$$

and further to

$$\int_0^\pi \sin x \, dx$$

This definite integral, as we determined in the chapter text, is equal to 2.

Question 17-7

What's the value of

$$\int_0^{\pi/2} (\cos x + \sin x) \, dx$$

representing the sum of the cosine and sine functions over the interval $(0,\pi/2)$?

Answer 17-7

As before, if we call our function $g(x)$, the basic antiderivative is

$$G(x) = \sin x - \cos x$$

When we evaluate this from $x = 0$ to $x = \pi/2$, we get

$$G(\pi/2) - G(0) = (\sin \pi/2 - \cos \pi/2) - (\sin 0 - \cos 0) = (1 - 0) - (0 - 1) = 2$$

We've just determined that

$$\int_0^{\pi/2} (\cos x + \sin x)\, dx = 2$$

We cut the width of the interval in half compared with the situation in Question and Answer 17-4, but the value of the definite integral is the same!

Question 17-8

Evaluate the definite integrals of the sine and cosine functions individually over the interval $(0, \pi/2)$. Show that the sum of these integrals is equal to the integral of the sum function obtained in Answer 17-7 over the same interval.

Answer 17-8

First, let's work out

$$\int_0^{\pi/2} \cos x\, dx$$

If we call the cosine function $f(x)$, then the basic antiderivative is

$$F(x) = \sin x$$

When we evaluate from $x = 0$ to $x = \pi/2$, we get

$$F(\pi/2) - F(0) = \sin \pi/2 - \sin 0 = 1 - 0 = 1$$

which shows that

$$\int_0^{\pi/2} \cos x\, dx = 1$$

Now, let's calculate

$$\int_0^{\pi/2} \sin x\, dx$$

If we call the sine function $h(x)$, then the basic antiderivative is

$$H(x) = -\cos x$$

When we evaluate from $x = 0$ to $x = \pi/2$, we get

$$H(\pi/2) - H(0) = -\cos \pi/2 - (-\cos 0) = 0 - (-1) = 1$$

which tells us that

$$\int_0^{\pi/2} \sin x \, dx = 1$$

Therefore

$$\int_0^{\pi/2} \cos x \, dx + \int_0^{\pi/2} \sin x \, dx = 1 + 1 = 2$$

This agrees with

$$\int_0^{\pi/2} (\cos x + \sin x) \, dx = 2$$

which we found before.

Question 17-9

Suppose we are given these two functions:

$$f(x) = e^x$$

and

$$g(x) = e$$

What's the area of the region bounded by the graphs of these functions and the dependent-variable axis?

Answer 17-9

Figure 20-7 shows this situation. The two curves intersect at the point $(1, e)$ because

$$f(1) = e^1 = e$$

and

$$g(1) = e$$

We must integrate the difference between the function for the line and the curve over the interval $(0,1)$. If we call our difference function $q(x)$, then

$$q(x) = g(x) - f(x) = e - e^x$$

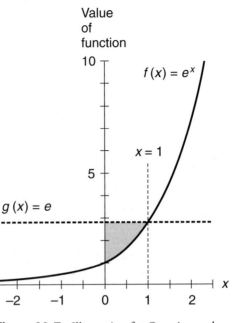

Figure 20-7   Illustration for Question and
Answer 17-9.

We must work out the integral

$$\int_0^1 (e - e^x)\, dx$$

The basic antiderivative is

$$Q(x) = ex - e^x$$

When we evaluate from $x = 0$ to $x = 1$, we get

$$Q(1) - Q(0) = (e \cdot 1 - e^1) - (e \cdot 0 - e^0) = (e - e) - (0 - 1) = 1$$

We've shown that

$$\int_0^1 (e - e^x)\, dx = 1$$

That's the area of the shaded region in Fig. 20-7.

Question 17-10

What's the area between the curve and the line shown in Fig. 20-8, over the interval for which $-2 < x < -1$? The function corresponding to the upper line is

$$f(x) = 1 - x$$

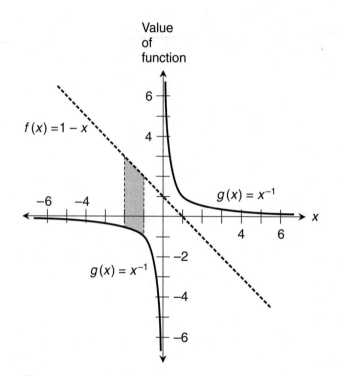

Figure 20-8 Illustration for Question and Answer 17-10.

and the function corresponding to the lower curve is

$$g(x) = x^{-1}$$

Answer 17-10

We can find the difference function by subtracting g from f. If we call that difference function $p(x)$, then

$$p(x) = f(x) - g(x) = 1 - x - x^{-1}$$

The area of the shaded region is

$$\int_{-2}^{-1} (1 - x - x^{-1})\, dx$$

The basic antiderivative is

$$P(x) = x - x^2/2 - \ln |x|$$

First, let's evaluate this for $x = -1$. We obtain

$$P(-1) = -1 - (-1)^2/2 - \ln |-1| = -3/2$$

Next, we evaluate for $x = -2$. We have

$$P(-2) = -2 - (-2)^2/2 - \ln |-2| = -4 - \ln 2$$

That's as simple as we can get this expression because ln 2 is irrational, so

$$P(-1) - P(-2) = -3/2 - (-4 - \ln 2) = 5/2 + \ln 2$$

We've figured out that

$$\int_a^b (1 - x - x^{-1})\, dx = 5/2 + \ln 2$$

That's the area between the line and the curve in Fig. 20-8. We can use a calculator and round off to three decimal places to get an approximation of 3.193 square units.

Chapter 18

Question 18-1

How can we informally describe the law of the mean?

Answer 18-1

Imagine two points on a smooth curve. Somewhere between these points, the derivative of the function is equal to the slope of the chord connecting the points.

Question 18-2

How can we formally state and illustrate the law of the mean for finding the length of an arc?

Answer 18-2

Suppose we choose two points on the graph of a function $y = f(x)$, and these points describe one of the chords in an approximation of the arc length between the points. Suppose that for the ith chord, the endpoints are where $x = x_{i-1}$ and $x = x_i$, as shown in Fig. 20-9. Then the slope of the chord is

$$\Delta_i y/\Delta x = [f(x_i) - f(x_{i-1})] / (x_i - x_{i-1})$$

If $f(x)$ is defined and continuous between the chord endpoints, then there is at least one point on the curve between the chord endpoints for which $f'(x)$ is equal to the slope of the chord. If we call the x-value of such a point x_i^*, then

$$x_{i-1} \le x_i^* \le x_i$$

and

$$f'(x_i^*) = \Delta_i y/\Delta x$$

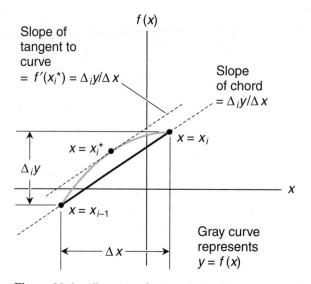

Figure 20-9   Illustration for Question and Answer 18-2.

Question 18-3

How does the formula for true arc length arise? What's the precise length L of the arc in the graph of a function $f(x)$ between two points where $x = a$ and $x = b$?

Answer 18-3

We get the formula by adding up the lengths of chords connecting points on the arc. As the number of chords increases without bound, we get a limit that defines

$$L = \int_a^b \{1 + [f'(x)]^2\}^{1/2} \, dx$$

Question 18-4

What's the arc-in-a-box method of finding the largest and smallest possible absolute values of the arc length between two points on the graph of a function?

Answer 18-4

The arc-in-a-box method works as follows.

- Construct a box with vertical and horizontal edges, such that the endpoints of the arc are at opposite corners (Fig. 20-10).
- Be sure that the arc is entirely contained within the box.
- Be sure that the arc curves in the same sense (clockwise or counterclockwise) everywhere within the box.

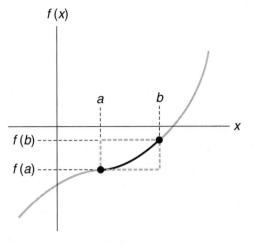

Figure 20-10   Illustration for Questions
and Answers 18-4 and 18-5.

Once we've done this, we can be sure that:

- The absolute value of the arc length is greater than or equal to the diagonal measure of the box.
- The absolute value of the arc length is less than or equal to half the perimeter of the box.

Question 18-5
In Fig. 20-10, suppose that $a = 3$, $b = 7$, $f(a) = -3$, and $f(b) = -1$. What are the minimum and maximum possible lengths of the arc in the box?

Answer 18-5
The box width is the absolute value of the difference between the x-values of the points:

$$|b - a| = |7 - 3| = 4$$

The box height is the absolute value of the difference in the function values at the points:

$$|f(b) - f(a)| = |(-1) - (-3)| = 2$$

The minimum possible arc length is the diagonal measure of the box:

$$(4^2 + 2^2)^{1/2} = (16 + 4)^{1/2} = 20^{1/2}$$

The maximum possible arc length is half the perimeter of the box:

$$(2 \cdot 4 + 2 \cdot 2) / 2 = (8 + 4) / 2 = 6$$

The arc length can't be less than $20^{1/2}$ units nor more than 6 units.

Question 18-6

How can we use the arc length formula to measure a line segment along the graph of

$$f(x) = (4/3)x$$

from $x = 0$ to $x = 3$?

Answer 18-6

To begin, we differentiate the function, getting

$$f'(x) = 4/3$$

The formula for the arc length L, stated again for convenience, is

$$L = \int_a^b \{1 + [f'(x)]^2\}^{1/2} \, dx$$

When we input 4/3 in place of $f'(x)$, set $a = 0$, and set $b = 3$, we get

$$L = \int_0^3 [1 + (4/3)^2]^{1/2} \, dx$$

Using arithmetic, this simplifies to

$$L = \int_0^3 5/3 \, dx$$

The indefinite integral without the constant of integration is

$$\int 5/3 \, dx = (5/3)x$$

When we evaluate this from 0 to 3, we get

$$L = (5/3) \cdot 3 - (5/3) \cdot 0 = 5$$

The line segment is 5 units long. If you sketch a graph of this situation, you'll see that we've used calculus to evaluate the dimensions of a so-called "3-4-5 right triangle."

Question 18-7

How can we use the arc length formula to measure a line segment along the graph of

$$f(x) = 6x - 5$$

from $x = 8$ to $x = 5$?

Answer 18-7

As before, we start by differentiating the function. In this case, we get

$$f'(x) = 6$$

When we substitute 6 for $f'(x)$ and include the bounds in the formula, we get

$$L = \int_8^5 (1 + 6^2)^{1/2} \, dx$$

which simplifies to

$$L = \int_8^5 37^{1/2} \, dx$$

The indefinite integral without the constant of integration is

$$\int 37^{1/2} \, dx = 37^{1/2} \, x$$

When we evaluate this from 8 to 5, we get

$$L = 37^{1/2} \cdot 5 - 37^{1/2} \cdot 8 = -3 \cdot 37^{1/2}$$

That's an irrational number equal to approximately −18.248.

Question 18-8

Why is the result in Answer 18-7 negative?

Answer 18-8

The arc length is negative because we move in the negative-x direction.

Question 18-9

Consider the following function, which represents a parabola:

$$f(x) = x^2/2$$

How can we measure the arc from $x = 0$ to $x = 1$, as shown in Fig. 20-11?

Answer 18-9

Our first step is to differentiate the function. The derivative is

$$f'(x) = x$$

When we substitute x for $f'(x)$ and include the bounds in the formula, we get

$$L = \int_a^b (1 + x^2)^{1/2} \, dx$$

From the table of integrals (App. G), we find the form

$$\int (x^2 + a^2)^{1/2} \, dx = (x/2)(x^2 + a^2)^{1/2} + (1/2) \, a^2 \ln |x + (x^2 + a^2)^{1/2}| + c$$

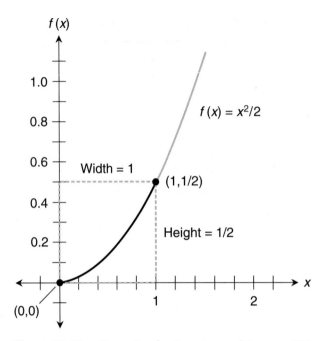

Figure 20-11   Illustration for Questions and Answers 18-9 and 18-10.

which can be rewritten as

$$\int (a^2 + x^2)^{1/2}\ dx = (x/2)(a^2 + x^2)^{1/2} + (1/2)\ a^2 \ln |x + (a^2 + x^2)^{1/2}| + c$$

If we let $a = 1$, we obtain

$$\int (1 + x^2)^{1/2}\ dx = (x/2)(1 + x^2)^{1/2} + (1/2) \ln |x + (1 + x^2)^{1/2}| + c$$

To find the arc length, we evaluate the following expression from $x = 0$ to $x = 1$:

$$(x/2)(1 + x^2)^{1/2} + (1/2) \ln |x + (1 + x^2)^{1/2}|$$

Plugging $x = 1$ into the above formula, we get

$$(1/2)(1 + 1)^{1/2} + (1/2) \ln |1 + (1 + 1)^{1/2}| = (1/2) \cdot 2^{1/2} + (1/2) \ln |1 + 2^{1/2}|$$

Plugging in $x = 0$, we get

$$(0/2)(1 + 0)^{1/2} + (1/2) \ln |0 + (1 + 0)^{1/2}| = 0$$

The arc length L is the difference between these results, or

$$L = (1/2) \cdot 2^{1/2} + (1/2) \ln |1 + 2^{1/2}|$$

Using a calculator and approximating to three decimal places gives us

$$L \approx 1.148$$

Question 18-10

How can we use the arc-in-a-box scheme to see that Answer 18-9 is reasonable?

Answer 18-10

Refer again Fig. 20-11. The width of the dashed gray box is $|1 - 0|$, or 1 unit. The height of the box is $|1/2 - 0|$, or 1/2 unit. The minimum possible arc length is the diagonal measure of the box:

$$[1^2 + (1/2)^2]^{1/2} = (1 + 1/4)^{1/2} = 5^{1/2}/2$$

That's an irrational number. A calculator tells us that it's roughly 1.118 units. The maximum possible arc length is half the perimeter of the box:

$$[2 \cdot 1 + 2 \cdot (1/2)] / 2 = (2 + 1) / 2 = 3/2$$

The arc length can't be less than $5^{1/2}/2 \approx 1.118$ nor more than $3/2 = 1.5$. In Answer 18-9, we found that the arc length is 1.148 units, rounded to three decimal places. That's between the constraints.

Chapter 19

Question 19-1

What's the principle of linearity for definite integrals, as it applies to a sum?

Answer 19-1

Let f_1 and f_2 be integrable functions of x over a continuous interval (a,b). Let k_1 and k_2 be constants. Then

$$\int_a^b \{k_1 [f_1(x)] + k_2 [f_2(x)]\} \, dx = k_1 \int_a^b f_1(x) \, dx + k_2 \int_a^b f_2(x) \, dx$$

Question 19-2

What's the principle of linearity for definite integrals, as it applies to a difference?

Answer 19-2

Let f_1 and f_2 be integrable functions of x over a continuous interval (a,b). Let k_1 and k_2 be constants. Then

$$\int_a^b \{k_1 [f_1(x)] - k_2 [f_2(x)]\} \, dx = k_1 \int_a^b f_1(x) \, dx - k_2 \int_0^1 f_2(x) \, dx$$

Question 19-3

Does the principle of linearity for sums work with any finite number of functions and constants?

Answer 19-3

Yes, if we stay with the same interval, stay with the same variable, and make sure that each function can be integrated over the entire interval.

Question 19-4

How can we evaluate the following definite integral using the principle of linearity?

$$\int_0^1 (e^x/e + 4 \cos x) \, dx$$

Answer 19-4

The principle allows us to rewrite this as

$$e^{-1} \int_0^1 e^x \, dx + 4 \int_0^1 \cos x \, dx$$

The basic antiderivative of the cosine is the sine. The exponential function is its own basic antiderivative. We can therefore resolve the above sum of integrals to

$$e^{-1} (e^1 - e^0) + 4 (\sin 1 - \sin 0)$$

which simplifies to

$$1 - e^{-1} + 4 \sin 1$$

Question 19-5

What's the formula for integration by parts?

Answer 19-5

For any two integrable functions $f(x)$ and $g(x)$,

$$\int f(x) \cdot g'(x) \, dx = f(x) \cdot g(x) - \int g(x) \cdot f'(x) \, dx$$

Question 19-6

How can we find the following indefinite integral using integration by parts?

$$\int (5x + 1)(2 \sin x) \, dx$$

Answer 19-6

To set up the parts, let's consider our functions f and g' to be

$$f(x) = 5x + 1$$

and

$$g'(x) = 2 \sin x$$

Then we have

$$g(x) = -2 \cos x$$

and

$$f'(x) = 5$$

Plugging these into the formula for integration by parts, we get

$$\int (5x + 1)(2 \sin x)\, dx = (5x + 1)(-2 \cos x) - \int (-2 \cos x) \cdot 5\, dx$$

$$= (5x + 1)(-2 \cos x) + 10 \int \cos x\, dx$$

The basic antiderivative of the cosine is the sine, so the last integral in the above equation becomes sin x plus a constant of integration c. Substituting, we get

$$\int (5x + 1)(2 \sin x)\, dx = (5x + 1)(-2 \cos x) + 10 \sin x + c$$

$$= -10x \cos x - 2 \cos x + 10 \sin x + c$$

Question 19-7

How can we use integration by parts to evaluate the following indefinite integral?

$$\int (\pi/2 \sin x + x \sin x)\, dx$$

Answer 19-7

As the first step in solving this problem, we must get the integrand into the proper form. We can factor it to obtain

$$\int (\pi/2 + x)(\sin x)\, dx$$

To set up the parts, let's say that

$$f(x) = \pi/2 + x$$

and

$$g'(x) = \sin x$$

Then we have

$$g(x) = -\cos x$$

and

$$f'(x) = 1$$

In this case,

$$\int (\pi/2 + x)(\sin x)\ dx = (\pi/2 + x)(-\cos x) - \int (-\cos x)\cdot 1\ dx$$

$$= (\pi/2 + x)(-\cos x) + \int \cos x\ dx$$

The last integral can be simplified to sin x plus a constant of integration c, because the basic antiderivative of the cosine is the sine. Substituting back, we get

$$\int (\pi/2 + x)(\sin x)\ dx = (\pi/2 + x)(-\cos x) + \sin x + c$$

$$= -\pi/2\ \cos x - x\cos x + \sin x + c$$

Question 19-8

Consider an integral of the form

$$\int k / (ax + b)\ dx$$

where x is the variable, and a, b, and k are constants. How can we resolve this?

Answer 19-8

We can rewrite the integral as

$$k \int (ax + b)^{-1}\ dx$$

Then we can use an integral table such as App. G to derive the formula

$$k \int (ax + b)^{-1}\ dx = ka^{-1}\ \ln |ax + b| + c$$

Question 19-9

How can we resolve the following integral?

$$\int -(5x + 1)^{-1}\ dx$$

Answer 19-9

We move the minus sign in front of the integral symbol, getting

$$- \int (5x + 1)^{-1} \, dx$$

Then we apply the formula stated in Answer 19-8, obtaining

$$-5^{-1} \ln |5x + 1| + c$$

which can also be written as

$$(-1/5) \ln |5x + 1| + c$$

Question 19-10

How can we resolve the following integral using the method of partial fractions?

$$\int (6x + 8)(x^2 + 3x + 2)^{-1} \, dx$$

Answer 19-10

The integrand breaks down into a sum of two constant multiples of reciprocals of linear functions. We get

$$\int [4\,(x + 2)^{-1} + 2\,(x + 1)^{-1}] \, dx$$

When we split this into a sum of integrals and pull out the constants, we get

$$4 \int (x + 2)^{-1} \, dx + 2 \int (x + 1)^{-1} \, dx$$

Applying the formula from Answer 19-8 to each addend gives us

$$4 \ln |x + 2| + 2 \ln |x + 1| + c$$

PART

3

Advanced Topics

- -

Differentiating Inverse Functions

The derivative of an inverse function can often be found directly, but sometimes an indirect scheme works better. Let's examine both methods.

A General Formula

To differentiate an inverse function, we can find the derivative of the original function, find the reciprocal of that derivative, substitute one variable for the other, and thereby get the derivative of the inverse function "through the back door."

What is an inverse function?

In Chap. 1, we learned about relations, functions, and their inverses. Let's review the meaning of *inverse* in this context. Suppose that x and y are variables, and f and f^{-1} are functions that are inverses of each other. Also suppose that

$$f(x) = y$$

and

$$f^{-1}(y) = x$$

Then we can be sure that

$$f^{-1}[f(x)] = x$$

and

$$f[f^{-1}(y)] = y$$

When we seek the inverse of a function, we might get a relation that's not a true function, because some values of the independent variable map to more than one value of the dependent variable. In many such cases, we can "force" the inverse of the original function to behave as another true function by excluding all values of either variable that map to more than one value of the other variable.

Differentiating "through the back door"

Let's recall the chain rule from Chap. 6. If f and g are differentiable functions of the same variable x, then

$$\{g\,[f\,(x)]\}' = g'\,[f\,(x)] \cdot f'\,(x)$$

If g happens to be the inverse of f, then

$$\{f^{-1}\,[f\,(x)]\}' = f^{-1\prime}\,[f\,(x)] \cdot f'\,(x)$$

where $f^{-1\prime}$ means the derivative of the inverse function. The inverse function and the original function "undo" each other on the left-hand side of this equation, so it becomes

$$x' = f^{-1\prime}\,[f\,(x)] \cdot f'\,(x)$$

If we say that $y = f\,(x)$, then we can simplify to obtain

$$dx/dx = f^{-1\prime}\,(y) \cdot dy/dx$$

which we can further simplify to

$$dx/dx = (dx/dy)\,(dy/dx)$$

Because $dx/dx = 1$, we have

$$1 = (dx/dy)\,(dy/dx)$$

which can be rearranged to get

$$(dx/dy) = (dy/dx)^{-1}$$

- -

Are you confused?

It's easy to be confused by the use of the superscript -1 after the name of a function such as f, as compared with its use after an expression such as (dy/dx). When it's written after the name of a function, a superscript -1 tells us to think about the *inverse function*. When it's written after a quantity, a superscript -1 means that we should work with the *reciprocal* of that quantity.

- -

Example: "front door"

Let's look at a function that has a well-defined inverse, and whose domain and range are both the entire set of real numbers. We can find the inverse and then differentiate it directly. Consider

$$f(x) = x^3$$

If we call $y = f(x)$, then

$$y = x^3$$

and

$$dy/dx = 3x^2$$

The inverse of f undoes the work of f. In this case,

$$x = f^{-1}(y) = y^{1/3}$$

When we differentiate x with respect to y, we get

$$dx/dy = y^{-2/3}/3$$

We can also write this as

$$f^{-1\prime}(y) = y^{-2/3}/3$$

This is the derivative of the inverse function, in which y is the independent variable and x is the dependent variable.

Same example: "back door"

Now let's differentiate the inverse function using the alternative method we've just learned. We already know that

$$dy/dx = 3x^2$$

We find dx/dy by taking the reciprocal of dy/dx, like this:

$$(dx/dy) = (dy/dx)^{-1} = (3x^2)^{-1} = x^{-2}/3$$

That's expressed in terms of x, but we want it in terms of y. We've seen that

$$x = y^{1/3}$$

Substituting, we get

$$x^{-2}/3 = (y^{1/3})^{-2}/3 = y^{-2/3}/3$$

-- --

Here's a challenge!

Differentiate the inverse of the following function, where the domain is the set of all real numbers x such that $x > 1$. First, find the inverse and then differentiate it directly. Then, use the "back door" method.

$$f(x) = e^x + 1$$

Solution

Let's introduce a dependent variable y, so we have

$$y = f(x)$$

and

$$x = f^{-1}(y)$$

We can derive x in terms of y from our knowledge of logarithms and exponentials. Here's how f can be morphed into f^{-1}, step-by-step:

$$f(x) = e^x + 1$$
$$y = e^x + 1$$
$$y - 1 = e^x$$
$$\ln(y - 1) = \ln(e^x)$$
$$\ln(y - 1) = x$$
$$x = \ln(y - 1)$$
$$f^{-1}(y) = \ln(y - 1)$$

When we differentiate using the chain rule, we get

$$f^{-1\prime}(y) = (y - 1)^{-1} \cdot 1 = (y - 1)^{-1}$$

That's the "front door" method. Now let's use the "back door" scheme. The derivative of y with respect to x is

$$dy/dx = e^x$$

When we take the reciprocal of this, we get

$$dx/dy = (e^x)^{-1}$$

We want to express this in terms of y, not in terms of x. As part of our algebraic exercise a short while ago, we saw that

$$x = \ln(y - 1)$$

By substitution,

$$dx/dy = [e^{\ln(y-1)}]^{-1} = (y-1)^{-1}$$

which can also be written as

$$f^{-1\prime}(y) = (y-1)^{-1}$$

This agrees with the result we got when we differentiated the inverse function directly through the "front door."

- -

Derivative of the Arcsine

Do you wonder what these "back door" formulas for differentiating inverse functions are useful for? Let's look at some examples of this method in action.

Restricting the domain

The sine function is a true function, regardless of its domain. We usually work only with input values between 0 and 2π, or perhaps $-\pi$ and π. But theoretically, we can input any real number. When we seek the inverse of the sine function, however, things get more complicated. We can "stand the wave on end" in Cartesian xy-coordinates and say that

$$y = f(x) = \sin x$$

and

$$x = f^{-1}(y) = \text{Arcsin } y$$

If we want the Arcsine relation to be a true function, we must restrict the domain of the sine function before we "stand the wave on end." The most common way to do this is to specify that

$$-\pi/2 \le x \le \pi/2$$

ensuring that the Arcsine relation is a true function of y. When we define the domain of the sine this way (Fig. 21-1), we capitalize the name of its inverse, writing

$$f^{-1}(y) = \text{Arcsin } y$$

- -

Are you confused?

Do you wonder why we restrict the domain of the sine function to values in the closed interval $[-\pi/2, \pi/2]$ to define the Arcsine function? Can't we use any closed interval that makes the mapping between the domain and the range a bijection? (If you don't remember what a bijection is, review its definition in Chap. 1.) The answer is "Yes and no." We can use certain other intervals. But whatever interval we choose, it must be

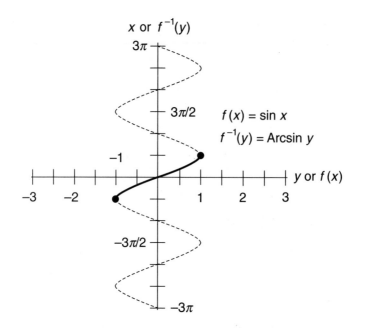

Figure 21-1   The Arcsine function is the inverse of the sine
function whose domain is restricted to values of
x in the closed interval $[-\pi/2, \pi/2]$.

exactly 1/2 cycle (π radians) wide, either from a positive peak to the next negative peak, or from a negative
peak to the next positive peak. The interval for which

$$-\pi/2 \le x \le \pi/2$$

is the one that most mathematicians prefer. It's called the *principal branch* of the sine function.

Getting the formula

The direct approach to differentiation doesn't work well with a function such as the Arcsine.
We'll have an easier time if we go through the "back door." To illustrate, we can differentiate
the inverse of

$$f(x) = \sin x$$

for $-\pi/2 \le x \le \pi/2$. Let's introduce the function f^{-1} and the variable y so that

$$y = f(x)$$

and

$$x = f^{-1}(y)$$

We can express x in terms of y as

$$x = \text{Arcsin } y$$

For the original sine function, the derivative of y with respect to x is

$$dy/dx = \cos x$$

When we take the reciprocal of each side, we get

$$dx/dy = (\cos x)^{-1}$$

This is the derivative of the Arcsine function, but we want it in terms of y. Before we attempt a substitution, let's remember an important identity from basic trigonometry. For all values of x,

$$(\cos x)^2 + (\sin x)^2 = 1$$

This well-known identity can be rewritten as

$$(\cos x)^2 = 1 - (\sin x)^2$$

Taking the 1/2 power of both sides, we get

$$\cos x = [1 - (\sin x)^2]^{1/2}$$

We know that $\sin x = y$, so we can substitute in the above equation to get

$$\cos x = (1 - y^2)^{1/2}$$

A while ago, we found that

$$dx/dy = (\cos x)^{-1}$$

We can now replace $\cos x$ by $(1 - y^2)^{1/2}$ to obtain

$$dx/dy = [(1 - y^2)^{1/2}]^{-1}$$

which can be simplified to

$$dx/dy = (1 - y^2)^{-1/2}$$

This is the derivative of the Arcsine function with the independent variable y. We've determined that

$$d/dy\,(\text{Arcsin } y) = (1 - y^2)^{-1/2}$$

Are you astute?

Here's an important fact that you might have already suspected. The derivative of the Arcsine function is defined only over the *open* interval $(-1,1)$, that is, for all real numbers y such that $-1 < y < 1$, even though the Arcsine function itself is defined over the *closed* interval $[-1,1]$. Can you see why? Here's a hint: Look at the endpoints of the solid curve in the graph of Fig. 21-1, and try to imagine the slope at those points.

Derivative of the Arccosine

Differentiating the inverse of the cosine function gives us a formula similar to the one for the derivative of the inverse of the sine. Let's define the inverse function, and then we'll derive the formula.

Restricting the domain

The cosine function, like its "cousin" the sine, is a true function even if we allow the domain to encompass the entire set of real numbers. But, as with the sine, things change when we seek the inverse (Fig. 21-2). We have

$$y = f(x) = \cos x$$

and

$$x = f^{-1}(y) = \text{Arccos } y$$

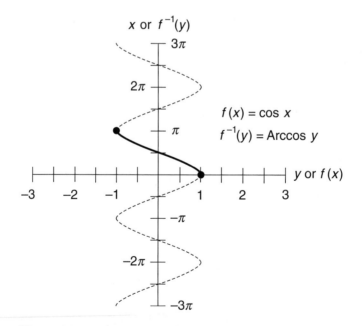

Figure 21-2   The Arccosine function is the inverse of the cosine function whose domain is restricted to values of x in the closed interval $[0, \pi]$.

but if we want the Arccosine relation to be a true function, we must restrict the domain of the cosine to the values $0 \leq x \leq \pi$. When we do this, we capitalize the name of its inverse, writing

$$f^{-1}(y) = \text{Arccos } y$$

- -

Are you confused?

You might wonder, as with the sine and Arcsine, why we restrict the domain of the cosine function to the interval $[0,\pi]$ to get the Arccosine function. The reason is the same as that for the sine and Arcsine, except that the interval is shifted by 1/4 cycle. The zone for which $0 \leq x \leq \pi$ is the principal branch.

- -

Getting the formula

Differentiating the Arccosine is similar to differentiating the Arcsine. Let's go through the process step-by-step. We'll start by letting

$$f(x) = \cos x$$

for $0 \leq x \leq \pi$. We define the function f^{-1} and the variable y so that

$$y = f(x)$$

and

$$x = f^{-1}(y)$$

We can express x in terms of y as

$$x = \text{Arccos } y$$

The derivative dy/dx is

$$dy/dx = -\sin x$$

When we take the reciprocal of this, we get

$$dx/dy = (-\sin x)^{-1}$$

We want to get the right-hand side in terms of y. That familiar trigonometric identity comes in handy again! For all values of x,

$$(\cos x)^2 + (\sin x)^2 = 1$$

which can be rewritten as

$$(\sin x)^2 = 1 - (\cos x)^2$$

Taking the 1/2 power of both sides gives us

$$\sin x = [1 - (\cos x)^2]^{1/2}$$

We can multiply through by −1 to obtain

$$-\sin x = -[1 - (\cos x)^2]^{1/2}$$

We know that $\cos x = y$, so we can substitute for x in the right-hand side to get

$$-\sin x = -(1 - y^2)^{1/2}$$

A while ago, we found that

$$dx/dy = (-\sin x)^{-1}$$

We can replace $-\sin x$ by $-(1 - y^2)^{1/2}$ in the above equation, obtaining

$$dx/dy = [-(1 - y^2)^{1/2}]^{-1}$$

which simplifies to

$$dx/dy = -(1 - y^2)^{-1/2}$$

We have just figured out that

$$d/dy\,(\text{Arccos } y) = -(1 - y^2)^{-1/2}$$

- -

Are you astute?

The derivative of the Arccosine function is defined only over the *open* interval (−1,1), even though the function itself is defined over the *closed* interval [−1,1]. The reason is the same as that for the Arcsine function and its derivative.

Here's a challenge!

Consider the following sum of composite functions:

$$h(z) = \text{Arcsin } (z/2) + \text{Arccos } (z^2/2)$$

Define the domain of h, and then differentiate it.

Solution (part 1)

The domain of the Arcsine function is the set of all inputs that produce defined outputs. That's the set of all real numbers in the closed interval $[-1,1]$. In our situation, the argument of the Arcsine is $z/2$, so the domain of that function is the set of all z for which $-2 \le z \le 2$. For the Arccosine function, the domain is, again, the set of all reals in the closed interval $[-1,1]$, so the domain of the right-hand term in the sum is the set of all z for which $-2^{1/2} \le z \le 2^{1/2}$. The domain of h is therefore the set of all z for which $-2^{1/2} \le z \le 2^{1/2}$.

Are you confused?

Do you wonder why the domain of h is the set of all z in $[-2^{1/2}, 2^{1/2}]$, and not the set of all z in $[-1,1]$ or the set of all z in $[-2,2]$? The reason is that the domain of h must be the *intersection* of the domains of the functions in the sum. In other words, if the output of h is to make any sense, we must get a defined output from the Arcsine function when we input $z/2$ to it, *and* we must get a defined output from the Arccosine function when we input $z^2/2$ to it. This happens for any real number z in $[-2^{1/2}, 2^{1/2}]$, but not for any z outside of that interval.

Solution (part 2)

Now that we know the domain of h, we can differentiate the two terms in the sum independently, and then add the results. Let's call those terms

$$p(z) = \text{Arcsin}(z/2)$$

and

$$q(z) = \text{Arccos}(z^2/2)$$

We've already determined the derivatives of the Arcsine and Arccosine functions. If we let the independent variable be z, then

$$d/dz\,(\text{Arcsin } z) = (1 - z^2)^{-1/2}$$

and

$$d/dz\,(\text{Arccos } z) = -(1 - z^2)^{-1/2}$$

Using the chain rule to differentiate p and q, we get

$$p'(z) = [1 - (z/2)]^{-1/2}\,(1/2) = [1 - (z/2)]^{-1/2}\,/\,2$$

and

$$q'(z) = -[1 - (z^2/2)]^{-1/2}\,(z) = -z\,[1 - (z^2/2)]^{-1/2}$$

The derivative h' is the sum $p' + q'$, so

$$h'(z) = [1 - (z/2)]^{-1/2}\,/\,2 - z\,[1 - (z^2/2)]^{-1/2}$$

Practice Exercises

This is an open-book quiz. You may (and should) refer to the text as you solve these problems. Don't hurry! You'll find worked-out answers in App. C. The solutions in the appendix may not represent the only way a problem can be figured out. If you think you can solve a particular problem in a quicker or better way than you see there, by all means try it!

1. Differentiate the inverse of the following function over the domain $x > 0$. First, find the inverse and then differentiate it directly. Then, use the "back door" method.

$$f(x) = x^2 + 2$$

2. Differentiate the inverse of the following function over the domain $x > 0$. First, find the inverse and then differentiate it directly. Then, use the "back door" method.

$$f(x) = \ln x$$

3. Differentiate the inverse of the following function. First, find the inverse and then differentiate it directly. Then, use the "back door" method.

$$f(x) = x^5 + 4$$

4. Define the domain of the following function. Then, differentiate it with respect to x.

$$f(x) = 5 \text{ Arcsin } x$$

5. Define the domain of the following function. Then, differentiate it with respect to t.

$$g(t) = -6t^2 \text{ Arcsin } t$$

6. Define the domain of the following function. Then, differentiate it with respect to z.

$$h(z) = \text{Arccos } z^2$$

7. Define the domain of the following function. Then, differentiate it with respect to v.

$$f(v) = 3v^3 \text{ Arccos } (v/3)$$

8. Define the domain of the following function. Then, differentiate it with respect to *s*.

$$h(s) = \text{Arcsin } s^3 - \text{Arccos } 2s$$

9. Define the domain of the following function. Then, differentiate it with respect to *w*.

$$g(w) = (\text{Arccos } w^2 - \text{Arcsin } w^2)/2$$

10. Define the domain of the following function. Then, differentiate it with respect to *x*.

$$f(x) = \text{Arcsin } e^x$$

Implicit Differentiation

Until now, we've seen functions with the dependent variable on the left-hand side of an equation, and an expression containing the independent variable on the right-hand side. In this chapter, we'll differentiate relations that aren't so clearly denoted.

Two-Way Relations

When we see a two-variable equation that isn't expressed as an obvious function of either variable, we can usually differentiate it if we accept some ambiguity and restrict the domain.

How shall we write it?

Any equation in two variables is a *two-way relation*. The mappings are inverses of each other, although one or both may fail to be true functions. Consider

$$y = x^2 + 1$$

Here, y is a true function of x, and the equation is written in the usual form for a function. But we can also write

$$y - 1 = x^2$$

If we take the positive-and-negative square root of each side and then transpose the sides, we get

$$x = \pm(y - 1)^{1/2}$$

This is a relation in which x is expressed in terms of y, but it's not a true function because it assigns more than one value of the dependent variable x to any real-number input value of y larger than 1. There's also another distinction. The domain of the function

$$y = x^2 + 1$$

is the entire set of real numbers. But the domain of the relation

$$x = \pm(y-1)^{1/2}$$

must be restricted to the set of reals y such that $y \geq 1$ if we want a real-number output.

Equations of circles

When working with graphs in the Cartesian xy-plane, we can write the general equation for a circle in the standard form

$$(x-x_0)^2 + (y-y_0)^2 = r^2$$

where x_0 and y_0 are real constants that tell us the coordinates (x_0, y_0) of the center of the circle, and r is a positive real constant that tells us the *radius* (Fig. 22-1). When the circle is centered at the origin, the formula is simpler because $x_0 = 0$ and $y_0 = 0$. Then we have

$$x^2 + y^2 = r^2$$

The simplest possible case is the *unit circle,* centered at the origin and having a radius equal to 1. Its equation is

$$x^2 + y^2 = 1$$

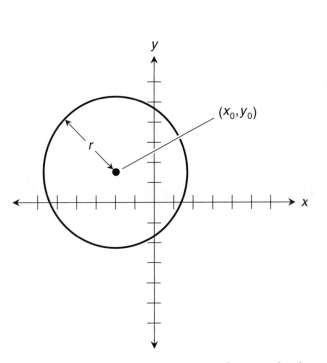

Figure 22-1   Graph of the circle for $(x-x_0)^2 + (y-y_0)^2 = r^2$.

Equations of ellipses

The general form for the equation of an *ellipse* in the Cartesian *xy*-plane, as shown in Fig. 22-2, is

$$(x - x_0)^2/a^2 + (y - y_0)^2/b^2 = 1$$

where x_0 and y_0 are real constants representing the coordinates (x_0,y_0) of the center of the ellipse, a is a positive real constant that represents the distance from (x_0,y_0) to the curve along a line parallel to the *x* axis, and b is a positive real constant that tells us the distance from (x_0,y_0) to the curve along a line parallel to the *y* axis. When we plot *x* on the horizontal axis and *y* on the vertical axis (the usual scheme), a is the length of the *horizontal semi-axis* or "horizontal radius" of the ellipse, and b is the length of the *vertical semi-axis* or "vertical radius." For ellipses centered at the origin, we have $x_0 = 0$ and $y_0 = 0$, so the general equation is

$$x^2/a^2 + y^2/b^2 = 1$$

Equations of hyperbolas

The general form for the equation of a *hyperbola* in the Cartesian *xy*-plane, as shown in Fig. 22-3, is

$$(x - x_0)^2/a^2 - (y - y_0)^2/b^2 = 1$$

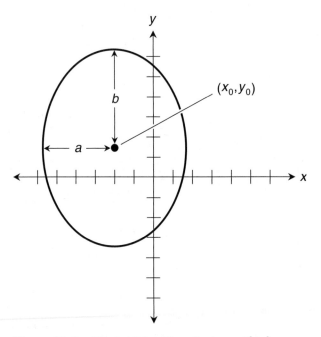

Figure 22-2   Graph of the ellipse for $(x - x_0)^2/a^2 +$ $(y - y_0)^2/b^2 = 1$.

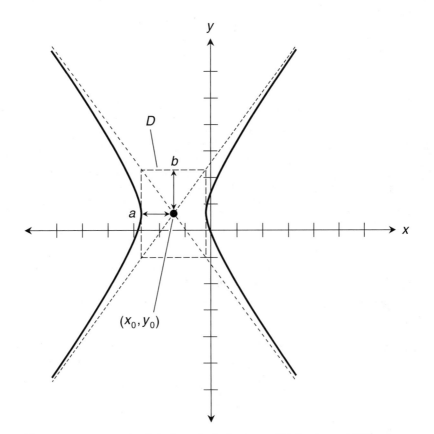

Figure 22-3 Graph of the hyperbola for $(x - x_0)^2/a^2 - (y - y_0)^2/b^2 = 1$.

where x_0 and y_0 are real constants that tell us the coordinates (x_0, y_0) of the center. The dimensions are more difficult to define than those of a circle or an ellipse. Suppose that D is a rectangle whose center is at (x_0, y_0), whose vertical edges are tangent to the hyperbola, and whose corners lie on the *asymptotes* of the hyperbola. (An asymptote, as we remember from precalculus, is a straight line that a curve approaches but never reaches as we move away from the origin.) When we define D this way, then a is the distance from (x_0, y_0) to D along a line parallel to the x axis, and b is the distance from (x_0, y_0) to D along a line parallel to the y axis. We call a the length of the horizontal semi-axis, and we call b the length of the vertical semi-axis. For hyperbolas centered at the origin, we have $x_0 = 0$ and $y_0 = 0$, so the general equation becomes

$$x^2/a^2 - y^2/b^2 = 1$$

The simplest possible case is the *unit hyperbola* whose equation is

$$x^2 - y^2 = 1$$

- -

Are you confused?

The equations of ellipses and hyperbolas don't always appear in the standard forms. Don't let this baffle you! Suppose you see the equation

$$x^2/3 + y^2/4 = 1$$

This is in the standard form for an ellipse. If you multiply through by 12, you get

$$4x^2 + 3y^2 = 12$$

You can then subtract $3y^2$ from each side, getting

$$4x^2 = 12 - 3y^2$$

This represents the same curve as the original equation. But if you saw it for the first time, you might not realize that the graph is an ellipse in the xy-plane.

- -

Two-Way Derivatives

We can differentiate a two-variable relation with respect to either variable, even if the equation does not look like it's in any standard form. This process is called *implicit differentiation*.

Example: unit circle

Let's scrutinize the equation of the unit circle, which has a radius of 1 and is centered at the origin. Again, that equation is

$$x^2 + y^2 = 1$$

The above equation can be rewritten as

$$y = \pm(1 - x^2)^{1/2}$$

or

$$x = \pm(1 - y^2)^{1/2}$$

The top equation produces a real-number output y only when $-1 \le x \le 1$. The bottom equation produces a real-number output x only when $-1 \le y \le 1$. When we find derivatives with respect to either variable, we must remember these limitations.

Let's find y' by differentiating both sides of the original equation with respect to x, and then solving the result for y'. We start with

$$x^2 + y^2 = 1$$

Differentiating through, we get

$$d/dx\,(x^2) + d/dx\,(y^2) = d/dx\,(1)$$

Term-by-term, this works out to

$$2x + 2yy' = 0$$

Do you wonder how we get $2yy'$ here? It comes from the chain rule. We must use that rule when we differentiate y^2 with respect to x, because we have a composite relation. First, x maps into y. Then, y maps into y^2. Think of the derivative like this:

$$d/dx\,(y^2) = [d/dy\,(y^2)]\,(dy/dx) = 2y\,(dy/dx) = 2yy'$$

We can subtract $2x$ from each side of the equation

$$2x + 2yy' = 0$$

to obtain

$$2yy' = -2x$$

We can divide through by $2y$ if we insist that $y \neq 0$. That gives us

$$y' = (-2x)\,/(2y)$$

which simplifies to

$$y' = -x/y$$

We've found the derivative dy/dx of the two way-relation. It's in terms of both x and y, but that's a common result in implicit differentiation. This derivative exists only for values of x between, but *not* including, -1 and 1. If $x = -1$ or $x = 1$, then we get $y = 0$ when we solve for it in the original equation, making y' undefined. If $x < -1$ or $x > 1$, then y isn't part of the real-number relation. (It would be part of the relation if we let it be a complex number, but we aren't going into the realm of complex numbers here.)

Now let's find x' by differentiating both sides with respect to y, and then solving the result for x'. We have

$$d/dy\,(x^2) + d/dy\,(y^2) = d/dy\,(1)$$

When we differentiate term-by-term, we get

$$2xx' + 2y = 0$$

As before, the chain rule is needed on $d/dy\,(x^2)$ to get $2xx'$. When we subtract $2y$ from each side, we obtain

$$2xx' = -2y$$

We can divide each side by $2x$ while insisting that $x \neq 0$, getting

$$x' = (-2y) / (2x)$$

which simplifies to

$$x' = -y/x$$

We've found the derivative dx/dy. But there's a restriction, just as there was with the other derivative. The value of x' is defined only for $-1 < y < 1$. When we have $y = -1$ or $y = 1$, then we get $x = 0$ when we solve for it in the original equation, making x' undefined. When we have $y < -1$ or $y > 1$, then x isn't in the relation, so x' is again undefined.

Example: ellipse

Consider the equation of an ellipse centered at $(x_0, y_0) = (1, -1)$, with a horizontal semi-axis of $a = 2$ units and a vertical semi-axis of $b = 3$ units. Figure 22-4 is a graph of this ellipse. Its equation is

$$(x - 1)^2 / 4 + (y + 1)^2 / 9 = 1$$

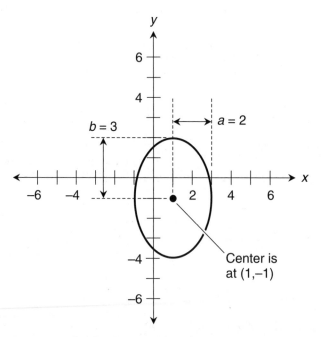

Figure 22-4   Graph of the ellipse for
$(x - 1)^2 / 4 + (y + 1)^2 / 9 = 1$.

Before we start differentiating, we should find the domains for both mappings in our two-way relation. The x-coordinate of the center is $x_0 = 1$. The horizontal semi-axis is 2 units long. Knowing these facts, we can see that the relation is defined only for values of x up to, and including, 2 units to the left or right of x_0. That is,

$$1 - 2 \leq x \leq 1 + 2$$

which means we must have

$$-1 \leq x \leq 3$$

The y-coordinate of the center is $y_0 = -1$. The vertical semi-axis measures 3 units. Therefore, the relation is defined only for values of y up to, and including, 3 units above and below y_0. That is,

$$-1 - 3 \leq y \leq -1 + 3$$

which means we must have

$$-4 \leq y \leq 2$$

To find y' (the derivative dy/dx), let's get the original equation into a form that's easier to work with. Here's the original equation for the ellipse:

$$(x - 1)^2 / 4 + (y + 1)^2 / 9 = 1$$

We can multiply through by 36 to get

$$9\,(x - 1)^2 + 4\,(y + 1)^2 = 36$$

Expanding the squared binomials gives us

$$9\,(x^2 - 2x + 1) + 4\,(y^2 + 2y + 1) = 36$$

which multiplies out to

$$9x^2 - 18x + 9 + 4y^2 + 8y + 4 = 36$$

Subtracting 9 from each side and then subtracting 4 from each side, we get an equation with the variables all on the left and a constant alone on the right:

$$9x^2 - 18x + 4y^2 + 8y = 23$$

Differentiating this term-by-term with respect to x, we get

$$d/dx\,(9x^2) - d/dx\,(18x) + d/dx\,(4y^2) + d/dx\,(8y) = d/dx\,(23)$$

This works out to

$$18x - 18 + 8yy' + 8y' = 0$$

Note, once again, that we must use the chain rule on $d/dx\,(4y^2)$ to get $8yy'$. When we add the quantity $(18 - 18x)$ to each side, we get

$$8yy' + 8y' = 18 - 18x$$

The left side can be rewritten to give us

$$(8y + 8)\,y' = 18 - 18x$$

Dividing through by the quantity $(8y + 8)$ with the restriction that $y \neq -1$, we get

$$y' = (18 - 18x) / (8y + 8) = (9 - 9x)/(4y + 4)$$

We've found dy/dx, but it has meaning only for $-1 < x < 3$. If $x = -1$ or $x = 3$, then the original equation for the ellipse tells us that $y = -1$, making y' undefined because the denominator of our ratio, the quantity $(4y + 4)$, becomes 0. If $x < -1$ or $x > 3$, then y isn't part of the real-number relation at all, so the derivative can't exist.

Now let's work out x' (the derivative dx/dy). We've already morphed the equation of the ellipse into the convenient form

$$9x^2 - 18x + 4y^2 + 8y = 23$$

When we differentiate term-by-term with respect to y, we have

$$d/dy\,(9x^2) - d/dy\,(18x) + d/dy\,(4y^2) + d/dy\,(8y) = d/dy\,(23)$$

Applying the rules of differentiation including the chain rule when necessary, we get

$$18xx' - 18x' + 8y + 8 = 0$$

Adding the quantity $(-8 - 8y)$ to each side, we get

$$18xx' - 18x' = -8 - 8y$$

The left side can be rewritten to give us

$$(18x - 18)\,x' = -8 - 8y$$

Dividing through by the quantity $(18x - 18)$ with the restriction that $x \neq 1$, we get

$$x' = (-8 - 8y) / (18x - 18) = (-4 - 4y) / (9x - 9) = (4y + 4) / (9 - 9x)$$

We've found dx/dy, but it's defined only when $-4 < y < 2$. If $y = -4$ or $y = 2$, then the original equation for the ellipse tells us that $x = 1$, making x' undefined because the denominator of our ratio, the quantity $(9 - 9x)$, becomes 0. If $y < -4$ or $y > 2$, then x isn't in the relation at all, so the derivative is meaningless.

Example: hyperbola

Consider the equation of a hyperbola centered at $(x_0, y_0) = (1, -1)$, with a horizontal semi-axis of $a = 2$ units and a vertical semi-axis of $b = 3$ units. This hyperbola is graphed in Fig. 22-5, and its equation is

$$(x - 1)^2 / 4 - (y + 1)^2 / 9 = 1$$

Let's find the domains for the two parts of this relation. The x-coordinate of the center is $x_0 = 1$. The length of the horizontal semi-axis is 2 units. Knowing these things and examining Fig. 22-5, we can see that the relation is defined only for values of x that are at least 2 units to the left of x_0, or at least 2 units to the right of x_0. That is,

$$x \leq 1 - 2 \ \text{ or } \ x \geq 1 + 2$$

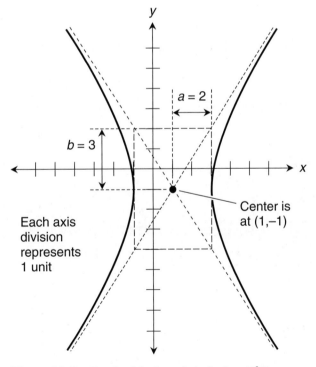

Figure 22-5 Graph of the hyperbola for $(x - 1)^2/4 -$ $(y + 1)^2/9 = 1$. Each axis division represents 1 unit.

which means we must have

$$x \leq -1 \ \text{ or } \ x \geq 3$$

The y-coordinate of the center is $y_0 = -1$. The vertical semi-axis measures 3 units. When we look at the graph, we can see that these facts, while interesting geometrically, don't place any restrictions on the domain of the relation when y is the independent variable. The relation that maps y to x is defined for all real numbers y.

To find y', let's get the equation into a more workable form. Here, again, is the original equation:

$$(x-1)^2 / 4 - (y+1)^2 / 9 = 1$$

Multiplying through by 36 gives us

$$9 (x-1)^2 - 4 (y+1)^2 = 36$$

Expanding the squared binomials produces

$$9 (x^2 - 2x + 1) - 4 (y^2 + 2y + 1) = 36$$

which multiplies out to

$$9x^2 - 18x + 9 - 4y^2 - 8y - 4 = 36$$

Subtracting 9 from each side and then adding 4 to each side, we get

$$9x^2 - 18x - 4y^2 - 8y = 31$$

Differentiating this term-by-term with respect to x, we get

$$d/dx\,(9x^2) - d/dx\,(18x) - d/dx\,(4y^2) - d/dx\,(8y) = d/dx\,(31)$$

This works out to

$$18x - 18 - 8yy' - 8y' = 0$$

Adding the quantity $(18 - 18x)$ to each side, we get

$$-8yy' - 8y' = 18 - 18x$$

The left side can be rewritten to give us

$$(-8y - 8)\,y' = 18 - 18x$$

Dividing through by the quantity $(-8y - 8)$ with the restriction that $y \neq -1$, we get

$$y' = (18 - 18x) / (-8y - 8) = (9 - 9x) / (-4y - 4) = (9x - 9) / (4y + 4)$$

This expression has meaning only when $x < -1$ or $x > 3$. If $x = -1$ or $x = 3$, then the original equation for the hyperbola tells us that $y = -1$, making y' undefined because the denominator of our ratio, $(4y + 4)$, is equal to 0. If $-1 < x < 3$, then y isn't part of the real-number relation, so there can be no derivative.

Now let's work out x'. We've already got the equation into a form that's easy to work with:

$$9x^2 - 18x - 4y^2 - 8y = 31$$

When we differentiate term-by-term with respect to y, we have

$$d/dy\,(9x^2) - d/dy\,(18x) - d/dy\,(4y^2) - d/dy\,(8y) = d/dy\,(31)$$

Working it out term-by-term yields

$$18xx' - 18x' - 8y - 8 = 0$$

Adding the quantity $(8y + 8)$ to each side, we get

$$18xx' - 18x' = 8y + 8$$

The left side can be rewritten to give us

$$(18x - 18)\,x' = 8y + 8$$

Dividing through by the quantity $(18x - 18)$ with the restriction that $x \neq 1$, we get

$$x' = (8y + 8)\,/\,(18x - 18)\ =\ (4y + 4)\,/\,(9x - 9)$$

We've found dx/dy, and it's defined over the entire set of real numbers y. We can see this by looking at Fig. 22-5. There's no point on the curve for which its slope is parallel to the x axis. The curve never passes through any point where $x = 1$, which would be necessary to get a zero denominator in dx/dy.

- -

Are you astute?

Let's look again at the derivatives we've obtained for the unit circle, the ellipse, and the hyperbola in this section. Have you noticed that they follow a pattern? For the unit circle, we got

$$y' = -x/y$$

and

$$x' = -y/x$$

For the ellipse, we found that

$$y' = (9 - 9x)\,/\,(4y + 4)$$

and

$$x' = (4y + 4)\,/\,(9 - 9x)$$

For the hyperbola, we derived

$$y' = (9x - 9) / (4y + 4)$$

and

$$x' = (4y + 4) / (9x - 9)$$

In each case, y' and x' are reciprocals of each other. These are not coincidences! Think about it. When we work out y', we're finding dy/dx. When we work out x', we're finding dx/dy. When we write the derivatives as ratios in this way, it's easy to see that if both derivatives are defined, we have

$$dy/dx = (dx/dy)^{-1}$$

and

$$dx/dy = (dy/dx)^{-1}$$

This principle is so reliable that you can use it to check your work whenever you're doing implicit differentiation of an equation in two variables. Work out the derivatives both ways. You should always get mutual reciprocals. If you don't, you've made a mistake somewhere.

Are you confused?

You might wonder if *any* two-way relation involving positive integer powers of the variables can be reduced to the standard form for a line, parabola, circle, ellipse, or hyperbola. The answer is no—at least, not if either variable is raised to an integer power larger than 2. Make up some equations in which one or both variables are raised to integer powers of 3 or larger, and then plot their graphs.

Ponder this!

Imagine the set of all possible two-way relations where the exponents of the variables are either 1 or 2, and never anything else. It's tempting to suppose that any such relation must represent a line, parabola, circle, ellipse, or hyperbola. What do you think? Is this proposition true, or not?

- -

Practice Exercises

This is an open-book quiz. You may (and should) refer to the text as you solve these problems. Don't hurry! You'll find worked-out answers in App. C. The solutions in the appendix may not represent the only way a problem can be figured out. If you think you can solve a particular problem in a quicker or better way than you see there, by all means try it!

1. What type of curve does the following equation represent? Put the equation in standard form for that type of curve. If the curve has a center, what are its coordinates? If the curve has semi-axes, how long are they?

$$3y^2 = 12x^2 - 48$$

2. What type of curve does the following equation represent? Put the equation in standard form. If the curve has a center, where is it? If it has semi-axes, how long are they?

$$4x^2 + y^2 + 8x + 2y + 1 = 0$$

3. What are the coordinates of the center of the circle represented by the following equation? What's the radius of that circle?

$$2x^2 = 288 - 2y^2$$

4. Determine the values of x and y for which the two-way relation stated in Prob. 1 is defined. Then find y' and x', and verify that they are reciprocals of each other.

5. Determine the values of x and y for which the two-way relation stated in Prob. 2 is defined. Then find y' and x', and verify that they are reciprocals of each other.

6. Determine the values of x and y for which the two-way relation stated in Prob. 3 is defined. Then find y' and x', and verify that they are reciprocals of each other.

7. Look again at the equation for the unit circle

$$x^2 + y^2 = 1$$

and the derivatives we found for it. What are the two values of y' for $x = 1/2$? What are the two values of x' for $y = 2^{-1/2}$?

8. Look again at the equation for the ellipse

$$(x - 1)^2/4 + (y + 1)^2/9 = 1$$

whose graph is shown in Fig. 22-4. Examine the derivatives we found. What is y' for $x = 1$? What is x' for $y = -1$?

9. Look again at the equation for the hyperbola graphed in Fig. 22-5:

$$(x - 1)^2 / 4 - (y + 1)^2/9 = 1$$

Examine the derivatives we found. What is y' when $x = -3$?

10. Prove that for any specific value of x, the derivatives y' for the two-way relation representing a circle centered at the origin, are always exact negatives of each other. We must assume, of course, that the value we choose for x is part of the real-number relation.

CHAPTER

23

The L'Hôpital Principles

In this chapter, we'll learn how to evaluate limits using the *l'Hôpital* (pronounced lo-pi-TALL) *principles,* named after the French mathematician Guillaume François Antoine de l'Hôpital (1661–1704), who described the rules in one of the first calculus textbooks ever written. Sometimes you'll see his name spelled *l'Hôspital.*

Expressions That Tend Toward 0/0

Imagine that we want to find the limit of a ratio in which both the numerator and the denominator approach 0. This is an example of an *indeterminate expression.* We can't evaluate 0/0 directly, because it's undefined. But we might be able to evaluate it indirectly.

How it works

If we want to find the limit of a ratio that tends toward 0/0, and if we can differentiate the numerator and the denominator, then the limit of the ratio of the derivatives is the same as the limit of the original ratio.

Imagine two functions $f(x)$ and $g(x)$ with three properties. First, f and g are both differentiable everywhere in some open interval around the point $x = a$ (except maybe not exactly at $x = a$). Second,

$$\underset{x \to a}{Lim}\ f(x) = 0$$

and

$$\underset{x \to a}{Lim}\ g(x) = 0$$

Third, $g'(x) \neq 0$ at every point within the defined interval where $x \neq a$. If all three of these conditions are met, then

$$\underset{x \to a}{Lim}\ f(x)/g(x) = \underset{x \to a}{Lim}\ f'(x)/g'(x)$$

Example

Let's figure out the following limit, where the numerator and denominator both approach 0 as x approaches 1:

$$\underset{x \to 1}{Lim} \ (3x - 3)/(x - 1)$$

If we name the numerator and the denominator

$$f(x) = 3x - 3$$

and

$$g(x) = x - 1$$

respectively, then our limit is of the form

$$\underset{x \to 1}{Lim} \ f(x)/g(x)$$

The derivatives are

$$f'(x) = 3$$

and

$$g'(x) = 1$$

The l'Hôpital rule for expressions that tend toward 0/0 tells us that

$$\underset{x \to 1}{Lim} \ f(x)/g(x) = \underset{x \to 1}{Lim} \ f'(x)/g'(x) = \underset{x \to 1}{Lim} \ 3/1$$

The value of the final (rightmost) expression is obviously 3, no matter what happens to x. Therefore, we know that

$$\underset{x \to 1}{Lim} \ (3x - 3)/(x - 1) = 3$$

Applying the rule twice

Let's scrutinize a more complicated limit in which the numerator and denominator both approach 0 as x approaches 1:

$$\underset{x \to 1}{Lim} \ (3x^2 - 6x + 3)/(x^2 - 2x + 1)$$

If we name the functions

$$f(x) = 3x^2 - 6x + 3$$

and

$$g(x) = x^2 - 2x + 1$$

then the derivatives are

$$f'(x) = 6x - 6$$

and

$$g'(x) = 2x - 2$$

The l'Hôpital rule for expressions that tend toward 0/0 tells us that

$$\lim_{x \to 1} f(x)/g(x) = \lim_{x \to 1} f'(x)/g'(x) = \lim_{x \to 1} (6x - 6)/(2x - 2)$$

This new expression, like the original one, approaches 0/0 as x approaches 1. The limit still defies direct evaluation, but we can apply l'Hôpital's principle again. Let's differentiate the numerator and denominator a second time, getting

$$f''(x) = 6$$

and

$$g''(x) = 2$$

The l'Hôpital rule now indicates that

$$\lim_{x \to 1} f'(x)/g'(x) = \lim_{x \to 1} f''(x)/g''(x) = \lim_{x \to 1} 6/2$$

This is equal to 3, no matter what happens to x. This third limit must be equal to the second one we found, which in turn must equal the original one. Therefore

$$\lim_{x \to 1} (3x^2 - 6x + 3)/(x^2 - 2x + 1) = 3$$

- -

Are you confused?

Do you wonder if the l'Hôpital rule *always* leads to a resolution for the limit of an expression that tends toward 0/0? The answer is no, it doesn't. Sometimes a limit "runs away" toward positive or negative infinity (or both!) when we try to apply the rule to it. In cases like that, we must conclude that the limit is infinite.

Here's a challenge!

Evaluate the following limit, where the numerator and denominator both approach 0 as x approaches 1. If the limit is infinite, then say so.

$$\lim_{x \to 1} (5x - 5)/(3x^2 - 6x + 3)$$

Solution

If we name the functions

$$f(x) = 5x - 5$$

and

$$g(x) = 3x^2 - 6x + 3$$

then the derivatives are

$$f'(x) = 5$$

and

$$g'(x) = 6x - 6$$

Using the l'Hôpital principle, we find that

$$\underset{x \to 1}{Lim} \ f(x)/g(x) = \underset{x \to 1}{Lim} \ f'(x)/g'(x) = \underset{x \to 1}{Lim} \ 5/(6x - 6)$$

The numerator in the final limit stays constant at 5, but the denominator approaches 0 as x approaches 1. The absolute value of the quantity $[f'(x)/g'(x)]$ therefore increases without bound as x approaches 1. If we approach 1 from the left (values smaller than 1, but increasing) the ratio "runs away" toward negative infinity, but if we approach from the right (values larger than 1, but decreasing) the ratio "runs away" toward positive infinity.

- -

An important restriction

"Well," you might say, "suppose we apply the l'Hôpital rule to the above limit a second time? Will that resolve it?" Let's try it and see. The second derivatives of the functions are

$$f''(x) = 0$$

and

$$g''(x) = 6$$

producing the apparent result

$$\underset{x \to 1}{Lim} \ f'(x)/g'(x) = \underset{x \to 1}{Lim} \ f''(x)/g''(x) = \underset{x \to 1}{Lim} \ 0/6$$

This limit appears to be equal to 0. But this isn't a legitimate answer, because we have violated a critical restriction on the use of l'Hôpital's rule. We must *never* use this tactic to look for the limit of an expression that *obviously* blows up as the variable approaches the limiting value.

It was okay to use the l'Hôpital rule the first time in the preceding "challenge," because we were faced with an expression that tended toward 0/0. But the second application of the rule was inappropriate.

Expressions That Tend Toward ±∞/±∞

Now think of a limit where the numerator and the denominator both "run away" toward positive or negative infinity. We can't work with infinity directly as if it were a number. But we might be able to find the limit of such a ratio by differentiation.

How it works

This l'Hôpital principle is almost identical to the 0/0 version, except for a couple of changes in the "numbers." As before, suppose that f and g are functions of a variable x, and both functions are differentiable over an open interval containing $x = a$, except possibly at $x = a$ itself. Also suppose that

$$\mathrm{Lim}_{x \to a} f(x) = \pm\infty$$

and

$$\mathrm{Lim}_{x \to a} g(x) = \pm\infty$$

Here, the symbol $\pm\infty$ means "positive infinity or negative infinity." Furthermore, suppose that $g'(x) \neq 0$ at every point in the defined interval where $x \neq a$. If all these things are true, then

$$\mathrm{Lim}_{x \to a} f(x)/g(x) = \mathrm{Lim}_{x \to a} f'(x)/g'(x)$$

Example

Let's look at the limit, as x approaches 0 from the right, of the natural logarithm function divided by the reciprocal function:

$$\mathrm{Lim}_{x \to 0+} (\ln x)/(x^{-1})$$

We can name the functions

$$f(x) = \ln x$$

and

$$g(x) = x^{-1}$$

In this case, we have

$$\mathrm{Lim}_{x \to 0+} f(x) = -\infty \quad \text{and} \quad \mathrm{Lim}_{x \to 0+} g(x) = +\infty$$

The derivatives are

$$f'(x) = x^{-1}$$

and

$$g'(x) = -x^{-2}$$

Using the l'Hôpital principle for expressions that tend toward $\pm\infty/\pm\infty$, we find that

$$\operatorname*{Lim}_{x\to0+} f(x)/g(x) = \operatorname*{Lim}_{x\to0+} f'(x)/g'(x) = \operatorname*{Lim}_{x\to0+} x^{-1}/(-x^{-2})$$

The last expression in the above equation is awkward, but it can be rewritten as

$$x^{-1}/(-x^{-2}) = (1/x)/(-1/x^2) = (1/x)\,(-x^2) = -x^2/x = -x$$

We can do these maneuvers without inadvertently dividing by 0. That's because x, while it might become vanishingly small, never actually equals 0. Now our limit is

$$\operatorname*{Lim}_{x\to0+} -x$$

This expression tends toward 0, so we can conclude that

$$\operatorname*{Lim}_{x\to0+} (\ln x)/(x^{-1}) = 0$$

Another variant of the rule

Here's a l'Hôpital principle that can help us find the limit of an expression as the value of the variable increases or decreases endlessly. Suppose that f and g are functions of a variable x, both functions are differentiable, and $g'(x) \ne 0$ as x approaches positive or negative infinity. Also, suppose that one of the following things is true:

$$\operatorname*{Lim}_{x\to+\infty} f(x) = 0 \text{ and } \operatorname*{Lim}_{x\to+\infty} g(x) = 0$$

$$\operatorname*{Lim}_{x\to-\infty} f(x) = 0 \text{ and } \operatorname*{Lim}_{x\to-\infty} g(x) = 0$$

$$\operatorname*{Lim}_{x\to+\infty} f(x) = \pm\infty \text{ and } \operatorname*{Lim}_{x\to+\infty} g(x) = \pm\infty$$

$$\operatorname*{Lim}_{x\to-\infty} f(x) = \pm\infty \text{ and } \operatorname*{Lim}_{x\to-\infty} g(x) = \pm\infty$$

In any situation like this,

$$\operatorname*{Lim}_{x\to a} f(x)/g(x) = \operatorname*{Lim}_{x\to a} f'(x)/g'(x)$$

Another example

Let's figure out the limit, as x is positive and increases without bound, of the ratio of the natural logarithm function to the exponential function:

$$\lim_{x \to +\infty} (\ln x)/e^x$$

We can name the functions

$$f(x) = \ln x$$

and

$$g(x) = e^x$$

The values of these functions both tend toward positive infinity as x increases without bound. That is,

$$\lim_{x \to +\infty} f(x) = +\infty \text{ and } \lim_{x \to +\infty} g(x) = +\infty$$

The derivatives are

$$f'(x) = x^{-1}$$

and

$$g'(x) = e^x$$

Applying the wisdom of l'Hôpital, we find that

$$\lim_{x \to +\infty} f(x)/g(x) = \lim_{x \to +\infty} f'(x)/g'(x) = \lim_{x \to +\infty} x^{-1}/e^x$$

The rightmost expression can be rewritten as

$$x^{-1}/e^x = (1/x)(1/e^x) = 1/(xe^x) = (xe^x)^{-1}$$

We must find

$$\lim_{x \to +\infty} (xe^x)^{-1}$$

As x approaches positive infinity, both x and e^x do the same, so the product xe^x tends toward positive infinity as well. The reciprocal therefore tends toward 0, so

$$\lim_{x \to +\infty} (\ln x)/e^x = 0$$

- -

Are you astute?

Occasionally, you'll want to find the limit of a ratio that can be converted into another form and then evaluated directly, even though it looks like it needs one of l'Hôpital's rules. If you think you can simplify an expression before working out the limit, go ahead and try it! You might get a ratio that's easier to manage, or an expression that doesn't involve a quotient at all. For example, if you have a ratio where a variable is raised to negative powers in both the numerator and the denominator, you can try to simplify the expression with algebra before looking for the limit.

- -

Other Indeterminate Limits

Occasionally, you'll encounter the limit of an indeterminate product, sum, or difference. If such an expression can be converted to a ratio of the form 0/0 or $\pm\infty/\pm\infty$, then you may be able to use one of l'Hôpital's principles to evaluate it.

Expressions that tend toward 0 · (+∞) or 0 · (−∞)

Let's examine the limit, as x approaches 0 from the left (that is, from the negative side), of the product of the sine function and the reciprocal function:

$$\lim_{x \to 0-} (\sin x)\,(x^{-1})$$

In this situation, the sine function approaches 0 while the reciprocal function tends toward negative infinity. We can rewrite the expression as a ratio to get

$$\lim_{x \to 0-} (\sin x)/x$$

Now both the numerator and the denominator approach 0, so we can apply the l'Hôpital rule for the form 0/0. Let's name the functions in our ratio

$$f(x) = \sin x$$

and

$$g(x) = x$$

The derivatives are

$$f'(x) = \cos x$$

and

$$g'(x) = 1$$

Therefore

$$Lim_{x \to 0-} f(x)/g(x) = Lim_{x \to 0-} f'(x)/g'(x) = Lim_{x \to 0-} (\cos x)/1 = Lim_{x \to 0-} (\cos x)$$

The rightmost expression approaches 1 as x approaches 0 from the left. We have determined that

$$Lim_{x \to 0-} (\sin x)(x^{-1}) = 1$$

Expressions that tend toward $+\infty - (+\infty)$

Let's work out the following limit of a difference in which both terms tend toward positive infinity as x approaches 0 from the right (that is, from the positive side):

$$Lim_{x \to 0+} 4x^{-1} - 4(e^x - 1)^{-1}$$

Using algebra, we can rearrange the expression to get a ratio so our limit becomes

$$Lim_{x \to 0+} (4e^x - 4 - 4x)/(xe^x - x)$$

As x approaches 0 from the right, both the numerator and the denominator in the above expression tend toward 0. We can apply the l'Hôpital rule for limits of this form. Let's call the numerator and denominator functions

$$f(x) = 4e^x - 4 - 4x$$

and

$$g(x) = xe^x - x$$

The derivatives are

$$f'(x) = 4e^x - 4$$

and

$$g'(x) = e^x + xe^x - 1$$

Therefore

$$Lim_{x \to 0+} f(x)/g(x) = Lim_{x \to 0+} f'(x)/g'(x) = Lim_{x \to 0+} (4e^x - 4)/(e^x + xe^x - 1)$$

As x approaches 0 from the right, the numerator and denominator in the above ratio both approach 0, so we must apply l'Hôpital's rule again. The second derivatives are

$$f''(x) = 4e^x$$

and

$$g''(x) = e^x + e^x + xe^x = 2e^x + xe^x$$

Now we have

$$\underset{x \to 0+}{Lim} \ f'(x)/g'(x) = \underset{x \to 0+}{Lim} \ f''(x)/g''(x) = \underset{x \to 0+}{Lim} \ (4e^x)/(2e^x + xe^x)$$

As x approaches 0 from either direction, $f''(x)$ approaches 4 and $g''(x)$ approaches 2. We have found that

$$\underset{x \to 0+}{Lim} \ 4x^{-1} - 4(e^x - 1)^{-1} = 4/2 = 2$$

Expressions tending toward $+\infty \cdot 0, -\infty \cdot 0, -\infty + (+\infty), +\infty + (-\infty),$ or $-\infty - (-\infty)$

Whenever we see a limit whose expression tends toward positive infinity times 0, negative infinity times 0, negative infinity plus positive infinity, positive infinity plus negative infinity, or negative infinity minus negative infinity, we can convert it to a form we already know how to work with, as follows:

- The form $+\infty \cdot 0$ is equivalent to $0 \cdot (+\infty)$
- The form $-\infty \cdot 0$ is equivalent to $0 \cdot (-\infty)$
- The form $-\infty + (+\infty)$ is equivalent to $+\infty - (+\infty)$
- The form $+\infty + (-\infty)$ is equivalent $+\infty - (+\infty)$
- The form $-\infty - (-\infty)$ is equivalent to $+\infty - (+\infty)$

- -

Are you confused?

Don't be flabbergasted by the way we bandy around the notions of positive and negative infinity (and the symbols $+\infty$ and $-\infty$) as if they're real numbers. They're not, of course! But expressions like these are convenient, even though they aren't technically rigorous.

Here's a challenge!

Evaluate the following limit, where the first term tends toward negative infinity and the second term tends toward positive infinity as x decreases endlessly (that is, becomes large negatively without bound):

$$\underset{x \to -\infty}{Lim} \ x^3 + x^2$$

Solution

Let's use algebra to rearrange the expression to get a ratio. We can morph it like this:

$$x^3 + x^2 = x^3 + (x^{-2})^{-1} = x^3 + (1/x^{-2}) = (x+1)/(x^{-2})$$

Our limit can now be written as

$$\underset{x \to -\infty}{Lim}\ (x+1)/(x^{-2})$$

As x becomes large negatively without bound, the numerator tends toward negative infinity while the denominator tends toward 0. That means the ratio tends toward negative infinity, so we can conclude that

$$\underset{x \to -\infty}{Lim}\ x^3 + x^2 = \underset{x \to -\infty}{Lim}\ (x+1)/(x^{-2}) = -\infty$$

Here's a lesson!

Perhaps the greatest challenge in regards to l'Hôpital's rules is knowing when we *can* apply them to advantage, and when we *can't*. These principles are useful only when we encounter, or can derive, the limit of a ratio that tends toward 0/0 or $\pm\infty/\pm\infty$. In other situations, l'Hôpital's rules rarely work and should not be used. If we apply them inappropriately, we will likely get invalid results.

- -

Practice Exercises

This is an open-book quiz. You may (and should) refer to the text as you solve these problems. Don't hurry! You'll find worked-out answers in App. C. The solutions in the appendix may not represent the only way a problem can be figured out. If you think you can solve a particular problem in a quicker or better way than you see there, by all means try it!

1. Evaluate the following limit, where the numerator and denominator both approach 0 as x approaches 0. (We can approach from the left or from the right; it doesn't matter.) If the limit is infinite, then say so.

$$\underset{x \to 0}{Lim}\ (\sin x)/(8x)$$

2. Evaluate the following limit, where the numerator and denominator both approach 0 as x approaches 1 (from either side). If the limit is infinite, then say so.

$$\underset{x \to 1}{Lim}\ (6x^2 - 12x + 6)/(x^2 - 2x + 1)$$

3. Evaluate the following limit, where the numerator and denominator both approach 0 as x approaches 0 (from either side). If the limit is infinite, then say so.

$$\underset{x \to 0}{Lim}\ (12 \sin x - 12x)/x^3$$

4. Evaluate the following limit, where the numerator and denominator both tend toward positive infinity as x increases endlessly. If the limit is infinite, then say so.

$$\underset{x \to +\infty}{Lim}\ 7x/(\ln x)$$

5. Evaluate the following limit, where the numerator and denominator both tend toward positive infinity as x increases endlessly. If the limit is infinite, then say so.

$$\lim_{x \to +\infty} e^x / x^2$$

6. Evaluate the following limit, where the numerator and denominator both tend toward positive infinity as x approaches 0 from the right. If the limit is infinite, then say so.

$$\lim_{x \to 0+} x^{-3} / x^{-2}$$

7. Evaluate the following limit. Compare it to the limit of the same expression as x approaches 0 from the negative side, which we found equal to 1 in the chapter text.

$$\lim_{x \to 0+} (\sin x) \, x^{-1}$$

8. Evaluate the following limit, where the first factor tends toward 0 and the second factor tends toward negative infinity as x approaches 0 from the left. If the limit is infinite, then say so.

$$\lim_{x \to 0-} x^5 \ln |x|$$

9. Evaluate the following limit, where both terms tend toward positive infinity as x approaches 0 from the right. If the limit is infinite, then say so.

$$\lim_{x \to 0+} 2x^{-1} - 3(e^x - 1)^{-1}$$

10. Evaluate the following limit, where both terms tend toward positive infinity as x approaches $\pi/2$ from the left. If the limit is infinite, then say so.

$$\lim_{x \to \pi/2-} \sec x - \tan x$$

24

Partial Derivatives

In this chapter, we'll learn how to differentiate *multi-variable functions* with two or three independent variables. When graphed, such functions require three or four dimensions, respectively. We can draw three-dimensional (3D) graphs on a flat page or computer screen if we include perspective. Four-dimensional (4D) graphs can't be drawn, so we must rely entirely on equations.

Multi-Variable Functions

As we've learned, a function is a special sort of mapping from the elements of one set (the domain) to the elements of another set (the range). Until now, the domain of every function we've encountered has been a set of real numbers. But we can also have functions whose domains are sets of ordered pairs or triples of real numbers.

Two inputs, one output

Imagine that we want to define the *topography* (terrain) of a rolling prairie in terms of a function. We need three dimensions: east-west position, north-south position, and elevation. We choose a starting point and call it the origin of *Cartesian three-space* in variables x, y, and z. At the origin, we assign $x = 0$, $y = 0$, and $z = 0$. That's the ordered triple $(x,y,z) = (0,0,0)$. We define the coordinate axes like this:

- Positive values of x are east of the origin
- Negative values of x are west of the origin
- Positive values of y are north of the origin
- Negative values of y are south of the origin
- Positive values of z are above the origin
- Negative values of z are below the origin

For convenience, let's say that $z = 0$ at sea level, so the xy-plane is at sea level within a few kilometers of the origin, where we can ignore the curvature of the earth. Unless we're in a strange part of the world or somewhere underwater in the ocean, we'll always find that the elevation *z* is positive.

Suppose that we wander around on the prairie and measure the elevation of the terrain above the *xy*-plane as a function of our horizontal location in terms of *x* and *y*. We spend a long time at this, taking down notes to make a detailed *topographical map* of the region. That map represents a function *f* that maps ordered pairs (*x*,*y*) into values of *z*, so we can write

$$z = f(x,y)$$

If we split the ordered pairs into individual real numbers, we can say that our function *f* has two independent (input) variables, *x* and *y*. The dependent variable, *z*, is the output.

If *f* is a true function, then there can never be more than one value of *z* for any particular ordered pair (*x*,*y*). In our prairie analogy, this translates into "No outcroppings, no overhangs, and no caves." In the real world, most prairie regions fit that description.

A pure-mathematics example

Figure 24-1 illustrates the graph of a simple function with two independent variables. (Actually, this drawing shows only a few curves on the true surface, which resembles a tall, thin bowl that extends infinitely upward.) This object is a *paraboloid*, which is what we get when we rotate a parabola on its axis to create a two-dimensional surface. The domain of the function is represented by the entire *xy*-plane. The range is the set of values on the *z* axis larger than a certain minimum, represented by the *apex* or *vertex* of the paraboloid, or its lowest point. We can envision curves on the paraboloid, formed by its intersection with various flat planes in space. If we cut the paraboloid with a vertical plane, such a curve is a parabola. If we cut the paraboloid with a horizontal plane, such a curve is a circle.

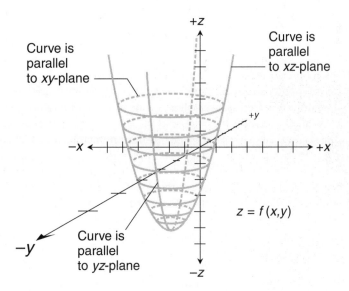

Figure 24-1   Graph of a two-variable function in Cartesian
xyz-space, showing some curves on the surface.

The vertical-line test

If the axis of the paraboloid in Fig. 24-1 is parallel to the z axis, then the surface represents a function of the variables x and y. If we call the function f, then

$$z = f(x,y)$$

because there is never more than one output value z for any input pair (x,y). Do you remember the vertical-line test for a single-variable function? (If not, look back at Chap. 1.) We can apply a similar rule to the graph of a two-variable function in xyz-space, as long as that graph appears as a surface.

Imagine a straight, infinitely long vertical line parallel to the z axis. We move this line around, so the point where it intersects the xy-plane sweeps through every possible ordered pair (x,y). Our graph represents a function if and only if the movable vertical line never cuts through the surface at more than one point. The paraboloid in Fig. 24-1 passes this test. Some surfaces, such as a sphere, fail the test. They represent relations, but not true functions, in Cartesian xyz-space.

Three inputs, one output

Let's perform a different experiment on the prairie. Imagine that we're no longer restricted to the surface. We choose a starting point and call it the origin, as before. But this time, we make all three of the variables x, y, and z independent. Their literal meanings are the same as before:

- Positive values of x are east of the origin
- Negative values of x are west of the origin
- Positive values of y are north of the origin
- Negative values of y are south of the origin
- Positive values of z are above the origin
- Negative values of z are below the origin

The difference between this situation and the previous one is that our elevation (or altitude) doesn't depend on the location. Altitude is now an independent variable, along with the east-west and north-south positions. Let's say that the dependent variable is the wind speed, and call it w. The value of w depends on where we are in space: the values of x, y, and z. Any change in x, y, or z is likely to produce a change in w.

Suppose that we launch weather balloons from numerous places all around the prairie. The balloons rise into the air, reaching various east-west positions, north-south positions, and altitudes. Each balloon has a weather-monitoring device with a radio transmitter that sends us wind-speed data for points (x,y,z) scattered around in the sky. We check all this data at 12:00 noon, local time, on a certain day. Then we compile the data to get points in a relation g such that

$$w = g(x,y,z)$$

This relation g maps ordered triples (x,y,z) to a fourth number, represented by the variable w. In the real world, g must be a true function. We can't have more than one wind speed at any single point at noon on the day of our experiment. There's never more than one value of w for any particular ordered triple (x,y,z).

What about time?

If we add another independent variable to the mix—time, represented by t—we'll get a function of four variables, with inputs consisting of *ordered quadruples* (x,y,z,t) and the dependent variable w. The inclusion of time as a variable makes our experiment more difficult, because we must take wind-speed readings from our weather instruments at frequent intervals, not only at noon. These added tasks complicate our function, and the larger amount of data complicates any graph that we try to create. Such a graph would occupy five dimensions: east-west position, north-south position, altitude, time, and wind speed!

- -

Are you confused?

We can't easily envision a vertical line test in the weather-balloon scenarios we've just described, because we're working in more than three dimensions. Any "vertical line" is perpendicular to each of the independent-variable axes, and parallel to the dependent-variable axis *Cartesian four-space* or *Cartesian five-space*.

To see if a relation with three or more independent variables is a true function, we must rely on mathematics and intuition. Sometimes this is easy, and sometimes it's hard. We know that it's impossible to have more than one wind speed at any place and time. But in a pure-mathematics situation with no physical analogy, it can be difficult to test a relation to see if it's a true function.

- -

Two Independent Variables

If we want to find the derivative of a function $f(x,y)$, we can try to differentiate f with respect to x alone or with respect to y alone. If we're successful, the results are called *partial derivatives*. If it's impossible to find both of the partial derivatives, then we must conclude that $f(x,y)$ isn't entirely differentiable.

"Slope" of a surface at a point

Examine Fig. 24-1 again. Imagine that we want to find the slope at a particular point on the surface. On most surfaces, there are infinitely many tangent lines at any specific point. All of those lines lie in a single *tangent plane*. The surface "rests on" the tangent plane the way a ball rests on a floor.

If we want to *fully* define the "slope" of a surface at a point, we must define the orientation of the plane tangent to the surface at that point. This can be done by calculating the *direction numbers* of a line passing through the point perpendicular to the tangent plane. That's a little complicated, but we don't have to worry about it here.

We can *partially* define the "slope" of the surface at a point by cutting through that surface with a flat plane perpendicular to any of the three coordinate axes. When we do that, we get curves. Note that:

- Any plane perpendicular to the y axis is parallel to the xz-plane.
- Any plane perpendicular to the x axis is parallel to the yz-plane.
- Any plane perpendicular to the z axis is parallel to the xy-plane.

If we have $z = f(x,y)$, then we're interested in planes of the first or second types. A plane of the first type lets us define z as a function of x for some fixed value of y. A plane of the second type allows us to define z as a function of y for some fixed value of x. If both of these *partial functions* are differentiable, we can find the partial derivatives of $f(x,y)$.

Derivative with respect to *x*

Let's slice the surface in Fig. 24-1 with a flat plane parallel to the *xz*-plane. There are infinitely many such planes to choose from, all perpendicular to the *y* axis and passing through some point on that axis. Any plane that intersects the surface does so along a curve. If we select one of the planes parallel to the *xz*-plane and stay within it, then we can find the derivative dz/dx for the intersection curve in that plane. Figure 24-2 illustrates an example. If we select another plane parallel to the *xz*-plane, then we can find dz/dx for the new curve.

We can find the *general partial derivative* of f with respect to x if we treat y as a constant. (This can be tricky, because we're used to thinking of y as a variable.) General partial derivatives are written like the general derivatives of single-variable functions, but using a "curly d" instead of an ordinary d to represent a differential. We can write the general partial derivative of f with respect to x in any of the following ways:

$$\partial z/\partial x$$
$$\partial f(x,y)/\partial x$$
$$\partial f/\partial x$$
$$\partial/\partial x \, f(x,y)$$
$$\partial/\partial x \, f$$

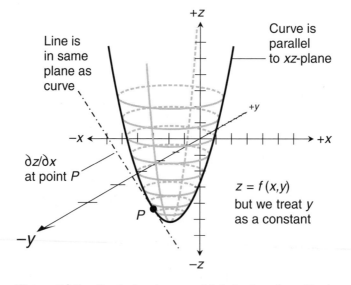

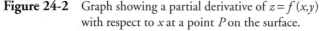

Figure 24-2   Graph showing a partial derivative of $z = f(x,y)$ with respect to x at a point P on the surface.

If we select a point P on the curve, then we can figure out the slope of the line through P tangent to the curve in a plane perpendicular to the y axis. This is a partial derivative of f with respect to x at the point P.

Derivative with respect to y

Now suppose we take the surface shown in Fig. 24-1 and slice it with a flat plane parallel to the *yz*-plane. Once again, there are infinitely many such planes. They are all perpendicular to the x axis, and they all go through some point on that axis. As before, any such plane that intersects the surface does so along a curve. If we choose one of these planes and work entirely in it, then we can find dz/dy for the curve where the plane and the surface intersect. If we move our reference plane back and forth, always keeping it parallel to the *yz*-plane and perpendicular to the x axis, we get other curves with other derivatives. Figure 24-3 shows one such scenario.

If we treat x as a constant, we can work out the general partial derivative of f with respect to y. In the situation we're dealing with now, we can write the partial derivative of f with respect to y in any of these formats:

$$\partial z/\partial y$$
$$\partial f(x,y)/\partial y$$
$$\partial f/\partial y$$
$$\partial/\partial y\, f(x,y)$$
$$\partial/\partial y\, f$$

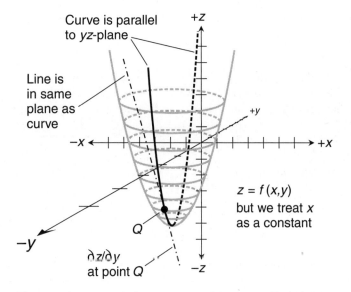

Figure 24-3   Graph showing a partial derivative of $z = f(x,y)$ with respect to y at a point Q on the surface.

If we choose a particular point Q on the curve, then we can figure out the slope of the line through Q tangent to the curve in a plane perpendicular to the x axis. This is a partial derivative of f with respect to y at P.

Example

Let's look at a simple function in which the domain is a set of real-number ordered pairs (x,y), and the range is a set of real numbers z. We will derive the general partial derivatives $\partial z/\partial x$ and $\partial z/\partial y$. Here's the function:

$$z = f(x,y) = x^3 y^5$$

To differentiate f with respect to x, we treat y as a constant and x as the independent variable. Because y is a constant, so is y^5. The partial derivative with respect to x is therefore

$$\partial z/\partial x = 3x^2 y^5$$

To differentiate f with respect to y, we treat x as a constant and y as the independent variable. Because x is a constant, so is x^3. The partial derivative with respect to y is therefore

$$\partial z/\partial y = 5x^3 y^4$$

- -

Are you confused?

Do you wonder which rules for single-variable differentiation also work for partial differentiation? Two of the rules that we learned in Part 1 can be expanded to create three tools that can help us find partial derivatives.

The *multiplication-by-constant rule for partial derivatives* says:

- If we take the partial derivative of a multi-variable function *after* it has been multiplied by a constant, we get the same result as we do if we take the partial derivative of the function and *then* multiply by the constant, as long as we differentiate with respect to the same variable both times.

The *sum rule for partial derivatives* tells us this:

- The partial derivative of the sum of two or more multi-variable functions is equal to the sum of their partial derivatives, as long as all the partial derivatives are found with respect to the same variable.

We can combine these two rules to obtain the *difference rule for two partial derivatives*, which says:

- The partial derivative of the difference between two multi-variable functions is equal to the difference between their partial derivatives, as long as both partial derivatives are found with respect to the same variable, and as long as we keep the subtraction in the same order.

- -

Another example

Let's derive the general partial derivatives $\partial z/\partial x$ and $\partial z/\partial y$ for the two-variable polynomial function

$$z = f(x,y) = 5x^5 + 2x^2 y^{-2} + 4xy^3$$

To differentiate with respect to x, we treat y as a constant. The derivative of the first term with respect to x is easy to find, because y doesn't appear there at all; it's $25x^4$. In the second term, x^2 is multiplied by the product of 2 and y^{-2}, so its derivative with respect to x is equal to $2x$ times $2y^{-2}$, which is $4xy^{-2}$. In the third term, x is multiplied by the product of 4 and y^3, so its derivative with respect to x is equal to 1 times $4y^3$, which is $4y^3$. When we add these three derivatives, we get

$$\partial z/\partial x = 25x^4 + 4xy^{-2} + 4y^3$$

To differentiate $f(x,y)$ with respect to y, we hold x constant. The derivative of the first term with respect to y is 0. That's because x is a constant, so $5x^5$ is also a constant, and the derivative of a constant is always 0. In second term, $2x^2$ is a constant, so the derivative with respect to y is equal to $2x^2$ times $-2y^{-3}$, which is $-4x^2 y^{-3}$. In the third term, $4x$ is a constant, so the derivative with respect to y is equal to $4x$ times $3y^2$, which is $12xy^2$. Adding these results, we get

$$\partial z/\partial y = 0 + (-4x^2 y^{-3}) + 12xy^2 = -4x^2 y^{-3} + 12xy^2$$

- -

Here's a challenge!

For the function stated in the last example, find the derivative with respect to x at the point $(x,y) = (1,-2)$. Mathematically, we write this as

$$\partial/\partial x \, (1,-2)$$

Then find the derivative with respect to y at the point $(x,y) = (2,3)$, which we formally write as

$$\partial/\partial y \, (2,3)$$

Solution

To find $\partial/\partial x \, (1,-2)$, we plug in the values $x = 1$ and $y = -2$ to the general derivative formula

$$\partial z/\partial x = 25x^4 + 4xy^{-2} + 4y^3$$

Working out the arithmetic, we obtain

$$\partial/\partial x \, (1,-2) = 25 \cdot 1^4 + 4 \cdot 1 \cdot (-2)^{-2} + 4 \cdot (-2)^3 = 25 + 1 + (-32) = -6$$

To find $\partial/\partial y$ (2,3), we plug in $x = 2$ and $y = 3$ to the general derivative formula

$$\partial z/\partial y = -4x^2 y^{-3} + 12xy^2$$

The arithmetic yields

$$\partial/\partial y\, (2,3) = -4 \cdot 2^2 \cdot 3^{-3} + 12 \cdot 2 \cdot 3^2 = -16/27 + 216 = 5{,}816\,/\,27$$

Three Independent Variables

If we want to differentiate a function that has three independent variables, say $w = g\,(x,y,z)$, we can try to differentiate with respect to x, with respect to y, or with respect to z. If we succeed at these tasks, we obtain partial derivatives. If it's impossible to find all three of the partial derivatives, then g is not fully differentiable.

Derivatives with respect to x, y, and z

We can find the general partial derivative of a function $w = g\,(x,y,z)$ with respect to x if we treat both y and z as constants. It's written in one of these ways:

$$\partial w/\partial x$$
$$\partial g\,(x,y,z)/\partial x$$
$$\partial g/\partial x$$
$$\partial/\partial x\, g\,(x,y,z)$$
$$\partial/\partial x\, g$$

The same thing can be done with the other two independent variables. If we hold x and z constant, we can find the partial derivative with respect to y. If we hold x and y constant, we can work out the partial derivative with respect to z. The same notations as those above can be used for these partial derivatives. Simply replace every occurrence of ∂x with either ∂y or ∂z, as applicable.

Example

Let's find $\partial w/\partial x$, $\partial w/\partial y$, and $\partial w/\partial z$ for the following function g. The domain is a set of real-number ordered triples (x,y,z), and the range is a set of real numbers w.

$$w = g\,(x,y,z) = x^3 y^{-1} z^2$$

To figure out $\partial w/\partial x$, we hold y and z constant and let x be the independent variable. When we do that, $y^{-1} z^2$ is a constant, because it's the result of arithmetic operations on other constants. The derivative with respect to x is therefore

$$\partial w/\partial x = 3x^2 y^{-1} z^2$$

To find $\partial w / \partial y$, we keep x and z constant and think of y as the independent variable. In that case, $x^3 z^2$ is a constant, so we have

$$\partial w / \partial y = -x^3 y^{-2} z^2$$

To determine $\partial w / \partial z$, we hold x and y constant and consider z as the independent variable. This means that $x^3 y^{-1}$ is a constant, so we obtain

$$\partial w / \partial z = 2x^3 y^{-1} z$$

Another example

Now we'll find $\partial w / \partial x$, $\partial w / \partial y$, and $\partial w / \partial z$ for a more complicated function h. The domain is, as before, a set of real-number ordered triples (x, y, z), and the range is a set of real numbers w.

$$w = h(x, y, z) = 2e^x y^{-2} e^z + 4xyz$$

To figure out $\partial w / \partial x$, we hold y and z constant and think of x as the independent variable. In the first term, $y^{-2} e^z$ is a constant, so the derivative with respect to x is $2e^x y^{-2} e^z$. In the second term, yz is a constant, so the derivative with respect to x is $4yz$. Adding these two results gives us

$$\partial w / \partial x = 2e^x y^{-2} e^z + 4yz$$

To find $\partial w / \partial y$, we keep x and z constant, while y is the independent variable. The first term has the constant $2e^x e^z$, so its derivative with respect to y is $-4e^x y^{-3} e^z$. In the second term, $4xz$ is a constant, so the derivative with respect to y is $4xz$. Adding, we get

$$\partial w / \partial y = -4e^x y^{-3} e^z + 4xz$$

To figure out $\partial w / \partial z$, we hold x and y constant and let z be the independent variable. In the first term, $2e^x y^{-2}$ is a constant, so the derivative with respect to z is $2e^x y^{-2} e^z$. In the second term, $4xy$ is a constant, so the derivative with respect to z is $4xy$. Adding yields

$$\partial w / \partial z = 2e^x y^{-2} e^z + 4xy$$

- -

Are you confused?

Do you wonder what happens when $y = 0$ in either of the previous examples? In both situations, there's a factor where y is raised to a negative-integer power, so if we input $y = 0$, we get an undefined quantity. Both functions have singularities. Because neither g nor h is defined at any point where $y = 0$, none of their partial derivatives are defined at such points.

Here's a challenge!

Find all three partial derivatives at the point $(x,y,z) = (2,3,4)$ for the function in the first example we worked out above:

$$w = g(x,y,z) = x^3 y^{-1} z^2$$

Solution

To work out $\partial/\partial x$ (2,3,4), we input $x = 2$, $y = 3$, and $z = 4$ to the general partial derivative $\partial w/\partial x$. That derivative is

$$\partial w/\partial x = 3x^2 y^{-1} z^2$$

Working out the arithmetic, we get

$$\partial/\partial x\,(2,3,4) = 3 \cdot 2^2 \cdot 3^{-1} \cdot 4^2 = 3 \cdot 4 \cdot (1/3) \cdot 16 = 64$$

To find $\partial/\partial y$ (2,3,4), we input $x = 2$, $y = 3$, and $z = 4$ to the general partial derivative $\partial w/\partial y$. That derivative is

$$\partial w/\partial y = -x^3 y^{-2} z^2$$

When we work out the arithmetic, we get

$$\partial/\partial y\,(2,3,4) = -(2^3) \cdot 3^{-2} \cdot 4^2 = -8 \cdot (1/9) \cdot 16 = -128/9$$

To calculate $\partial/\partial z$ (2,3,4), we input $x = 2$, $y = 3$, and $z = 4$ to the formula for $\partial w/\partial z$. That derivative is

$$\partial w/\partial z = 2x^3 y^{-1} z$$

Doing the arithmetic, we obtain

$$\partial/\partial z\,(2,3,4) = 2 \cdot 2^3 \cdot 3^{-1} \cdot 4 = 2 \cdot 8 \cdot (1/3) \cdot 4 = 64/3$$

- -

Practice Exercises

This is an open-book quiz. You may (and should) refer to the text as you solve these problems. Don't hurry! You'll find worked-out answers in App. C. The solutions in the appendix may not represent the only way a problem can be figured out. If you think you can solve a particular problem in a quicker or better way than you see there, by all means try it!

1. Look back at Fig. 24-1. Suppose that we call y the dependent variable, so the surface defines

$$y = g(x,z)$$

where x and z are the independent variables. Is g a true function? Why or why not?

2. In Fig. 24-1, suppose that we call x the dependent variable, so the surface defines

$$x = h(y,z)$$

where y and z are the independent variables. Is h a true function? Why or why not?

3. Consider the following function, in which the domain is a set of real ordered pairs (x,y), and the range is a set of real z's. Find the general partial derivative $\partial z / \partial x$.

$$z = f(x,y) = -x^2 + 2xy + 4y^3$$

4. Find the general partial derivative $\partial z / \partial y$ of the function stated in Prob. 3.

5. For the function stated in Prob. 3, find $\partial / \partial x\,(1,1)$ and $\partial / \partial y\,(1,1)$.

6. For the function stated in Prob. 3, find $\partial / \partial x\,(-2,-3)$ and $\partial / \partial y\,(-2,-3)$.

7. Consider the following function, in which the domain is a set of real ordered triples (x,y,z), and the range is a set of reals w. Find the general partial derivative $\partial w / \partial x$.

$$w = g(x,y,z) = xy \ln |z| - 2x^2 y^3 z^4$$

8. Find the general partial derivative $\partial w / \partial y$ of the function stated in Prob. 7.

9. Find the general partial derivative $\partial w / \partial z$ of the function stated in Prob. 7.

10. For the function stated in Prob. 7, find these three specific partial derivatives:

$$\partial / \partial x\,(-1,2,e)$$
$$\partial / \partial y\,(-2,-1,2e)$$
$$\partial / \partial z\,(1,-2,0)$$

CHAPTER

25

Second Partial Derivatives

We can differentiate single-variable functions more than once, as we have seen. The same is true with multi-variable functions. In this chapter, we'll learn how to partially differentiate two-variable and three-variable functions twice in succession.

Two Variables, Second Partials

Imagine a function f of two real-number input variables x and y, giving us a real-number output z:

$$z = f(x,y)$$

We can differentiate f twice with respect to x or twice with respect to y, getting new functions of those variables. The new functions are called *simple second partial derivatives*, or *second partials*.

Second partials relative to x or y

To determine the second partial of f with respect to x, we must treat y as a constant, differentiate with respect to x as we learned to do in Chap. 24, and then differentiate with respect to x again. The second partial of f with respect to x can be denoted in any of the following ways:

$$\partial^2 z/\partial x^2$$
$$\partial^2 f(x,y)/\partial x^2$$
$$\partial^2 f/\partial x^2$$
$$\partial^2/\partial x^2 \, f(x,y)$$
$$\partial^2/\partial x^2 \, f$$

If we treat x as a constant, then we can generate the second partial of f with respect to y. First, we work out $\partial z/\partial y$, and then we differentiate that function with respect to y. This second partial can be represented by any of the following expressions:

$$\partial^2 z/\partial y^2$$
$$\partial^2 f(x,y)/\partial y^2$$
$$\partial^2 f/\partial y^2$$
$$\partial^2/\partial y^2 \, f(x,y)$$
$$\partial^2/\partial y^2 \, f$$

We'll use the first notations in these lists, assuming that the independent variables are x and y, and the dependent variable is z.

Example

Let's find both of the second partials, $\partial^2 z/\partial x^2$ and $\partial^2 z/\partial y^2$, of the monomial function

$$z = f(x,y) = x^3 y^5$$

We worked out the first partial with respect to x in Chap. 24 by holding y constant, getting

$$\partial z/\partial x = 3x^2 y^5$$

To find the second partial with respect to x, we keep holding y constant, and then we differentiate with respect to x again. Because y is a constant, so is y^5. Therefore

$$\partial^2 z/\partial x^2 = 6x y^5$$

The first partial with respect to y, as we found in Chap. 24, was derived by holding x constant and differentiating with respect to y. We obtained

$$\partial z/\partial y = 5x^3 y^4$$

To figure out the second partial with respect to y, we keep treating x as a constant and differentiate relative to y again. Because x is a constant, so is $5x^3$. Therefore

$$\partial^2 z/\partial y^2 = 20x^3 y^3$$

- -

Are you confused?

It's reasonable to wonder whether the multiplication-by-constant rule, the sum rule, and the difference rule for first partial derivatives also apply to second partials. They do, under certain conditions.

The multiplication-by-constant rule for second partials says:

- If we take the second partial of a multi-variable function *after* it has been multiplied by a constant, we get the same result as we do if we take the second partial of the function and *then* multiply by the constant, as long as we differentiate with respect to the same variable throughout the process.

The sum rule for second partials says:

- The second partial of the sum of two or more multi-variable functions is equal to the sum of their second partials, as long as all the partial derivatives are found with respect to the same variable.

The difference rule for second partials says:

- The second partial of the difference between two multi-variable functions is equal to the difference between their second partials, as long as we differentiate with respect to the same variable throughout the process, and as long as we keep the subtraction in the same order.

Another example

Let's derive the second partials $\partial^2 z/\partial x^2$ and $\partial^2 z/\partial y^2$ of the polynomial function

$$z = f(x,y) = 5x^5 + 2x^2y^{-2} + 4xy^3$$

In Chap. 24, we differentiated this function term-by-term with respect to x to get the first partial

$$\partial z/\partial x = 25x^4 + 4xy^{-2} + 4y^3$$

Let's continue to hold y constant and differentiate with respect to x again. In the first term, the constant is 25. In the second term, the constant is $4y^{-2}$. The third term is a pure constant, $4y^3$, so it will disappear when we differentiate it with respect to x. Working through, we get

$$\partial^2 z/\partial x^2 = 100x^3 + 4y^{-2} + 0 = 100x^3 + 4y^{-2}$$

The first partial of the original function f with respect to y, as we found in Chap. 24, is

$$\partial z/\partial y = -4x^2y^{-3} + 12xy^2$$

We keep treating x as a constant, and we differentiate the first partial relative to y. In the first term, $-4x^2$ is a constant. In the second term, $12x$ is a constant. We therefore have

$$\partial^2 z/\partial y^2 = 12x^2y^{-4} + 24xy$$

Here's a challenge!

For the original function f in the example we just finished, find the *specific second partial* with respect to x at the point $(x,y) = (1,-2)$. We denote this as

$$\partial^2/\partial x^2 \, (1,-2)$$

Then find the specific second partial with respect to y at the point $(x,y) = (2,1)$. We denote this as

$$\partial^2/\partial y^2 \ (2,1)$$

Solution

To find the second partial with respect to x at $(1,-2)$, we input $x = 1$ and $y = -2$ to the general formula

$$\partial^2 z/\partial x^2 = 100x^3 + 4y^{-2}$$

Calculating, we get

$$\partial^2/\partial x^2 \ (1,-2) = 100 \cdot 1^3 + 4 \cdot (-2)^{-2} = 100 + 1 = 101$$

To find $\partial^2/\partial y^2 \ (2,1)$, we input $x = 2$ and $y = 1$ to the general formula

$$\partial^2 z/\partial y^2 = 12x^2 y^{-4} + 24xy$$

The arithmetic gives us

$$\partial^2/\partial y^2 \ (2,1) = 12 \cdot 2^2 \cdot 1^{-4} + 24 \cdot 2 \cdot 1 = 48 + 48 = 96$$

- -

Two Variables, Mixed Partials

Once again, let's consider a function f of two real-number inputs x and y, giving us a real-number output z, so

$$z = f(x,y)$$

We've seen what happens when we differentiate f twice with respect to x or y. But we can differentiate f with respect to x and then differentiate $\partial z/\partial x$ relative to y. Alternatively, we can differentiate f relative to y and then differentiate $\partial z/\partial y$ relative to x. If we do either of these things, we get a *general mixed second partial derivative*, also called a *mixed partial*.

Differentiating with respect to x and then y

When we work out the mixed partial of f with respect to x and then y, we can write the result as

$$\partial/\partial y \ (\partial z/\partial x)$$

assuming, of course, that we use the names of the function and the variables so that

$$z = f(x,y)$$

This same mixed partial can be denoted in other ways, too, such as:

$$\partial^2 z / \partial y \partial x$$
$$\partial^2 f(x,y) / \partial y \partial x$$
$$\partial^2 f / \partial y \partial x$$
$$\partial^2 / \partial y \partial x \, f(x,y)$$
$$\partial^2 / \partial y \partial x \, f$$

- -

Are you confused?

In the denominators of the above differential ratios, the variables appear in *reverse order* from the way the differentiation is done. Although we differentiate with respect to x and then with respect to y, the denominator says $\partial y \partial x$. This is nothing more than a notational convention, but it can be confusing.

- -

Differentiating with respect to y and then x

We can go the other way as well. First, we work out $\partial z / \partial y$, and then we differentiate that function with respect to x. This second partial can be represented by any of the following expressions:

$$\partial / \partial z \, (\partial x / \partial y)$$
$$\partial^2 z / \partial x \partial y$$
$$\partial^2 f(x,y) / \partial x \partial y$$
$$\partial^2 f / \partial x \partial y$$
$$\partial^2 / \partial x \partial y \, f(x,y)$$
$$\partial^2 / \partial x \partial y \, f$$

Again, the variables in the denominators of the differential ratios appear backward from the way we do the differentiation. We differentiate with respect to y and then x, but the denominator says $\partial x \partial y$.

Example

Let's find the two mixed partials, $\partial^2 z / \partial y \partial x$ and $\partial^2 z / \partial x \partial y$, of a function that we've worked with before:

$$z = f(x,y) = x^3 y^5$$

To find $\partial^2 z / \partial y \partial x$, we differentiate f with respect to x, and then differentiate the resulting function with respect to y. The first partial derivative with respect to x is

$$\partial z / \partial x = 3x^2 y^5$$

To differentiate relative to y, we hold x constant in $\partial z/\partial x$. That makes $3x^2$ a constant, so we have

$$\partial^2 z/\partial y \partial x = 15x^2 y^4$$

To find $\partial^2 z/\partial x \partial y$, we differentiate f relative to y, and then differentiate the resulting function relative to x. We've already seen that

$$\partial z/\partial y = 5x^3 y^4$$

To differentiate with respect to x, we hold y constant in $\partial z/\partial y$. That means y^4 is a constant, so

$$\partial^2 z/\partial x \partial y = 15x^2 y^4$$

The two mixed partials are the same. Do you think this is mere happenstance? Let's try another example and see if the coincidence occurs again.

Another example

Let's find the two mixed partials, $\partial^2 z/\partial y \partial x$ and $\partial^2 z/\partial x \partial y$, of the trinomial function we've worked with before:

$$z = f(x,y) = 5x^5 + 2x^2 y^{-2} + 4xy^3$$

The first partial relative to x, which we've already found, is

$$\partial z/\partial x = 25x^4 + 4xy^{-2} + 4y^3$$

Now we hold x constant in $\partial z/\partial x$ and differentiate with respect to y, getting

$$\partial^2 z/\partial y \partial x = 0 - 8xy^{-3} + 12y^2 = -8xy^{-3} + 12y^2$$

To find $\partial^2 z/\partial x \partial y$, we differentiate relative to y and then relative to x. We've seen that

$$\partial z/\partial y = -4x^2 y^{-3} + 12xy^2$$

To find the derivative of this function with respect to x, we hold y constant, getting

$$\partial^2 z/\partial x \partial y = -8xy^{-3} + 12y^2$$

The two mixed partials are the same again!

A theorem

If you like, make up several two-variable functions and find their mixed partials "both ways." If you work out all the derivatives correctly, you'll see that the mixed partials are always the same, regardless of which variable you differentiate against (that is, with respect to) first.

Mathematicians have proved that this is always the case. It's an important theorem in the calculus of two-variable functions.

- -

Here's a challenge!

For the original function f in the example we just finished, calculate the value of the mixed partial $\partial^2 z/\partial y \partial x$ at the point $(x,y) = (3,1)$. We denote this as

$$\partial^2/\partial y \partial x\,(3,1)$$

Then find the value of the mixed partial $\partial^2 z/\partial x \partial y$ at the point $(x,y) = (2,0)$, written as

$$\partial^2/\partial x \partial y\,(2,0)$$

Solution

To find $\partial^2/\partial y \partial x\,(3,1)$, we input $x = 3$ and $y = 1$ to the general formula

$$\partial^2 z/\partial y \partial x = -8xy^{-3} + 12y^2$$

The arithmetic yields

$$\partial^2/\partial y \partial x\,(3,1) = -8 \cdot 3 \cdot 1^{-3} + 12 \cdot 1^2 = -24 + 12 = -12$$

To find $\partial^2/\partial x \partial y\,(2,0)$, we plug in $x = 2$ and $y = 0$ to the general formula

$$\partial^2 z/\partial x \partial y = -8xy^{-3} + 12y^2$$

The arithmetic gives us

$$\partial^2/\partial x \partial y\,(2,0) = -8 \cdot 2 \cdot 0^{-3} + 12 \cdot 0^2$$

The first term contains 0 to the -3 power. That's undefined, so the entire expression is undefined at the point $(x,y) = (2,0)$. In fact, the mixed partials are undefined at any point where $y = 0$.

- -

Three Variables, Second Partials

Let's see what happens when we differentiate a three-variable function twice. As we did in Chap. 24, let's name the function and variables

$$w = g\,(x,y,z)$$

We can try to differentiate twice with respect to x, twice with respect to y, or twice with respect to z. If all the functions are differentiable, we will obtain second partials.

Second partials with respect to *x*, *y*, or *z*

We can find the second partial of *g* with respect to *x* if we treat both *y* and *z* as constants, and then differentiate against *x* twice in succession. The expression for this second partial can be written in any of these forms:

$$\partial^2 w / \partial x^2$$
$$\partial^2 g\,(x,y,z) / \partial x^2$$
$$\partial^2 g / \partial x^2$$
$$\partial^2 / \partial x^2\ g\,(x,y,z)$$
$$\partial^2 / \partial x^2\ g$$

If we hold *x* and *z* constant, we can find the second partial with respect to *y*, and denote it in any of these ways:

$$\partial^2 w / \partial y^2$$
$$\partial^2 g\,(x,y,z) / \partial y^2$$
$$\partial^2 g / \partial y^2$$
$$\partial^2 / \partial y^2\ g\,(x,y,z)$$
$$\partial^2 / \partial y^2\ g$$

If we hold *x* and *y* constant, we can find the second partial relative to *z*. As you can guess, we can denote this in any of the following formats:

$$\partial^2 w / \partial z^2$$
$$\partial^2 g\,(x,y,z) / \partial z^2$$
$$\partial^2 g / \partial z^2$$
$$\partial^2 / \partial z^2\ g\,(x,y,z)$$
$$\partial^2 / \partial z^2\ g$$

As we did earlier in this chapter, we will use the first notations in these lists, remembering that we're calling *x*, *y*, and *z* the independent variables, while *w* is the dependent variable.

Example

Let's find $\partial^2 w / \partial x^2$, $\partial^2 w / \partial y^2$, and $\partial^2 w / \partial z^2$ for the function we worked with in the first three-variable example in Chap. 24:

$$w = g\,(x,y,z) = x^3 y^{-1} z^2$$

We found $\partial w / \partial x$ by treating *y* and *z* as constants and differentiating the function relative to *x*. That gave us

$$\partial w / \partial x = 3x^2 y^{-1} z^2$$

Continuing to hold y and z constant, we can differentiate with respect to x again, getting

$$\partial^2 w / \partial x^2 = 6xy^{-1}z^2$$

We determined $\partial w / \partial y$ by holding x and z constant. Differentiating against y, we got

$$\partial w / \partial y = -x^3 y^{-2} z^2$$

When we keep treating x and z as constants, differentiating relative to y again yields

$$\partial^2 w / \partial y^2 = 2x^3 y^{-3} z^2$$

We figured out $\partial w / \partial z$ by holding x and y constant and differentiating against z, getting

$$\partial w / \partial z = 2x^3 y^{-1} z$$

Maintaining x and y as constants, we can differentiate against z again to get

$$\partial^2 w / \partial z^2 = 2x^3 y^{-1}$$

Another example

Now we'll find $\partial^2 w / \partial x^2$, $\partial^2 w / \partial y^2$, and $\partial^2 w / \partial z^2$ for the function we worked with in the second three-variable example in Chap. 24:

$$w = h(x,y,z) = 2e^x y^{-2} e^z + 4xyz$$

The first partial of the original function relative to x is

$$\partial w / \partial x = 2e^x y^{-2} e^z + 4yz$$

We keep holding y and z constant, differentiating with respect to x again to get

$$\partial^2 w / \partial x^2 = 2e^x y^{-2} e^z + 0 = 2e^x y^{-2} e^z$$

The first partial of the original function relative to y is

$$\partial w / \partial y = -4e^x y^{-3} e^z + 4xz$$

Continuing to hold x and z constant, we differentiate relative to y again, obtaining

$$\partial^2 w / \partial y^2 = 12e^x y^{-4} e^z + 0 = 12e^x y^{-4} e^z$$

The first partial of the original function relative to z is

$$\partial w / \partial z = 2e^x y^{-2} e^z + 4xy$$

If we keep treating x and y as constants, differentiating against z again gives us

$$\partial^2 w/\partial z^2 = 2e^x y^{-2} e^z + 0 = 2e^x y^{-2} e^z$$

- -

Are you confused?

It's easy to become frustrated by the fancy notation in expressions such as those we've been working with in this chapter. The situation is rarely as complicated as it looks. Remember these things, and you'll have a minimum of trouble.

- Once you decide which variable you're differentiating against, then any combination of the other variables should be treated as a constant, just as if it were a real number.
- The derivative of a constant times the variable you're differentiating against is always equal to that constant, all by itself.
- The derivative of a constant alone is always equal to 0.

Here's a challenge!

Find all three simple second partials at the point $(x,y,z) = (2,3,4)$ for the function in the first example we worked out above:

$$w = g(x,y,z) = x^3 y^{-1} z^2$$

Solution

To work out $\partial^2/\partial x^2$ (2,3,4), we input $x = 2$, $y = 3$, and $z = 4$ to

$$\partial^2 w/\partial x^2 = 6xy^{-1}z^2$$

Doing the arithmetic, we get

$$\partial^2/\partial x^2 \, (2,3,4) = 6 \cdot 2 \cdot 3^{-1} \cdot 4^2 = 6 \cdot 2 \cdot (1/3) \cdot 16 = 64$$

To find $\partial^2/\partial y^2$ (2,3,4), we input $x = 2$, $y = 3$, and $z = 4$ to

$$\partial^2 w/\partial y^2 = 2x^3 y^{-3} z^2$$

Calculating, we obtain

$$\partial^2/\partial y^2 \, (2,3,4) = 2 \cdot 2^3 \cdot 3^{-3} \cdot 4^2 = 2 \cdot 8 \cdot (1/27) \cdot 16 = 256/27$$

To determine $\partial^2/\partial z^2$ (2,3,4), we input $x = 2$, $y = 3$, and $z = 4$ to

$$\partial^2 w/\partial z^2 = 2x^3 y^{-1}$$

Working it out, we get

$$\partial^2/\partial z^2 \,(2,3,4) = 2 \cdot 2^3 \cdot 3^{-1} = 2 \cdot 8 \cdot (1/3) = 16/3$$

Three Variables, Mixed Partials

When we differentiate a three-variable function twice, we can differentiate against one variable and then another. In that case, we get a mixed partial.

Six ways to mix

There are six different ways we can find mixed second partials of a three-variable function. Here they are, along with the notations. Let's call the independent variables *x, y,* and *z,* and let's call the dependent variable *w.*

- We can differentiate against *x* and then *y* to get $\partial^2 w/\partial y \partial x$.
- We can differentiate against *x* and then *z* to get $\partial^2 w/\partial z \partial x$.
- We can differentiate against *y* and then *x* to get $\partial^2 w/\partial x \partial y$.
- We can differentiate against *y* and then *z* to get $\partial^2 w/\partial z \partial y$.
- We can differentiate against *z* and then *x* to get $\partial^2 w/\partial x \partial z$.
- We can differentiate against *z* and then *y* to get $\partial^2 w/\partial y \partial z$.

Are you confused?

As with mixed partials in two-variable functions, the sequence of differentials in the denominators can be misleading. Don't be fooled! They go in the opposite order from the way we usually read text. For example, $\partial^2 w/\partial z \partial y$ means that we differentiate with respect to *y* first, and then with respect to *z*.

Example

Let's look again at the monomial function *g* for which we found the simple second partials. This time, we'll determine all six of the mixed second partials. Here's the original function for reference:

$$w = g\,(x,y,z) = x^3 y^{-1} z^2$$

The first partial of the original function relative to *x* is

$$\partial w/\partial x = 3x^2 y^{-1} z^2$$

To differentiate $\partial w/\partial x$ against *y,* we hold *x* and *z* constant, getting

$$\partial^2 w/\partial y \partial x = -3x^2 y^{-2} z^2$$

To differentiate $\partial w/\partial x$ against z, we hold x and y constant, getting

$$\partial^2 w/\partial z \partial x = 6x^2 y^{-1} z$$

The first partial of the original function relative to y is

$$\partial w/\partial y = -x^3 y^{-2} z^2$$

To differentiate $\partial w/\partial y$ against x, we hold y and z constant, getting

$$\partial^2 w/\partial x \partial y = -3x^2 y^{-2} z^2$$

To differentiate $\partial w/\partial y$ against z, we hold x and y constant, getting

$$\partial^2 w/\partial z \partial y = -2x^3 y^{-2} z$$

The first partial of the original function relative to z is

$$\partial w/\partial z = 2x^3 y^{-1} z$$

To differentiate $\partial w/\partial z$ against x, we hold y and z constant, getting

$$\partial^2 w/\partial x \partial z = 6x^2 y^{-1} z$$

To differentiate $\partial w/\partial z$ against y, we hold x and z constant, getting

$$\partial^2 w/\partial y \partial z = -2x^3 y^{-2} z$$

- -

Here's a challenge!

Find the six mixed second partials for the binomial three-variable function h that we've worked with a couple of times already. Here's the function again, for reference:

$$w = h(x,y,z) = 2e^x y^{-2} e^z + 4xyz$$

Solution

This problem is a little bit tedious, but it's tricky. We have many opportunities to make a mistake, but only one way to get it right! The first partial of the original function relative to x is

$$\partial w/\partial x = 2e^x y^{-2} e^z + 4yz$$

To differentiate $\partial w/\partial x$ against y, we hold x and z constant to get

$$\partial^2 w/\partial y \partial x = -4e^x y^{-3} e^z + 4z$$

To differentiate $\partial w/\partial x$ against z, we hold x and y constant to get

$$\partial^2 w/\partial z\partial x = 2e^x y^{-2}e^z + 4y$$

The first partial of the original function relative to y is

$$\partial w/\partial y = -4e^x y^{-3}e^z + 4xz$$

To differentiate $\partial w/\partial y$ against x, we hold y and z constant to get

$$\partial^2 w/\partial x\partial y = -4e^x y^{-3}e^z + 4z$$

To differentiate $\partial w/\partial y$ against z, we hold x and y constant to get

$$\partial^2 w/\partial z\partial y = -4e^x y^{-3}e^z + 4x$$

The first partial of the original function relative to z is

$$\partial w/\partial z = 2e^x y^{-2}e^z + 4xy$$

To differentiate $\partial w/\partial z$ against x, we hold y and z constant to get

$$\partial^2 w/\partial x\partial z = 2e^x y^{-2}e^z + 4y$$

To differentiate $\partial w/\partial z$ against y, we hold x and z constant to get

$$\partial^2 w/\partial y\partial z = -4e^x y^{-3}e^z + 4x$$

- -

Practice Exercises

This is an open-book quiz. You may (and should) refer to the text as you solve these problems. Don't hurry! You'll find worked-out answers in App. C. The solutions in the appendix may not represent the only way a problem can be figured out. If you think you can solve a particular problem in a quicker or better way than you see there, by all means try it!

1. Consider the following function, in which the domain is a set of real ordered pairs (x,y), and the range is a set of reals z. Find the second partial $\partial^2 z/\partial x^2$.

$$z = f(x,y) = -x^2 + 2xy + 4y^3$$

2. Find the second partial $\partial^2 z/\partial y^2$ of the function stated in Prob. 1.

3. Find the specific second partials $\partial^2/\partial x^2$ (3,2) and $\partial^2/\partial y^2$ (3,2) of the function stated in Prob. 1.

4. Find the mixed second partial $\partial^2 z/\partial y \partial x$ of the function stated in Prob. 1.

5. Find the mixed second partial $\partial^2 z/\partial x \partial y$ of the function stated in Prob. 1.

6. Find the specific mixed partials $\partial^2/\partial y \partial x$ (3,2) and $\partial^2/\partial x \partial y$ (3,2) of the function stated in Prob. 1.

7. Consider the following function, in which the domain is a set of real ordered triples (x,y,z), and the range is a set of reals w. Find $\partial^2 w/\partial x^2$, $\partial^2 w/\partial y^2$, and $\partial^2 w/\partial z^2$.

$$w = g(x,y,z) = xy \ln |z| - 2x^2 y^3 z^4$$

8. For the function stated in Prob. 7, find these three specific second partials:

$$\partial^2/\partial x^2 \ (-1,2,e)$$
$$\partial^2/\partial y^2 \ (-2,-1,2e)$$
$$\partial^2/\partial z^2 \ (1,-2,0)$$

9. Find all six mixed second partials for the function stated in Prob. 7.

10. Find all six specific mixed second partials for the function stated in Prob. 7, at the point where $(x,y,z) = (1,2,e)$.

Surface-Area and Volume Integrals

In Part 2, we used integration to find areas of flat surfaces. Integration can also help us determine surface areas and volumes of solids in three dimensions (3D). In this chapter, we'll see how this technique works with cylinders, cones, spheres, and prisms.

A Cylinder

Figure 26-1 shows a cylinder in Cartesian three-space. The cylinder is lying on its side, so its left-hand face is in the yz-plane, with the center at the origin. The axis of the cylinder lies along the x axis. The right-hand face is h units to the right of the left-hand face. The radius is r. This is a *right circular cylinder*, meaning that the end faces are circles oriented at right angles to the axis.

Circumference vs. displacement

Imagine that we uniformly slice the outer shell into cylindrical cross-sectional bands, each of width Δx. We increase the number of bands indefinitely, so their widths become infinitesimals dx and they approach true geometric circles. These circles all lie in planes parallel to the end faces, and they all have radius r. Their circumferences are therefore equal to $2\pi r$. The circumference of any particular cross-sectional circle is a function f of the displacement x to the right of the cylinder's left-hand face. It's a constant function:

$$f(x) = 2\pi r$$

because it doesn't depend on where the center of the circle is located on the x axis.

Lateral-surface area integral

To find the *lateral-surface area* A of the cylinder (the area of the sleeve only, not including either end face), we can integrate f along the length of the cylinder from its left-hand

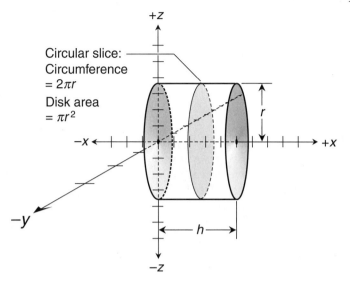

Figure 26-1 Integration can be used to find the lateral-surface area and volume of a cylinder.

face where $x = 0$ to its right-hand face where $x = h$. When we set up the integral, it comes out as

$$A = \int_0^h 2\pi r \, dx$$

The basic antiderivative is

$$F(x) = 2\pi r x$$

When we evaluate F from $x = 0$ to $x = h$, we get

$$A = 2\pi r x \Big]_0^h = 2\pi r h - 2\pi r \cdot 0 = 2\pi r h$$

This is familiar! We recognize it from basic geometry as the formula for the lateral-surface area of a cylinder in terms of its radius r and its length h.

Cross-sectional area vs. displacement

Let's cut the cylinder uniformly into cross-sectional slices again. As before, we make the number of slices approach infinity. But now, instead of slicing through the sleeve, we cut through the interior. This produces coin-like slabs parallel to the cylinder faces. Every slab has equal thickness Δx. As the number of slabs approaches infinity, their thicknesses become differentials dx which become arbitrarily close to 0. The slabs approach two-dimensional (2D) disks of radius r and area πr^2. As we did with the circular slices, we can think of the area of any particular disk as a function g of its displacement x to the right of the cylinder's left-hand face. This is a constant function:

$$g(x) = \pi r^2$$

Volume integral

To find the volume V of the cylinder, we integrate g along the entire length, just as we did with the function f for the circles. We start at the left-hand face where $x = 0$, and travel along the x axis until we reach the right-hand face where $x = h$. That gives us

$$V = \int_0^h \pi r^2 \, dx$$

The basic antiderivative is

$$G(x) = \pi r^2 x$$

When we evaluate G from $x = 0$ to $x = h$, we get

$$V = \pi r^2 x \Big]_0^h = \pi r^2 h - \pi r^2 \cdot 0 = \pi r^2 h$$

This is the formula for the volume of a cylinder that we learned in 3D geometry.

- -

Are you confused?

Do you wonder how we can add up circumferences to get area, or add up areas to get volume? How do we get the extra dimensions? The answer is that we're not adding circumferences or areas in the conventional sense. We're taking advantage of the same mathematical trick that Riemann used with rectangles to figure out the areas defined by curves in the Cartesian plane.

To determine the surface area of a cylinder, we stack *infinitely many infinitely thin bands* to get the cylinder shell. To get the volume, we stack *infinitely many infinitely thin slabs* to get the cylinder and its interior. We call the bands "circles" and the slabs "disks" because, as their widths or thicknesses approach 0, that's what they more and more closely resemble.

- -

A Cone

Figure 26-2 shows a cone tipped on its side, so its apex is at the origin of Cartesian *xyz*-space. Its base (the flat face at the right) is parallel to the *yz*-plane. The axis of the cone is a portion of the coordinate x axis. The base of the cone is h units from the apex. The radius of the base is r units. This is a *right circular cone,* meaning that the base is a circle oriented at a right angles to the axis.

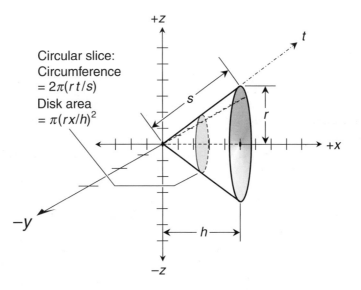

Figure 26-2   Integration can be used to find the slant-surface area and volume of a cone.

Circumference vs. displacement

We can slice the cone's shell into cross sections, and then increase the number of slices indefi-nitely. These slices approach circles parallel to the *yz*-plane, perpendicular to the *x* axis, and centered on that axis. Unlike the situation with the cylinder, these circles vary in radius, depending on where on the *x* axis we center them.

The circumference of the cone's circular base is $2\pi r$. Any point on the edge of the base is *s* units from the apex, where *s* is the *slant height* of the cone as shown in the figure. The slant height is

$$s = (r^2 + h^2)^{1/2}$$

The radius of any particular cross section is directly proportional to its distance from the apex as measured along the shell of the cone. Let's call this distance *t*, and invent an axis for it as shown. This *t* axis can be any straight line connecting the cone's apex with some point on the edge of its base. The radius of a circular slice whose edge is *t* units from the apex is therefore equal to *r* times *t/s*, so its circumference is $2\pi(rt/s)$.

We can describe the circumference of any particular cross-sectional circle as a function *f* of the distance *t* between its edge and the cone's apex. This isn't a constant function, as with the cylinder. Instead, we have

$$f(t) = 2\pi(rt/s) = (2\pi r/s)t$$

where $2\pi r/s$ is a constant, and *t* is the independent variable.

- -

Are you confused?

Do you wonder why we define the circumferences of our cross-sectional circles in terms of the displacement t along the cone's shell, rather than in terms of the displacement x along the cone's axis? When we want to find the surface area of an object, we must integrate a function that's defined *on that surface.* If we cut through the cone's interior from the apex to the center of the base, the resulting distance h is less than the slant height s. If we try to derive the surface-area formula that way, we'll get answers that are only h/s of what they should be.

- -

Slant-surface area integral

Let A be the *slant-surface area* of the cone (that is, the surface area not including the base). We can integrate $f(t)$ along a straight line in the cone's shell from the apex, where $t = 0$, to any point on the outer edge of the base, where $t = s$. This gives us the definite integral

$$A = \int_0^s (2\pi r/s)t \, dt$$

The basic antiderivative is

$$F(t) = [(2\pi r/s)t^2] \, / \, 2 = (\pi r/s)t^2$$

When we evaluate F from $t = 0$ to $t = s$, we get

$$A = (\pi r/s)t^2 \Big]_0^s = (\pi r/s)s^2 - (\pi r/s) \cdot 0^2 = \pi rs$$

This formula is taught in some basic geometry courses, but not all. If you go to the Internet and do a search on the phrase "surface area of a cone," you'll find that this is indeed the formula for the slant-surface area of a cone in terms of its base radius r and its slant height s. If you want the formula in terms of the radius and the true height h, then you can substitute the quantity $(r^2 + h^2)^{1/2}$ in place of s, getting

$$A = \pi r(r^2 + h^2)^{1/2}$$

Cross-sectional area vs. displacement

Now let's slice up the cone through its interior, getting disks parallel to the base. Each disk has a radius that's directly proportional to its distance x from the apex. The largest such disk is the base itself. The radius of any particular disk is equal to r times x/h, so its area is $\pi(rx/h)^2$.

We can describe the area of a cross-sectional disk as a function g of its distance x between its center and the cone's apex. In this situation, we travel straight through the cone rather than on its shell, because we're concerned with the cone's interior, not its surface. We have

$$g(x) = \pi(rx/h)^2 = (\pi r^2/h^2)x^2$$

where $\pi r^2/h^2$ is a constant, and x is the independent variable.

Volume integral

To find the volume V, we integrate our function g along the axis of the cone from its apex where $x = 0$ to its base where $x = h$. This gives us

$$V = \int_0^h (\pi r^2/h^2)x^2 \, dx$$

The basic antiderivative is

$$G(x) = [(\pi r^2/h^2)x^3] / 3$$

When we evaluate G from $x = 0$ to $x = h$, we get

$$V = [(\pi r^2/h^2)x^3] / 3 \Big]_0^h = [(\pi r^2/h^2)h^3] / 3 - [(\pi r^2/h^2) \cdot 0^3] / 3 = \pi r^2 h/3$$

This is the familiar geometric formula for the volume of a cone in terms of its base radius r and its height h.

- -

Here's a challenge!

Using integration, find the slant-surface area and volume of a cone with a base radius r of 3 units and a height h of 4 units. Then compare these results with the values obtained using the standard geometric formulas.

Solution

To begin, we must find the slant height of the cone. We're told that $r = 3$ and $h = 4$. Remembering the formula for slant height s in terms of r and h, we can calculate

$$s = (r^2 + h^2)^{1/2} = (3^2 + 4^2)^{1/2} = (9 + 16)^{1/2} = 25^{1/2} = 5$$

The circumference of a circular slice as a function f of the distance t between its edge and the cone's apex is

$$f(t) = (2\pi r/s)t = (2\pi \cdot 3/5)t = (6\pi/5)t$$

To find the slant-surface area A, we must calculate

$$A = \int_0^5 (6\pi/5)t \, dt$$

The basic antiderivative is

$$F(t) = [(6\pi/5)t^2] / 2 = (3\pi/5)t^2$$

When we evaluate F from $t = 0$ to $t = 5$, we get

$$A = (3\pi/5)t^2 \bigg]_0^5 = (3\pi/5) \cdot 5^2 - (3\pi/5) \cdot 0^2 = (3\pi/5) \cdot 25 = 15\pi$$

Using the geometric formula, we get the same result:

$$A = \pi r s = \pi \cdot 3 \cdot 5 = 15\pi$$

Now let's calculate the volume. The area of a cross-sectional disk as a function g of the distance x between its center and the cone's apex is

$$g(x) = (\pi r^2/h^2)x^2 = (\pi \cdot 3^2/4^2)x^2 = (9\pi/16)x^2$$

To find the volume V, we integrate g along the axis of the cone from its apex where $x = 0$ to its base where $x = 4$, getting

$$V = \int_0^4 (9\pi/16)x^2 \, dx$$

The basic antiderivative is

$$G(x) = [(9\pi/16)x^3] / 3 = (3\pi/16)x^3$$

Evaluating G from $x = 0$ to $x = 4$, we get

$$V = [(3\pi/16)x^3 \bigg]_0^4 = (3\pi/16) \cdot 4^3 - (3\pi/16) \cdot 0^3 = (3\pi/16) \cdot 64 = 12\pi$$

The formula from 3D geometry tells us that

$$V = \pi r^2 h/3 = \pi \cdot 3^2 \cdot 4/3 = 12\pi$$

A Sphere

Figure 26-3 shows a sphere in Cartesian *xyz*-space. The center of the sphere is at the origin, and its radius is *r*. Let's use integration to derive general formulas for the surface area and the volume of this sphere.

Circumference vs. arc displacement

We can slice the sphere's outer surface into cross sections parallel to the *yz*-plane, and then force the number of slices to increase endlessly. When we have infinitely many such slices, each one is a circle on the sphere's shell. Suppose we start at the far right-hand pole as shown in Fig. 26-3, where the sphere intersects the positive *x* axis. We move toward the left until we reach the opposite pole where the sphere intersects the negative *x* axis. The radius of a circular

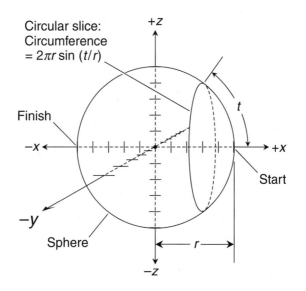

Figure 26-3   Integration can be used to find the
surface area of a sphere.

slice depends on its distance *t*, as measured over the sphere's surface, from the starting point.
The smallest possible distance is $t = 0$, and the largest possible value is $t = \pi r$, or half the
sphere's circumference.

We can write down a function that describes the radius of any particular cross-sectional
circle based on the arc distance *t*. If we call the function f^*, we have

$$f^*(t) = r \sin(t/r)$$

Here, t/r is the angle, in radians, between any point on the circle's edge and the far right-hand
pole of the sphere. The circumference of such a circle is 2π times its radius. If *f* is the function
that describes this circumference, then

$$f(t) = 2\pi f^*(t) = 2\pi r \sin(t/r)$$

Surface-area integral

To find the surface area *A* of the sphere, we must integrate *f* halfway around the sphere from
$t = 0$ to $t = \pi r$. We can write that integral as

$$A = \int_0^{\pi r} 2\pi r \sin(t/r)\ dt = 2\pi r \int_0^{\pi r} \sin(t/r)\ dt$$

When we look through App. G, the table of indefinite integrals, we find the formula

$$\int \sin ax\ dx = -a^{-1} \cos ax + c$$

From this, letting $a = r^{-1}$, we can infer that the basic antiderivative of the last integrand above is

$$F^*(t) = -r \cos(t/r)$$

The basic antiderivative of the complete function f is $2\pi r$ times this, or

$$F(t) = -2\pi r^2 \cos(t/r)$$

When we evaluate F from $t = 0$ to $t = \pi r$, we get

$$A = -2\pi r^2 \cos(t/r) \Big]_0^{\pi r} = -2\pi r^2 \cos \pi - (-2\pi r^2 \cos 0) = 2\pi r^2 + 2\pi r^2 = 4\pi r^2$$

This is the well-known geometric formula for the surface area of a sphere in terms of its radius r.

- -

Are you confused?

You might wonder why we follow a *geodesic* on the sphere's surface instead of cutting through the sphere along the x axis to define the circumferences of the cross-sectional circles. (On any sphere, a geodesic is a circle whose center is at the center of the sphere. A geodesic, also called a *great circle*, is as large as any circle on the sphere can be. On the earth's surface, for example, all meridians are geodesics because they pass through both poles.)

The situation here is similar to the case with the cone. When we use integration to figure out the surface area of a solid, remember that we must integrate on, or over, that surface. If we cut straight through the sphere, then we travel a distance equal to the sphere's diameter, or twice the radius. But we want to go halfway around the sphere's circumference. That distance is greater than the diameter by a factor of $\pi/2$.

- -

Cross-sectional area vs. displacement

Imagine that we cut up the sphere into solid disks instead of circles, all parallel to the yz-plane with their edges on the sphere's shell (Fig. 26-4). The radius r_0 of such a disk is a function of the x-value of the point through which its center passes. Imagine x as the base of a right triangle (shown in dashed gray), r_0 as its height, and r as its hypotenuse. From the Pythagorean theorem in geometry, we know that

$$x^2 + r_0^2 = r^2$$

so therefore

$$r_0 = (r^2 - x^2)^{1/2}$$

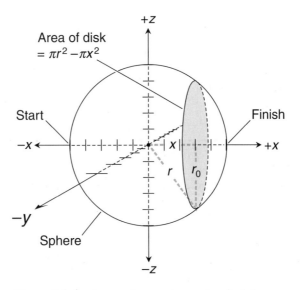

Figure 26-4 Integration can be used to find the volume of a sphere.

With this information, we can describe the area of any particular cross-sectional disk as a function of x. We use the formula for the area of a disk in terms of its radius. If we call our function g, then

$$g(x) = \pi r_0^2 = \pi [(r^2 - x^2)^{1/2}]^2 = \pi(r^2 - x^2) = \pi r^2 - \pi x^2$$

In this situation, we travel straight through the sphere rather than over its surface. When thinking about volume, we're concerned with the whole sphere, interior included, and not only the surface. This makes our job easier than it was when we derived the formula for the surface area. This time, as we integrate, we can travel a straight path instead of having to follow an arc.

Volume integral

We can define the volume V of the sphere by integrating our function g along the x axis from the left-hand pole of the sphere to the right-hand pole. The lower bound is therefore at the point where $x = -r$, and the upper bound is at $x = r$. We have

$$V = \int_{-r}^{r} (\pi r^2 - \pi x^2)\, dx$$

Keep in mind that r, the radius of the sphere, is a constant, so the basic antiderivative is

$$G(x) = \pi r^2 x - \pi x^3/3$$

When we evaluate G from $x = -r$ to $x = r$, we get

$$V = (\pi r^2 x - \pi x^3/3) \Big]_{-r}^{r} = (\pi r^3 - \pi r^3/3) - (-\pi r^3 + \pi r^3/3) = 4\pi r^3/3$$

We recognize this from our courses in geometry as the formula for the volume of a sphere in terms of its radius r.

- -

Here's a challenge!

Using integration, find the surface area and volume of a sphere with a radius of 2 units. Compare the results with the values obtained using the formulas from 3D geometry.

Solution

As we find the surface area of the sphere, Fig. 26-3 can be helpful for reference. Remember that the circumference of a cross-sectional circle as a function of the arc displacement t is described by

$$f(t) = 2\pi r \sin(t/r)$$

To find the surface area A, we integrate f halfway around the sphere from $t = 0$ to $t = \pi r$. We know that $r = 2$, so

$$A = \int_0^{\pi r} 2\pi r \sin(t/r)\, dt = 4\pi \int_0^{2\pi} \sin(t/2)\, dt$$

The table of indefinite integrals (App. G) gives us the formula

$$\int \sin ax\, dx = -a^{-1} \cos ax + c$$

Letting $a = 1/2$ and leaving out the constant of integration, we obtain

$$\int \sin(t/2)\, dt = -2\cos(t/2)$$

The basic antiderivative F of the complete function f is 4π times this, or

$$F(t) = -8\pi \cos(t/2)$$

When we evaluate F from $t = 0$ to $t = 2\pi$, we get

$$A = -8\pi \cos(t/2) \Big]_0^{2\pi} = -8\pi \cos \pi - (-8\pi \cos 0) = 8\pi + 8\pi = 16\pi$$

The formula from solid geometry gives us

$$A = 4\pi r^2 = 4\pi \cdot 2^2 = 16\pi$$

Now, let's refer to Fig. 26-4 as we work out the volume. The area of any particular cross-sectional disk, based on the location of its center along the x axis, is

$$g(x) = \pi r^2 - \pi x^2$$

We've been told that $r = 2$, so this function becomes

$$g(x) = 4\pi - \pi x^2$$

The sphere's volume V is obtained by integrating the function g along the x axis from the point where $x = -2$ to the point where $x = 2$. Therefore,

$$V = \int_{-2}^{2} (4\pi - \pi x^2)\, dx$$

The basic antiderivative is

$$G(x) = 4\pi x - \pi x^3/3$$

When we evaluate G from $x = -2$ to $x = 2$, we get

$$V = (4\pi x - \pi x^3/3) \Big]_{-2}^{2} = (4\pi \cdot 2 - \pi \cdot 2^3/3) - [4\pi \cdot (-2) - \pi \cdot (-2)^3/3]$$

$$= (8\pi - 8\pi/3) - [-8\pi - (-8\pi/3)] = 16\pi/3 + 16\pi/3 = 32\pi/3$$

From solid geometry, we can calculate the volume as

$$V = 4\pi r^3/3 = 4\pi \cdot 2^3/3 = 32\pi/3$$

- -

Practice Exercises

This is an open-book quiz. You may (and should) refer to the text as you solve these problems. Don't hurry! You'll find worked-out answers in App. C. The solutions in the appendix may not represent the only way a problem can be figured out. If you think you can solve a particular problem in a quicker or better way than you see there, by all means try it!

1. Using the integration techniques in this chapter, find the lateral-surface area of a cylinder with a radius of 2 units and a length of 5 units. Compare this with the result using the formula from 3D geometry.

2. Find the volume of the cylinder described in Prob. 1, first using integration, and then using the formula from 3D geometry.

3. Using integration, find the slant-surface area of a cone with a radius of 12 units and a slant height of 13 units. Compare this with the result using the formula from 3D geometry.

4. Find the volume of the cone described in Prob. 3, first using integration, and then using the formula from 3D geometry.

5. Using integration, find the surface area of a sphere with a radius of 10 units. Compare this with the result using the formula from 3D geometry.

6. Find the volume of the sphere described in Prob. 3, first using integration, and then using the formula from 3D geometry.

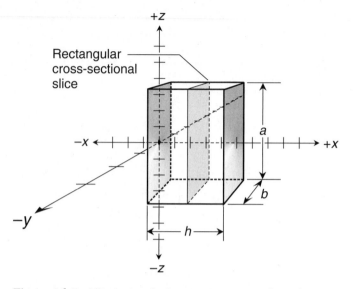

Figure 26-5 Illustration for Practice Exercises 7 through 10.

7. Derive a formula for the lateral-surface area (not including either flat end face) of a rectangular prism using integration. Assume that the length of the prism is *h*, and the end faces are rectangles of dimensions *a* and *b* as shown in Fig. 26-5.

8. From 3D geometry, verify that the formula derived in the solution to Prob. 7 is correct.

9. Derive a formula for the volume of a rectangular prism using integration. Assume that the length of the prism is *h*, and the end faces are rectangles of dimensions *a* and *b* as shown in Fig. 26-5.

10. From 3D geometry, verify that the formula derived in the solution to Prob. 9 is correct.

CHAPTER
27

Repeated, Double, and Iterated Integrals

In this chapter, we'll develop a method of integrating single-variable functions twice in succession. Then we'll integrate two-variable functions twice, first with respect to one variable and then with respect to the other.

Repeated Integrals in One Variable

Just as we can differentiate a single-variable function twice, we can integrate a single-variable function twice. We must set the constants of integration to 0 with each repetition to avoid ambiguity.

Multiple definite integrals

Let's start out with a constant function. If we call the function f, the variable x, and the constant k, then

$$f(x) = k$$

The indefinite integral with respect to x is

$$\int k \, dx = kx + c$$

where c is the constant of integration. Suppose that we set $c = 0$ and then integrate again with respect to x. That gives us

$$\int \left(\int k \, dx \right) dx = \int kx \, dx = kx^2/2 + d$$

where d is another constant of integration. If we set $d = 0$ and then specify an interval with bounds $x = a$ and $x = b$ for both integrals, we get

$$kx^2/2 \, \Big]_a^b = kb^2/2 - ka^2/2$$

455

We can write out the complete repeated integral like this:

$$\int_a^b \int_a^b k \, dx \, dx = kx^2/2 \Big]_a^b = kb^2/2 - ka^2/2$$

We must write two differentials dx here, one for each integral.

- -

Are you confused?

In this scheme, we find basic antiderivatives twice in succession, and then evaluate the final expression *at the end of the process*. We don't find the first definite integral and then integrate again. (We will operate that way later in this chapter with functions of two variables, but that approach won't work with functions of one variable.)

- -

Example

Let's try our newfound skill on a monomial linear function (a constant times a variable). Suppose we want to find

$$\int_1^2 \int_1^2 12t \, dt \, dt$$

First, we find the indefinite integral without the constant of integration. That's the basic antiderivative

$$\int 12t \, dt = 6t^2$$

Now we integrate again, leaving out the constant of integration to get

$$\int 6t^2 \, dt = 2t^3$$

The repeated definite integral is therefore

$$\int_1^2 \int_1^2 12t \, dt \, dt = 2t^3 \Big]_1^2 = 2 \cdot 2^3 - 2 \cdot 1^3 = 16 - 2 = 14$$

Another example

The sine and the cosine functions behave in interesting ways when we integrate them over and over. Let's consider the example

$$\int_{\pi/2}^{\pi} \int_{\pi/2}^{\pi} \sin z \, dz \, dz$$

The first basic antiderivative is

$$\int \sin z \, dz = -\cos z$$

Integrating again without the constant gives us

$$\int -\cos z \, dz = -\sin z$$

The repeated definite integral is therefore

$$\int_{\pi/2}^{\pi} \int_{\pi/2}^{\pi} \sin z \, dz \, dz = -\sin z \Big]_{\pi/2}^{\pi} = -\sin \pi - (-\sin \pi/2) = 0 - (-1) = 1$$

- -

Here's a challenge!

In Chap. 8, we worked out a problem involving a fictitious episode with Sir Isaac Newton, an apple, and the cliffs of Dover. According to our make-believe story, Sir Isaac took an apple to the cliffs, tossed it off, and watched it plunge to the beach. He (and we) knew the function that defined the apple's vertical fallen distance vs. time. He timed the fall, and saw that the apple struck the beach after 4 seconds. Based on that data, we calculated the height h_c of the cliff.

Now let's change the story. Suppose that Sir Isaac knew only that the apple would accelerate vertically downward at a constant rate of 10 meters per second per second. He knew that this would be true, neglecting air resistance, no matter how heavy the apple was. He timed the fall and saw that the apple struck the beach after 4 seconds (Fig. 27-1). Calculate the height h_c of the cliff on this basis. Remember that the apple's vertical speed is the derivative of its vertical fallen distance over time, and its vertical acceleration is the derivative of its vertical speed at any instant in time.

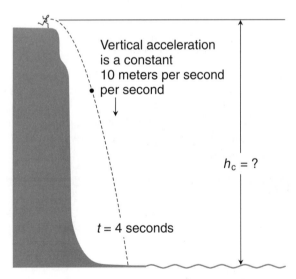

Figure 27-1   Sir Isaac hurls an apple from the cliffs of Dover. The vertical acceleration is 10 meters per second per second, and the apple takes 4 seconds to fall. What is the altitude h_c of the apple above the beach at the instant it leaves Sir Isaac's hand?

Solution

To work this out, we integrate the acceleration function twice, from $t = 0$ seconds (when the apple leaves Sir Isaac's hand) to $t = 4$ seconds (when the apple lands). The acceleration function is a constant 10, so our repeated integral is

$$\int_0^4 \int_0^4 10 \ dt \ dt$$

The first basic antiderivative is

$$\int 10 \ dt = 10t$$

When we take the basic antiderivative again, we get

$$\int 10t \ dt = 5t^2$$

The repeated definite integral is

$$\int_0^4 \int_0^4 10 \ dt \ dt = 5t^2 \ \bigg]_0^4 = 5 \cdot 4^2 - 5 \cdot 0^2 = 80 - 0 = 80$$

The height of the cliff, h_c, is 80 meters. This is the same answer that we got when we worked out the solution in Chap. 8.

- -

Double Integrals in Two Variables

While repeated integration can be done with single-variable functions, we're more likely to encounter it with functions of two variables. An interesting application is calculating the *mathematical volume* of an object in 3D coordinates. In Chap. 26, we were concerned with the *true geometric volume* or "real-world" volume of a solid object. "Real-world" volume can never be negative. Mathematical volume can be positive, negative, or zero. We'll see how this works in this chapter and the next.

Prisms and slabs

Imagine a function f of variables x and y in Cartesian xyz-space. Suppose that the graph of the function is a surface such as the one as shown in Fig. 27-2. For any given point (x,y,z) on the surface, we have

$$z = f(x,y)$$

We can approximate the mathematical volume of the object defined by this surface with respect to any rectangular region in the xy-plane, as long as the function f is defined for every point (x,y) in that region. The object is a solid whose "base" lies in the xy-plane.

Imagine breaking up the solid into square prisms whose heights are defined by the value of z in a selected corner of each prism's top face. (In Fig. 27-2, that point is the one where the

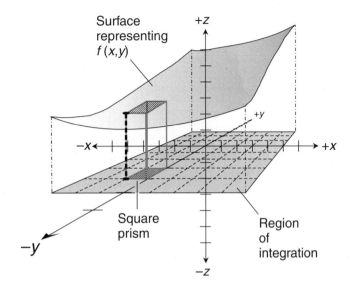

Figure 27-2 We can approximate the volume defined by a surface with respect to a rectangular region in the *xy*-plane by breaking the solid into square prisms whose heights are defined at selected points in their top faces.

x and *y* values are smallest.) The defined height of the prism is shown as a heavy, dashed vertical line that runs along one of its corners. In this example, the height of the prism is positive ($z > 0$), so its mathematical volume is positive. If the height were negative ($z < 0$), then the prism's mathematical volume would be negative. If the height of the prism were zero, then its mathematical volume would be zero.

Now imagine that we make the *xy*-plane grid for the prisms increasingly fine while staying inside the same rectangular region. In Fig. 27-2, the grid has eight squares along the *x* axis and eight squares along the *y* axis. That means there are 8^2, or 64, prisms in total. We could increase this to 80 by 80, getting 6,400 prisms; then to 800 by 800, getting 640,000 prisms; then to 8,000 by 8,000, getting 64,000,000 prisms; and so on without end. As the number of prisms increases while the rectangular region stays the same, our approximation of the solid's mathematical volume improves. As the number of prisms approaches infinity, the sum of their individual mathematical volumes approaches the actual mathematical volume of the solid defined by the surface relative to the rectangular region.

In practice, it's difficult to add up the mathematical volumes of rectangular prisms, as portrayed in Fig. 27-2, to find the mathematical volume of a solid defined by a surface with respect to a plane region. But there's an easier way. We can integrate $f(x,y)$ with respect to one of the variables, in effect slicing the solid crosswise into flat, thin slabs. Then we can integrate the resulting function with respect to the other variable. When we do this, we stack up the slabs, rather than assembling an array of prisms, to get the mathematical volume of the solid.

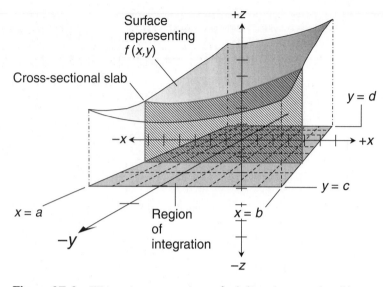

Figure 27-3 We can integrate twice to find the volume enclosed by a surface with respect to a rectangular region in the *xy*-plane by integrating with respect to *x* and then with respect to *y*.

Slabs parallel to the *xz*-plane

Let's slice the solid into arbitrarily thin vertical slabs parallel to the *xz*-plane, find their areas by integration, and then integrate the function defining those areas along the *y* axis. This scheme is diagrammed in Fig. 27-3. We must work through two separate integrals.

First, we integrate $f(x,y)$ with respect to *x*, treating *y* as a constant. We can think of this as a "partial antiderivative." But we must consider the limits of integration, too. Suppose that the *x*-value edges of the rectangle in the *xy*-plane are $x = a$ and $x = b$, as shown in the drawing. That means we must determine

$$\int_a^b f(x,y) \, dx$$

Once we've worked out this definite integral, we get a function of *y* alone. We integrate that function with respect to *y*, and evaluate the result from $y = c$ to $y = d$. The bounds of this interval represent the *y*-value edges of the rectangle in the *xy*-plane, as shown. When we finish this process, we have the value of

$$\int_c^d \left[\int_a^b f(x,y) \, dx \right] dy$$

If we give the *xy*-plane region a name such as *R*, we can rewrite the above expression in the shortened form

$$\iint_R f(x,y) \, dx \, dy$$

Slabs parallel to the yz-plane

There's nothing special about the way we sliced up the solid in the above example. We can just as easily slice it into slabs parallel to the *yz*-plane. As before, we find the slabs' areas by integration, but this time we first integrate $f(x,y)$ with respect to *y*. Then we integrate the function defining those areas as we move along the *x* axis. This scheme is shown in Fig. 27-4.

As we integrate $f(x,y)$ with respect to *y*, we hold *x* constant. Imagine that the *y*-value edges of the rectangle in the *xy*-plane are $y = c$ and $y = d$. We figure out the definite integral

$$\int_{c}^{d} f(x,y) \, dy$$

When we do this, we get a function of *x*. We integrate this new function with respect to *x* over the interval from $x = a$ to $x = b$, obtaining

$$\int_{a}^{b} \left[\int_{c}^{d} f(x,y) \, dy \right] \, dx$$

As before, we can call the *xy*-plane region *R*. The above expression can then be written as

$$\iint_{R} f(x,y) \, dy \, dx$$

- -

Are you confused?

There's an obvious difference between the two processes described above, and a subtle difference in the integral expressions we write at the end. In the first case, we get differentials in the order *dx dy*. In the

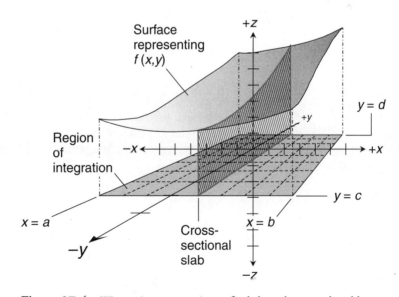

Figure 27-4 We can integrate twice to find the volume enclosed by a surface with respect to a rectangular region in the *xy*-plane by integrating with respect to *y* and then with respect to *x*.

second process, they come out as *dy dx.* It's reasonable to ask, "Does the order of the differentials matter?" From a pure mathematics point of view, there's a theoretical difference between

$$\iint_{R} f(x,y)\ dx\ dy$$

and

$$\iint_{R} f(x,y)\ dy\ dx$$

But in practice, the quantity we're describing—the mathematical volume of the solid—is the same either way. If these schemes really work (and they do, as long as the function is integrable over the entire region *R*), then they *must* give us identical results. After all, it's the same object, no matter how we slice it! This fact comes out of a theorem proved early in the twentieth century by the Italian mathematician Guido Fubini.

- -

Iterated Integrals in Two Variables

When we calculate a double definite integral for a two-variable function as described above, first with respect to one independent variable and then with respect to the other, the process is called *iterated integration.*

Example

Consider the surface described by this two-variable function *f* in Cartesian *xyz*-space:

$$f(x,y) = 3x^2 + 3y^2 + 1$$

We will calculate the mathematical volume of the solid defined by *f* with respect to the square region in the *xy*-plane whose edges are $x = -1$, $x = 1$, $y = -1$, and $y = 1$.

First, let's find the mathematical volume of the solid using double integration, slicing through the solid to create slabs parallel to the *xz*-plane by integrating with respect to *x,* and then stacking up the slabs by integrating with respect to *y.* Written in the shorthand form, our double integral is

$$\iint_{R} (3x^2 + 3y^2 + 1)\ dx\ dy$$

where *R* is the square in the *xy*-plane with respect to which we want to find the mathematical volume. Written out "the long way," which more completely shows us what we must do, the iterated integral is

$$\int_{-1}^{1} \left[\int_{-1}^{1} (3x^2 + 3y^2 + 1)\ dx \right] dy$$

The integral inside the large square brackets is

$$\int_{-1}^{1} (3x^2 + 3y^2 + 1)\, dx$$

To integrate with respect to x, we treat y as a constant, just as if we wanted to find the partial derivative with respect to x. But here, we're finding a "partial antiderivative." We obtain

$$x^3 + 3xy^2 + x \Big]_{-1}^{1}$$

Evaluating the expression from $x = -1$ to $x = 1$, we get

$$(1 + 3y^2 + 1) - (-1 - 3y^2 - 1) = 6y^2 + 4$$

We substitute this new one-variable equation in place of the integral inside the large square brackets above, obtaining

$$\int_{-1}^{1} (6y^2 + 4)\, dy$$

The basic antiderivative should be evaluated from -1 to 1, so we must calculate

$$2y^3 + 4y \Big]_{-1}^{1}$$

Working out the arithmetic, we get

$$(2 \cdot 1^3 + 4 \cdot 1) - [2 \cdot (-1)^3 + 4 \cdot (-1)] = 12$$

Now let's go the other way by slicing up the object in planes parallel to the *yz*-plane, and then integrating with respect to x. Written in the shorthand form, our double integral is

$$\iint_{R} (3x^2 + 3y^2 + 1)\ dy\, dx$$

The iterated integral is

$$\int_{-1}^{1} \Big[\int_{-1}^{1} (3x^2 + 3y^2 + 1)\, dy \Big] dx$$

The integral inside the large square brackets is

$$\int_{-1}^{1} (3x^2 + 3y^2 + 1)\, dy$$

To integrate with respect to y, we hold x constant, getting the antiderivative

$$3x^2y + y^3 + y \Big]_{-1}^{1}$$

Evaluating the expression from $y = -1$ to $y = 1$, we get

$$(3x^2 + 1 + 1) - (-3x^2 - 1 - 1) = 6x^2 + 4$$

We substitute this for the integral inside the large square brackets above, obtaining

$$\int_{-1}^{1} (6x^2 + 4) \ dx$$

This resolves to

$$2x^3 + 4x \Big]_{-1}^{1}$$

Working out the arithmetic, we get the same result as we did when we found the mathematical volume the other way:

$$(2 \cdot 1^3 + 4 \cdot 1) - [2 \cdot (-1)^3 + 4 \cdot (-1)] = 12$$

Another example

Let's try our iterated-integration skills with a region that isn't centered at the origin. We should expect to get the same answer either way we do the integration, but the routes will differ. Here's a surface described by a two-variable function g in xyz-space:

$$g(x,y) = 2x - y$$

Our task is to calculate the mathematical volume of the solid defined by g with respect to the rectangle in the xy-plane whose edges are at $x = 0$, $x = 2$, $y = 0$, and $y = 4$.

This problem is interesting because, if we graph the function g in xyz-space (or get a computer to do it for us), we'll see that in the rectangular region we've specified, some of the surface is "above" the xy-plane ($z > 0$) and some of the surface is "below" the xy-plane ($z < 0$). Any portions of the solid's mathematical volume where $z < 0$ will contribute negatively to our final answer. We should not be surprised if this object has a zero or negative net mathematical volume.

First, let's slice through the solid in planes parallel to the xz-plane by integrating with respect to x, and then we'll integrate with respect to y. Written in the shorthand form, our double integral is

$$\iint_R (2x - y) \ dx \ dy$$

The iterated integral is

$$\int_0^4 \left[\int_0^2 (2x - y)\ dx \right]\ dy$$

The inside integral is

$$\int_0^2 (2x - y)\ dx$$

We treat y as a constant, obtaining the antiderivative

$$x^2 - xy \Big]_0^2$$

Evaluating the expression from $x = 0$ to $x = 2$, we get

$$(2^2 - 2y) - (0^2 - 0y) = 4 - 2y$$

We substitute this single-variable equation in place of the integral inside the large square brackets above, obtaining

$$\int_0^4 (4 - 2y)\ dy$$

The antiderivative is

$$4y - y^2 \Big]_0^4$$

Working out the arithmetic, we get

$$(4 \cdot 4 - 4^2) - (4 \cdot 0 - 0^2) = 0$$

The object has zero mathematical volume, even though it's three-dimensional! The average position of the surface within the rectangular region happens to coincide exactly with the *xy*-plane, although the surface is oriented at a slant. With respect to the region we've defined in the *xy*-plane, the positive and negative portions of the solid's mathematical volume are equal and opposite, so they cancel each other out.

Now let's slice up the solid in planes parallel to the *yz*-plane, and then integrate with respect to *x*. If we do everything right, we should get a final answer of 0 again. In shorthand, we have

$$\iint_R (2x - y)\ dy\ dx$$

The iterated integral is

$$\int_0^2 \left[\int_0^4 (2x - y)\ dy \right]\ dx$$

The inside integral is

$$\int_0^4 (2x - y)\ dy$$

To integrate with respect to y, we hold x constant, getting

$$2xy - y^2/2 \Big]_0^4$$

Evaluating the expression from $y = 0$ to $y = 4$, we get

$$(2x \cdot 4 - 4^2/2) - (2x \cdot 0 - 0^2/2) = 8x - 8$$

We substitute this for the integral inside the large square brackets above, obtaining

$$\int_0^2 (8x - 8) \, dx$$

This resolves to

$$4x^2 - 8x \Big]_0^2$$

Evaluating, we get

$$(4 \cdot 2^2 - 8 \cdot 2) - (4 \cdot 0^2 - 8 \cdot 0) = 0$$

Practice Exercises

This is an open-book quiz. You may (and should) refer to the text as you solve these problems. Don't hurry! You'll find worked-out answers in App. C. The solutions in the appendix may not represent the only way a problem can be figured out. If you think you can solve a particular problem in a quicker or better way than you see there, by all means try it!

1. Imagine that we're in a hovering dirigible above the surface of an otherwise uninhabited planet, where gravity causes falling objects to accelerate at 24 feet per second per second. We drop a brick, which reaches the surface after 14 seconds. Assuming that air resistance is not a factor, what's our altitude?

2. We test-drive a car on a flat, straight road. Before starting the experiment, we bring the car to a stop and mark our position. Then we accelerate at an increasing rate of $6t$ meters per second per second, where t represents the elapsed time, in seconds, after the start. Find the function that expresses our distance in meters from the starting position as time passes.

3. In the situation of Prob. 3, how many meters will we travel in the first second? The first 2 seconds? The first 3 seconds? The first 4 seconds? If we can keep up this increasing rate of acceleration for 10 seconds, how far from the starting point will we end up?

4. Consider the flat surface described by the following constant function in Cartesian *xyz*-space:

$$f(x,y) = 4$$

Draw a simple 3D graph of this situation. Then, using the formula from 3D geometry, calculate the mathematical volume of the rectangular box defined by f with respect to the region whose edges are segments of the lines $x = -3$, $x = 5$, $y = -5$, and $y = 3$ in the xy-plane.

5. Find the mathematical volume of the box described in Prob. 4 by integrating f relative to the region, first with respect to x and then with respect to y.

6. Find the mathematical volume of the box described in Prob. 4 by integrating f relative to the region, first with respect to y and then with respect to x.

7. Consider the surface described by the following function in Cartesian xyz-space:

$$f(x,y) = 4x + 4y$$

Integrate against x and then against y to find the mathematical volume of the solid that this surface defines with respect to the region whose edges are segments of the lines $x = 1$, $x = 3$, $y = 0$, and $y = 5$ in the xy-plane.

8. Find the mathematical volume of the region described in Prob. 7 by integrating against y and then against x.

9. Consider the surface described by the following function in Cartesian xyz-space:

$$f(x,y) = x^2 + 2xy + y^2$$

Integrate against x and then against y to find the mathematical volume of the solid that this surface defines with respect to the region whose edges are segments of the lines $x = 1$, $x = 3$, $y = 0$, and $y = 5$ in the xy-plane.

10. Find the mathematical volume of the region described in Prob. 9 by integrating against y and then against x.

28

More Volume Integrals

We could devote a dozen chapters to double integration. Obviously, we don't have room for that, but let's look at a few examples that are a little more complicated than the ones we saw in the last chapter.

Slicing and Integrating

Imagine that we're working in Cartesian *xyz*-space, and we want to find the mathematical volume of a solid defined by a surface with respect to some flat, but non-rectangular, integration region R in the *xy*-plane. Suppose that R has a finite, positive area, and is enclosed by the graphs of two functions

$$y = g(x)$$

and

$$y = h(x)$$

We can break up the solid into slabs, just as we can with rectangular regions of integration. Figure 28-1 shows a situation of this sort.

First, we slice

When the boundaries of R are functions of *x*, the best way to find the mathematical volume of the solid is to slice it into slabs perpendicular to the *x* axis, and therefore parallel to the *yz*-plane. This is the way it's done in Fig. 28-1. The illustrated slab is bounded by the region of integration, the surface, and two vertical lines representing values of *y*. One vertical line passes through the graph of $g(x)$, and the other vertical line goes through the graph of $h(x)$. Here, "vertical" means "parallel to the *z* axis."

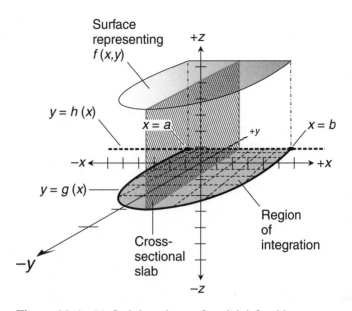

Figure 28-1   To find the volume of a solid defined by an
irregular region, we can sometimes integrate
cross-sectional slabs as shown here. We're
located close to the negative *y*-axis, looking in
toward the origin.

Next, we integrate

For any given value of *x*, we have a unique slab as defined above. We can find the area *A* of the
slab by integrating the surface function $f(x,y)$ with respect to *y*, from $g(x)$ to $h(x)$ for that
particular value of *x*. That is,

$$A = \int_{g(x)}^{h(x)} f(x,y)\ dy$$

If we choose a different value of *x*, we get a different slab, and its area is likely to be different.
We can choose slabs that intersect the *x* axis anywhere from $x = a$ to $x = b$ as shown in Fig. 28-1,
and the value of the above integral varies accordingly. We integrate with respect to *y*, so we
must treat *x* as a constant when working out the antiderivative.

Finally, we integrate again

Imagine that we're at the point where $x = a$ in the situation of Fig. 28-1. Here, the cross-
sectional slab is actually a line segment parallel to the *z* axis. As such, it has no area. As we
start moving toward the right along the *x* axis, the slab is slender, but growing. As we keep
moving, the area of the slab increases until it reaches a maximum at some value of *x* between
a and *b*. (Figure 28-1 shows the slab near that maximum-area point.) As we get close to $x =$
b, the slab becomes slimmer until, when *x* finally attains the value *b*, the slab collapses into
another line segment parallel to the *z* axis.

If we integrate the function that expresses the Area A of a slab with respect to our position on the x axis, we obtain the mathematical volume V of the solid defined by the region of integration R and the surface:

$$V = \int_a^b A \, dx$$

This can be written out in full as

$$V = \int_a^b \left[\int_{g(x)}^{h(x)} f(x,y) \, dy \right] dx$$

or as

$$V = \int_a^b \int_{g(x)}^{h(x)} f(x,y) \, dy \, dx$$

If we're sure we have explicitly defined the region of integration R, we can write the above expression in the shorthand form

$$V = \iint_R f(x,y) \, dy \, dx$$

Base Bounded by Curve and x Axis

Imagine a flat region R with a finite, positive area in the xy-plane of Cartesian xyz-space. Suppose that R is bounded by the x axis and the graph of a nonlinear function $h(x)$. If the graph of $h(x)$ doesn't intersect the x axis twice, then the region must also be bounded by one or two lines parallel to the y axis to ensure that it's completely enclosed. Figure 28-2 shows an example of a situation like this. Here, the functions whose graphs define R are

$$g(x) = 0$$

and

$$h(x) = 3x^2$$

On the right, R is bounded by the line $x = 2$ to ensure that the base of the solid is a completely enclosed region.

- -

Are you confused?

Have you noticed that the coordinate axes appear different in Fig. 28-2 than they have looked in earlier examples? We're taking a new perspective on xyz-space. In all the 3D graphs portrayed until now, we've seen the situation from out in space near the negative y-axis, looking in toward the origin. Now, we're out in space near the positive z-axis, looking in toward the origin. Things look different, but the relative orientations of the positive and negative x, y, and z axes are the same as they've always been.

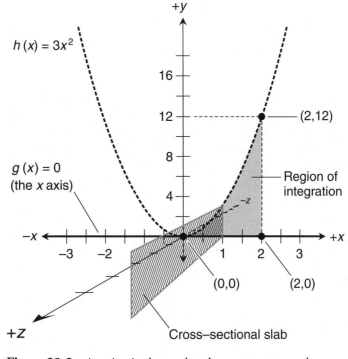

Figure 28-2 A region in the *xy*-plane between a curve and
the *x* axis. We can integrate twice to find the
mathematical volume defined by a surface (not
shown) with respect to this region.

- -

This new vantage point gives us a good broadside view of the *xy*-plane, so we can clearly see the shape of
the region of integration. Unfortunately, this perspective makes it difficult to show the surface represented
by $f(x,y)$. But if that surface is flat and level, we can imagine it as a sheet parallel to the page on which
Fig. 28-2 is printed. Such a sheet might be "in front of" the page or "behind" it, depending on whether
$f(x,y)$ is positive or negative.

- -

The proper structure of Cartesian *xyz*-space

Whenever you see Cartesian *xyz*-space, the axes always are (or should be) oriented in a certain
way relative to each other, regardless of the point of view. You learned this relationship in precalculus. This is a good time to review it. Think in terms of compass directions and altitude.

Draw an *x* axis so that the positive values go "east" and the negative values go "west." Then
draw a *y* axis going "north" and "south," at a right angle to the *x* axis. These two axes intersect
at the point where $x = 0$ and $y = 0$. The positive *y*-values go "north," and the negative *y*-
values go "south." Now imagine a third axis, the *z* axis, which is "vertical" and perpendicular
to both the *x* axis and the *y* axis. The positive *z*-values go "upward," and the negative *z*-values
go "downward." The *z* axis passes through the point where the *x* axis and *y* axis intersect. At
this point, $z = 0$.

Now imagine this set of three mutually perpendicular lines, which all intersect at a single point where $x = 0$, $y = 0$, and $z = 0$, as a rigid structure in space, as if each axis were an inflexible rod. No matter where you go, the structure stays the same, even though its appearance changes as your perspective changes. If you have a 3D computer graphing program and you experiment with it for awhile, you'll be able to create some beautiful examples of this type of coordinate structure.

If you like, look back at all the drawings you've seen so far in this book showing Cartesian *xyz*-space. You'll notice that the axes are oriented in the same relative way, regardless of your point of view. It's important that Cartesian *xyz*-space always be set up in exactly this fashion. Otherwise, sooner or later, you'll get *seriously* confused because you'll end up with graphs that are "inside-out," "upside-down," or "backward"!

Flat, level surface

Let's calculate the mathematical volume of a solid with respect to the region in Fig. 28-2, as defined by a surface whose function is

$$f(x,y) = 4$$

This surface is a plane parallel to the *xy*-plane and 4 units "in front of" it, as seen from the perspective of Fig. 28-2. Written in the shorthand form, our double integral is

$$\iint\limits_R 4 \, dy \, dx$$

The *x*-value interval for the region of integration is (0,2), so the iterated integral is

$$\int_0^2 \left[\int_{g(x)}^{h(x)} 4 \, dy \right] dx$$

The integral inside the large square brackets is

$$\int_{g(x)}^{h(x)} 4 \, dy$$

which resolves to

$$4y \, \Big]_{g(x)}^{h(x)}$$

We remember that the functions whose graphs define the region of integration R are

$$g(x) = 0$$

and

$$h(x) = 3x^2$$

Evaluating $4y$ from $y = 0$ to $y = 3x^2$, we get

$$4 \cdot 3x^2 - 4 \cdot 0 = 12x^2$$

We substitute this for the integral inside the square brackets above, obtaining

$$\int_0^2 12x^2 \, dx$$

which resolves to

$$4x^3 \Big]_0^2$$

Working out the arithmetic to derive the mathematical volume V, we get

$$V = 4 \cdot 2^3 - 4 \cdot 0^3 = 32$$

Flat, sloping surface

Now let's see what happens with a surface that's flat, but not parallel to the xy-plane. Imagine a solid with respect to the region of integration R in Fig. 28-2 as defined by

$$f(x,y) = 2x + 4y$$

In shorthand form, the double integral is

$$\iint_R (2x + 4y) \, dy \, dx$$

The x-value interval enclosing R is, as before, $(0,2)$. Therefore, we can write the iterated integral as

$$\int_0^2 \left[\int_{g(x)}^{h(x)} (2x + 4y) \, dy \right] dx$$

The integral inside the large square brackets represents the area of a cross-sectional slab with x held constant:

$$\int_{g(x)}^{h(x)} (2x + 4y) \, dy$$

We integrate against y to obtain

$$2xy + 2y^2 \Big]_{g(x)}^{h(x)}$$

Again, we remember that the *xy*-plane functions whose graphs define the region of integration *R* are

$$g(x) = 0$$

and

$$h(x) = 3x^2$$

Evaluating the quantity $(2xy + 2y^2)$ from $y = 0$ to $y = 3x^2$ yields

$$[2x \cdot 3x^2 + 2(3x^2)^2] - (2x \cdot 0 + 2 \cdot 0^2) = 6x^3 + 18x^4$$

Substituting this for the integral inside the large square brackets above, we get

$$\int_0^2 (6x^3 + 18x^4) \, dx$$

which resolves to

$$3x^4/2 + 18x^5/5 \Big]_0^2$$

Working out the arithmetic to get the mathematical volume *V,* we obtain

$$V = (3 \cdot 2^4/2 + 18 \cdot 2^5/5) - (3 \cdot 0^4/2 + 18 \cdot 0^5/5) = 696/5$$

Warped surface

Imagine a solid defined by the following function with respect to the region of integration *R* shown in Fig. 28-2:

$$f(x,y) = 3x^2 + y$$

This surface is not flat, but warped, making our problem a little more complicated than the situation in the previous examples. Written in shorthand form, the double integral is

$$\iint_R (3x^2 + y) \, dy \, dx$$

As before, the span of *x* values for the region of integration is (0,2). The iterated integral is therefore

$$\int_0^2 \left[\int_{g(x)}^{h(x)} (3x^2 + y) \, dy \right] dx$$

The integral inside the large square brackets is

$$\int_{g(x)}^{h(x)} (3x^2 + y)\, dy$$

Holding x constant and integrating against y, we get

$$3x^2 y + y^2/2 \,\Big]_{g(x)}^{h(x)}$$

The functions whose graphs define the base of the solid are

$$g(x) = 0$$

and

$$h(x) = 3x^2$$

Evaluating the quantity $(3x^2 y + y^2/2)$ from $y = 0$ to $y = 3x^2$, we get

$$[3x^2 \cdot 3x^2 + (3x^2)^2/2] - (3x^2 \cdot 0 + 0^2/2) = 27x^4/2$$

We substitute this for the integral inside the large square brackets above to obtain

$$\int_0^2 27x^4/2\, dx$$

which resolves to

$$27x^5/10 \,\Big]_0^2$$

Working out the arithmetic, we get a mathematical volume of

$$V = 27 \cdot 2^5/10 - 27 \cdot 0^5/10 = 432/5$$

Base Bounded by Curve and Line

Consider a region in the xy-plane, bounded by a parabola representing $g(x)$ and a straight line representing $h(x)$ as shown in Fig. 28-3. In this situation, the functions whose graphs define the region R are

$$g(x) = x^2 - 4$$

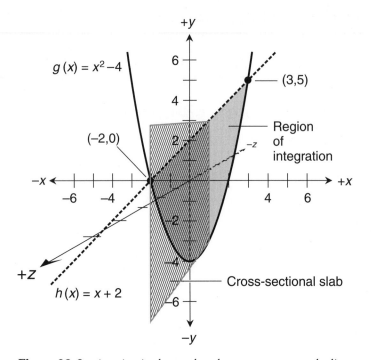

Figure 28-3 A region in the *xy*-plane between a curve and a line. We can integrate twice to find the mathematical volume defined by a surface (not shown) with respect to this region.

and

$$h(x) = x + 2$$

Let's calculate the mathematical volumes of solids with respect to *R*, as defined by three different surfaces.

Flat, level surface

Imagine a flat surface 4 units "in front of" the *xy*-plane. The function for this surface is

$$f(x,y) = 4$$

Written in shorthand, the double integral for the mathematical volume of the solid is

$$\iint_R 4 \ dy \ dx$$

The interval representing the span of x values for R is $(-2,3)$. These are the x values of the points where the parabola and the line intersect in Fig. 28-3. The iterated integral is therefore

$$\int_{-2}^{3} \left[\int_{g(x)}^{h(x)} 4 \, dy \right] dx$$

The integral inside the large square brackets is

$$\int_{g(x)}^{h(x)} 4 \, dy$$

which resolves to

$$4y \Big]_{g(x)}^{h(x)}$$

Again, the functions whose graphs define R are

$$g(x) = x^2 - 4$$

and

$$h(x) = x + 2$$

When we evaluate $4y$ from $y = x^2 - 4$ to $y = x + 2$, we obtain

$$4 \cdot (x + 2) - 4 \cdot (x^2 - 4) = -4x^2 + 4x + 24$$

We substitute this for the integral inside the square brackets above, obtaining

$$\int_{-2}^{3} (-4x^2 + 4x + 24) \, dx$$

which resolves to

$$-4x^3/3 + 2x^2 + 24x \Big]_{-2}^{3}$$

Working out the arithmetic, we get

$$V = (-4 \cdot 3^3/3 + 2 \cdot 3^2 + 24 \cdot 3) - [-4 \cdot (-2)^3/3 + 2 \cdot (-2)^2 + 24 \cdot (-2)] = 250/3$$

Flat, sloping surface

Let's calculate the mathematical volume of a solid relative to the region of integration R in Fig. 28-3 as defined by

$$f(x,y) = 4x$$

The graph of this function is a flat, sloping surface. In shorthand, we have

$$\iint\limits_{R} 4x \, dy \, dx$$

The x-value interval enclosing R is $(-2,3)$, so can write the iterated integral as

$$\int_{-2}^{3} \left[\int_{g(x)}^{h(x)} 4x \, dy \right] dx$$

The integral inside the large square brackets is

$$\int_{g(x)}^{h(x)} 4x \, dy$$

When we integrate against y, holding x constant, we obtain

$$4xy \left. \right]_{g(x)}^{h(x)}$$

Again, the xy-plane functions whose graphs define R are

$$g(x) = x^2 - 4$$

and

$$h(x) = x + 2$$

When we evaluate $4xy$ from $y = x^2 - 4$ to $y = x + 2$, we get

$$4x(x+2) - 4x(x^2 - 4) = -4x^3 + 4x^2 + 24x$$

Substituting this for the integral inside the square brackets above, we get

$$\int_{-2}^{3} (-4x^3 + 4x^2 + 24x) \, dx$$

which resolves to

$$-x^4 + 4x^3/3 + 12x^2 \left. \right]_{-2}^{3}$$

Working out the arithmetic, we obtain

$$V = (-3^4 + 4 \cdot 3^3/3 + 12 \cdot 3^2) - [-(-2)^4 + 4 \cdot (-2)^3/3 + 12 \cdot (-2)^2] = 125/3$$

Warped surface

Now imagine a warped surface in Cartesian *xyz*-space that represents

$$f(x,y) = x^2$$

Let's find the mathematical volume of the solid defined by this surface and the region illustrated in Fig. 28-3. The shorthand form of the double integral is

$$\iint_R x^2 \, dy \, dx$$

The interval of *x* values is (−2,3), so the iterated integral is

$$\int_{-2}^{3} \left[\int_{g(x)}^{h(x)} x^2 \, dy \right] dx$$

The integral inside the large square brackets is

$$\int_{g(x)}^{h(x)} x^2 \, dy$$

Holding *x* constant and integrating with respect to *y*, we get

$$x^2 y \; \Big]_{g(x)}^{h(x)}$$

As in the previous two examples, the functions whose graphs define the region of integration in the *xy*-plane are

$$g(x) = x^2 - 4$$

and

$$h(x) = x + 2$$

Evaluating $x^2 y$ from $y = x^2 - 4$ to $y = x + 2$ produces

$$x^2(x + 2) - x^2(x^2 - 4) = -x^4 + x^3 + 6x^2$$

When we substitute this for the integral inside the large square brackets above, we get

$$\int_{-2}^{3} (-x^4 + x^3 + 6x^2) \, dx$$

which resolves to

$$-x^5/5 + x^4/4 + 2x^3 \; \Big]_{-2}^{3}$$

Working out the arithmetic to derive the mathematical volume, we get

$$V = (-3^5/5 + 3^4/4 + 2 \cdot 3^3) - [-(-2)^5/5 + (-2)^4/4 + 2 \cdot (-2)^3] = 125/4$$

- -

Here's a challenge!

Calculate the mathematical volume of a solid with respect to the region of integration in Fig. 28-3 as defined by

$$f(x,y) = -x$$

This function's graph is a flat, sloping surface that cuts through the *xy*-plane along the *y* axis.

Solution

The shorthand form of the double integral for the mathematical volume is

$$\iint_R -x \, dy \, dx$$

The *x*-value interval enclosing the region of integration is $(-2,3)$, so can write the iterated integral as

$$\int_{-2}^{3} \left[\int_{g(x)}^{h(x)} -x \, dy \right] dx$$

The integral inside the large square brackets is

$$\int_{g(x)}^{h(x)} -x \, dy$$

When we hold *x* constant and integrate against *y*, we obtain

$$-xy \; \Big]_{g(x)}^{h(x)}$$

Once again, the *xy*-plane functions whose graphs enclose the region of integration are

$$g(x) = x^2 - 4$$

and

$$h(x) = x + 2$$

When we evaluate $-xy$ from $y = x^2 - 4$ to $y = x + 2$, we get

$$-x(x + 2) - [-x(x^2 - 4)] = x^3 - x^2 - 6x$$

Substituting this for the integral inside the large square brackets above, we get

$$\int_{-2}^{3} (x^3 - x^2 - 6x)\, dx$$

which resolves to

$$x^4/4 - x^3/3 - 3x^2 \Big]_{-2}^{3}$$

Working out the arithmetic, we obtain

$$V = (3^4/4 - 3^3/3 - 3 \cdot 3^2) - [(-2)^4/4 - (-2)^3/3 - 3 \cdot (-2)^2] = -125/12$$

- -

Base Bounded by Two Curves

Now let's look at a region in the xy-plane that's bounded by two parabolas as shown in Fig. 28-4. Here, the functions whose graphs define the region of integration are

$$g(x) = x^2/2 - 2$$

and

$$h(x) = -x^2/2 + 2$$

We will calculate the mathematical volumes of solids with respect to this region, as defined by three different surfaces in xyz-space.

Flat, level surface

Let's find the mathematical volume of a solid relative to the region of integration R in Fig. 28-4, as defined by

$$f(x,y) = 4$$

The short form of the double integral is

$$\iint_{R} 4\ dy\, dx$$

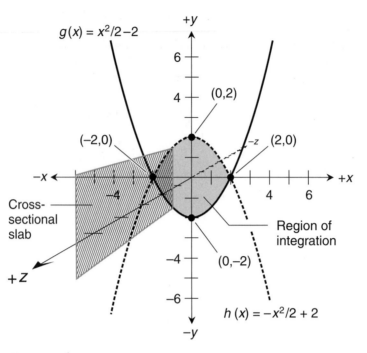

g(x) = x²/2−2

(0,2)

(−2,0)

(2,0)

−z

(0,−2)

h(x) = −x²/2 + 2

Cross-
sectional
slab

+Z

Region of
integration

Figure 28-4   A region in the *xy*-plane between two curves. We can
integrate twice to find the mathematical volume defined
by a surface (not shown) with respect to this region.

The span of *x* values for *R* is (−2,2). The iterated integral is therefore

$$\int_{-2}^{2} \left[\int_{g(x)}^{h(x)} 4 \, dy \right] dx$$

The integral inside the large square brackets is

$$\int_{g(x)}^{h(x)} 4 \, dy$$

Integrating with respect to *y*, we get

$$4y \Big]_{g(x)}^{h(x)}$$

We recall that the functions whose graphs enclose the region of integration are

$$g(x) = x^2/2 - 2$$

and

$$h(x) = -x^2/2 + 2$$

Evaluating $4y$ from $y = x^2/2 - 2$ to $y = -x^2/2 + 2$, we get

$$4 \cdot (-x^2/2 + 2) - 4 \cdot (x^2/2 - 2) = -4x^2 + 16$$

We substitute this for the integral inside the large square brackets above, obtaining

$$\int_{-2}^{2} (-4x^2 + 16)\, dx$$

which resolves to

$$-4x^3/3 + 16x \,\Big]_{-2}^{2}$$

Working out the arithmetic to derive the solid's mathematical volume, we get

$$V = (-4 \cdot 2^3/3 + 16 \cdot 2) - [-4 \cdot (-2)^3/3 + 16 \cdot (-2)] = 128/3$$

Flat, sloping surface

Let's find the mathematical volume of a solid relative to the region in Fig. 28-4 as defined by the following function, which represents a sloping surface that cuts through the xy-plane along the x axis:

$$f(x,y) = 2y$$

In shorthand form, the double integral is

$$\iint_R 2y\, dy\, dx$$

The span of x values over the region of integration is $(-2,2)$. The iterated integral is therefore

$$\int_{-2}^{2} \Big[\int_{g(x)}^{h(x)} 2y\, dy \Big]\, dx$$

The integral inside the large square brackets is

$$\int_{g(x)}^{h(x)} 2y\, dy$$

Integrating with respect to y, we get

$$y^2 \, \Big]_{g(x)}^{h(x)}$$

Once again, the functions whose graphs enclose our xy-plane region of integration are

$$g(x) = x^2/2 - 2$$

and

$$h(x) = -x^2/2 + 2$$

Evaluating y^2 from $y = x^2/2 - 2$ to $y = -x^2/2 + 2$, we get

$$(-x^2/2 + 2)^2 - (x^2/2 - 2)^2 = 0$$

Substituting this for the integral inside the large square brackets above, we get

$$\int_{-2}^{2} 0 \, dx$$

which resolves to

$$0 \, \Big]_{-2}^{2}$$

Working out the arithmetic is a trivial task. We have simply

$$V = 0 - 0 = 0$$

The surface lies partly above the xy-plane and partly below it. The mathematical volume relative to the region of integration R on the positive y-side of (or "above") the x axis is positive, but the mathematical volume relative to R on the negative y-side of (or "below") the x axis is negative to the same extent. Therefore, they cancel each other out, giving us a *mathematical* volume of 0—even though the solid, if we could construct it in the "real world," would have a finite and positive *geometric* volume!

Warped surface

Finally, let's see what happens when we have a warped surface that cuts through the xy-plane exactly along the y axis. This time, we'll figure out the mathematical volume of a solid defined by the region of integration in Fig. 28-4 and the function

$$f(x,y) = x^3$$

In shorthand form, the double integral is

$$\iint_R x^3 \, dy \, dx$$

The span of x values is, once again, $(-2,2)$, so the iterated integral is

$$\int_{-2}^{2} \left[\int_{g(x)}^{h(x)} x^3 \, dy \right] dx$$

The integral inside the large square brackets is

$$\int_{g(x)}^{h(x)} x^3 \, dy$$

Integrating with respect to y, we get

$$x^3 y \left.\right]_{g(x)}^{h(x)}$$

The functions whose graphs define our xy-plane region are

$$g(x) = x^2/2 - 2$$

and

$$h(x) = -x^2/2 + 2$$

Evaluating $x^3 y$ from $y = x^2/2 - 2$ to $y = -x^2/2 + 2$, we get

$$x^3(-x^2/2 + 2) - x^3(x^2/2 - 2) = -x^5 + 4x^3$$

Substituting this for the integral inside the large square brackets above, we get

$$\int_{-2}^{2} (-x^5 + 4x^3) \, dx$$

which resolves to

$$-x^6/6 + x^4 \left.\right]_{-2}^{2}$$

Working out the arithmetic, we obtain

$$V = (-2^6/6 + 2^4) - [-(-2)^6 + (-2)^4] = 0$$

The mathematical volume on the positive x-side (or "to the right") of the y axis is positive, and the mathematical volume on the negative x-side (or "to the left") of the y axis is negative to an equal extent. The positive and negative mathematical volumes precisely cancel, leaving us with a mathematical volume of zero for the entire solid.

- -

Are you confused?

In this chapter, we always slice the solid regions so that the slabs are perpendicular to the x axis and parallel to the yz-plane. We never go the other way, slicing the solid into slabs perpendicular to the y axis and parallel to the xz-plane. If you wonder why, go ahead and try to solve one of the above problems that way! It will be difficult, because in all the examples here, we define the base region of the solid using functions of x, not functions of y.

Once in awhile, we'll encounter base regions in the xy-plane that are defined as functions of y, with x as the dependent variable. For example, suppose that we have

$$x = g(y)$$

and

$$x = h(y)$$

In situations like this, it's easier to find the solid's mathematical volume by slicing it into slabs perpendicular to the y axis and parallel to the xz-plane, and then integrating from $y = a$ to $y = b$ as shown in Fig. 28-5. When we compare this with Fig. 28-1, it's apparent that we've changed our point of view. The mathematical calculations are a little different; but geometrically, the process works in the same way. We get

$$V = \int_{a}^{b} \left[\int_{g(y)}^{h(y)} f(x,y) \; dx \right] dy$$

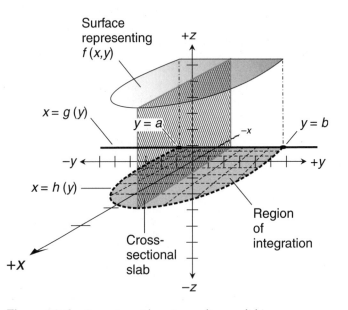

Figure 28-5   Sometimes it's easier to slice a solid into slabs perpendicular to the y axis, rather than perpendicular to the x axis. Here, we see the situation from a point of view near the positive x-axis, looking in toward the origin.

which can also be written as

$$V = \int_{a}^{b} \int_{g(y)}^{h(y)} f(x,y) \; dx \; dy$$

If we call the region of integration *R*, we can shorten this further to

$$V = \iint_{R} f(x,y) \; dx \; dy$$

In the volume integral, we now have *dx dy* instead of *dy dx*. This tells us to integrate first with respect to *x* to get the areas of the slabs for constant values of *y*, and then to integrate with respect to *y* to get the mathematical volume of the solid.

- -

Practice Exercises

This is an open-book quiz. You may (and should) refer to the text as you solve these problems. Don't hurry! You'll find worked-out answers in App. C. The solutions in the appendix may not represent the only way a problem can be figured out. If you think you can solve a particular problem in a quicker or better way than you see there, by all means try it!

1. Refer to Fig. 28-6. Imagine that we're doing the second integral in the iteration, looking at the areas of the cross-sectional slabs as we move from *x* = *a* to *x* = *b*. The surface

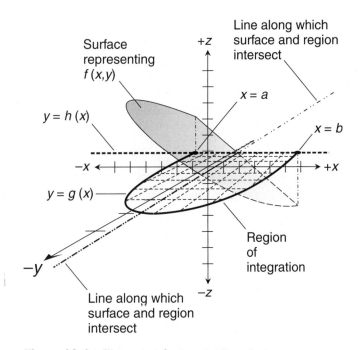

Figure 28-6   Illustration for Practice Exercise 1.

representing $f(x,y)$ is flat, and cuts through the region of integration. The surface lies partly "above" the xy-plane and partly "below" it. Describe how the area of the cross-sectional slab varies as we move from $x = a$ to $x = b$. (This exercise is difficult, but is intended to stimulate your imagination!)

2. For the solid defined by the flat, level surface $f(x,y) = 4$ and the shaded region in Fig. 28-2, we found $V = 32$. This mathematical volume can also be found by determining the true geometric area of the region of integration and then multiplying by 4, which is the height of the solid. Find the mathematical volume this way, and verify that the result agrees with what we got in the chapter text.

3. Refer again to Fig. 28-2. Consider a solid defined by the following function and the region of integration shown:

$$f(x,y) = 2x - 6y$$

Calculate the mathematical volume of this solid using the method we've learned in this chapter.

4. Refer to Fig. 28-2 one more time. Consider a solid defined by the following function and the region of integration shown:

$$f(x,y) = 3x^2 - 4y$$

Calculate the mathematical volume of this solid using the method we've learned in this chapter.

5. For the solid defined by the flat, level surface $f(x,y) = 4$ and the shaded region in Fig. 28-3, we found $V = 250/3$. This mathematical volume can also be found by determining the true geometric area of the region of integration and then multiplying by 4, which is the height of the solid. Find the mathematical volume this way, and verify that the result agrees with what we got in the chapter text.

6. Refer again to Fig. 28-3. Consider a solid defined by the following function and the region of integration shown:

$$f(x,y) = -4x$$

Calculate the mathematical volume of this solid using the method we've learned in this chapter.

7. Refer to Fig. 28-3 one more time. Consider a solid defined by the following function and the region of integration shown:

$$f(x,y) = -x^2$$

Calculate the mathematical volume of this solid using the method we've learned in this chapter.

8. For the solid defined by the flat, level surface $f(x,y) = 4$ and the shaded region in Fig. 28-4, we found $V = 128/3$. This mathematical volume can also be found by determining the true geometric area of the region of integration and then multiplying by 4, which is the height of the solid. Find the mathematical volume this way, and verify that the result agrees with what we got in the chapter text.

9. Refer again to Fig. 28-4. Consider a solid defined by the following function and the region of integration shown:

$$f(x,y) = 2x + 1$$

Calculate the mathematical volume of this solid using the method we've learned in this chapter.

10. Refer to Fig. 28-4 one more time. Consider a solid defined by the following function and the region of integration shown:

$$f(x,y) = 2y - 1$$

Calculate the mathematical volume of this solid using the method we've learned in this chapter.

29

What's a Differential Equation?

A differential equation contains one or more derivatives or differentials along with variables and constants. In this chapter, we'll look at a few extremely basic examples of *ordinary differential equations* (ODEs), which don't have partial derivatives.

Elementary First-Order ODEs

An *elementary first-order ODE* can be manipulated into an equation with only a derivative on the left-hand side, and only a single-variable function on the right-hand side.

How to recognize one

Imagine an independent variable x and a dependent variable y along with a continuous function $f(x)$, bundled into an equation. We have an elementary first-order ODE if we can morph it into the standard form

$$dy/dx = f(x)$$

To solve this, our objective is to eliminate the derivative and obtain an equation of the form

$$y = H(x)$$

where H is a *family of functions*. We call it a family because it contains a constant whose value can change, giving rise to a set of functions that are similar but not identical. The constant arises when we take an indefinite integral as part of the solution process.

Example 1

Let's start with something simple that's in the standard form for an elementary first-order ODE to begin with:

$$dy/dx = -2x$$

When we integrate both sides with respect to x, we get

$$\int (dy/dx) \ dx = \int -2x \ dx$$

Because these are indefinite integrals, each has its own constant of integration when we work it out. We obtain the general antiderivatives

$$\int (dy/dx) \ dx = y + c_1$$

and

$$\int -2x \ dx = -x^2 + c_2$$

where c_1 and c_2 are the constants. When we take the right-hand sides of these two equations and combine them, we get

$$y + c_1 = -x^2 + c_2$$

Subtracting c_1 from each side gives us

$$y = -x^2 - c_1 + c_2$$

We can consolidate c_1 and c_2 into a single constant c, like this:

$$-c_1 + c_2 = c$$

That simplifies our solution equation to

$$y = H(x) = -x^2 + c$$

where H is the family of solution functions.

Example 2

Let's work out an elementary first-order ODE where the left-hand side is a little more complicated:

$$dy/dx - 6x^2 = -e^x$$

Adding $6x^2$ to both sides, we get

$$dy/dx = -e^x + 6x^2$$

Integrating both sides with respect to x, we get

$$\int (dy/dx) \ dx = \int (-e^x + 6x^2) \ dx$$

The general antiderivatives are

$$\int (dy/dx) \ dx = y + c_1$$

and

$$\int (-e^x + 6x^2) \ dx = -e^x + 2x^3 + c_2$$

Putting these antiderivatives together into a single equation, we obtain

$$y + c_1 = -e^x + 2x^3 + c_2$$

When we subtract c_1 from each side, we get

$$y = -e^x + 2x^3 - c_1 + c_2$$

As before, we can consolidate c_1 and c_2 into a single constant c, writing the solution as

$$y = H(x) = -e^x + 2x^3 + c$$

Example 3

Now we'll tackle something more messy, in which we find a polynomial along with a trigonometric function. Let's solve

$$dy/dx + \cos x = 3x^2 + 8x - 7$$

If we subtract $\cos x$ from both sides, we get

$$dy/dx = 3x^2 + 8x - 7 - \cos x$$

This is in the standard elementary first-order ODE form. When we integrate both sides, we get

$$\int (dy/dx) \ dx = \int (3x^2 + 8x - 7 - \cos x) \ dx$$

The general antiderivatives are

$$\int (dy/dx) \ dx = y + c_1$$

and

$$\int (3x^2 + 8x - 7 - \cos x) \ dx = x^3 + 4x^2 - 7x - \sin x + c_2$$

where c_1 and c_2 are the constants of integration. Combining the right-hand sides of these two equations, we get

$$y + c_1 = x^3 + 4x^2 - 7x - \sin x + c_2$$

Subtracting c_1 from both sides gives us

$$y = x^3 + 4x^2 - 7x - \sin x - c_1 + c_2$$

Once again, we can consolidate the constants and call the combination c, getting

$$y = H(x) = x^3 + 4x^2 - 7x - \sin x + c$$

- -

Are you confused?

You might wonder if the constant in the solution to an elementary first-order ODE be eliminated or resolved into a specific real number. Sometimes it can, but not always. In many physics and engineering situations, the constant of integration cancels itself out. We saw an example of how that sort of thing can happen when we calculated the height of Sir Isaac's cliff in Chap. 27. For the constant to disappear or attain a specific real-number value, we must define certain initial conditions, such as when to start a timer or where to start a race. The purely mathematical solution to an elementary first-order ODE always includes a constant whose value we don't necessarily know, because it's the result of taking indefinite integrals.

- -

Elementary Second-Order ODEs

A differential equation can contain a second derivative along with a function. Then we have an *elementary second-order ODE.*

How to recognize one

Consider an equation with an independent variable x and a dependent variable y, along with a continuous function $f(x)$. We have an elementary second-order ODE if we can morph it into the form

$$d^2y/dx^2 = f(x)$$

To solve this type of equation, our objective is to get something of the form

$$y = H(x)$$

where H is a family of functions that contains two constants, which result from repeated indefinite integration.

Example 4

Let's solve an elementary second-order ODE that consists only of a second derivative and a monomial linear function:

$$d^2y/dx^2 = -2x$$

When we take the indefinite integrals of both sides with respect to x, we get

$$\int (d^2y/dx^2)\, dx = \int -2x\, dx$$

Because these are indefinite integrals, each has its own constant of integration. We obtain the general antiderivatives

$$\int (d^2y/dx^2)\, dx = dy/dx + c_1$$

and

$$\int -2x\, dx = -x^2 + c_2$$

where c_1 and c_2 are constants. When we combine these two antiderivatives into a single equation, we get

$$dy/dx + c_1 = -x^2 + c_2$$

Subtracting c_1 from each side, we get

$$dy/dx = -x^2 - c_1 + c_2$$

Letting the quantity $(-c_1 + c_2)$ be a consolidated constant p, we can simplify to

$$-c_1 + c_2 = p$$

That gives us

$$dy/dx = -x^2 + p$$

We recognize this as an elementary first-order ODE. When we integrate both sides, we get

$$\int (dy/dx)\, dx = \int (-x^2 + p)\, dx$$

The general antiderivatives are

$$\int (dy/dx)\, dx = y + c_3$$

and

$$\int (-x^2 + p)\, dx = -x^3/3 + px + c_4$$

where c_3 and c_4 are new constants. When we combine the right-hand sides of these two equations, we get

$$y + c_3 = -x^3/3 + px + c_4$$

Subtracting c_3 from each side produces

$$y = -x^3/3 + px - c_3 + c_4$$

We can consolidate c_3 and c_4 and call the combination q, like this:

$$-c_3 + c_4 = q$$

That simplifies our solution equation to

$$y = H(x) = -x^3/3 + px + q$$

where H is the family of solution functions, and p and q are constants whose values we don't necessarily know.

Example 5

Now let's solve an elementary second-order ODE that isn't quite so neat:

$$d^2y/dx^2 - 12x^2 = 2e^x$$

Adding $12x^2$ to both sides, we get

$$d^2y/dx^2 = 2e^x + 12x^2$$

Integrating through with respect to x gives us

$$\int (d^2y/dx^2)\ dx = \int (2e^x + 12x^2)\ dx$$

The general antiderivatives are

$$\int (d^2y/dx^2)\ dx = dy/dx + c_1$$

and

$$\int (2e^x + 12x^2)\ dx = 2e^x + 4x^3 + c_2$$

Putting the general antiderivatives together into a single equation, we obtain

$$dy/dx + c_1 = 2e^x + 4x^3 + c_2$$

Subtracting c_1 from each side gives us

$$dy/dx = 2e^x + 4x^3 - c_1 + c_2$$

Letting the quantity $(-c_1 + c_2)$ be a consolidated constant p, we can simplify to

$$dy/dx = 2e^x + 4x^3 + p$$

When we integrate both sides of this elementary first-order ODE, we get

$$\int (dy/dx)\, dx = \int (2e^x + 4x^3 + p)\, dx$$

The general antiderivatives are

$$\int (dy/dx)\, dx = y + c_3$$

and

$$\int (2e^x + 4x^3 + p)\, dx = 2e^x + x^4 + px + c_4$$

where c_3 and c_4 are new constants. Combining the right-hand sides of these equations, we get

$$y + c_3 = 2e^x + x^4 + px + c_4$$

Subtracting c_3 from each side yields

$$y = 2e^x + x^4 + px - c_3 + c_4$$

When we add $-c_3$ to c_4 and call the sum q, we can simplify to obtain the solution

$$y = H(x) = 2e^x + x^4 + px + q$$

Example 6

Now let's solve the following elementary second-order ODE:

$$d^2y/dx^2 - \cos x = 3x^2 + 6x + 10$$

Adding $\cos x$ to both sides, we get

$$d^2y/dx^2 = 3x^2 + 6x + 10 + \cos x$$

Integrating both sides gives us

$$\int (d^2y/dx^2)\, dx = \int (3x^2 + 6x + 10 + \cos x)\, dx$$

The general antiderivatives are

$$\int (d^2y/dx^2)\, dx = dy/dx + c_1$$

and

$$\int (3x^2 + 6x + 10 + \cos x)\, dx = x^3 + 3x^2 + 10x + \sin x + c_2$$

Combining these results into a single equation, we obtain

$$dy/dx + c_1 = x^3 + 3x^2 + 10x + \sin x + c_2$$

Subtracting c_1 from each side yields

$$dy/dx = x^3 + 3x^2 + 10x + \sin x - c_1 + c_2$$

Adding $-c_1$ and c_2 to get a single constant p, we obtain

$$dy/dx = x^3 + 3x^2 + 10x + \sin x + p$$

Integrating through with respect to x gives us

$$\int (dy/dx)\, dx = \int (x^3 + 3x^2 + 10x + \sin x + p)\, dx$$

The general antiderivatives are

$$\int (dy/dx)\, dx = y + c_3$$

and

$$\int (x^3 + 3x^2 + 10x + \sin x + p)\, dx = x^4/4 + x^3 + 5x^2 - \cos x + px + c_4$$

where c_3 and c_4 are new constants. Combining the right-hand sides of the above equations into a single equation yields

$$y + c_3 = x^4/4 + x^3 + 5x^2 - \cos x + px + c_4$$

When we subtract c_3 from each side, we get

$$y = x^4/4 + x^3 + 5x^2 - \cos x + px - c_3 + c_4$$

We can add $-c_3$ to c_4, getting a single constant q to produce the solution

$$y = H(x) = x^4/4 + x^3 + 5x^2 - \cos x + px + q$$

- -

Are you confused?

Do you wonder why we can't consolidate p and q in the solutions to the elementary second-order ODEs in Examples 4 through 6, the way we combined c_1 and c_2 in the solutions to the elementary first-order ODEs in Examples 1 through 3? The reason goes back to the algebra of polynomials. Within an expression or equation, we can only consolidate constants that apply to the same power or function of a single variable or set of variables.

In the elementary first-order ODEs, the constants c_1 and c_2 both stood alone. We could treat them both as multiples of 1, which is x^0, so we could merge them into a single constant. But in the solutions to the elementary second-order ODEs, the constant p multiplies x or x^1, while q multiplies 1 or x^0. The different powers of x can't be merged into a monomial, so we have no convenient way to combine their multiples.

Are you astute?

You might ask, "After we've derived a solution to an ODE, how do we know if it's correct?" To check the solution to a first-order ODE, we differentiate our answer once. To check the solution to a second-order ODE, we differentiate our answer twice. If we've done everything right, we should either get the original ODE straightaway, or else get something that we can morph into the original ODE.

Here's a challenge!

Differentiate the solutions in each of the examples we worked out in this chapter. If necessary, morph the derivatives to get back the original equations, thereby showing that our solutions were correct.

Solution

We must check each of the examples in the order they appeared. The first three were elementary first-order ODEs, and the second three were elementary second-order ODEs.

Checking example 1. We finished with the solution

$$y = -x^2 + c$$

Differentiating both sides gives us

$$dy/dx = -2x$$

That was the original equation.

Checking example 2. We derived the solution

$$y = -e^x + 2x^3 + c$$

Differentiating through, we get

$$dy/dx = -e^x + 6x^2$$

Subtracting $6x^2$ from each side gives us the original equation, which was

$$dy/dx - 6x^2 = -e^x$$

Checking example 3. Our solution equation was

$$y = x^3 + 4x^2 - 7x - \sin x + c$$

When we differentiate through, we obtain

$$dy/dx = 3x^2 + 8x - 7 - \cos x$$

Adding cos *x* to both sides produces the ODE we began with, which was

$$dy/dx + \cos x = 3x^2 + 8x - 7$$

Checking example 4. We ended with the solution

$$y = -x^3/3 + px + q$$

Differentiating both sides gives us

$$dy/dx = -x^2 + p$$

Differentiating again yields

$$d^2y/dx^2 = -2x$$

That's the equation we started with.

Checking example 5. We derived the solution equation

$$y = 2e^x + x^4 + px + q$$

Differentiating both sides, we obtain

$$dy/dx = 2e^x + 4x^3 + p$$

When we differentiate a second time, we get

$$d^2y/dx^2 = 2e^x + 12x^2$$

Subtracting $12x^2$ from each side produces the original differential equation

$$d^2y/dx^2 - 12x^2 = 2e^x$$

Checking example 6. The solution we obtained was

$$y = x^4/4 + x^3 + 5x^2 - \cos x + px + q$$

We differentiate both sides to get

$$dy/dx = x^3 + 3x^2 + 10x + \sin x + p$$

Differentiating again yields

$$d^2y/dx^2 = 3x^2 + 6x + 10 + \cos x$$

We can subtract cos *x* from both sides to get the original equation

$$d^2y/dx^2 - \cos x = 3x^2 + 6x + 10$$

- -

For further study

Differential equations are an outgrowth of calculus. This chapter was written only to give you a glimpse into the topic by showing you how to solve some of the simplest possible ODEs. *Differential Equations Demystified* by Steven G. Krantz (McGraw-Hill, 2005) offers an excellent introduction to the subject.

Practice Exercises

This is an open-book quiz. You may (and should) refer to the text as you solve these problems. Don't hurry! You'll find worked-out answers in App. C. The solutions in the appendix may not represent the only way a problem can be figured out. If you think you can solve a particular problem in a quicker or better way than you see there, by all means try it!

1. Solve the differential equation

$$dy/dx = \sin x + 3$$

2. Solve the differential equation

$$dy/dx + \sin x = \cos x$$

3. Solve the differential equation

$$2\,dy/dx - 4e^x = 16x^3$$

4. Solve the differential equation

$$d^2y/dx^2 = \cos x + 5x$$

5. Solve the differential equation

$$d^2y/dx^2 + 2 \sin x = 3 \cos x$$

6. Solve the differential equation

$$2d^2y/dx^2 - 2e^x = 24x^2$$

7. Solve the differential equation

$$dy/(3\,dx) + 6x^{-2} = e^x$$

Here's a hint: As the first step, multiply through by 3.

8. Solve the differential equation

$$dy - 4e^x \, dx = x^2 \, dx + 2x \, dx$$

   Here's a hint: As the first step, divide through by dx.

9. Solve the differential equation stated in Prob. 8 by adding the quantity $(4e^x \, dx)$ to each side, separating out the dx multipliers from the sum on the right-hand side using the distributive law, and finally integrating straight through.

10. Check each of the solutions to Exercises 1 through 8 by differentiating and then morphing, if necessary, to get the original ODE back.

CHAPTER 30

Review Questions and Answers

Part Three

This is not a test! It's a review of important general concepts you learned in the previous nine chapters. Read it though slowly and let it "sink in." If you're confused about anything here, or about anything in the section you've just finished, go back and study that material some more.

Chapter 21

Question 21-1

What does a superscript −1 mean when it's written after the name of a function? What does it mean when it's written after dy/dx?

Answer 21-1

When we see a superscript −1 after the name of a function, it tells us to work with the inverse of that function. When we see a superscript −1 written after dy/dx, it tells us to work with the reciprocal derivative, dx/dy.

Question 21-2

Consider the following equation. What does the expression $f^{-1\prime}(y)$ mean?

$$dx/dx = f^{-1\prime}(y) \cdot dy/dx$$

Answer 21-2

The expression $f^{-1\prime}(y)$ refers to the derivative of x with respect to y, or dx/dy. That's the derivative of the inverse of $f(x)$.

Question 21-3

Suppose that we have a function f of an independent variable x, along with a dependent variable y such that

$$y = f(x)$$

How can we find $f^{-1\prime}(y)$ without first figuring out the inverse function itself?

Answer 21-3

We can differentiate $f(x)$ with respect to x, getting dy/dx. Then we can take the reciprocal of dy/dx, which gives us dx/dy. That's the derivative of the inverse of f.

Question 21-4

How can we express the above mentioned principle in words alone, in a way that's easy to remember?

Answer 21-4

We can say, "The derivative of the inverse equals the inverse of the derivative." But we must be careful about the context. The first time we say "inverse," we mean "inverse function." The second time, we mean "multiplicative inverse" or "reciprocal."

Question 21-5

When we want to define the inverse of the sine function, we must first define the principal branch to ensure that the inverse is a true function. What's the principal branch of the sine function?

Answer 21-5

The principal branch of the sine function is the set of all input values in the closed interval $[-\pi/2, \pi/2]$, giving us output values in the closed interval $[-1, 1]$. For example, if

$$y = \sin x$$

then the principal branch has a domain of

$$-\pi/2 \leq x \leq \pi/2$$

and a range of

$$-1 \leq y \leq 1$$

Question 21-6

What's the principal branch of the cosine function?

Answer 21-6

The principal branch of the cosine function is the set of all input values in the closed interval $[0,\pi]$, giving us output values in the closed interval $[-1,1]$. If we have

$$y = \cos x$$

then the principal branch has a domain of

$$0 \le x \le \pi$$

and a range of

$$-1 \le y \le 1$$

Question 21-7

The derivatives of the Arcsine and Arccosine functions are defined only over the *open* interval $(-1,1)$, even though the functions themselves are defined over the *closed* interval $[-1,1]$. Why?

Answer 21-7

At the extreme endpoints of the interval, the functions have defined values, but the slopes of the graphs are undefined ("vertical"). We can see this by looking back at pages 382 and 384 (Figs. 21-1 and 21-2). Because the derivative of a function is equivalent to the slope of its graph, the derivative must be undefined at any point where the slope of the graph is undefined.

Question 21-8

What happens if we take an input variable x, let a function operate on it to get an output variable y, and then let the inverse of the function operate on y?

Answer 21-8

We get x back again, assuming the function and its inverse are both defined, and are in fact true functions, for x and y.

Question 21-9

Imagine a function f that operates on an input x to get an output y. That is,

$$f(x) = y$$

Now suppose that at a particular input value x_1, we have an output value y_1 and a derivative of 2. That is,

$$f(x_1) = y_1$$

and

$$f'(x_1) = 2$$

How can we figure out the value of $f^{-1\prime}(y_1)$?

Answer 21-9

We can take the reciprocal of the derivative at x_1 to obtain the derivative of the inverse at y_1, as follows:

$$f^{-1\prime}(y_1) = [f'(x_1)]^{-1} = 2^{-1} = 1/2$$

Question 21-10

Imagine a function g that operates on an input u to get an output v. That is,

$$g(u) = v$$

Suppose that when the input is a certain value u_1, the output is v_1 and the derivative is 0. That is,

$$g(u_1) = v_1$$

and

$$g'(u_1) = 0$$

How can we determine $g^{-1\prime}(v_1)$?

Answer 21-10

We can't, because it's not defined. If we take the reciprocal of the derivative at u_1 in an attempt to find the derivative of the inverse at v_1, we get

$$g^{-1\prime}(v_1) = [g'(u_1)]^{-1} = 0^{-1} = 1/0$$

Chapter 22

Question 22-1

What's the equation of a unit circle centered at the origin of the Cartesian xy-plane? What's the equation of a circle of radius r centered at the origin? What's the equation of a circle of radius r centered at the point (x_0, y_0)? Can the equation of a circle in the Cartesian xy-plane ever represent a true function of either variable? If so, how do we know? If not, why not?

Answer 22-1

The equation of a unit circle centered at the origin is

$$x^2 + y^2 = 1$$

The equation of a circle of radius r centered at the origin is

$$x^2 + y^2 = r^2$$

The equation of a circle of radius r centered at (x_0, y_0) is

$$(x - x_0)^2 + (y - y_0)^2 = r^2$$

The equation of a circle in the xy-plane is never a true function of either variable. Whichever variable is defined as independent, the relation produces two outputs for some inputs. A true function is not "allowed" to map into more than one output value for any single input value.

Question 22-2

What's the general form for the equation of an ellipse in the Cartesian xy-plane? Is such a relation a true function of either variable? If so, how do we know? If not, why not?

Answer 22-2

The equation of an ellipse in xy-coordinates can be written in the form

$$(x - x_0)^2 / a^2 + (y - y_0)^2 / b^2 = 1$$

where x_0 and y_0 are the coordinates of the center, a is the distance from (x_0, y_0) to the curve along a line parallel to the x axis, and b is the distance from (x_0, y_0) to the curve along a line parallel to the y axis. Such a relation is never a true function of either variable, for the same reason as the equation of a circle in the xy-plane is never a true function.

Question 22-3

What's the general form for the equation of a hyperbola in xy-coordinates? Is the relation a true function of either variable? If so, how do we know? If not, why not?

Answer 22-3

The equation of a hyperbola can be written in the form

$$(x - x_0)^2 / a^2 - (y - y_0)^2 / b^2 = 1$$

where x_0 and y_0 are the coordinates of the center, and a and b are positive real-number constants that determine the dimensions and shape of the curve. Such a relation is never a function of either variable. If we define x as the independent variable, then the relation produces two outputs for most inputs. If we define y as the independent variable, then the relation produces two outputs for all inputs.

Question 22-4

Suppose we're told to use implicit differentiation to find y' and x' for the following equation. How does that process work?

$$x^2 + y^2 = 36$$

Answer 22-4

We can find y' by differentiating both sides of the original equation with respect to x, and then solving the result for y'. We can find x' by differentiating both sides with respect to y, and then solving the result for x'.

Question 22-5

What are the steps in the implicit differentiation process to find y' for the equation stated in Question 22-4?

Answer 22-5

Our objective is to find dy/dx. We start with

$$x^2 + y^2 = 36$$

Differentiating through with respect to x, we get

$$d/dx\,(x^2) + d/dx\,(y^2) = d/dx\,(36)$$

Term-by-term, this works out to

$$2x + 2yy' = 0$$

Subtracting $2x$ from each side, we obtain

$$2yy' = -2x$$

We can divide through by $2y$ if we insist that $y \neq 0$. That gives us

$$y' = (-2x) / (2y)$$

which simplifies to

$$y' = dy/dx = -x/y$$

Question 22-6

What are the steps in the implicit differentiation process to find x' for the equation stated in Question 22-4?

Answer 22-6

This process follows a "parallel track" to Answer 22-5. This time, our goal is to find dx/dy. We start with

$$x^2 + y^2 = 36$$

Differentiating through with respect to y, we get

$$d/dy\,(x^2) + d/dy\,(y^2) = d/dy\,(36)$$

Term-by-term, this works out to

$$2xx' + 2y = 0$$

Subtracting $2y$ from each side, we obtain

$$2xx' = -2y$$

We can divide through by $2x$ if we insist that $x \neq 0$. That gives us

$$x' = (-2y)\,/\,(2x)$$

which simplifies to

$$x' = dx/dy = -y/x$$

Question 22-7

What type of curve is represented by the following general equation?

$$px^2 + qy^2 = r$$

Here, x and y are variables, and p, q, and r are positive real-number constants. What are the steps in the implicit differentiation process to find y' for this equation?

Answer 22-7

This equation represents an ellipse centered at the origin. Our objective is to find a general formula for dy/dx. We start with

$$px^2 + qy^2 = r$$

Differentiating through with respect to x, we get

$$d/dx\,(px^2) + d/dx\,(qy^2) = d/dx\,(r)$$

Term-by-term, this works out to

$$2px + 2qyy' = 0$$

Subtracting $2px$ from each side, we obtain

$$2qyy' = -2px$$

We can divide through by $2qy$ if we insist that $y \neq 0$. That gives us

$$y' = (-2px) / (2qy)$$

which simplifies to

$$y' = dy/dx = (-px) / (qy)$$

Question 22-8

What are the steps in the implicit differentiation process to find x' for the equation stated in Question 22-7?

Answer 22-8

This time, we want to find a general formula for dx/dy. As before, we start with

$$px^2 + qy^2 = r$$

Differentiating through with respect to y, we get

$$d/dy\,(px^2) + d/dy\,(qy^2) = d/dy\,(r)$$

Term-by-term, this works out to

$$2pxx' + 2qy = 0$$

Subtracting $2qy$ from each side, we obtain

$$2pxx' = -2qy$$

We can divide through by $2px$ if we insist that $x \neq 0$. That gives us

$$y' = (-2qy) / (2px)$$

which simplifies to

$$y' = dy/dx = (-qy) / (px)$$

Question 22-9

What type of curve is represented by the following general equation?

$$px^2 - qy^2 = r$$

Here, x and y are variables, and p, q, and r are positive real-number constants. What are the steps in the implicit differentiation process to find y' for this equation?

Answer 22-9

This equation represents a hyperbola centered at the origin. We want to find a general formula for dy/dx. We start with

$$px^2 - qy^2 = r$$

Differentiating through with respect to x, we get

$$d/dx\,(px^2) - d/dx\,(qy^2) = d/dx\,(r)$$

Term-by-term, this works out to

$$2px - 2qyy' = 0$$

Subtracting $2px$ from each side, we obtain

$$-2qyy' = -2px$$

We can divide through by $-2qy$ if we insist that $y \neq 0$. That gives us

$$y' = (-2px)\,/\,(-2qy)$$

which simplifies to

$$y' = dy/dx = (px)\,/\,(qy)$$

Question 22-10

What are the steps in the implicit differentiation process to find x' for the equation stated in Question 22-9?

Answer 22-10

This time, we want to find a general formula for dx/dy. As before, we start with

$$px^2 - qy^2 = r$$

Differentiating through with respect to y, we get

$$d/dy\,(px^2) - d/dy\,(qy^2) = d/dy\,(r)$$

Term-by-term, this works out to

$$2pxx' - 2qy = 0$$

Adding $2qy$ to each side, we obtain

$$2pxx' = 2qy$$

We can divide through by $2px$ if we insist that $x \neq 0$. That gives us

$$x' = (2qy) / (2px)$$

which simplifies to

$$x' = dx/dy = (qy) / (px)$$

Chapter 23

Question 23-1

How can we informally state l'Hôpital's rule for finding limits of expressions that tend toward 0/0?

Answer 23-1

If we want to find the limit of a ratio that tends toward 0/0, and if we can differentiate both the numerator and the denominator, then the limit of the ratio of the derivatives is the same as the limit of the original ratio.

Question 23-2

How can we state the above principle formally?

Answer 23-2

Consider two functions $f(x)$ and $g(x)$ with three properties. First, f and g are both differentiable everywhere in some open interval containing the point where $x = a$ (but not necessarily differentiable at $x = a$ itself). Second,

$$\lim_{x \to a} f(x) = 0$$

and

$$\lim_{x \to a} g(x) = 0$$

Third, $g'(x) \neq 0$ at every point within the defined interval where $x \neq a$. If all three of these conditions are met, then

$$\lim_{x \to a} f(x) / g(x) = \lim_{x \to a} f'(x) / g'(x)$$

Question 23-3

There's an important restriction on the use of the l'Hôpital rule for finding the limits of expressions that tend toward 0/0. What is that restriction?

Answer 23-3

We must not use this rule to seek the limit of an expression that *obviously* blows up (that is, "runs away" toward positive or negative infinity) as the variable approaches the limiting value.

Question 23-4

How can we informally state l'Hôpital's rule for finding limits of expressions that tend toward $\pm\infty/\pm\infty$?

Answer 23-4

If we want to find the limit of a ratio that tends toward $\pm\infty/\pm\infty$ and if we can differentiate both the numerator and the denominator, then the limit of the ratio of the derivatives is the same as the limit of the original ratio.

Question 23-5

How can we state the above rule formally?

Answer 23-5

Suppose that f and g are functions of a variable x, and both functions are differentiable over an open interval containing the point where $x = a$ (but not necessarily differentiable at $x = a$ itself). Also suppose that

$$\underset{x \to a}{Lim}\ f(x) = \pm\infty$$

and

$$\underset{x \to a}{Lim}\ g(x) = \pm\infty$$

Finally, suppose that $g'(x) \neq 0$ at every point in the interval where $x \neq a$. If all three of these conditions are met, then

$$\underset{x \to a}{Lim}\ f(x) / g(x) = \underset{x \to a}{Lim}\ f'(x) / g'(x)$$

Question 23-6

Another of l'Hôpital's rules can help us evaluate the limit of an expression as the value of the variable approaches $+\infty$ or $-\infty$. What's that rule, stated formally?

Answer 23-6

Suppose that $f(x)$ and $g(x)$ are differentiable functions, and $g'(x) \neq 0$ as the value of x increases or decreases without bound. Also, suppose that one of these four things is true:

$$\underset{x \to +\infty}{Lim}\ f(x) = 0 \quad \text{and} \quad \underset{x \to +\infty}{Lim}\ g(x) = 0$$

$$\underset{x \to -\infty}{Lim}\ f(x) = 0 \quad \text{and} \quad \underset{x \to -\infty}{Lim}\ g(x) = 0$$

$$\underset{x \to +\infty}{Lim}\ f(x) = \pm\infty \quad \text{and} \quad \underset{x \to +\infty}{Lim}\ g(x) = \pm\infty$$

$$\underset{x \to -\infty}{Lim}\ f(x) = \pm\infty \quad \text{and} \quad \underset{x \to -\infty}{Lim}\ g(x) = \pm\infty$$

Then

$$Lim_{x \to a} f(x) / g(x) = Lim_{x \to a} f'(x) / g'(x)$$

Question 23-7

Suppose that we want to find the limit of the product of two functions, one of which approaches 0 and the other of which approaches $+\infty$. How can we rearrange this expression so we can use l'Hôpital's principles in an attempt to evaluate it?

Answer 23-7

We can take the function that approaches 0 and divide it by the reciprocal of the function that approaches $+\infty$. This gives us a ratio that approaches 0/0, so we can use l'Hôpital's rule for limits of that type.

Question 23-8

Suppose that we want to find the limit of the difference between two functions, both of which approach $+\infty$. How can we rearrange this expression so we can try to use l'Hôpital's principles to evaluate it?

Answer 23-8

Using algebra, we can try to rewrite the expression as a ratio that approaches an expression of the form 0/0 or $\pm\infty/\pm\infty$, and then use the appropriate l'Hôpital rule in an attempt to find the limit.

Question 23-9

Imagine that we've been told to figure out the limit of an expression that tends toward one of these forms:

$$+\infty \cdot 0$$
$$-\infty \cdot 0$$
$$-\infty + (+\infty)$$
$$+\infty + (-\infty)$$
$$-\infty - (-\infty)$$

How can we rearrange expressions such as these so we can use l'Hôpital's principles in an attempt to evaluate them?

Answer 23-9

Whenever we see a limit whose expression tends toward one of these forms, we can convert it to a form we already know how to deal with, as follows:

- The form $+\infty \cdot 0$ is equivalent to $0 \cdot (+\infty)$
- The form $-\infty \cdot 0$ is equivalent to $0 \cdot (-\infty)$

- The form $-\infty + (+\infty)$ is equivalent to $+\infty - (+\infty)$
- The form $+\infty + (-\infty)$ is equivalent $+\infty - (+\infty)$
- The form $-\infty - (-\infty)$ is equivalent to $+\infty - (+\infty)$

Question 23-10

How can we evaluate the following limit, where the first factor tends toward 0 and the second factor tends toward negative infinity as x approaches 0 from the left?

$$\lim_{x \to 0-} x \ln(-x)$$

Answer 23-10

Let's rewrite this as the limit of a ratio by putting x^{-1} in the denominator. When we do that, we obtain

$$\lim_{x \to 0-} [\ln(-x)] / x^{-1}$$

Both the numerator and the denominator in this ratio tend toward negative infinity as x approaches 0 from the left, so we can apply the l'Hôpital rule for limits of expressions of the form $-\infty/(-\infty)$. Let's call the functions

$$f(x) = \ln(-x)$$

and

$$g(x) = x^{-1}$$

Because we approach 0 from the left, x is always negative, so

$$\ln(-x) = \ln|x|$$

Therefore, we can rewrite our functions as

$$f(x) = \ln|x|$$

and

$$g(x) = x^{-1}$$

The derivatives are

$$f'(x) = x^{-1}$$

and

$$g'(x) = -x^{-2}$$

Therefore

$$\underset{x \to 0-}{Lim}\ f(x)\ /\ g(x)\ =\ \underset{x \to 0-}{Lim}\ f'(x)\ /\ g'(x)\ =\ \underset{x \to 0-}{Lim}\ x^{-1}\ /\ (-x^{-2})$$

The rightmost expression can be simplified with algebra:

$$x^{-1}\ /\ (-x^{-2}) = (1/x)\ /\ (-1/x^2) = (1/x)\ (-x^2) = -x$$

Now we have the manageable limit

$$\underset{x \to 0-}{Lim}\ -x$$

This expression approaches 0 as x approaches 0 from the left. We've determined that

$$\underset{x \to 0-}{Lim}\ x \ln(-x) = 0$$

Chapter 24

Question 24-1

Suppose that we have a surface in Cartesian xyz-space, and we can define a relation H so that

$$z = H(x,y)$$

is represented by this surface. How can we apply the vertical-line test to see if H is a true function of x and y?

Answer 24-1

Imagine a straight, infinitely long vertical line parallel to the z axis. We move this line around, so the point where it intersects the xy-plane sweeps through every possible ordered pair (x,y). The graph of H represents a true function if and only if the movable vertical line never cuts through the surface at more than one point, as shown in Fig. 30-1.

Question 24-2

Imagine a function $f(x,y)$ in Cartesian xyz-space, whose graph shows up as a surface. If we cut through this surface with a plane perpendicular to the x axis, we get a curve. What sort of function does this curve represent? If we take the derivative of this function, what do we get?

Answer 24-2

The curve represents a function of y, with x treated as a constant. If we take the derivative of this function, we get a partial derivative of $f(x,y)$ with respect to y.

Question 24-3

Imagine the same function $f(x,y)$ as the one stated in Question 24-2. If we cut through this surface with a plane perpendicular to the y axis, we get a curve. What sort of function does this curve represent? If we take the derivative of this function, what do we get?

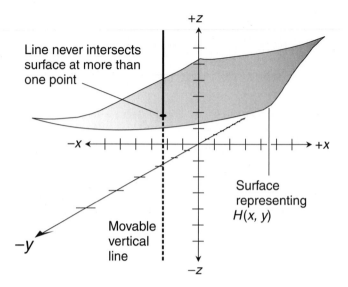

Figure 30-1   Illustration for Question and Answer 24-1.

Answer 24-3

The curve represents a function of x, with y held constant. If we take the derivative, we get a partial derivative of $f(x,y)$ with respect to x.

Question 24-4

What's the multiplication-by-constant rule for partial differentiation?

Answer 24-4

If we take the partial derivative of a multi-variable function *after* it has been multiplied by a constant, we get the same result as we do if we take the partial derivative of the function and *then* multiply by the constant, as long as we differentiate with respect to the same variable on both occasions.

Question 24-5

What's the sum rule for partial differentiation?

Answer 24-5

The partial derivative of the sum of two or more multi-variable functions is equal to the sum of their partial derivatives with respect to the same variable.

Question 24-6

What's the difference rule for partial differentiation?

Answer 24-6

The partial derivative of the difference between two multi-variable functions is equal to the difference between their partial derivatives, as long as both partial derivatives are found with respect to the same variable, and as long as we keep the subtraction in the same order.

Question 24-7

Consider the following function, in which the domain is a set of real-number ordered triples (x,y,z) with $z \neq 0$, and the range is a set of real numbers w. What's the general partial derivative $\partial w / \partial x$?

$$w = g\,(x,y,z)\ = e^x + y^2 + \ln\,|z|$$

Answer 24-7

To find the partial derivative with respect to x, we hold y and z constant. That makes the second and third terms constants, so their derivatives are both 0. The derivative of the first term with respect to x is e^x, because the exponential function is its own derivative. Therefore

$$\partial w / \partial x = e^x + 0 + 0 = e^x$$

Question 24-8

What's the general partial derivative $\partial w / \partial y$ of the function stated in Question 24-7?

Answer 24-8

To find the partial derivative with respect to y, we hold x and z constant. That makes the first and third terms constants, so their derivatives are both 0. The derivative of the second term with respect to y is equal to $2y$. Therefore

$$\partial w / \partial y = 0 + 2y + 0 = 2y$$

Question 24-9

What's the general partial derivative $\partial w / \partial z$ of the function stated in Question 24-7?

Answer 24-9

To find the partial derivative with respect to z, we hold x and y constant. That makes the first and second terms constants, so their derivatives are both 0. The derivative of the third term with respect to z is equal to z^{-1}. Therefore

$$\partial w / \partial z = 0 + 0 + z^{-1} = z^{-1}$$

Question 24-10

For the function stated in Question 24-7, how can we find these three specific partial derivatives?

$$\partial/\partial x\,(0,2,4)$$
$$\partial/\partial y\,(0,2,4)$$
$$\partial/\partial z\,(0,2,4)$$

Answer 24-10

We simply plug in the numbers to the results we have obtained. When we do that for each partial derivative in turn, we obtain

$$\partial/\partial x\,(0,2,4) = e^0 = 1$$
$$\partial/\partial y\,(0,2,4) = 2 \cdot 2 = 4$$
$$\partial/\partial z\,(0,2,4) = 4^{-1} = 1/4$$

Chapter 25

Question 25-1

Imagine a function f of two variables x and y, giving us an output z, like this:

$$z = f(x,y)$$

How do we determine the second partial of f with respect to x, denoted $\partial^2 z/\partial x^2$? How do we determine the second partial of f with respect to y, denoted $\partial^2 z/\partial y^2$?

Answer 25-1

To determine $\partial^2 z/\partial x^2$, we treat y as a constant and differentiate with respect to x twice. To find $\partial^2 z/\partial y^2$, we treat x as a constant and differentiate with respect to y twice.

Question 25-2

What is the multiplication-by-constant rule for second partials?

Answer 25-2

If we take the second partial of a multi-variable function *after* it has been multiplied by a constant, we get the same result as we do if we take the second partial of the function and *then* multiply by the constant, as long as we differentiate with respect to the same variable throughout the process.

Question 25-3

What is the sum rule for second partials?

Answer 25-3

The second partial of the sum of two or more multi-variable functions is equal to the sum of their second partials, as long as all the partial derivatives are found with respect to the same variable.

Question 25-4

What is the difference rule for second partials?

Answer 25-4

The second partial of the difference between two multi-variable functions is equal to the difference between their second partials, as long as we differentiate with respect to the same variable throughout the process, and as long as we keep the subtraction in the same order.

Question 25-5

Once again, suppose that we have a function f of two variables x and y, giving us an output z, as follows:

$$z = f(x,y)$$

How do we determine the mixed partial of f with respect to x and then y, denoted $\partial^2 z / \partial y \partial x$?

Answer 25-5

To figure out $\partial^2 z / \partial y \partial x$, we hold y constant and differentiate the original function f with respect to x, obtaining $\partial z / \partial x$. Then we take the function $\partial z / \partial x$, hold x constant, and differentiate it with respect to y.

Question 25-6

How do we determine the mixed partial of f (as stated in Question 25-5) with respect to y and then x, denoted $\partial^2 z / \partial x \partial y$?

Answer 25-6

To figure out $\partial^2 z / \partial x \partial y$, we hold x constant and differentiate f against y, getting $\partial z / \partial y$. Then we take $\partial z / \partial y$, hold y constant, and differentiate against x.

Question 25-7

If we start with the two-variable function f stated in Question 25-5 and work out both mixed partials, and if we haven't made any mistakes in our work, how will the function $\partial^2 z / \partial y \partial x$ compare with the function $\partial^2 z / \partial x \partial y$?

Answer 25-7

They will be identical.

Question 25-8

Imagine that we have a function g of three variables x, y, and z, along with an output variable w, so that

$$w = g(x,y,z)$$

How can we find the second partial of g relative to x? Relative to y? Relative to z? Let's denote these second partials as $\partial^2 w / \partial x^2$, $\partial^2 w / \partial y^2$, and $\partial^2 w / \partial z^2$ respectively.

Answer 25-8

The processes are similar, but we must be careful which variables we hold constant. It's easy to get confused!

- To find the second partial with respect to x, we treat y and z as constants, differentiate the original function against x to get $\partial w / \partial x$, and then differentiate $\partial w / \partial x$ against x to get $\partial^2 w / \partial x^2$.
- To find the second partial with respect to y, we treat x and z as constants, differentiate the original function against y to get $\partial w / \partial y$, and then differentiate $\partial w / \partial y$ against y to get $\partial^2 w / \partial y^2$.
- To find the second partial with respect to z, we treat x and y as constants, differentiate the original function against z to get $\partial w / \partial z$, and then differentiate $\partial w / \partial z$ against z to get $\partial^2 w / \partial z^2$.

Question 25-9

Suppose that we want to find all of the mixed second partials of a three-variable function g such that

$$w = g(x,y,z)$$

where x, y, and z are the independent variables, and w is the dependent variable. How many different mixed second partials are there? How do we find them?

Answer 25-9

There are six different mixed second partials, representing the fact that there are six different permutations of two objects (x and y) taken out of three (x, y, and z).

- We can differentiate against x and then y to get $\partial^2 w / \partial y \partial x$.
- We can differentiate against x and then z to get $\partial^2 w / \partial z \partial x$.
- We can differentiate against y and then x to get $\partial^2 w / \partial x \partial y$.
- We can differentiate against y and then z to get $\partial^2 w / \partial z \partial y$.
- We can differentiate against z and then x to get $\partial^2 w / \partial x \partial z$.
- We can differentiate against z and then y to get $\partial^2 w / \partial y \partial z$.

Question 25-10

How can we find the six mixed second partials for the following three-variable function?

$$w = h\,(x,y,z) = x^{-3}y^2 e^z$$

Answer 25-10

Let's find the mixed partials in the order listed in Answer 25-9. The first partial of the original function relative to x is found by holding y and z constant, obtaining

$$\partial w/\partial x = -3x^{-4}y^2 e^z$$

To differentiate $\partial w/\partial x$ against y, we hold x and z constant to get

$$\partial^2 w/\partial y\partial x = -6x^{-4}y e^z$$

To differentiate $\partial w/\partial x$ against z, we hold x and y constant to get

$$\partial^2 w/\partial z\partial x = -3x^{-4}y^2 e^z$$

The first partial of the original function relative to y is found by holding x and z constant, obtaining

$$\partial w/\partial y = 2x^{-3}y e^z$$

To differentiate $\partial w/\partial y$ against x, we hold y and z constant to get

$$\partial^2 w/\partial x\partial y = -6x^{-4}y e^z$$

To differentiate $\partial w/\partial y$ against z, we hold x and y constant to get

$$\partial^2 w/\partial z\partial y = 2x^{-3}y e^z$$

The first partial of the original function relative to z is found by holding x and y constant, obtaining

$$\partial w/\partial z = x^{-3}y^2 e^z$$

To differentiate $\partial w/\partial z$ against x, we hold y and z constant to get

$$\partial^2 w/\partial x\partial z = -3x^{-4}y^2 e^z$$

To differentiate $\partial w/\partial z$ against y, we hold x and z constant to get

$$\partial^2 w/\partial y\partial z = 2x^{-3}y e^z$$

Chapter 26

Question 26-1

Figure 30-2 shows a right circular cylinder in Cartesian *xyz*-space. The left-hand face of the cylinder lies in the *yz*-plane. The axis of the cylinder is along the coordinate *x* axis. The radius is *r* and the height (shown here as length) is *h*. How can we integrate to find the lateral-surface area *A* of this cylinder?

Answer 26-1

We integrate the circumference of a circular slice along the length of the cylinder's surface from its left-hand end where $x = 0$ to its right-hand end where $x = h$, getting

$$A = \int_0^h 2\pi r \, dx$$

Question 26-2

Look at Fig. 30-2 again. How can we integrate to find the volume *V* of the cylinder?

Answer 26-2

We integrate the area of a cross-sectional disk along the length of the cylinder, starting at the left-hand face where $x = 0$, and moving along the *x* axis until we reach the right-hand face where $x = h$. That gives us

$$V = \int_0^h \pi r^2 \, dx$$

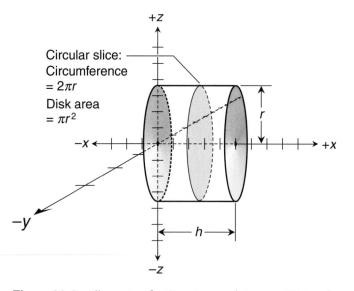

Figure 30-2   Illustration for Questions and Answers 26-1 and 26-2.

Question 26-3

How can we integrate to find the volume of a right circular cylinder whose radius r is 10 units and whose length h is 7 units?

Answer 26-3

To find the volume V, we use the integral formula

$$V = \int_0^h \pi r^2 \, dx$$

We know that $r = 10$ and $h = 7$, so the integral becomes

$$V = \int_0^7 \pi \cdot 10^2 \, dx = \int_0^7 100\pi \, dx = 100\pi x \Big]_0^7 = 100\pi \cdot 7 - 100\pi \cdot 0 = 700\pi$$

Question 26-4

Figure 30-3 shows a right circular cone in *xyz*-space. The apex is at the origin and the base is parallel to the *yz*-plane. The cone's axis lies along the coordinate *x* axis. The radius at the base is r, the height (shown here as length) is h, and the slant height is s. The distance between the apex and any particular point on the cone's shell is t. How can we integrate to find the slant-surface area A of this cone?

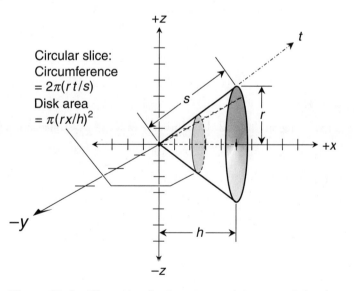

Figure 30-3   Illustration for Questions and Answers 26-4 and 26-6.

Answer 26-4

We can integrate the function representing the circumference of a circular cross-sectional slice along a straight line from the apex, where $t = 0$, to any point on the outer edge of the base, where $t = s$. When we do this, we get

$$A = \int_0^s (2\pi r t/s)\, dt = \int_0^s (2\pi r/s)t\, dt$$

Question 26-5

How can we integrate to find the slant-surface area A of a right circular cone whose radius r is 1 unit and whose slant height s is 2 units?

Answer 26-5

To find A, we use the integral formula

$$A = \int_0^s (2\pi r/s)t\, dt$$

We've been told that $r = 1$ and $s = 2$, so we have

$$A = \int_0^2 (2\pi \cdot 1/2)t\, dt = \int_0^2 \pi t\, dt = \left. \pi t^2/2 \right]_0^2 = 2\pi - 0 = 2\pi$$

Question 26-6

Look again at Fig. 30-3. How can we integrate to find V, the cone's interior volume?

Answer 26-6

We can integrate the function representing the area of a cross-sectional disk along the axis of the cone from its apex where $x = 0$ to its base where $x = h$, getting

$$V = \int_0^h \pi (rx/h)^2\, dx = \int_0^h (\pi r^2/h^2)x^2\, dx$$

Question 26-7

How can we integrate to find the volume V of a right circular cone whose radius r is 1 unit and whose height h is also 1 unit?

Answer 26-7

To find V, we use the integral formula

$$V = \int_0^h (\pi r^2/h^2)x^2\, dx$$

We've been told that $r = 1$ and $h = 1$, so we have

$$V = \int_0^1 (\pi \cdot 1^2/1^2)x^2\, dx = \int_0^1 \pi x^2\, dx = \left. \pi x^3/3 \right]_0^1 = \pi/3 - 0 = \pi/3$$

Question 26-8

Figure 30-4 shows a sphere in *xyz*-space. The center is at the origin. The radius is *r*. As we travel over the sphere's surface, *t* is the distance between the right-hand pole and any particular point on the surface, as shown. How can we integrate to find the surface area *A* of the sphere?

Answer 26-8

We integrate the function representing the circumference of a circular slice by traveling half-way around the sphere from $t = 0$ to $t = \pi r$. That's

$$A = 2\pi r \int_0^{\pi r} \sin(t/r)\, dt$$

Question 26-9

Figure 30-5 shows another sphere in *xyz*-space. The center is at the origin. The radius is *r*. As we travel straight through the sphere, *x* is the distance between the origin and the center of a disk whose radius is r_0. How can we integrate to find the volume *V* of the sphere?

Answer 26-9

We integrate the function representing the area of a cross-sectional disk along the *x* axis from the left-hand pole of the sphere (where $x = -r$) to the right-hand pole (where $x = r$). Therefore

$$V = \int_{-r}^{r} (\pi r^2 - \pi x^2)\, dx$$

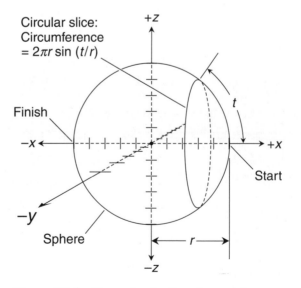

Figure 30-4 Illustration for Question and Answer 26-8.

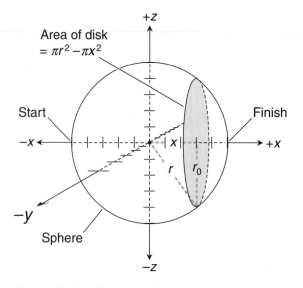

Figure 30-5   Illustration for Question and Answer 26-9.

Question 26-10

How can we integrate to find the surface area and volume of a sphere whose radius r is 1 unit?

Answer 26-10

To determine the surface area A, we must find

$$A = 2\pi r \int_0^{\pi r} \sin(t/r)\, dt$$

We know that $r = 1$, so the integral becomes

$$A = 2\pi \int_0^{\pi} \sin t\, dt = -2\pi \cos t\, \Big]_0^{\pi} = -2\pi \cos \pi - (-2\pi \cos 0) = 2\pi + 2\pi = 4\pi$$

The volume V can be found by using the formula

$$V = \int_{-r}^{r} (\pi r^2 - \pi x^2)\, dx$$

Because $r = 1$, we have

$$V = \int_{-1}^{1} (\pi - \pi x^2)\, dx = \pi x - \pi x^3/3 \, \Big]_{-1}^{1} = (\pi - \pi/3) - (-\pi + \pi/3) = 2\pi/3 + 2\pi/3 = 4\pi/3$$

Chapter 27

Question 27-1

How do we find a repeated definite integral of a single-variable function $f(x)$?

Answer 27-1

First, we take the indefinite integral with respect to x, which gives us

$$\int f(x)\, dx = F_1(x)$$

where F_1 is the basic antiderivative of f. We ignore the constant of integration, because we intend to find a definite integral. Next, we integrate again to get

$$\int F_1(x)\, dx = \int \left[\int f(x)\, dx\right] dx = F_2(x)$$

where F_2 is the second basic antiderivative of f. As with the first integral, we ignore the constant of integration. If we specify an interval with bounds $x = a$ and $x = b$ for both integrals, we get

$$F_2(x)\Big]_a^b = F_2(b) - F_2(a)$$

We can write out the complete repeated integral like this:

$$\int_a^b \int_a^b f(x)\, dx\, dx = F_2(b) - F_2(a)$$

Question 27-2

What precautions must we take when we calculate a repeated definite integral of a single-variable function?

Answer 27-2

We must wait until we have integrated both times before we evaluate the expression over the defined interval. The bounds of integration on both integral symbols must be identical.

Question 27-3

How can we find the repeated definite integral of the cosine function over the interval from 0 to π?

Answer 27-3

Let's call the variable z. Then we can state our problem as

$$\int_0^\pi \int_0^\pi \cos z\, dz\, dz$$

The first basic antiderivative is

$$\int \cos z\, dz = \sin z$$

Integrating again without the constant gives us

$$\int \sin z\, dz = -\cos z$$

The repeated definite integral is therefore

$$\int_0^\pi \int_0^\pi \cos z \, dz \, dz = -\cos z \Big]_0^\pi = -\cos \pi - (-\cos 0) = 1 - (-1) = 2$$

Question 27-4

What's the difference between the mathematical volume and the true geometric volume of a solid in 3D space?

Answer 27-4

Mathematical volume is the direct result of the integration process. As such, it can be positive, negative, or zero. True geometric volume, also called "real-world volume," can never be negative.

Question 27-5

Imagine a function $z = f(x,y)$ whose graph is a surface in Cartesian xyz-space. We want to find the mathematical volume of the solid object defined by this surface and a rectangle in the xy-plane enclosed by the lines $x = a$, $x = b$, $y = c$, and $y = d$. How can we do this by integrating with respect to x and then with respect to y?

Answer 27-5

We integrate $f(x,y)$ against x from a to b, getting

$$\int_a^b f(x,y) \, dx$$

This gives us a function of y that we integrate from c to d. The entire double integral can be written as

$$\int_c^d \left[\int_a^b f(x,y) \, dx \right] dy$$

or as

$$\int_c^d \int_a^b f(x,y) \, dx \, dy$$

Question 27-6

How can we find the mathematical volume of the solid described in Question 27-5 by integrating with respect to y and then with respect to x?

Answer 27-6

We integrate $f(x,y)$ against y from c to d, getting

$$\int_c^d f(x,y) \, dy$$

This gives us a function of x that we integrate from a to b to get

$$\int_a^b \left[\int_c^d f(x,y) \, dy \right] dx$$

which can also be written as

$$\int_a^b \int_c^d f(x,y) \, dy \, dx$$

Question 27-7

Suppose that we calculate both integrals as denoted in Answers 27-5 and 27-6 for a particular solid object in *xyz*-space. How should they compare, assuming we haven't made any mistakes?

Answer 27-7

They should be equal to each other. That is, we should get

$$\int_c^d \int_a^b f(x,y) \, dx \, dy = \int_a^b \int_c^d f(x,y) \, dy \, dx$$

Question 27-8

Imagine a function f that is defined at every point inside a rectangle R in the *xy*-plane of Cartesian *xyz*-space. Let's say that

$$z = f(x,y)$$

The graph of f is a surface; let's call it S. With respect to R, S defines a solid object that we call T, as shown in Fig. 30-6. How can we approximate the mathematical volume of T using a grid of squares in R?

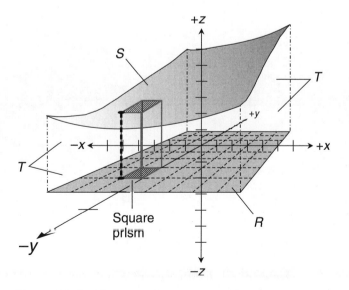

Figure 30-6   Illustration for Question and Answer 27-8.

Answer 27-8

Imagine breaking up the solid into square prisms whose heights are defined by the value of z in a selected corner of each prism's top face. Suppose that we make the xy-plane grid for the prisms increasingly fine while staying inside R. This causes the number of prisms to increase. If we keep doing this indefinitely, the sum of the prisms' mathematical volumes approaches the mathematical volume of T.

Question 27-9

Imagine the same function, region, and surface as described in Question 27-8. How can we illustrate the process of finding the mathematical volume of T by integrating with respect to x and then with respect to y, as we did in Answer 27-5 above?

Answer 27-9

Figure 30-7 shows how this works. We slice T into slabs perpendicular to the y axis and parallel to the xz-plane. Then we integrate against x from $x = a$ to $x = b$, determining the area of a slab with y held constant. Finally, we integrate the function that describes how the slab area varies as we move along the y axis from $y = c$ to $y = d$.

Question 27-10

Imagine the same function, region, and surface as described in Questions 27-8 and 27-9. How can we illustrate the process of finding the mathematical volume of T by integrating with respect to y and then with respect to x, as we did in Answer 27-6 above?

Answer 27-10

Figure 30-8 is a diagram of this process. We slice T into slabs perpendicular to the x axis and parallel to the yz-plane. Then we integrate against y from $y = c$ to $y = d$, finding the area of a

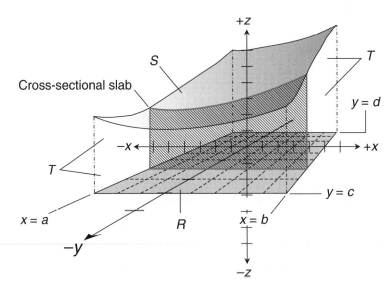

Figure 30-7   Illustration for Question and Answer 27-9.

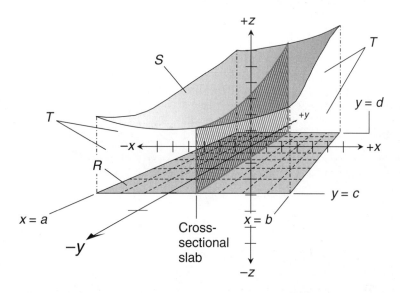

Figure 30-8   Illustration for Question and Answer 27-10.

slab with x held constant. Finally, we integrate the function that describes how the slab area varies as we move along the x axis from $x = a$ to $x = b$.

Chapter 28

Question 28-1

Imagine a solid object T defined by a surface S in Cartesian xyz-space that represents a function

$$z = f(x,y)$$

We want to double integrate to find the mathematical volume of T with respect to a non-rectangular region of integration R in the xy-plane, enclosed by the graphs of

$$y = g(x)$$

and

$$y = h(x)$$

What's the first step in the process?

Answer 28-1

We slice T into slabs perpendicular to the x axis, and therefore parallel to the yz-plane. Each slab is bounded by R, S, and two vertical lines (parallel to the z axis), one passing through the graph of $g(x)$, and the other passing through the graph of $h(x)$. This situation is illustrated in Fig. 30-9.

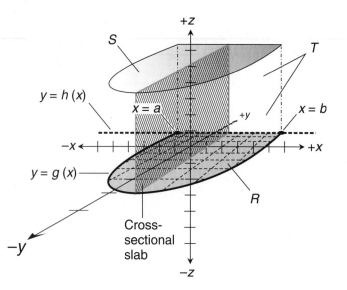

Figure 30-9   Illustration for Questions and Answers 28-1 through 28-4.

Question 28-2

What's the second step in the process for finding the mathematical volume of the solid object *T* described in Question 28-1 and illustrated in Fig. 30-9?

Answer 28-2

We can define the area *A* of a cross-sectional slab for a constant value of *x* by finding

$$A = \int_{g(x)}^{h(x)} f(x,y) \, dy$$

Question 28-3

What's the final step in finding the mathematical volume *V* of the solid *T* described in Question 28-1 and illustrated in Fig. 30-9?

Answer 28-3

We note the *x* values of the points where the graphs of *g* and *h* intersect in the *xy*-plane. These are *x* = *a* and *x* = *b* as shown in Fig. 30-9. Next, we integrate the function that expresses the area of a slab vs. the value of *x*, over the interval from *x* = *a* to *x* = *b*. That integral is

$$V = \int_{a}^{b} A \, dx = \int_{a}^{b} \left[\int_{g(x)}^{h(x)} f(x,y) \, dy \right] dx$$

This double integral can also be written out as

$$V = \int_{a}^{b} \int_{g(x)}^{h(x)} f(x,y) \, dy \, dx$$

Question 28-4

Refer to Fig. 30-9 once again. Imagine that we're integrating the function that defines the area A of a cross-sectional slab as we move parallel to the x axis. The surface S represents a positive constant function $f(x,y)$, so S lies entirely "above" the xy-plane. How does A change as we move from $x = a$ to $x = b$?

Answer 28-4

Let's start out at the point where $x = a$. The cross-sectional slab is a line segment parallel to the z axis and "above" the xy-plane. As such, it has no mathematical area. As we move toward the right along the x axis, the slab is slender at first, but it becomes progressively wider. As we keep moving, the width, and therefore the mathematical area, of the slab increases until it reaches a maximum at some value of x between a and b. As we keep moving toward the right along the x axis, the slab slims back down and its mathematical area decreases until, when we reach $x = b$, it collapses into a line segment with no mathematical area.

Question 28-5

What is the proper structure of Cartesian xyz-space? That is, in what relative way should the three axes be oriented?

Answer 28-5

Imagine the x axis so that the positive values go "east" and the negative values go "west." Then imagine the y axis so that the positive values go "north" and the negative values go "south." Finally imagine the z axis so that the positive values go "straight up" and the negative values go "straight down." All three axes should intersect at a single point, the origin, where

$$x = y = z = 0$$

Once this relative orientation has been defined for the axes, we should consider the whole structure as if it were rigid, made of stiff metal rods, for example. We can then turn, tumble, or roll the entire structure in any way we want, and the nature of the system will not change. All of the Cartesian xyz-space graphs in this book show the axes in this relative orientation, although the points of view differ.

Question 28-6

Imagine a solid object T defined by a surface S in Cartesian xyz-space that represents a function

$$z = f(x,y)$$

We want to double integrate to find the mathematical volume of T with respect to a non-rectangular region of integration R in the xy-plane, enclosed by the graphs of

$$x = g(y)$$

and

$$x = h(y)$$

What's the first step in the process?

Answer 28-6

We slice T into slabs perpendicular to the y axis. Each slab is bounded by R, S, and two vertical lines (parallel to the z axis). One vertical line passes through the graph of $g(y)$, and the other vertical line goes through the graph of $h(y)$, as shown in Fig. 30-10. Note the subtle difference between this graph and Fig. 30-9. We've changed our point of view. Previously, we looked toward the origin from somewhere near the negative y-axis. Now, we're looking toward the origin from somewhere near the positive x-axis.

Question 28-7

What's the second step in the process for finding the mathematical volume of T described in Question 28-6 and illustrated in Fig. 30-10?

Answer 28-7

We can define the area A of a cross-sectional slab for a constant value of y by finding

$$A = \int_{g(y)}^{h(y)} f(x,y) \ dx$$

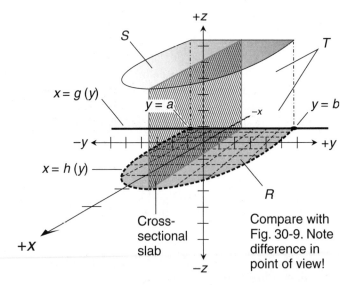

Figure 30-10   Illustration for Questions and Answers 28-6
through 28-10.

Question 28-8

What's the final step in finding the mathematical volume of the solid described in Question 28-6 and illustrated in Fig. 30-10?

Answer 28-8

We integrate, over the interval from $y = a$ to $y = b$, the function that tells us the area of a slab depending on the value of y. That integral is

$$V = \int_a^b A \, dy = \int_a^b \left[\int_{g(y)}^{h(y)} f(x,y) \, dx \right] dy$$

This double integral can also be written as

$$V = \int_a^b \int_{g(y)}^{h(y)} f(x,y) \, dx \, dy$$

Question 28-9

Refer again to Fig. 30-10. Imagine that we're integrating the function that defines the area A of a cross-sectional slab as we move from $y = a$ to $y = b$. The surface S represents a positive constant function $f(x,y)$. How does A vary as we move from $y = a$ to $y = b$?

Answer 28-9

This situation is similar to the one described in Answer 28-4. The only significant difference is that this time, we're going along the y axis instead of along the x axis. We begin at $y = a$, where the cross-sectional slab is really nothing more than a line segment parallel to the z axis. As we move along the y axis in the positive direction, the slab starts out slim, but widens. The mathematical area increases until it reaches a maximum somewhere between $y = a$ and $y = b$. As we keep moving along the y axis, the slab's mathematical area decreases until, when we reach $y = b$, it becomes a line segment again.

Question 28-10

What would happen to the mathematical volume of the solid T shown in Fig. 30-10 if we were to integrate the function defining the area of the movable slab from $y = b$ to $y = a$, instead of from $y = a$ to $y = b$? What would happen to the true geometric volume?

Answer 28-10

The mathematical volume would be multiplied by -1, because we would have reversed the direction of one of the integrals. The true geometric volume of the solid would not change, however. We'd still be working with the same "real-world" object!

Chapter 29

Question 29-1

How can we recognize an elementary first-order ordinary differential equation (ODE)?

Answer 29-1

Imagine an equation consisting of an independent variable x, a dependent variable y, and a continuous function f. We have an elementary first-order ODE if we can morph this equation into the standard form

$$dy/dx = f(x)$$

Question 29-2

How can we solve an elementary first-order ODE once we have it in the standard form described in Answer 29-1?

Answer 29-2

To solve this type of equation, we integrate both sides with respect to x, getting

$$y = H(x)$$

where H is a family of functions, any two of which differ by some real-number constant, but are otherwise identical. The constant arises because we must take the indefinite integral to get rid of the derivative.

Question 29-3

How can we solve the following elementary first-order ODE?

$$dy/dx - \cos x = -e^x$$

Answer 29-3

We begin by manipulating the equation into the standard form. Let's add cos x to both sides. Then we get

$$dy/dx = -e^x + \cos x$$

That's in the standard form for an elementary first-order ODE. Taking the indefinite integral of both sides with respect to x gives us

$$\int (dy/dx)\ dx = \int (-e^x + \cos x)\ dx$$

The general antiderivatives are

$$\int (dy/dx)\ dx = y + c_1$$
$$\int (-e^x + \cos x)\ dx = -e^x + \sin x + c_2$$

where c_1 and c_2 are the constants of integration. Putting the antiderivatives together into a single equation, we obtain

$$y + c_1 = -e^x + \sin x + c_2$$

When we subtract c_1 from each side, we get

$$y = -e^x + \sin x - c_1 + c_2$$

We can add $-c_1$ to c_2, call the sum c, and then write the solution as

$$y = H(x) = -e^x + \sin x + c$$

where $H(x)$ represents the entire family of functions, all of which would be identical except for the fact that they differ by real-number constants.

Question 29-4

Can the constant in the solution to an elementary first-order ODE, such as the one we found in Answer 29-3, be eliminated or resolved into a specific real number?

Answer 29-4

Sometimes it can, but not always. If the constant of integration cancels itself out, as often happens in physical-science problems, we can eliminate it. But in the purely mathematical solution, we must leave the constant in. That's because it can't be resolved into any particular real number.

Question 29-5

How can we get the following elementary first-order ODE into the standard form, allowing us to solve it by taking the indefinite integral of both sides with respect to x?

$$dy + \cos x \, dx = 3x^2 \, dx + 8x \, dx - 7 \, dx$$

Answer 29-5

We can divide the entire equation through by dx, obtaining

$$dy/dx + \cos x = 3x^2 + 8x - 7$$

When we subtract $\cos x$ from both sides, we get the standard form

$$dy/dx = 3x^2 + 8x - 7 - \cos x$$

Question 29-6

How can we recognize an elementary second-order ODE?

Answer 29-6

Imagine that we encounter an equation with an independent variable x and a dependent variable y, along with a continuous function f. We have an elementary second-order ODE, if we can manipulate it into the form

$$d^2y/dx^2 = f(x)$$

Question 29-7

What strategy can we use to solve an equation in the form shown in Answer 29-6?

Answer 29-7

We must take the indefinite integral of both sides, retain the constant of integration, and then take the indefinite integral of both sides again. When we do that, we end up with an equation of the form

$$y = H(x)$$

where H is a family of functions that contains two constants. These constants result from the repeated indefinite integration. One of the constants stands all by itself, and the other multiplies the independent variable x.

Question 29-8

How can we solve the following elementary second-order ODE?

$$d^2y/dx^2 - 15x^3 = 2 \sin x$$

Answer 29-8

Let's begin by getting the equation into the standard form. We can add $15x^3$ to both sides, obtaining

$$d^2y/dx^2 = 2 \sin x + 15x^3$$

Integrating through with respect to x gives us

$$\int (d^2y/dx^2)\, dx = \int (2 \sin x + 15x^3)\, dx$$

The general antiderivatives are

$$\int (d^2y/dx^2)\, dx = dy/dx + c_1$$

and

$$\int (2 \sin x + 15x^3)\, dx = -2 \cos x + 15x^4/4 + c_2$$

where c_1 and c_2 are the constants of integration. Putting the general antiderivatives together into a single equation, we obtain

$$dy/dx + c_1 = -2 \cos x + 15x^4/4 + c_2$$

Subtracting c_1 from each side gives us

$$dy/dx = -2 \cos x + 15x^4/4 - c_1 + c_2$$

Letting the quantity $(-c_1 + c_2)$ be a consolidated constant p, we can simplify the above equation to

$$dy/dx = -2 \cos x + 15x^4/4 + p$$

When we integrate both sides, we get

$$\int (dy/dx) \, dx = \int (-2 \cos x + 15x^4/4 + p) \, dx$$

The general antiderivatives are

$$\int (dy/dx) \, dx = y + c_3$$

and

$$\int (-2 \cos x + 15x^4/4 + p) \, dx = -2 \sin x + 3x^5/4 + px + c_4$$

where c_3 and c_4 are new constants of integration (not necessarily the same as c_1 and c_2, which we got when we integrated the first time). Combining the right-hand sides of these equations, we obtain

$$y + c_3 = -2 \sin x + 3x^5/4 + px + c_4$$

Subtracting c_3 from each side yields

$$y = -2 \sin x + 3x^5/4 + px - c_3 + c_4$$

When we let the quantity $(-c_3 + c_4)$ be a single constant and call it q, we can simplify to obtain the solution

$$y = H(x) = -2 \sin x + 3x^5/4 + px + q$$

Question 29-9

Can we consolidate the constants p and q in the solution to the elementary second-order ODE that we found in Answer 29-8?

Answer 29-9

No, we can't. Within an expression or equation, we can only consolidate constants that apply to the same power or function of a single variable or set of variables. Here, p multiplies x to the first power, while q stands alone (that is, it multiplies x to the zeroth power).

Question 29-10

Suppose that we've derived a solution to an elementary first-order or second-order ODE. How can we test our solution to be sure that it's correct?

Answer 29-10

To check the solution to a first-order ODE, we can differentiate once. To check the solution to a second-order ODE, we can differentiate twice. If our solution is correct, the differentiation process will produce the original ODE, or else an equation that can be manipulated into the original ODE.

Final Exam

This exam is designed to test your general knowledge of calculus theory, not to measure how fast you can perform calculations. A good score is at least 120 answers correct. The answers are listed in App. D. This test is long, so don't try to take it in a single session. Feel free to draw diagrams, sketch graphs, or use a calculator. But don't look back at the text or refer to outside references.

1. Consider the following generalized cubic function:

$$f(x) = ax^3$$

where a is a nonzero real number. What is the derivative of this function?
(a) $f'(x) = ax^2/3$
(b) $f'(x) = ax^2$
(c) $f'(x) = 3ax^2$
(d) $f'(x) = 3ax$
(e) $f'(x) = ax/3$

2. If f and g are differentiable functions of a real variable x, which of the following statements is true in general?
(a) $\{g[f(x)]\}' = g'[f(x)] \cdot f'[g(x)]$
(b) $\{g[f(x)]\}' = g'[f(x)] \cdot [f'(x)]^2$
(c) $\{g[f(x)]\}' = g'[f(x)]/[f'(x)]^2$
(d) $\{g[f(x)]\}' = g'[f(x)] \cdot f'(x)$
(e) $\{g[f(x)]\}' = g'(x) \cdot f'[g(x)]$

3. Imagine a function whose graph shows up as a curve in rectangular xy-coordinates. Suppose we call the function

$$y = p(x)$$

Now suppose that we find the derivative of the function p at a particular point (x_0, y_0), and it turns out to be equal to -4. Which of the following statements describes this situation?

(a) The slope of a line tangent to the curve at (x_0, y_0) is equal to -4.

(b) The y-intercept of a line tangent to the curve at (x_0, y_0) is equal to -4.

(c) The value of y_0 is -4 times the value of x_0.

(d) The x-intercept of a line tangent to the curve at (x_0, y_0) is equal to -4.

(e) The value of x_0 is -4 times the value of y_0.

4. Which of the following types of functions is not continuous?

(a) A step function.

(b) A linear function.

(c) A constant function.

(d) An absolute-value function.

(e) A quadratic function.

5. Consider the function

$$f(x) = -x^2 + 3x - 2$$

graphed in rectangular coordinates. Which of the following ordered pairs, if any, reflects an inflection point in the curve?

(a) $(0,-2)$

(b) $(1,0)$

(c) $(3/2,1/4)$

(d) $(2,0)$

(e) None of the above

6. Which of these statements are true, and which are false?

- The limit of the sum of two expressions is equal to the sum of the limits of the expressions.
- The limit of the difference between two expressions is equal to the difference between the limits of the expressions (in the same order).
- A constant times the limit of an expression is equal to the limit of the expression times the constant.

(a) All three of the statements are true.

(b) The first statement is false, but the other two are true.

(c) The second statement is false, but the other two are true.

(d) The third statement is false, but the other two are true.

(e) All three of the statements are false.

7. In the situation shown by Fig. FE-1, the limit of $g(x)$, as x approaches any fixed nonzero real number k from the right, is equal to

(a) the limit of $g(x)$ as x approaches k from the left.

(b) twice the limit of $g(x)$ as x approaches k from the left.

(c) half the limit of $g(x)$ as x approaches k from the left.

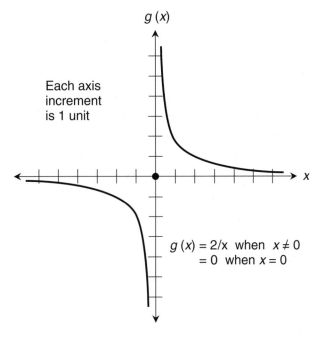

$g(x)$

Each axis
increment
is 1 unit

x

$g(x) = 2/x$ when $x \neq 0$
$= 0$ when $x = 0$

Figure FE-1 Illustration for Final Exam Questions 7
and 8.

(d) -1 times the limit of $g(x)$ as x approaches k from the left.

(e) twice the reciprocal of the limit of $g(x)$ as x approaches k from the left.

8. In the situation shown by Fig. FE-1, $g(x)$ is

(a) a function of x, defined and continuous for all real numbers x.

(b) a function of x, undefined at the point where $x = 0$, but continuous for all real numbers x.

(c) a function of x, defined for all real numbers x, but discontinuous at the point where $x = 0$.

(d) a function of x, but undefined and discontinuous for all real numbers x.

(e) not a function of x, because it fails the vertical-line test.

9. If we call the independent variable x and the we call the dependent variable y, then the slope of a line passing through two points is equal to

(a) the difference in the y values divided by the difference in the x values.

(b) the sum of the y values divided by the sum of the x values.

(c) the average of the y values divided by the average of the x values.

(d) the ratio of the y values times the ratio of the x values.

(e) the product of the y values times the product of the x values.

10. Suppose we're confronted with the following composite function, and we're told to differentiate it:

$$f(x) = \cos\left[\sin\left(\ln x\right)\right]$$

What strategy makes the most sense for tackling this problem?

(a) Apply the product rule for derivatives twice, and the chain rule once.

(b) Apply the product rule three times.

(c) Apply the chain rule twice.

(d) Apply the quotient rule for derivatives once, and the chain rule once.

(e) Give up! This function is nondifferentiable for all real numbers!

11. Which of the following statements is true for differentiable functions f and g of a single real variable z?

(a) $d(f-g)/dz = dg/dz - df/dz$

(b) $d(fg)/dz = (dg/dz)(df/dz)$

(c) $d(f+g)/dz = dg/dz + df/dz$

(d) $d(f/g)/dz = (dg/dz)/(df/dz)$

(e) $d[f(g)]/dz = d[g(f)]/dz$

12. Fill in the blank to make the following statement true: "To find the derivative of the reciprocal of a differentiable function, we must first find the derivative of the original function, then _____, and finally divide by the square of the original function."

(a) take the reciprocal

(b) multiply by -1

(c) add the original function

(d) subtract the original function

(e) take the absolute value

13. Consider the following function f of a real variable x, where a is a real-number constant:

$$f(x) = a/(3x^5)$$

The derivative of this function is

(a) $f'(x) = (-5a)/(3x^6)$, provided $x \neq 0$

(b) $f'(x) = 15ax^6$, provided $x \neq 0$

(c) $f'(x) = -15ax^{-6}$, provided $x \neq 0$

(d) $f'(x) = (-5/3)/(ax^4)$, provided $x \neq 0$

(e) impossible to calculate without more information.

14. Suppose that we evaluate certain limits involving a function $f(x)$ at the point where $x = 5$. When we do this, we find that the function is continuous at the point, while

$$\underset{\Delta x \to 0+}{Lim}\ [f(5 + \Delta x) - f(5)]/\Delta x = 2$$

and

$$\underset{\Delta x \to 0-}{Lim} \; [f(5 + \Delta x) - f(5)] \, / \, \Delta x = 8$$

What can we conclude about $f'(5)$?

(a) It's not defined.

(b) It's equal to 2.

(c) It's equal to 8.

(d) It's the arithmetic mean of 2 and 8, which is $(2 + 8)/2$, or 5.

(e) It's the geometric mean of 2 and 8, which is $(2 \cdot 8)^{1/2}$, or 4.

15. Consider the following derivative, where p, q, and r are nonzero real constants, and y is a real variable:

$$d/dy \, [p(qy + r)]$$

This derivative works out as

(a) $pqy + pr$

(b) $pq + pr$

(c) pqy

(d) pq

(e) q

16. Consider the following linear relation between two variables x and y:

$$3x + 6y = 4$$

When this relation is graphed as a function of x, the slope of the resulting straight line is

(a) 3.

(b) 6.

(c) 2/3.

(d) −1/2.

(e) −3/2.

17. Consider again the linear relation stated in Question 16. When this relation is graphed as a function of x, the y-intercept is

(a) 3.

(b) 6.

(c) 2/3.

(d) −1/2.

(e) −3/2.

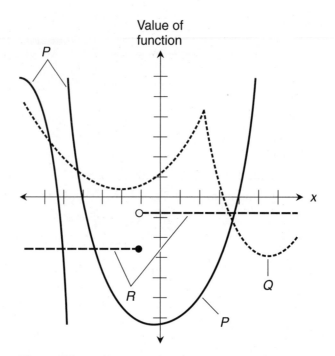

Figure FE-2   Illustration for Final Exam Questions 18 and 19.

18. Which graph or graphs in Fig. FE-2 contain a point at which the associated function is clearly discontinuous within the portions of the domain and range we can see?
    (a) Curve *P* only.
    (b) Curves *P* and *Q*, but not curve *R*.
    (c) Curve *Q* only.
    (d) Curves *P* and *R*, but not curve *Q*.
    (e) All three curves *P*, *Q*, and *R*.

19. Which graph or graphs in Fig. FE-2 contain a point at which the associated function is continuous but nondifferentiable within the portions of the domain and range we can see?
    (a) Curve *P* only.
    (b) Curves *P* and *Q*, but not curve *R*.
    (c) Curve *Q* only.
    (d) Curves *P* and *R*, but not curve *Q*.
    (e) All three curves *P*, *Q*, and *R*.

20. Which of the following functions is its own derivative?
    (a) $f_1(x) = \sin x$
    (b) $f_2(x) = -e^x$

(c) $f_3(x) = 2 \ln x$

(d) $f_4(x) = x^{-1}$

(e) $f_5(x) = x^4/4$

21. How can the following expression be precisely translated into words?

$$\underset{\Delta x \to 0+}{Lim}\ x^{-2} = +\infty$$

(a) As x approaches 0 from the positive direction, the limit of x^{-2} becomes equal to positive infinity if x gets small enough.

(b) As x approaches and then reaches 0 from the positive direction, x^{-2} approaches and then reaches positive infinity.

(c) As x starts out at 0 and then becomes positive, x^{-2} starts out at positive infinity and gets smaller.

(d) As x gets smaller and smaller but remains positive, x^{-2} gets larger and larger positively and endlessly.

(e) None of the above verbal expressions is technically correct.

22. Consider the following quadratic function:

$$f(x) = -8x^2 + 6x - 5$$

What is the x value of the extremum? Is the extremum an absolute maximum or an absolute minimum?

(a) The x value is 3/8 at an absolute maximum.

(b) The x value is 3/8 at an absolute minimum.

(c) The x value is −8 at an absolute maximum.

(d) The x value is −8 at an absolute minimum.

(e) The x value is 5, but it's a local extremum, and we can't tell whether it's a minimum or a maximum without more information.

23. Consider the following function of x, where the domain is the entire set of reals:

$$f(x) = (1/x^2) - 4$$

For what values of x, if any, is this function nondifferentiable because of a discontinuity?

(a) $x = 0$.

(b) $x = 2$.

(c) $x = 2$ and $x = 2$.

(d) $x = 4$.

(e) There are no discontinuities, so there is no reason to suspect that the function is nondifferentiable on that basis.

24. Consider the following function of x, where the domain is the entire set of reals:

$$f(x) = 1 / (x^2 - 4)$$

For what values of x, if any, is this function nondifferentiable because of a discontinuity?
(a) $x = 0$.
(b) $x = 2$.
(c) $x = -2$ and $x = 2$.
(d) $x = 4$.
(e) There are no discontinuities, so there is no reason to suspect that the function is nondifferentiable on that basis.

25. Consider the following function of x, where the domain is the entire set of reals:

$$f(x) = 1 / (x - 4)^2$$

For what values of x, if any, is this function nondifferentiable because of a discontinuity?
(a) $x = 0$.
(b) $x = 2$.
(c) $x = -2$ and $x = 2$.
(d) $x = 4$.
(e) There are no discontinuities, so there is no reason to suspect that the function is nondifferentiable on that basis.

26. Consider the graph of the function

$$y = -3 \tan x$$

with the domain restricted to $-\pi/2 < x < \pi/2$. What is the slope at the inflection point?
(a) 0
(b) 1
(c) 3
(d) −1
(e) −3

27. A relation is a mapping that can be expressed in terms of
(a) irrational numbers.
(b) rational numbers.
(c) quotients of integers.
(d) ordered pairs.
(e) extrema and inflection points.

28. Consider the following function of a real variable x:

$$f(x) = -87x^4 + 22x^3 - 11x^2 + 5x - 4$$

If we keep differentiating this function over and over, we will eventually get

(a) an exponential function.

(b) a logarithmic function.

(c) the sine function.

(d) the cosine function.

(e) the zero function.

29. Consider the following generalized linear function:

$$f(x) = ax$$

where a is a nonzero real number. What is the derivative of this function?

(a) $f'(x) = 0$

(b) $f'(x) = a$

(c) $f'(x) = a^2$

(d) $f'(x) = a^2/2$

(e) $f'(x) = x$

30. Suppose we have a function h that consists of a real variable z raised to a real-number power p, and then multiplied by a rational-number constant b. The derivative of this function can be expressed as

$$h'(z) = bpz^{(p-1)}$$

There is an important restriction to keep in mind, however. This derivative is not defined if $z = 0$ and

(a) $p > 2$.

(b) $p > 1$.

(c) $p \le 1$.

(d) $b = 0$.

(e) $b < 0$.

31. Consider the following derivative, where p and q are nonzero real constants, and z is a real variable:

$$d/dz \, [(pz + q)(qz - p)]$$

This derivative works out as

(a) $z(pq - q + p)$

(b) $2pqz - p^2 + q^2$

(c) $z^2 + p^2 - q^2$

(d) $p^2 + q^2 + z$

(e) $p + q - z$

32. Consider the following function of a real variable x:

$$f(x) = 7x^4 + 12x^3 - 71x^2 + 65x + 84 + 42e^x$$

If we keep differentiating this function over and over, we will eventually get

(a) an exponential function.

(b) a logarithmic function.

(c) the sine function.

(d) the cosine function.

(e) the zero function.

33. Suppose we have two differentiable functions g and h of a single variable y. If we add the functions and then take the derivative of their sum with respect to y, we get the same result as when we

(a) differentiate the functions individually with respect to y, and then add the derivatives.

(b) differentiate the functions individually with respect to y, then add the derivatives, and finally divide by 2.

(c) differentiate the functions individually with respect to y, then add the derivatives, and finally divide by the derivative of the sum of the original functions.

(d) differentiate the functions individually with respect to y, and then multiply the derivatives.

(e) differentiate the functions individually with respect to y, then multiply the derivatives, and finally divide by the derivative of the sum of the original functions.

34. In Fig. FE-3, each vertical axis division is 1 unit. Both curves are sinusoids. Knowing these things, we can reasonably conclude that the dashed curve shows

(a) the first derivative of the function portrayed by the solid curve.

(b) the second derivative of the function portrayed by the solid curve.

(c) the third derivative of the function portrayed by the solid curve.

(d) the fourth derivative of the function portrayed by the solid curve.

(e) None of the above

35. For an object in free fall, neglecting air resistance, the second derivative of the vertical fallen distance with respect to time is

(a) the vertical speed.

(b) the vertical jerk.

(c) the first derivative of the vertical speed with respect to time.

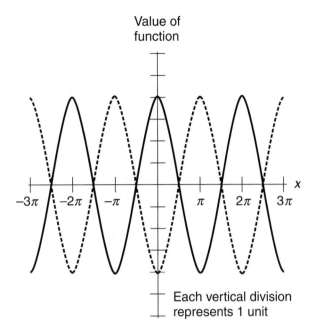

Figure FE-3 Illustration for Final Exam Question 34.

    (d) the first derivative of the vertical acceleration with respect to time.

    (e) the first derivative of the vertical jerk with respect to time.

36. Suppose you have a function whose graph shows up as a straight line in the rectangular xy-plane, where x is the independent variable and y is the dependent variable. Further, suppose that the line does not pass through the origin (0,0). The derivative of the function at a specific point (x_0, y_0) on the line is equal to

    (a) the y-intercept divided by the x-intercept.

    (b) the x-intercept divided by the y-intercept.

    (c) y_0 / x_0.

    (d) x_0 / y_0.

    (e) None of the above

37. Suppose we encounter a function h that takes a variable z to a positive-integer power n larger than 3, and then multiplies the result by a nonzero real constant k. How can we express the first derivative of h with respect to z?

    (a) $h'(z) = nk^n z^{(n-1)}$

    (b) $h'(z) = nkz^{(n-1)}$

    (c) $h'(z) = nz^{(n-1)}$

    (d) $h'(z) = nkz^n$

    (e) $h'(z) = nz^n$

38. Suppose we encounter a function g that multiplies a variable y by a nonzero real constant c, and then take the product cy to a positive-integer power m larger than 3. How can we express the first derivative of g with respect to z?

    (a) $dg/dy = mc^m y^{(m-1)}$

    (b) $dg/dy = mcy^{(m-1)}$

    (c) $dg/dy = my^{(m-1)}$

    (d) $dg/dy = mcy^m$

    (e) $dg/dy = cy^m$

39. Mathematical induction is a technique that allows us to

    (a) prove a general statement true if we can prove a large enough number of specific examples true.

    (b) prove, in effect, infinitely many statements in a few steps.

    (c) prove that if A implies B, then "not A" implies "not B."

    (d) prove that if a theorem holds for all the rational numbers, then it holds for all the real numbers.

    (e) prove that if a theorem holds for all the rational numbers, then it holds for all the integers.

40. A function is a relation in which every element in the domain maps to

    (a) one or more elements in the range.

    (b) at most one element in the range.

    (c) another element in the domain.

    (d) itself.

    (e) a single, constant element in another relation.

41. Consider the following generalized quadratic function:

    $$f(x) = ax^2$$

    where a is a nonzero real number. What is the derivative of this function?

    (a) $f'(x) = a/2$

    (b) $f'(x) = a$

    (c) $f'(x) = 2a$

    (d) $f'(x) = a^2/2$

    (e) $f'(x) = 2ax$

42. Which of the following functions has an inverse relation that is not a function? Assume in each case that the domain of the original function is the entire set of reals.

    (a) $f_1(x) = 6x + 4$

    (b) $f_2(x) = x^3$

    (c) $f_3(x) = -5x^5$

    (d) $f_4(x) = \ln x$

    (e) $f_5(x) = 4x^2$

43. Which of the following expressions describes how the chain rule can be applied twice when evaluating a function *h* of a function *g* of a function *f*, assuming that all three functions are differentiable over their domains?

(a) $\{h[g(f)]\}' = h'[g(f)]$

(b) $\{h[g(f)]\}' = h'[g(f)] \cdot g'(f)$

(c) $\{h[g(f)]\}' = h'(g) \cdot g'(f) \cdot h'(f)$

(d) $\{h[g(f)]\}' = h'[g(f)] \cdot f'$

(e) $\{h[g(f)]\}' = h'(g) \cdot g'(f) \cdot f'$

44. To find the derivative of the quotient of two differentiable functions, we multiply the derivative of the first function by the second function, then multiply the derivative of the second function by the first function, then subtract the second product from the first product, and finally

(a) multiply by the second function.

(b) multiply by the square of the second function.

(c) divide by the second function.

(d) divide by the square of the second function.

(e) divide by the derivative of the second function.

45. Figure FE-4 shows the graph of a cubic function. At which points, or in which regions, is dy/dx positive?

(a) Region *P* (to the left of point *Q*).

(b) The region between points *Q* and *S*.

(c) Point *R* only.

(d) Points *Q* and *S* only.

(e) Region *T* (to the right of point *S*).

46. In Fig. FE-4, at which points, or in which regions, is dy/dx equal to 0?

(a) Region *P* (to the left of point *Q*).

(b) The region between points *Q* and *S*.

(c) Point *R* only.

(d) Points *Q* and *S* only.

(e) Region *T* (to the right of point *S*).

47. In Fig. FE-4, at which points, or in which regions, is d^2y/dx^2 equal to 0?

(a) Region *P* (to the left of point *Q*).

(b) The region between points *Q* and *S*.

(c) Point *R* only.

(d) Points *Q* and *S* only.

(e) Region *T* (to the right of point *S*).

48. Consider the following function of a real variable *x*:

$$f(x) = -87x^4 + 22x^3 - 11x^2 + 5x - 4 + 3x^{-1}$$

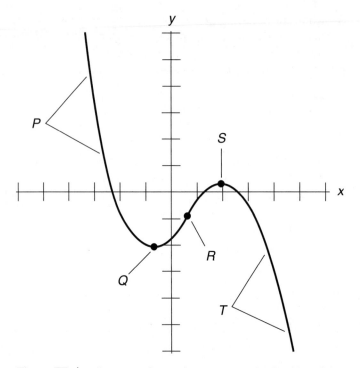

Figure FE-4   Illustration for Final Exam Questions 45, 46, and 47.

This function is differentiable over

(a) the entire set of real numbers.

(b) the entire set of reals except 0.

(c) only the set of positive reals.

(d) only the set of nonnegative reals.

(e) only the set of negative reals.

49. Once again, consider the function

$$f(x) = -87x^4 + 22x^3 - 11x^2 + 5x - 4 + 3x^{-1}$$

What is the fifth derivative of f with respect to x?

(a) $f^{(5)}(x) = 3e^x$

(b) $f^{(5)}(x) = 3 \ln x$

(c) $f^{(5)}(x) = 0$

(d) $f^{(5)}(x) = -360/x^6$

(e) Nothing. This function can't be differentiated five times.

50. If f and g are differentiable functions of the same variable, and if a and b are constants, then which of the following statements is *not* true in general?
    (a) $(af)' = a(f')$
    (b) $(f+g)' = (g+f)'$
    (c) $(f-g)' = (g-f)'$
    (d) $(af+bg)' = a(f') + b(g')$
    (e) $(fg)' = (gf)'$

51. What is the general resolution of the following indefinite integral? Call the constant of integration c.

$$\int 4t^{-1}\, dt$$

    (a) $-4t^{-2} + c$
    (b) $-2t^{-2} + c$
    (c) $4 \ln |t| + c$
    (d) $-4 \ln |t| + c$
    (e) This integral can't be resolved.

52. Suppose that f_1 and f_2 are integrable functions of x over a continuous interval (a,b). Let k_1 and k_2 be real-number constants. Then

$$\int_a^b \{k_1\,[f_1(x)] + k_2\,[f_2(x)]\}\, dx = k_1 \int_a^b f_1(x)\, dx + k_2 \int_a^b f_2(x)\, dx$$

    What, if anything, is wrong with this formula?
    (a) Nothing is wrong.
    (b) The constants to the right of the equals sign, appearing in front of the integral symbols, should both be $k_1 k_2$.
    (c) The constants to the right of the equals sign, appearing in front of the integral symbols, should both be $(k_1 + k_2)$.
    (d) The constants to the right of the equals sign, appearing in front of the integral symbols, should both be $(k_1^2 + k_2^2)^{1/2}$.
    (e) The integrands to the right of the equals sign should be $f_1'(x)$ in the first integral and $f_2'(x)$ in the second integral.

53. The principle of linearity for definite integrals works with any finite linear combination of functions and constants, as long as
    (a) we stay with the same variable.
    (b) we stay with the same interval.
    (c) each function can be integrated over the entire interval.
    (d) we're sure that (a), (b), and (c) are all true.
    (e) all the constants are the same.

54. Fill in the blank to make the following sentence true: "If we integrate a function over an interval from the minimum to the maximum, we get _____ the integral of the same function from the maximum to the minimum."

    (a) twice

    (b) half

    (c) the same result as

    (d) −1 times

    (e) the −1 power of

55. Refer to Fig. FE-5. What does the following expression represent?

$$\sum_{i=1}^{n} \Delta x \cdot f(a + i\Delta x)$$

    (a) The actual area defined by the curve.

    (b) The sum of the areas of the rectangles.

    (c) The sum of the perimeters of the rectangles.

    (d) The perimeter of the ith rectangle.

    (e) The area of the ith rectangle.

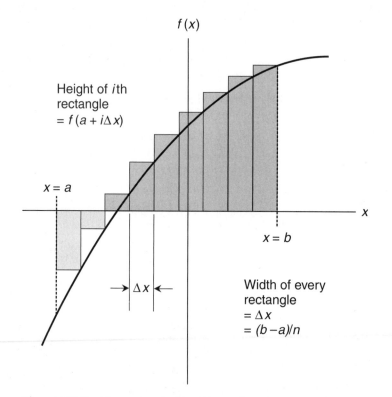

Figure FE-5   Illustration for Final Exam Questions 55 and 56.

56. We can rewrite the expression stated in Question 55 to get

$$\sum_{i=1}^{n} [(b-a)/n] \cdot f[a+i(b-a)/n]$$

What fact, evident from Fig. FE-5, allows us to do this?

(a) The area defined by the curve is equal to the sum of the areas of the rectangles.

(b) The height of the ith rectangle is equal to $f(a+i\Delta x)$.

(c) The number of rectangles approaches infinity.

(d) The width of each rectangle is equal to $b-a$.

(e) The increment Δx is equal to $(b-a)/n$.

57. The principle of integration by parts tells us that if f and g are integrable and differentiable functions of a variable x, then

$$\int f(x) \cdot g'(x)\ dx = f(x) \cdot g(x) - \int g(x) \cdot f'(x)\ dx$$

Now look at the following equation, which is meant to show an example of integration by parts. Is anything wrong here? If so, what can be done to make it right?

$$\int xe^x\ dx = xe^x - \int x\ dx$$

(a) Nothing is wrong.

(b) The integrand after the second integral symbol should be e^x, not x.

(c) The integrand after the second integral symbol should be $x^2/2$, not x.

(d) The expression after the equals sign should be $x^2 e^x$, not xe^x.

(e) The expression after the equals sign should be $x \ln x$, not xe^x.

58. What is the value of the following sum of definite integrals?

$$\int_0^1 2x\ dx + \int_1^0 2x\ dx$$

(a) 0

(b) 1

(c) −1

(d) j, the positive square root of −1

(e) −j, the negative square root of −1

59. Suppose we encounter the indefinite integral

$$\int [(5x-5)(x^2+x-6)^{-1}]\ dx$$

We can use the technique of partial fractions to split the above integrand into a sum of two manageable integrands. Those two integrands are

(a) $4(x-2)^{-1}$ and $2(x+3)^{-1}$.

(b) $(x-3/4)^{-1}$ and $16(x+2)^{-1}$.

(c) $(x-2)^{-1}$ and $4(x+3)^{-1}$.

(d) $4(x+2/3)^{-1}$ and $2(x-3/2)^{-1}$.

(e) $4(x-2/3)^{-1}$ and $4(x-3)^{-1}$.

60. If we want to evaluate the integral in Question 59 over the interval (0,5), we must take note of the fact that

(a) there's a singularity at $x=3$.

(b) there's a singularity at $x=3/2$.

(c) there's a singularity at $x=3/4$.

(d) there's a singularity at $x=2/3$.

(e) there's a singularity at $x=2$.

61. Suppose we want to find the true geometric area between the graphs of the following two functions:

$$f(x) = \sin x$$

and

$$g(x) = 2x/\pi$$

We can do this by integrating a certain function with respect to x, over the interval between the points where the graphs intersect. What are the x values of those two points?

(a) -1 and 1.

(b) $-\pi$ and π.

(c) 0 and 1.

(d) 0 and π.

(e) 1 and π.

62. If $h(z) = 0$, then which of the following is an antiderivative of h?

(a) 5

(b) $5z$

(c) $5z^2$

(d) $5z^{-1}$

(e) All of the above

63. Suppose that $g(z)$ is a function that is integrable over a certain interval. Someone tells us that the following statement is true for all possible values of p, q, and r in the interval:

$$\int_p^q g(z)\, dz + \int_p^r g(z)\, dz = \int_q^r g(z)\, dz$$

What, if anything, is wrong with this statement? If something is wrong with it, how can it be changed to make it correct?

(a) Nothing is wrong with this statement. It's true as it stands.

(b) The bounds attached to the integral symbols are wrong. We can make the statement true by changing the lower bound of the second integral to q and changing the lower bound of the third integral to p.

(c) The bounds attached to the integral symbols are wrong. We can make the statement true by reversing the bounds of the third integral.

(d) The bounds attached to the integral symbols are wrong. We can make the statement true by reversing the bounds of the second integral.

(e) The addition operation is wrong. We can make the statement true by changing it to subtraction.

64. Suppose we have a function $f(x)$ that contains no singularities, and is defined and continuous over an interval from $x = a$ to $x = b$. The formula for the length of the arc between these two points is

$$L = \int_{a}^{b} \{1 + [f'(x)]^2\}^{1/2} \, dx$$

If we want to apply this formula in any particular situation, which of the following would we most likely want to do first?

(a) Differentiate f with respect to x.

(b) Antidifferentiate f with respect to x.

(c) Square the function f.

(d) Add 1 to the function f.

(e) Take the positive square root of the function f.

65. Consider the function

$$g(x) = (x^2 - 3x + 2)^{-1}$$

Suppose we want to calculate the arc lengths along the graph of this function over various intervals. Which, if any, of the following intervals won't work with the formula for arc length stated in Question 64?

(a) $-2 < x < 0$

(b) $-2 < x < 2$

(c) $3 < x < 4$

(d) $1/2 < x < 3/4$

(e) $-1/2 < x < 1/2$

66. Suppose we encounter this mathematical expression.

$$\sum_{i=1}^{n} a_i = b$$

How would we say or write this in words?

(a) The sequence of terms a_i, from $i = 1$ to n, is equal to b.

(b) The average of the terms a_i, from $i = 1$ to n, is equal to b.

(c) The sum of the terms a_i, from $i = 1$ to n, is equal to b.

(d) The integral of the terms a_i, from $i = 1$ to n, is equal to b.

(e) None of the above

67. Let's look again at a situation we saw a little while ago (Question 61). We want to find the true geometric area between the graphs of

$$f(x) = \sin x$$

and

$$g(x) = 2x/\pi$$

We can do this by integrating a certain function with respect to x, over the interval between the points where the graphs intersect. In this situation,

(a) the integrand is $\sin x - 2x/\pi$.

(b) the integrand is $\sin x + 2x/\pi$.

(c) the integrand is $-\cos x - 2x/\pi$.

(d) the integrand is $-\cos x + 2x/\pi$.

(e) the integrand is $\cos x - 2x/\pi$.

68. If $g(t) = 7$, then which of the following is an antiderivative of g?

(a) $7t$

(b) $7t - 7$

(c) $7t + 7$

(d) $(9 + 5)(t/2) - 16$

(e) All of the above

69. If we consult a comprehensive table of indefinite integrals, we'll find

$$\int (ax + b)^{-1} \, dx = a^{-1} \ln |ax + b| + c$$

where a and b are constants, and x is the variable. This formula can come in handy when we want to use the technique of

(a) integration by parts.

(b) integration by linear differentiation.

(c) Riemann integration.

(d) improper integration.

(e) integration by partial fractions.

70. Figure FE-6 shows the graph of a function $h(x)$, along with a region defined by the curve within the interval $(-1,1)$ that contains a singularity at $x = 0$. Suppose we want to evaluate the definite integral

$$\int_{-1}^{1} h(x) \, dx$$

What's the correct way to do this?

(a) Find $H(x)$, and then subtract $H(-1)$ from $H(1)$.

(b) Find, if possible, the improper integral of $H(x)$ from $x = -1$ to $x = 0$. Then find, if possible, the improper integral of $H(x)$ from $x = 0$ to $x = 1$. Finally, assuming both integrals are defined and finite, add them.

(c) Find, if possible, the improper integral of $H(x)$ from $x = -1$ to $x = 0$. Then find, if possible, the improper integral of $H(x)$ from $x = 0$ to $x = 1$. Finally, assuming both integrals are defined and finite, subtract them from each other and take the absolute value of the result.

(d) Find, if possible, the improper integral of $H(x)$ from $x = -1$ to $x = 0$. Then find, if possible, the improper integral of $H(x)$ from $x = 0$ to $x = 1$. Finally, assuming both integrals are defined and finite, multiply them.

(e) Give up! It's impossible to evaluate this integral.

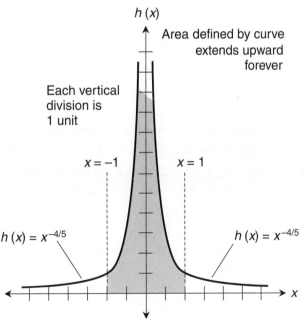

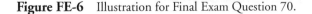

Figure FE-6   Illustration for Final Exam Question 70.

71. If $f(v) = -4v + 8$, which of the following is *not* an antiderivative of f?

    (a) $-2v^2 + 8v$

    (b) $-2v^2 + 8v + 8$

    (c) $-2v^2 + 8v - \pi$

    (d) $-2v^2 + 8$

    (e) $-2v^2 + 8v + \ln \pi$

72. Consider the definite integral

$$\int_a^b f(x)\, dx$$

Suppose that f blows up at the upper bound b. To approximate the value of this integral, we can imagine a tiny positive number ε and evaluate

$$\int_a^{b-\varepsilon} f(x)\, dx$$

We can make ε smaller and smaller, always keeping it positive, and we'll be able to get as close to the actual value of the integral from a to b as we want, but only if

    (a) the actual integral from a to b is positive.

    (b) $f(b)$ is positive.

    (c) the actual integral from a to b is finite.

    (d) the actual integral from a to b is equal to 0.

    (e) $f(b)$ is defined.

73. Imagine a car in a drag race. We start a timer as soon as the car begins the race. The total distance that the car has traveled from the starting point at any given instant in time can be found by

    (a) integrating the acceleration with respect to time.

    (b) integrating the speed with respect to time.

    (c) differentiating the acceleration with respect to time.

    (d) differentiating the speed with respect to time.

    (e) no known means.

74. How can we simplify the last expression in the following equation and have it remain valid in theoretical terms? In this example, y is the variable, and c_1, c_2, and c_3 are constants of integration.

$$\int (5y^{2/3} - y^{-1/2} - 4y^{-2})\, dy$$
$$= \int 5y^{2/3}\, dy + \int -y^{-1/2}\, dy + \int -4y^{-2}\, dy$$
$$= 3y^{5/3} + c_1 - 2y^{1/2} + c_2 + 4y^{-1} + c_3$$

(a) We can consolidate c_1, c_2, and c_3 into a single constant of integration c that's equal to $c_1/3 + c_2/3 + c_3/3$.

(b) We can consolidate c_1, c_2, and c_3 into a single constant of integration c that's equal to $c_1 c_2 c_3$.

(c) We can consolidate c_1, c_2, and c_3 into a single constant of integration c that's equal to $c_1 + c_2 + c_3$.

(d) We can get rid of the constants altogether.

(e) We can't simplify the last expression any further.

75. The sum rule for definite integrals works when we add up any finite number of functions, but only under certain conditions. Which conditions?

(a) All the functions must be integrable over the interval.

(b) All the integrals must be done in the same direction.

(c) The interval of integration must be the same for each function.

(d) All three conditions (a), (b), and (c) must hold true.

(e) None of the above conditions have to be true.

76. We want to evaluate the following integral. Why can't we use the Fundamental Theorem of Calculus directly over the interval (2,4) in an attempt to solve this problem?

$$\int_2^4 [(3z^2 + 2z - 6 + (z-3)^{-1/3}] \, dz$$

(a) The interval (2,4) is too narrow.

(b) The integrand function is not continuous for $z < 2$.

(c) The integrand function is not continuous for $z > 4$.

(d) The integrand function contains a singularity in the interval (2,4).

(e) We can!

77. Let's look, for the third time, at the situation we saw in Questions 61 and 67. We want to find the true geometric area between the graphs of

$$f(x) = \sin x$$

and

$$g(x) = 2x/\pi$$

We can do this by evaluating a certain expression over the interval between the x values of the points where the graphs intersect. In this situation,

(a) the expression is $\sin x - x^2/\pi$, evaluated from -1 to 1.

(b) the expression is $\sin x + x^2/\pi$, evaluated from $-\pi$ to π.

(c) the expression is $-\cos x - x^2/\pi$, evaluated from 0 to 1.

(d) the expression is $-\cos x + x^2/\pi$, evaluated from 0 to π.

(e) the expression is $\cos x - x^2/\pi$, evaluated from 1 to π.

78. Which of the following functions contains a singularity?
    (a) $f_1(x) = x^2$
    (b) $f_2(x) = e^x$
    (c) $f_3(x) = 3x^2 + 2x$
    (d) $f_4(x) = (5 + x)^{-1}$
    (e) All four of the above

79. Consider the sum of definite integrals

$$\int_3^5 7x^2 \, dx + \int_5^3 x^{2/3} \, dx$$

Can we rewrite this expression so that the sum rule can be used to convert it into a single definite integral of a polynomial function? If so, how?
    (a) We can't do it.
    (b) We can transpose the bounds of the first integral, and leave everything else the same.
    (c) We can transpose the bounds of the second integral, and leave everything else the same.
    (d) We can transpose the bounds of the first integral, and change the second integrand to $-x^{2/3}$.
    (e) We can transpose the bounds of the second integral, and change the second integrand to $-x^{2/3}$.

80. Suppose that we have two integrable functions f and g such that f operates on x, while g operates on $f(x)$. If we see an integral of the form

$$\int_a^b g[f(x)] \cdot f'(x) \, dx$$

where a and b are real numbers, then the rule of substitution lets us rewrite it as

$$\int_{f(a)}^{f(b)} g(y) \, dy$$

In the second integral, y is the equivalent of
    (a) $f(x)$ in the first integral.
    (b) $f'(x)$ in the first integral.
    (c) $f'(x) \, dx$ in the first integral.
    (d) $g'(x)$ in the first integral.
    (e) $g'(x) \, dx$ in the first integral.

81. In the second integral above (Question 80), dy is the equivalent of
    (a) $f(x)$ in the first integral.
    (b) $f'(x)$ in the first integral.

(c) $f'(x)\,dx$ in the first integral.

(d) $g'(x)$ in the first integral.

(e) $g'(x)\,dx$ in the first integral.

82. Suppose a, b, c, and d are real numbers and $a < b < c < d$. Consider the following sum of definite integrals:

$$\int_a^b 3x^5\,dx + \int_c^d 3e^x\,dx$$

Both of these integrals are defined over any interval in the set of reals, because both of the integrands are defined and continuous over the entire set of reals. Now suppose that we want to consolidate the above expression into a single definite integral in which the integrand is $(3x^5 + 3e^x)$. How should we set the bounds of integration?

(a) We can't, based on the information given here.

(b) The lower bound should be a, and the upper bound should be c.

(c) The lower bound should be a, and the upper bound should be d.

(d) The lower bound should be b, and the upper bound should be c.

(e) The lower bound should be b, and the upper bound should be d.

83. Suppose we want to find the true geometric area between the graphs of the following two functions:

$$f(x) = -e^x$$

and

$$g(x) = (1 - e)x - 1$$

We can do this by integrating a certain function with respect to x, over the interval between the points where the graphs intersect. What are the x values of those two points?

(a) -1 and 1.

(b) $-e$ and e.

(c) 1 and e.

(d) 0 and e.

(e) 0 and 1.

84. Consider a function f, a variable x, a constant a, and an integer n. Imagine that

$$f(x) = ax^n$$

Now suppose we're told that the general antiderivative of f with respect to x is

$$F(x) = ax^{(n+1)}/(n+1) + c$$

where c is the constant of integration. This formula won't work

(a) if $n = -1$.

(b) if $n < -1$.

(c) if n is not a rational number.

(d) if n is not an integer.

(e) in any case, because it's wrong!

85. Figure FE-7 shows the graphs of two functions in Cartesian coordinates. The true geometric area of the region enclosed by the line and the curve is

(a) positive, because most of it lies above the independent-variable axis.

(b) positive, because most of it lies to the right of the dependent-variable axis.

(c) positive, because true geometric areas are never negative.

(d) partly positive and partly negative.

(e) impossible to define, because it does not all lie in a single quadrant of the coordinate plane.

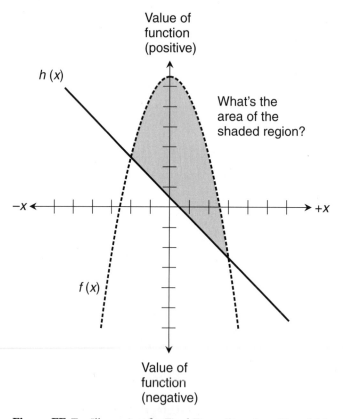

Figure FE-7   Illustration for Final Exam Questions 85 and 86.

86. To determine the true geometric area enclosed by the graphs in Fig. FE-7, we can
    (a) integrate the function $f(x) - h(x)$ over the interval between the points where the line and the curve intersect.
    (b) integrate the function $h[f(x)]$ over the interval between the points where the line and the curve intersect.
    (c) integrate the function $f(x) + h(x)$ over the interval between the points where the line and the curve intersect.
    (d) integrate the function $[f(x) + h(x)]/2$ over the interval between the points where the line and the curve intersect.
    (e) do nothing, because the region does not all lie in a single quadrant of the coordinate plane.

87. Suppose we want to evaluate the definite integral

$$\int_0^1 (x^{-1/7} + x^{-1/3})\, dx$$

We decide to go through the following four steps to solve this problem.

- Split the original integral into two separate integrals of monomial functions.
- Integrate those two monomial functions independently.
- If either of the monomial-function integrals is undefined, conclude that the entire integral is undefined.
- If both of the monomial-function integrals are defined, add them to get the value of the entire integral.

What, if anything, makes this process incorrect, incomplete, or otherwise unworkable? If anything is wrong with it, what can we do to make it work?

    (a) The first and second steps are unworkable. We can't split this integral into a sum of two other integrals, because the exponents aren't integers. We must therefore tackle the integrand as a single polynomial.
    (b) The third step is wrong. The entire integral is undefined only if both of the monomial-function integrals are undefined.
    (c) The fourth step is wrong. We can't add the two monomial-function integrals; we must find their arithmetic mean.
    (d) The fourth step is wrong. We can't add the two monomial-function integrals; we must find their geometric mean.
    (e) Nothing is wrong with this process.

88. Let's look again at a situation we saw a little while ago (Question 83). We want to find the true geometric area between the graphs of

$$f(x) = -e^x$$

and

$$g(x) = (1 - e)x - 1$$

We can do this by integrating a certain function with respect to x, over the interval between the points where the graphs intersect. In this situation,

(a) the integrand is $e^x - x + ex + 1$.

(b) the integrand is $-e^x + x + ex + 1$.

(c) the integrand is $e^x - x + ex - 1$.

(d) the integrand is $-e^x - x + ex + 1$.

(e) the integrand is $e^x - x - ex - 1$.

89. Suppose we encounter an integral of the form

$$\int_{-\infty}^{0} h(x)\, dx$$

Which of the following statements can we make about the value of this integral without knowing anything about h except the fact that $h(0)$ is defined?

(a) It's equal to 0.

(b) It's negative.

(c) It's positive.

(d) It's a real number.

(e) None of the above

90. Suppose we encounter an integral of the form

$$\int_{-\infty}^{\infty} f(z)\, dz$$

Which of the following statements can we make about the value of this integral without knowing anything about f except the fact that $f(0)$ is defined?

(a) It's equal to 0.

(b) It's infinite.

(c) It's finite.

(d) It's not a real number.

(e) None of the above

91. Suppose we encounter an integral of the form

$$\int_{-1}^{1} g(t)\, dt$$

Which of the following statements can we make about the value of this integral without knowing anything about g except the fact that $g(-1)$ and $g(1)$ are both defined?

(a) It's equal to 0.

(b) It's finite.

(c) It's infinite.

(d) It's an integer.

(e) None of the above

92. Suppose we encounter an integral of the form

$$\int_{-1}^{-1} P(v)\ dv$$

Which of the following statements can we make about the value of this integral without knowing anything about P except the fact that $P(-1)$ is defined?

(a) It's equal to 0.

(b) It's a positive real number.

(c) It's undefined.

(d) It's a negative real number.

(e) None of the above

93. Consider a normal probability distribution $P(x)$ whose graph has a left-hand extreme at $x = p$ and a right-hand extreme at $x = q$. Suppose we find $x = \varphi$ such that $p < \varphi < q$. Suppose we also have

$$\int_{p}^{\varphi} P(x)\ dx = 1/2$$

and

$$\int_{\varphi}^{q} P(x)\ dx = 1/2$$

What does φ represent?

(a) The least upper bound of P.

(b) The standard deviation of P.

(c) The deviation of P.

(d) The greatest lower bound of P.

(e) None of the above

94. In the situation shown by the graph of Fig. FE-8, it's tempting to think that we can integrate $f(x)$ and $g(x)$ separately over the interval between the points P and Q where the curves intersect, take the absolute values of both integrals, and then add those two absolute values to get the true geometric area between the curves. Can we do this and be sure we'll get the correct result?

(a) Yes, but only if P and Q are equally far from the dependent-variable axis.

(b) Yes, but only if P and Q are both exactly on the independent-variable axis.

(c) Yes, but only if P and Q are equally far from the independent-variable axis.

(d) Yes, but only if $f(x)$ and $g(x)$ are both quadratic functions.

(e) No! Never.

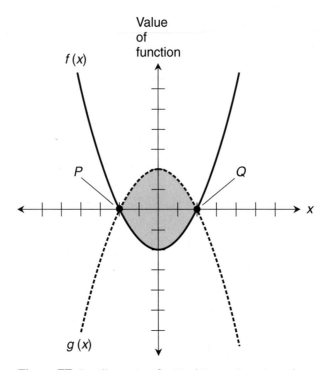

Figure FE-8   Illustration for Final Exam Question 94.

95. Let's look, for the third time, at the situation we saw in Questions 83 and 88. We want to find the true geometric area between the graphs of

$$f(x) = -e^x$$

and

$$g(x) = (1 - e)x - 1$$

We can do this by evaluating a certain expression over the interval between the *x* values of the points where the graphs intersect. In this situation,

(a) the expression is $e^x - x^2/2 + ex^2/2 - x$, evaluated from 0 to *e*.

(b) the expression is $-e^x - x^2/2 + ex^2/2 + x$, evaluated from 0 to 1.

(c) the expression is $-e^x - x^2/2 - ex^2/2 + x$, evaluated from $-e$ to *e*.

(d) the expression is $-e^x - x^2/2 + ex^2/2 + x$, evaluated from -1 to 1.

(e) the expression is $-e^x - x^2/2 - ex^2/2 - x$, evaluated from 1 to *e*.

96. Imagine that we have various functions of a variable *x*, and all of those functions can be integrated with respect to *x*. Which of the following statements is false?

    (a) The indefinite integral of the negative of a function is equal to the negative of the indefinite integral of the function.

    (b) The indefinite integral of the difference between two functions is equal to the difference between the indefinite integrals of the functions, as long as the order of subtraction stays the same.

    (c) The indefinite integral of the product of two functions is equal to the product of the indefinite integrals of the functions.

    (d) The indefinite integral of the sum of two functions is equal to the sum of the indefinite integrals of the functions, as long as the order of addition stays the same.

    (e) The indefinite integral of the sum of two functions is equal to the sum of the indefinite integrals of the functions, whether the order of addition stays the same or not.

97. In the graph of Fig. FE-9, the length of the chord labeled with Δx and Δy is

    (a) $\Delta y / \Delta x$.

    (b) $(\Delta y + \Delta x)^{1/2}$.

    (c) $(\Delta y - \Delta x)^{1/2}$.

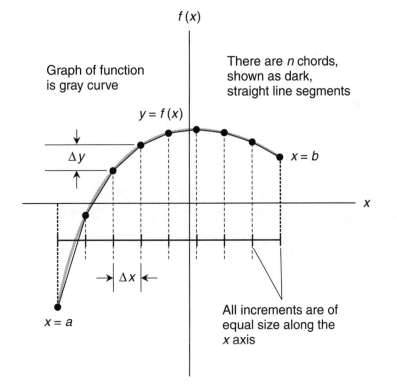

Figure FE-9   Illustration for Final Exam Questions 97, 98, and 99.

(d) $(\Delta y^2 + \Delta x^2)^{1/2}$.

(e) $(\Delta y^2 - \Delta x^2)^{1/2}$.

98. In Fig. FE-9, the law of the mean tells us *specifically* that

   (a) the length of the arc segment labeled with Δx and Δy is greater than or equal to the length of the corresponding chord.

   (b) the length of the arc over the interval from $x = a$ to $x = b$ is greater than or equal to the sum of the lengths of the chords connecting all the points over that interval.

   (c) there is a point on the chord labeled with Δx and Δy at which the slope of the chord is equal to the length of the corresponding arc segment.

   (d) there is a point on the arc segment labeled with Δx and Δy at which the derivative of the function representing the arc equals the slope of the corresponding chord.

   (e) as the number of chords n increases, the sum of the chord lengths over the interval from $x = a$ to $x = b$ approaches the true arc length over that interval.

99. Which of the following statements concerning Fig. FE-9 is *not necessarily true?*

   (a) the length of the arc segment labeled with Δx and Δy is greater than or equal to the length of the corresponding chord.

   (b) the length of the arc over the interval from $x = a$ to $x = b$ is greater than or equal to the sum of the lengths of the chords connecting all the points over that interval.

   (c) there is a point on the chord labeled with Δx and Δy at which the slope of the chord is equal to the length of the corresponding arc segment.

   (d) there is a point on the arc segment labeled with Δx and Δy at which the derivative of the function representing the arc equals the slope of the corresponding chord.

   (e) as the number of chords n increases, the sum of the chord lengths over the interval from $x = a$ to $x = b$ approaches the true arc length over that interval.

100. If we antidifferentiate the zero function twice with respect to an independent variable x, we can, in theory, get any of the following *except*

   (a) 0

   (b) 2

   (c) $2x$

   (d) $2x + 2$

   (e) $2x^2 + 2$

101. What type of curve does the following equation represent?

$$11x^2 + 2y^2 = 22$$

   (a) We need more information to answer this.

   (b) It's a parabola.

   (c) It's a circle.

   (d) It's an ellipse.

   (e) It's a hyperbola.

102. Look again at the equation in Question 101. What is y' when $x = 11$?

    (a) It's not defined.
    (b) It's equal to 0.
    (c) It's equal to $\pm 2^{1/2}$.
    (d) It's equal to $\pm 11^{1/2}$.
    (e) It's equal to $-(11/2)^{1/2}$.

103. Look again at the equation in Question 101. What is y' when $y = 11^{1/2}$?

    (a) It's not defined.
    (b) It's equal to 0.
    (c) It's equal to $\pm 2^{1/2}$.
    (d) It's equal to $\pm 11^{1/2}$.
    (e) It's equal to $-(11/2)^{1/2}$.

104. Consider a function f of two variables x and y. Suppose that we treat y as a constant, differentiate with respect to x, keep holding y constant, and differentiate with respect to x again. The result of this process is

    (a) the zero function.
    (b) a constant function.
    (c) a polynomial function.
    (d) a mixed second partial.
    (e) None of the above

105. Fill in the blank to make the following statement express one of l'Hôpital's principles. "If we want to find the limit of a ratio that tends toward _____, and if we can differentiate the numerator and the denominator, then the limit of the ratio of the derivatives is the same as the limit of the original ratio."

    (a) $1/0$
    (b) $\pm\infty/0$
    (c) $1/\pm\infty$
    (d) $0/0$
    (e) $\pm 2/\pi$

106. Fill in the blank to make the following statement express another of l'Hôpital's principles. "If we want to find the limit of a ratio that tends toward _____, and if we can differentiate the numerator and the denominator, then the limit of the ratio of the derivatives is the same as the limit of the original ratio."

    (a) $\pm\infty/\pm\infty$
    (b) $0/1$
    (c) $\pm\infty/0$
    (d) $\pm\infty/1$
    (e) $\pm\pi/2$

107. Imagine variables x and y along with true functions f and f^{-1} such that

$$f(x) = y$$

and

$$f^{-1}(y) = x$$

Which of the following statements can help us differentiate f^{-1} with respect to y?
(a) $[f'(y)]/[f^{-1\prime}(x)] = 1$
(b) $(dx/dy) = (dy/dx)^{-1}$
(c) $xy = 1$
(d) $f(x)\,f(y) = 1$
(e) $(dy/dx)^2 = 1$

108. Consider a function $z = f(x,y)$. Imagine that the 3D graph of f is a surface that intersects all three planes shown in Fig. FE-10, forming a curve in each plane. Suppose that the intersection curve in the plane parallel to the xz plane represents a function that can be differentiated. When we differentiate it, we have one of the partial derivatives of

(a) y with respect to x.
(b) y with respect to z.
(c) x with respect to y.
(d) z with respect to x.
(e) z with respect to y.

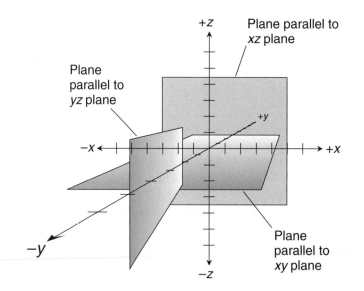

Figure FE-10   Illustration for Final Exam Questions 108 and 109.

109. Consider the same function $z = f(x,y)$ as we did in Question 108. Suppose that the intersection curve in the plane parallel to the *xz* plane represents a function that can be differentiated. When we differentiate it,
    (a) we hold *y* constant.
    (b) we hold *z* constant.
    (c) we hold *x* and *y* constant.
    (d) we hold *x* and *z* constant.
    (e) we hold *y* and *z* constant.

110. Suppose we use l'Hôpital's rule in an attempt to find an indeterminate limit, and we end up with a limit of the form $+\infty/-\infty$. What should we do?
    (a) Give up. We can't go any further to find this limit.
    (b) Conclude that the limit is not defined.
    (c) Apply l'Hôpital's rule again.
    (d) Conclude that the limit is equal to $-\infty$.
    (e) Conclude that the limit is equal to -1.

111. Consider a function *h* of two variables *u* and *v*. Suppose that we treat *u* as a constant, differentiate with respect to *v*, then treat *v* as a constant, and differentiate with respect to *u*. The result of this process is
    (a) the zero function.
    (b) a constant function.
    (c) a polynomial function.
    (d) a mixed second partial.
    (e) None of the above

112. Imagine that we want to find the mathematical volume of a solid defined by a surface in Cartesian *xyz*-space with respect to a region *R* in the *xy*-plane. Suppose that the surface is defined by the function $h(x,y)$. When we work out the double integrals we discover that

$$\iint_R h(x,y)\ dx\ dy = \iint_R h(x,y)\ dy\ dx = 0$$

What can we conclude, with certainty, from this result?
    (a) The surface is flat, and lies entirely in the *xy*-plane within the region *R*.
    (b) The function *f* contains a singularity somewhere in the region *R*.
    (c) The surface is flat and horizontal, parallel to the *xy*-plane within the region *R*.
    (d) The surface is perpendicular to the *xy*-plane at every point where it passes through the region *R*.
    (e) None of the above

113. Consider the following function in which *x* is the independent variable and *y* is the dependent variable:

$$y = f(x) = (x+1)^3$$

Suppose that we find the inverse relation f^{-1}, so we have

$$x = f^{-1}(y)$$

where y is the independent variable and x is the dependent variable. How must we restrict the domain of f to ensure that f^{-1} is a true function and is defined for all input values?

(a) We must restrict it to the set of all reals x such that $x \geq 1$.

(b) We must restrict it to the set of all reals x such that $x \geq 0$.

(c) We must restrict it to the set of all reals x such that $x \geq -1$.

(d) We must restrict it to the set of all reals x such that $-1 \leq x \leq 1$.

(e) We don't have to restrict it at all; x can be any real number.

114. Assuming we have restricted the domain of f in the situation of Question 113, what is the derivative of its inverse?

(a) $f^{-1\prime}(y) = 3(y+1)^2$

(b) $f^{-1\prime}(y) = 3y(y+1)^2$

(c) $f^{-1\prime}(y) = y^{-2/3}/3$

(d) $f^{-1\prime}(y) = y^{1/3} - 1$

(e) $f^{-1\prime}(y) = (y+1)^{1/3}$

115. Consider the following function g of two real independent variables s and t, where the dependent variable is a real number u:

$$u = g(s,t) = s^2 + 2st + t^2$$

The value of $\partial u / \partial t$ at the point where $(s,t) = (3,-2)$ is

(a) 2.

(b) −2.

(c) 4.

(d) 0.

(e) impossible to determine without more information.

116. Consider two functions $f(x)$ and $g(x)$ with three properties. First, f and g are both differentiable with respect to x over some open interval containing $x = a$, except possibly at $x = a$ itself. Second,

$$\lim_{x \to a} f(x) = 0$$

and

$$\lim_{x \to a} g(x) = 0$$

Third, $g'(x) = 0$. If all three of these conditions are met, then

$$\lim_{x \to a} f(x) / g(x) = \lim_{x \to a} f'(x) / g'(x)$$

Is anything wrong with this principle as stated here? If so, how can the principle be rewritten to make it correct?

(a) The first condition is misstated. It should require that f and g be continuous and differentiable over the entire set of real numbers.

(b) The second condition is misstated. It should require that the limits of both f and g should be nonzero real numbers as x approaches a.

(c) The third condition is misstated. It should require that $g'(x)$ be nonzero whenever $x \neq a$.

(d) The ratio of the derivatives is upside-down in the limit on the right-hand side of the equation. We should divide $g'(x)$ by $f'(x)$.

(e) Nothing is wrong with the principle as it is written here.

117. Consider the following function in which the domain is a set of real-number ordered pairs (x,y), and the range is a set of real numbers z:

$$z = f(x,y) = 7x^2 \ln |y|$$

The value of $\partial z / \partial x$ at the point $(x,y) = (2,1)$ is

(a) 0.

(b) 7.

(c) 14.

(d) 28.

(e) undefined.

118. In the situation of Question 117, $\partial z / \partial y$ at the point $(x,y) = (2,1)$ is

(a) 0.

(b) 7.

(c) 14.

(d) 28.

(e) undefined.

119. What's the value of the following repeated integral?

$$\int_{-3}^{1} \int_{-3}^{1} 24y \, dy \, dy$$

(a) We can't evaluate it, because it's not in a meaningful form.

(h) −112

(c) 112

(d) −96

(e) 96

120. Fill in the blank to make the following statement true: "To differentiate the inverse of a function (assuming that the inverse is a true function), we can differentiate the original function, then find the _____ of that derivative, and finally substitute one variable for the other."

   (a) negative

   (b) antiderivative

   (c) reciprocal

   (d) natural logarithm

   (e) absolute value

121. Suppose that we want to use calculus to derive a formula for the slant-surface area of the right circular cone shown in Fig. FE-11. To do this, we can integrate the function that defines

   (a) the circumference of each circular slice vs. the vertical distance y from the cone's apex.

   (b) the circumference of each circular slice vs. the distance t of its edge from the cone's apex.

   (c) the area of each disk-shaped slice vs. the distance x of its center from the cone's apex.

   (d) the area of each disk-shaped slice vs. the distance t of its edge from the cone's apex.

   (e) the surface area of each "sub-cone" vs. the distance x of the center of its base from the apex of the "main cone."

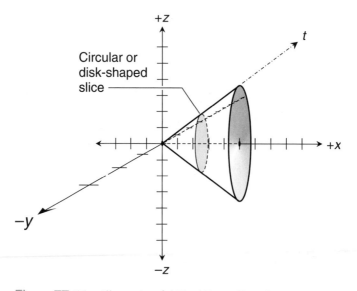

Figure FE-11   Illustration for Final Exam Questions 121 and 122.

122. If we want to use calculus to derive a formula for the volume of the right circular cone shown in Fig. FE-11, we can integrate the function that defines

(a) the circumference of each circular slice vs. the vertical distance y from the cone's apex.

(b) the circumference of each circular slice vs. the distance t of its edge from the cone's apex.

(c) the area of each disk-shaped slice vs. the distance x of its center from the cone's apex.

(d) the area of each disk-shaped slice vs. the distance t of its edge from the cone's apex.

(e) the volume of each "sub-cone" vs. the distance x of the center of its base from the apex of the "main cone."

123. Imagine a solid defined by a surface relative to a flat "base" region of finite, positive area that lies in the xy-plane of Cartesian xyz-space. Suppose the surface is defined by a function $f(x,y)$ such that, for every point within the "base" region, $f(x,y) > 0$. Then the mathematical volume of the solid

(a) must be positive.

(b) must be negative.

(c) must be 0.

(d) can be positive, negative, or 0.

(e) is undefined, because f contains a singularity.

124. Consider the following expression involving a function of two variables. What does this expression specifically tell us to do?

$$\int_{c}^{d} \left[\int_{a}^{b} f(x,y)\ dx \right]\ dy$$

(a) Integrate f against x, and evaluate the resulting expression from $y = a$ to $y = b$, getting another function. Then integrate that function against y, and evaluate the resulting expression from $x = c$ to $x = d$.

(b) Integrate f against x, and evaluate the resulting expression from $x = a$ to $x = b$, getting another function. Then integrate that function against y, and evaluate the resulting expression from $y = c$ to $y = d$.

(c) Integrate f against y, and evaluate the resulting expression from $y = a$ to $y = b$, getting another function. Then integrate that function against x, and evaluate the resulting expression from $x = c$ to $x = d$.

(d) Integrate f against y, and evaluate the resulting expression from $x = a$ to $x = b$, getting another function. Then integrate that function against x, and evaluate the resulting expression from $y = c$ to $y = d$.

(e) Integrate f against x, getting a second function. Then integrate against y, getting a third function. Then evaluate that function from $y = a$ to $y = b$. Finally, evaluate the remaining expression from $x = c$ to $x = d$.

125. What type of equation is this?

$$dy/dx + \cos x = 3x^2 + 8x - 7$$

(a) A second-order differential equation.

(b) A third-order differential equation.

(c) A partial differential equation.

(d) An extraordinary differential equation.

(e) An ordinary differential equation.

126. If we want to solve the equation in Question 125, what should we try to end up with?

    (a) We should try to get x all by itself on the left-hand side, and all the derivatives on the right-hand side.

    (b) We should try to get the derivative all by itself on the left-hand side, and all the variables and constants on the right-hand side.

    (c) We should try to get the derivative of x with respect to y on the left-hand side, and a function of y on the right-hand side.

    (d) We should try to get y all by itself on the left-hand side, and a function of x plus a constant on the right-hand side.

    (e) We should try to get the derivative all by itself on the left-hand side, and nothing but a real-number constant on the right-hand side.

127. When we define the lateral-surface area of a right circular cylinder by integrating along its length, what are we in effect doing?

    (a) Stacking up infinitely many circular cross-sections.

    (b) Stacking up infinitely many disk-shaped cross-sections.

    (c) Adding up the lengths of infinitely many line segments.

    (d) Adding up the circumferences of infinitely many concentric circles.

    (e) Adding up the lateral-surface areas of infinitely many concentric cylinders.

128. What type of equation is this?

$$d^2y/(3dx^2) - 2x = e^x + 5$$

    (a) A partial differential equation.

    (b) A cubic differential equation.

    (c) An exponential differential equation.

    (d) A second-order differential equation.

    (e) An improper differential equation.

129. If we want to solve the equation in Question 128, what is the most reasonable thing to do first?

    (a) Multiply the equation through by 3.

    (b) Differentiate both sides with respect to x.

    (c) Take the natural logarithm of both sides.

    (d) Consolidate the right-hand side into a single exponential function.

    (e) Integrate through with respect to y.

130. Imagine a solid defined by a surface relative to a flat "base" region of finite, positive area that lies in the *xy*-plane of Cartesian *xyz*-space. Suppose the surface is defined by a function $g(x,y)$ such that, for every point within the "base" region, $g(x,y) = 0$. Then the mathematical volume of the solid

(a) must be positive.

(b) must be negative.

(c) must be 0.

(d) can be positive, negative, or 0.

(e) is undefined, because *g* contains a singularity.

131. Imagine that we're calculating the mathematical volume of a solid defined by a surface in Cartesian *xyz*-space over a region *R* in the *xy*-plane. The function $h(x,y)$ that represents this surface is continuous and integrable within *R*. We find that

$$\iint_R h(x,y)\ dx\ dy < \iint_R h(x,y)\ dy\ dx$$

What can we conclude, with certainty, from this result?

(a) We've made a mistake somewhere in our calculations.

(b) The average value of *h* within *R* is negative.

(c) The average value of *h* within *R* is positive.

(d) Every value of *h* within *R* is negative.

(e) Every value of *h* within *R* is positive.

132. Imagine a solid defined by a surface relative to a flat "base" region of finite, positive area that lies in the *xy*-plane of Cartesian *xyz*-space. Suppose the surface is defined by a function $q(x,y)$ such that, for every point within the "base" region, $q(x,y) < 0$. Then the mathematical volume of the solid

(a) must be positive.

(b) must be negative.

(c) must be 0.

(d) can be positive, negative, or 0.

(e) is undefined, because *q* contains a singularity.

133. Suppose that we want to use calculus to derive a formula for the surface area of the sphere shown in Fig. FE-12. To do this, we can integrate the function that defines

(a) the circumference of each circular slice vs. the vertical distance *y* of its edge from the sphere's center.

(b) the circumference of each circular slice vs. the arc displacement *t* of its edge from the point where the positive *x*-axis passes through the sphere.

(c) the area of each disk-shaped slice vs. the distance *x* from the sphere's center.

(d) the area of each disk-shaped slice vs. the arc displacement *t* of its edge from the point where the positive *x*-axis passes through the sphere.

(e) The area of each disk-shaped slice from the point where the negative *x*-axis passes through the sphere to the point where the positive *x*-axis passes through the sphere.

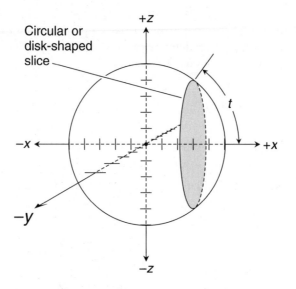

Figure FE-12   Illustration for Final Exam
Questions 133 and 134.

134. If we want to use calculus to derive a formula for the mathematical volume of the sphere shown in Fig. FE-12, we can integrate the function that defines
   (a) the circumference of each circular slice vs. the distance x of its center from the sphere's center.
   (b) the circumference of each circular slice vs. the arc displacement t of its edge from the point where the positive x-axis passes through the sphere.
   (c) the area of each disk-shaped slice vs. its radius.
   (d) the area of each disk-shaped slice vs. its circumference.
   (e) The area of each disk-shaped slice from the point where the negative x-axis passes through the sphere to the point where the positive x-axis passes through the sphere.

135. What's the value of the following repeated integral?

$$\int_{0}^{1} \int_{-3}^{0} 24y \, dy \, dy$$

   (a) We can't evaluate it, because it's not in the right form.
   (b) −112
   (c) 112
   (d) −108
   (e) 108

136. Imagine a solid defined by a surface relative to a flat "base" region of finite, positive area that lies in the *xy*-plane of Cartesian *xyz*-space. Suppose the surface is defined by a function $p\,(x,y)$ such that, for at least one point within the "base" region, $p\,(x,y) = 0$. Then the mathematical volume of the solid

    (a) must be positive.

    (b) must be negative.

    (c) must be 0.

    (d) can be positive, negative, or 0.

    (e) is undefined because *p* contains a singularity.

137. Figure FE-13 illustrates a process by which we can calculate the mathematical volume *V* that's defined by a 3D solid with respect to a flat region. Which of the following double integrals tells us how this process works?

    (a) $V = \displaystyle\int_{a}^{b} \int_{g(x)}^{h(x)} f(x,y)\, dy\, dx$

    (b) $V = \displaystyle\int_{a}^{b} \int_{g(x)}^{h(x)} f(x,y)\, dx\, dy$

    (c) $V = \displaystyle\int_{g(x)}^{h(x)} f(y)\, dy \int_{a}^{b} f(x)\, dx$

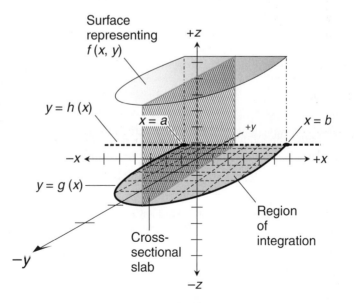

Figure FE-13   Illustration for Final Exam Question 137.

(d) $V = \int\limits_{g(x)}^{h(x)} \int\limits_{a}^{b} f(x,y) \, dx \, dy$

(e) $V = \int\limits_{a}^{b} f(x) \, dx \int\limits_{g(x)}^{h(x)} f(y) \, dy$

138. What is the solution to the following equation?

$$dy/dx - e^x = 0$$

(a) It can't be solved, because it contains a derivative.

(b) $y = e^x + c$, where c is a real-number constant.

(c) $y = 1$.

(d) $\ln (dy/dx) = x$.

(e) $dy = \ln x \, dx$.

139. Imagine a 3D coordinate system set up as Cartesian *xyz*-space. If the positive *x*-axis runs toward us and the positive *y*-axis runs off to our right, then the positive *z*-axis should run

(a) away from us.

(b) off to our left.

(c) straight up.

(d) straight down.

(e) None of the above

140. For a freely falling object in the earth's gravitational field, neglecting air resistance, the double integral of acceleration is

(a) the rate of acceleration per unit time, also called jerk.

(b) downward speed.

(c) downward displacement.

(d) altitude.

(e) undefined.

141. Consider the following function h, where the domain is a set of real-number ordered triples (x,y,z), and the range is a set of real numbers w:

$$w = h(x,y,z) = 2e^x y^{-2} e^z + 4xyz$$

What is $\partial w/\partial x$ at the point $(x,y,z) = (0,1,2)$?

(a) $-4e^2$

(b) $-e$

(c) 0

(d) $2e$

(e) $2e^2 + 8$

142. For the function *h* stated in Question 141, what's $\partial w/\partial y$ at $(x,y,z) = (0,1,2)$?
    (a) $-4e^2$
    (b) $-e^2$
    (c) -1
    (d) $-1 + e^2$
    (e) $-4e + e^2$

143. For the function *h* stated in Question 141, $\partial w/\partial z$ at $(x,y,z) = (0,1,2)$ is
    (a) $-2e$.
    (b) 0.
    (c) $2e^2$.
    (d) $4e + e^2$.
    (e) $8e + 2e^2$.

144. Consider the following function in which *x* is the independent variable and *y* is the dependent variable:

$$y = f(x) = \sin(x/2)$$

Suppose that we find the inverse relation f^{-1}, so we have

$$x = f^{-1}(y)$$

where *y* is the independent variable and *x* is the dependent variable. For what values of *y* is this inverse relation defined?
    (a) The set of all reals *y* such that $0 \le y \le \pi$.
    (b) The set of all reals *y* such that $y \ge 0$.
    (c) The set of all reals *y* such that $y \ge -1$.
    (d) The set of all reals *y* such that $-1 \le y \le 1$.
    (e) The set of all reals *y* such that $-\pi/2 \le y \le \pi/2$.

145. Suppose that we have established the domain of f^{-1} in the situation of Question 144. Also recall that for variables *x* and *y*,

$$d/dx (\sin x) = \cos x$$

and

$$d/dy (\text{Arcsin } y) = (1 - y^2)^{-1/2}$$

Based on this knowledge, what is the derivative of f^{-1}?
    (a) $f^{-1\prime}(y) = 2(1 - y^2)^{-1/2}$
    (b) $f^{-1\prime}(y) = (1 - y^2)^{-1/2} / 2$

(c) $f^{-1\prime}(y) = 2 \cos y$

(d) $f^{-1\prime}(y) = (\cos y) / 2$

(e) $f^{-1\prime}(y) = 2 (\cos y) (1 - y^2)^{-1/2}$

146. Consider the following three-variable function h, where x, y, and z are real-number independent variables and w is the real-number dependent variable:

$$w = h(x,y,z) = 2xy^2z^3$$

What is the second partial of this function with respect to x?

(a) $\partial^2 w/\partial x^2 = 2xz^3$

(b) $\partial^2 w/\partial x^2 = 2y^2z$

(c) $\partial^2 w/\partial x^2 = 2z$

(d) $\partial^2 w/\partial x^2 = 0$

(e) It is undefined.

147. Consider the following two-way relation:

$$9(x+1)^2 + 4(y-3)^2 = 36$$

Where, if anywhere, is the derivative dy/dx equal to 0 on the curve representing this relation?

(a) At the points $(-1,0)$ and $(-1,6)$.

(b) At the points $(-1,-6)$ and $(-1,12)$.

(c) At the points $(-3,3)$ and $(1,3)$.

(d) At the points $(-5,3)$ and $(3,3)$.

(e) There's no point where $dy/dx = 0$.

148. Consider again the two-way relation stated in Question 147. Where, if anywhere, is the derivative dy/dx undefined on the curve representing this relation?

(a) At the points $(-1,0)$ and $(-1,6)$.

(b) At the points $(-1,-6)$ and $(-1,12)$.

(c) At the points $(-3,3)$ and $(1,3)$.

(d) Only at the point $(-1,0)$.

(e) The derivative dy/dx is defined at every point on the curve.

149. Consider the following three-variable function h, where x, y, and z are real-number independent variables and w is the real-number dependent variable:

$$w = h(x,y,z) = 2xy^2z^3$$

What is the second mixed partial that we get when we differentiate h with respect to x, and then differentiate the result with respect to z?

(a) $\partial^2 w/\partial z\partial x = 3yz^3$

(b) $\partial^2 w/\partial z\partial x = 6y^2z^2$

(c) $\partial^2 w / \partial z \partial x = 12 y^2 z$

(d) $\partial^2 w / \partial z \partial x = 12 y z$

(e) $\partial^2 w / \partial z \partial x = 12 z$

150. Suppose we encounter

$$\lim_{x \to 0-} (x^{-2}) (\sin x)$$

What should be our very first step if we want to evaluate this limit?

(a) Convert the expression after the limit symbol to a ratio.

(b) Apply l'Hôpital's rule.

(c) Apply the chain rule.

(d) Differentiate the expression after the limit symbol.

(e) Give up. The expression is indeterminate, so we can't evaluate it.

A

Worked-Out Solutions to Exercises: Chapters 1 to 9

These solutions do not necessarily represent the only way a problem can be figured out. If you think you can solve a problem faster or better than the way it's done here, by all means try it! Always check your work to be sure your "alternative" answer is correct.

Chapter 1

1. This relation is a function. The domain is the set of all integers, and the range is the set of all nonnegative integers. Every element of the domain maps into one, but only one, element of the range. The inverse, which we get by reversing each ordered pair in the original relation, is

$$\{(0,0), (1,1), (2,-1), (3,2), (4,-2), (5,3), (6,-3), (7,4), (8,-4), \ldots\}$$

This is also a function. The domain is the set of all nonnegative integers, and the range is the set of all integers. Every element of the domain maps into one, but only one, element of the range.

2. This relation is a function. Once again, the domain is the set of all integers, and the range is the set of all nonnegative integers. Every element in the domain maps into one, but only one, element of the range. The inverse, which we get by reversing each ordered pair in the original relation, is

$$\{(0,0), (1,1), (1,-1), (2,2), (2,-2), (3,3), (3,-3), (4,4), (4,-4), \ldots\}$$

This is not a function! The domain is the set of all nonnegative integers, and the range is the set of all integers. With the exception of $(0,0)$, every element of the domain maps into two elements of the range. A function isn't allowed to do that.

3. Let's substitute a letter variable for $f(x)$. If we use y, the function is written

$$y = 4x - 5$$

To find the inverse, we must solve for x in terms of y. First, we add 5 to each side of the above equation, getting

$$y + 5 = 4x$$

Dividing through by 4 and transposing the sides of the equation left-to-right, we get

$$x = (y + 5)/4$$

which can be rewritten as

$$x = y/4 + 5/4$$

This means that

$$f^{-1}(y) = y/4 + 5/4$$

This is the inverse we seek. It's a linear function of y.

4. This constant function, $g(x)$, has the value 7 regardless of the value of x. If we graph this function and call the dependent variable y, we get a horizontal line through the point $y = 7$. The graph of the inverse relation is a vertical line, which fails the vertical-line test, so we know that this inverse is not a function.

5. The equation of the unit circle can be manipulated to get y in terms of x. Here is the original equation for reference:

$$x^2 + y^2 = 1$$

If we subtract x^2 from each side, we get

$$y^2 = 1 - x^2$$

Taking the positive-or-negative square root of each side gives us

$$y = \pm(1 - x^2)^{1/2}$$

This is a relation whose domain is the set of real numbers between, and including, -1 and 1. The range is also the set of reals between, and including, -1 and 1. The relation is not a function. With the exception of $x = -1$ and $x = 1$, every element of the domain maps into two elements in the range. This fact can be seen by looking at a graph of the unit circle (Fig. A-1). The graph fails the vertical-line test as shown by the dashed line, which intersects the circle twice.

6. The equation of the unit circle can be manipulated to get x in terms of y. Once again, here's the original equation for reference:

$$x^2 + y^2 = 1$$

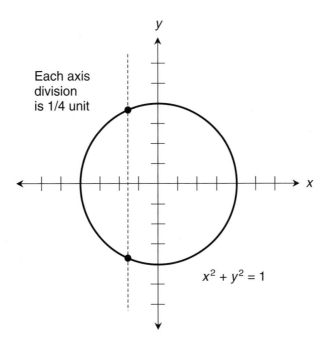

Figure A-1 Illustration for the solution to Prob. 5 in Chap. 1.

If we subtract y^2 from each side, we get

$$x^2 = 1 - y^2$$

Taking the positive-or-negative square root of each side gives us

$$x = \pm(1 - y^2)^{1/2}$$

This is the same situation as we encountered in the solution to Prob. 5. The names of the variables have been changed, that's all! We have, once again, a relation whose domain and range are both the set of real numbers between, and including, −1 and 1. As before, we don't have a function here because, with the exception of $y = -1$ and $y = 1$, every element of the domain maps into two elements in the range.

7. The complete graph of the inverse relation, g^{-1}, is a parabola that opens directly toward the right. If we restrict the range of this relation to the set of positive real numbers, we cut off the lower half of the parabola along with the point (0,0). This leaves only the upper half of the curve, which passes the vertical-line test as shown in Fig. A-2. Therefore, g^{-1} becomes a function when the range is restricted to the set of positive reals. It remains a function if we allow the range to include 0 as well as all the positive reals, because the graph still passes the vertical-line test. But we can't extend the range into the negative reals unless we remove at least some of the positive reals.

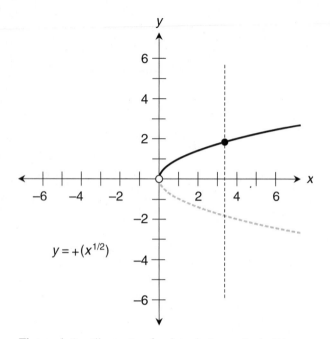

$$y = +(x^{1/2})$$

Figure A-2   Illustration for the solution to Prob. 7 in Chap. 1.

8. If we restrict the range of g^{-1} to the set of negative reals, we cut off the upper half of the parabola along with $(0,0)$. This leaves only the lower half, which passes the vertical-line test as shown in Fig. A-3. Therefore, g^{-1} becomes a function when the range is restricted to the set of negative reals. It remains a function if we also allow the range to include 0, but we can't extend the range into the positive reals unless we remove at least some of the negative reals.

9. If we fill in the gap in the domain for the relation shown in Fig. 1-9 by setting $y = 0$ when $x = 0$, we are, in effect, saying that $1/0 = 0$, because the relation is

$$y = 1/x$$

The mathematical truth of this can be debated! But it provides us with a y-value for every possible real number x. The resulting relation is a function, because the graph passes the vertical-line test. The same thing happens with the relation

$$y = \tan x$$

graphed in Fig. 1-10. If we set $y = 0$ whenever x is an odd-integer multiple of $\pi/2$, we still have a function of x, because the graph still passes the vertical-line test. We should expect to see some raised eyebrows, however, if we claim that the tangent of $\pi/2$ is actually equal to 0!

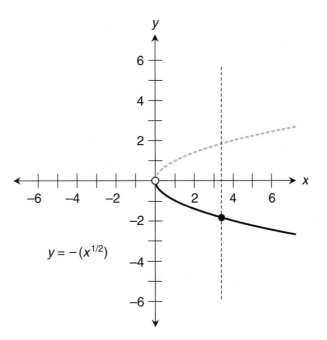

$$y = -(x^{1/2})$$

Figure A-3   Illustration for the solution to Prob. 8 in
          Chap. 1.

10. The inverse of the function shown in Fig. 1-9 is also a function. We can see that this is
    true by doing the horizontal-line test on the graph. We can take

$$y = 1/x$$

and multiply through by x provided that $x \neq 0$, getting

$$xy = 1$$

Then we can divide through by y, again with the constraint that $y \neq 0$, getting

$$x = 1/y$$

Chapter 2

1. This sequence starts out at 1/10, and then each succeeding number gets smaller by
   a factor of 10. The numbers get closer and closer to 0, but never actually reach 0.
   Therefore, the limit of the sequence is 0. We can write this as

$$\text{Lim}_{n \to \infty} 1/10^n = 0$$

2. In decimal form, $1/10 = 0.1$, $1/10^2 = 0.01$, $1/10^3 = 0.001$, and so on. Let's write the numbers in decimal form and then arrange the sum in as an accountant would do, with each term underneath its predecessor:

$$
\begin{array}{l}
0.1 \\
0.01 \\
0.001 \\
0.0001 \\
0.00001 \\
\cdots \\
\hline
0.11111\ldots
\end{array}
$$

Looking at the series this way, we can see that it ultimately adds up to the nonterminating, repeating decimal $0.11111\ldots$. From our algebra or number theory courses, we recall that this endless decimal number is equal to $1/9$. That's the limit of the sequence of partial sums in the series

$$1/10 + 1/10^2 + 1/10^3 + 1/10^4 + 1/10^5 + \cdots$$

3. This limit exists, and it is equal to 0. To see why, suppose that we start out with x at some positive real number for which the function is defined. As we increase the value of x, the value of $1/x^2$ decreases, but it always remains positive. If we choose some tiny positive real number r, no matter how close to 0 it might be, we can always find some large value of x for which $0 < 1/x^2 < r$. But no matter how large we make x, the number $1/x^2$ never becomes negative. Therefore,

$$\operatorname*{Lim}_{x \to \infty} 1/x^2 = 0$$

4. This limit does not exist. To see why, suppose that we start out with x at some positive real number for which the function is defined and then decrease x, letting it get arbitrarily close to 0 but always remaining positive. As we decrease the value of x, the value of $1/x^2$ remains positive and increases. If we choose some large positive real number s, no matter how gigantic, we can always find some small, positive value of x for which $1/x^2 > s$. Therefore,

$$\operatorname*{Lim}_{x \to 0+} 1/x^2$$

is not defined.

5. Figure A-4 is a graph of the function $f(x) = \log_{10} x$ for values of x from 0.1 up to 10, and for values of f from -1 to 1. The function varies smoothly throughout this span. If we start at values of x a little smaller than 3 and work our way toward 3, the value of f approaches $\log_{10} 3$. Therefore,

$$\operatorname*{Lim}_{x \to 3-} \log_{10} x = \log_{10} 3$$

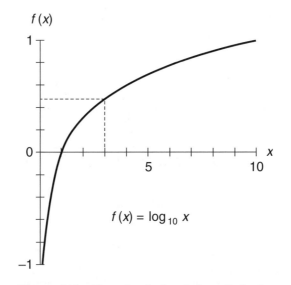

Figure A-4 Illustration for the solution to Probs. 5
through 8 in Chap. 2.

6. If we start at values of x slightly larger than 3 in Fig. A-4 and then work our way toward the point where $x = 3$, the value of f approaches $\log_{10} 3$. It's apparent from the graph that

$$\underset{n \to 3+}{Lim} \log_{10} x = \log_{10} 3$$

7. From the solution to Prob. 5, we can conclude that the function f is left-hand continuous at the point where $x = 3$. From the solution to Prob. 6, we know that f is right-hand continuous at that same point. By definition, therefore, f is continuous at the point where $x = 3$.

8. The function $f(x) = \log_{10} x$ is continuous if we restrict the domain to the set of positive reals. But it's not continuous if the domain is the set of all nonnegative reals. The base-10 logarithm of 0 is not defined, and that creates a discontinuity in f at $x = 0$. Not only that, but the function has no defined limit as x approaches 0 from the right.

9. Figure A-5 is a graph of the absolute-value function $f(x) = |x|$ for values of the domain between approximately -6 and 6. This function is continuous over the entire set of reals. Clearly, it is continuous over the set of positive reals because, when $x > 0$,

$$f(x) = x$$

which is a linear function. It is also continuous over the set of negative reals because, when $x < 0$,

$$f(x) = -x$$

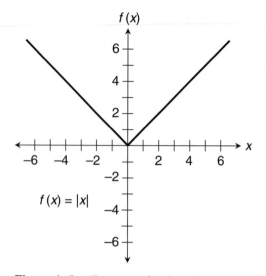

Figure A-5   Illustration for the solution to
Prob. 9 in Chap. 2.

which, again, is a linear function. The only doubt might arise when we look at the point where $x = 0$. But

$$\underset{x \to 0+}{Lim} \ |x|$$

and

$$\underset{x \to 0-}{Lim} \ |x|$$

are both defined and equal to 0, which is the value of $f(0)$. Therefore, by definition, the function $f(x) = |x|$ is continuous at the point where $x = 0$.

10. Figure A-6 is a graph of the function $f(x) = \csc x$ for the domain between, and including, -3π radians and 3π radians. The function is not continuous in that interval, because it's undefined when x is an integer multiple of π radians. That's not the only problem! The function has no defined limit as x approaches any integer multiple of π radians, either from the left or the right. In the interval we've specified, discontinuities occur when $x = -3\pi$, $x = -2\pi$, $x = -\pi$, $x = 0$, $x = \pi$, $x = 2\pi$, and $x = 3\pi$.

Chapter 3

1. According to the rule we worked out for the derivative of a basic quadratic function,

$$f(x) = x^2$$

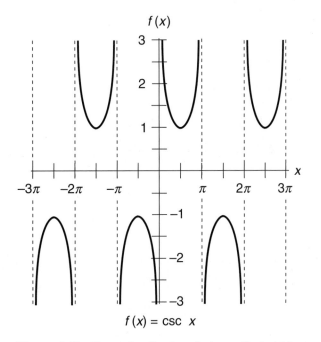

$$f(x) = \csc x$$

Figure A-6 Illustration for the solution to Prob. 10 in Chap. 2.

has the derivative function

$$f'(x) = 2x$$

The *x*-values of interest are $x = -3$, $x = -2$, and $x = -1$. We have

$$f'(-3) = 2 \times (-3) = -6$$
$$f'(-2) = 2 \times (-2) = -4$$
$$f'(-1) = 2 \times (-1) = -2$$

2. The function and its derivative are the same as in Prob. 1. The *x*-values of interest are $x = 0$, $x = 1$, $x = 2$, and $x = 3$. We have

$$f'(0) = 2 \times 0 = 0$$
$$f'(1) = 2 \times 1 = 2$$
$$f'(2) = 2 \times 2 = 4$$
$$f'(3) = 2 \times 3 = 6$$

3. According to the rule we worked out for the derivative of a monomial quadratic function,

$$f(x) = -2x^2$$

has the derivative function

$$f'(x) = 2 \times (-2)x = -4x$$

The x-values of interest are $x = -3$, $x = -2$, and $x = -1$. We have

$$f'(-3) = -4 \times (-3) = 12$$
$$f'(-2) = -4 \times (-2) = 8$$
$$f'(-1) = -4 \times (-1) = 4$$

4. The function and its derivative are the same as in Prob. 3. The x-values of interest are $x = 0$, $x = 1$, $x = 2$, and $x = 3$. We have

$$f'(0) = -4 \times 0 = 0$$
$$f'(1) = -4 \times 1 = -4$$
$$f'(2) = -4 \times 2 = -8$$
$$f'(3) = -4 \times 3 = -12$$

5. We found the general derivative of this function in the chapter text. The original function, again, is

$$f(x) = -7x^2 + 2x$$

The derivative function is

$$f'(x) = -14x + 2$$

The x-values of interest are $x = -3$, $x = -2$, and $x = -1$. We have

$$f'(-3) = -14 \times (-3) + 2 = 44$$
$$f'(-2) = -14 \times (-2) + 2 = 30$$
$$f'(-1) = -14 \times (-1) + 2 = 16$$

6. The function and its derivative are the same as in Prob. 5. The x-values of interest are $x = 0$, $x = 1$, $x = 2$, and $x = 3$. We have

$$f'(0) = -14 \times 0 + 2 = 2$$
$$f'(1) = -14 \times 1 + 2 = -12$$
$$f'(2) = -14 \times 2 + 2 = -26$$
$$f'(3) = -14 \times 3 + 2 = -40$$

7. According to the rule we worked out for the derivative of a monomial cubic function,

$$f(x) = 5x^3$$

has the derivative function

$$f'(x) = 3 \times 5x^2 = 15x^2$$

The x-values of interest are $x = -3$, $x = -2$, and $x = -1$. We have

$$f'(-3) = 15 \times (-3)^2 = 135$$
$$f'(-2) = 15 \times (-2)^2 = 60$$
$$f'(-1) = 15 \times (-1)^2 = 15$$

8. The function and its derivative are the same as in Prob. 7. The x-values of interest are $x = 0$, $x = 1$, $x = 2$, and $x = 3$. We have

$$f'(0) = 15 \times 0^2 = 0$$
$$f'(1) = 15 \times 1^2 = 15$$
$$f'(2) = 15 \times 2^2 = 60$$
$$f'(3) = 15 \times 3^2 = 135$$

9. We found the derivative of this function in the chapter text. The original function is

$$f(x) = 2x^3 - 5x$$

which has the derivative function

$$f'(x) = 6x^2 - 5$$

The x-values of interest are $x = -3$, $x = -2$, and $x = -1$. We have

$$f'(-3) = 6 \times (-3)^2 - 5 = 49$$
$$f'(-2) = 6 \times (-2)^2 - 5 = 19$$
$$f'(-1) = 6 \times (-1)^2 - 5 = 1$$

10. The function and its derivative are the same as in Prob. 9. The x-values of interest are $x = 0$, $x = 1$, $x = 2$, and $x = 3$. We have

$$f'(0) = 6 \times 0^2 - 5 = -5$$
$$f'(1) = 6 \times 1^2 - 5 = 1$$
$$f'(2) = 6 \times 2^2 - 5 = 19$$
$$f'(3) = 6 \times 3^2 - 5 = 49$$

Chapter 4

1. Figure A-7 is a graph of the function

$$f(x) = 1/x \quad \text{when } x < -1$$
$$= -1 \quad \text{when } x \geq -1$$

   This graph suggests that the function is defined and continuous over the entire set of real numbers, but the slope abruptly changes at the point where $x = -1$. We should suspect that the function is nondifferentiable at that point.

2. We can attempt to find the derivative at the point where $x = -1$ by evaluating

$$\underset{\Delta x \to 0}{Lim} \; [f(-1 + \Delta x) - f(-1)] / \Delta x$$

   from both the right and the left. Let's do it from the right first. To denote this, we add a plus sign after the 0 beneath "*Lim*" to get

$$\underset{\Delta x \to 0+}{Lim} \; [f(-1 + \Delta x) - f(-1)] / \Delta x$$

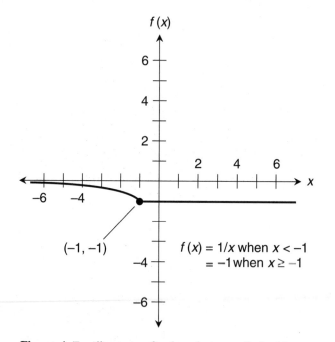

Figure A-7   Illustration for the solution to Prob. 1 in Chap. 4.

When Δx is small and positive, we're in the straight-line part of the function with a constant value of -1. We're told that $f(-1) = -1$. Let's substitute -1 for $f(-1 + \Delta x)$, and substitute -1 for $f(-1)$ in the above expression. Then we get

$$\underset{\Delta x \to 0+}{Lim} \; [-1 - (-1)] / \Delta x$$

which simplifies to

$$\underset{\Delta x \to 0+}{Lim} \; 0/\Delta x$$

and, because Δx is nonzero, finally to

$$\underset{\Delta x \to 0+}{Lim} \; 0$$

This limit is clearly equal to 0.

Now let's move to the left of the point where $x = -1$, into the curved part of the graph. When Δx is small and negative, we're in the zone where the function takes the reciprocal of the input. This time, we write

$$\underset{\Delta x \to 0-}{Lim} \; [f(-1 + \Delta x) - f(-1)] / \Delta x$$

As before, $f(-1) = -1$. When we substitute $1/(-1 + \Delta x)$ for $f(-1 + \Delta x)$, and substitute -1 for $f(-1)$ in the above expression, it becomes

$$\underset{\Delta x \to 0-}{Lim} \; [1/(-1 + \Delta x) - (-1)] / \Delta x$$

which can be simplified to

$$\underset{\Delta x \to 0-}{Lim} \; [1/(-1 + \Delta x) + 1] / \Delta x$$

and further to

$$\underset{\Delta x \to 0-}{Lim} \; [\Delta x / (\Delta x - 1)] / \Delta x$$

and finally to

$$\underset{\Delta x \to 0-}{Lim} \; 1/(\Delta x - 1)$$

As Δx approaches 0 from the left, the expression after "*Lim*" approaches $1/(-1)$, which is equal to -1. That's not the same value as we got when we evaluated the right-hand limit. Having found a disagreement between the right-hand and left-hand limits, we know that this function is nondifferentiable at the point where $x = -1$.

3. Figure A-8 is a graph of the function

$$f(x) = x \quad \text{when } x \leq 3$$
$$= 1 \quad \text{when } x > 3$$

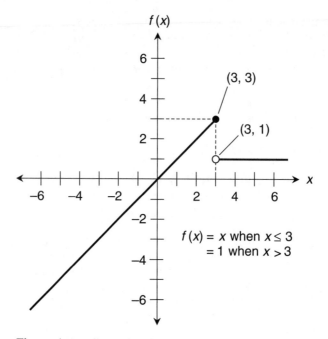

Figure A-8   Illustration for the solution to Prob. 3 in Chap. 4.

This graph suggests that the function is defined over the entire set of reals, but it is discontinuous at the point where $x = 3$. We should suspect that the function is nondifferentiable at that point.

4. Let's evaluate the limit as we approach the point where $x = 3$ from the right:

$$\underset{\Delta x \to 0+}{Lim} \; [f(3 + \Delta x) - f(3)] \, / \, \Delta x$$

The value of the function is 1 if we add any small positive quantity Δx to the input value of 3. We also know that $f(3) = 3$. Therefore, the above limit can be simplified to

$$\underset{\Delta x \to 0+}{Lim} \; (1 - 3) \, / \, \Delta x$$

and further to

$$\underset{\Delta x \to 0+}{Lim} \; -2/\Delta x$$

This limit does not exist. As Δx approaches 0 from the positive side, $-2/\Delta x$ blows up negatively. This fact alone is sufficient to prove that the function is nondifferentiable at the point where $x = 3$.

5. Figure A-9 is a graph of the function

$$f(x) = x \quad \text{when } x < 3$$
$$= 1 \quad \text{when } x \geq 3$$

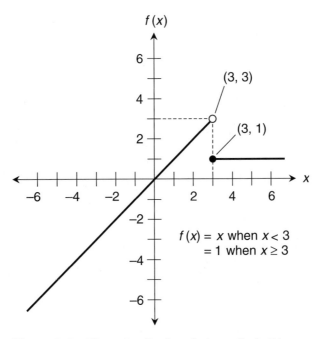

Figure A-9   Illustration for the solution to Prob. 5 in Chap. 4.

This graph, like Fig. A-8, suggests that the function is defined over the entire set of reals, but it is discontinuous at the point where $x = 3$. We should suspect that the function is nondifferentiable at that point.

6. Let's figure out the limit as we approach the point where $x = 3$ once again, but this time from the left:

$$\underset{\Delta x \to 0-}{Lim}\ [f(3 + \Delta x) - f(3)]\ /\ \Delta x$$

The output value of this function is equal to the input value if we add any tiny negative quantity Δx to the input value of 3. We also know that $f(3) = 1$. Therefore, the above limit can be simplified to

$$\underset{\Delta x \to 0-}{Lim}\ (3 + \Delta x - 1)\ /\ \Delta x$$

and further to

$$\underset{\Delta x \to 0-}{Lim}\ (2 + \Delta x)\ /\ \Delta x$$

As Δx approaches 0 from the negative side, the numerator, $2 + \Delta x$, gets arbitrarily close to 2 while the denominator gets arbitrarily close to 0. Therefore, the limit blows up negatively. That's all we need to prove that the function is nondifferentiable at the point where $x = 3$.

7. Figure A-10 is a graph of the function

$$f(x) = x^2 \quad \text{when } x < 1$$
$$= x^3 \quad \text{when } x \geq 1$$

This graph suggests that the function is defined and continuous over the entire set of real numbers. It's impossible to be sure, by visual inspection alone, whether or not the slope of this graph changes at the point where $x = 1$. We should check to see if the function is differentiable there.

8. Let's evaluate the limit as we approach the point where $x = 1$ from the right. We have

$$\underset{\Delta x \to 0+}{Lim} \; [f(1 + \Delta x) - f(1)] / \Delta x$$

When Δx is positive, we're in the "cubic zone." We know that $f(1) = 1^3 = 1$. When we substitute $(1 + \Delta x)^3$ for $f(1 + \Delta x)$, and substitute 1 for $f(1)$ in the above expression, it becomes

$$\underset{\Delta x \to 0+}{Lim} \; [(1 + \Delta x)^3 - 1] / \Delta x$$

When we cube the binomial, we get

$$\underset{\Delta x \to 0+}{Lim} \; [1 + 3\Delta x + 3(\Delta x)^2 + (\Delta x)^3 - 1] / \Delta x$$

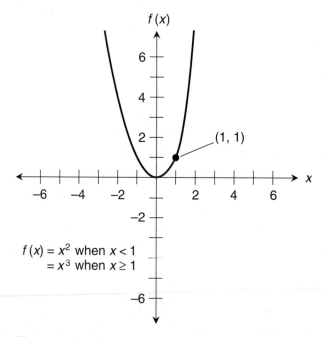

Figure A-10   Illustration for the solution to Prob. 7 in Chap. 4.

which simplifies to

$$\underset{\Delta x \to 0+}{Lim} \ [3\Delta x + 3(\Delta x)^2 + (\Delta x)^3] \ / \ \Delta x$$

and further to

$$\underset{\Delta x \to 0+}{Lim} \ 3 + 3\Delta x + (\Delta x)^2$$

As Δx tends toward 0, the second and third addends in the expression also approach 0. The limit is therefore equal to $3 + 0 + 0$, which is 3.

Now let's move to the left of the point where $x = 1$, into the "quadratic zone." We must work with the limit

$$\underset{\Delta x \to 0+}{Lim} \ [f(1 + \Delta x) - f(1)] \ / \ \Delta x$$

When we substitute $(1 + \Delta x)^2$ for $f(1 + \Delta x)$, and substitute 1 for $f(1)$ in the above expression, it becomes

$$\underset{\Delta x \to 0+}{Lim} \ [(1 + \Delta x)^2 - 1] \ / \ \Delta x$$

We can square the binomial to get

$$\underset{\Delta x \to 0-}{Lim} \ (1 + 2\Delta x + (\Delta x)^2 - 1] \ / \ \Delta x$$

which can be simplified to

$$\underset{\Delta x \to 0-}{Lim} \ [2\Delta x + (\Delta x)^2] \ / \ \Delta x$$

and further to

$$\underset{\Delta x \to 0-}{Lim} \ 2 + \Delta x$$

As Δx approaches 0 from the left, this expression approaches $2 + 0$ from the left, telling us that the limit is 2. That's not the same as the limit we got when we approached from the right. Therefore, the function is nondifferentiable at the point where $x = 1$.

9. Figure A-11 is a graph of the function

$$f(x) = \ x^2 \quad \text{when } x < 1$$
$$= 2x - 1 \quad \text{when } x \geq 1$$

This graph suggests that the function is defined and continuous over the entire set of real numbers. As in Fig. A-10, we can't be sure, by visual inspection alone, whether or not the slope of this graph changes at the point where $x = 1$. We should check to see if the function is differentiable there.

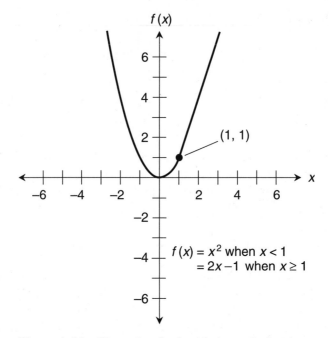

Figure A-11   Illustration for the solution to Prob. 9 in Chap. 4.

10. Let's evaluate the limit from the right. We have

$$\underset{\Delta x \to 0+}{Lim} \; [f(1 + \Delta x) - f(1)] / \Delta x$$

When Δx is small and positive, we're in the straight-line portion of the graph. We can calculate

$$f(1) = 2 \times 1 - 1 = 1$$

When we substitute $2(1 + \Delta x) - 1$ for $f(1 + \Delta x)$, and substitute 1 for $f(1)$ in the above expression, it becomes

$$\underset{\Delta x \to 0+}{Lim} \; [2 \times (1 + \Delta x) - 1 - 1] / \Delta x$$

When we multiply this out, we get

$$\underset{\Delta x \to 0+}{Lim} \; [2 + 2\Delta x - 1 - 1] / \Delta x$$

which simplifies to

$$\underset{\Delta x \to 0+}{Lim} \; 2\Delta x / \Delta x$$

and finally to

$$\underset{\Delta x \to 0+}{Lim} \ 2$$

Clearly, this limit is equal to 2.

When we move to the left of the point where $x = 1$ and let Δx approach 0 from the negative side, we can repeat the second half of the solution to Prob. 8 step-for-step, because we're dealing with the same curve and approaching the same point. When we calculate the limit

$$\underset{\Delta x \to 0+}{Lim} \ [f(1 + \Delta x) - f(1)] \ / \ \Delta x$$

we'll find that it is equal to 2, which is same as the limit we got when we approached the point where $x = 1$ from the right. We can therefore say with confidence that this two-part function is differentiable at the point where $x = 1$.

The curve in Fig. A-10 looks as if it goes smoothly through the transition point where $x = 1$, but there's actually a "jog in the road." As we move from the quadratic part of the curve into the cubic part, the slope changes abruptly from 2 to 3. In Fig. A-11, there's no abrupt change in the slope. It's like a well-engineered road, where a curve ends and a straightaway begins without any abrupt change in the direction.

Chapter 5

1. The derivatives of these functions can be found by using the power rule. When this rule is applied, the results are as follows:

   (a) $f'(x) = -40x^4$

   (b) $g'(z) = 84z^6$

   (c) $h'(t) = -441t^{20}$

2. We can apply the power rule because $f_3(x)$ is a monomial power function. We get

   $$f_3'(x) = 3x^2$$

   We want to find x_0 in the interval $0 \le x \le 1$, such that the value of this derivative is equal to 1. To do that, we set

   $$3x_0^2 = 1$$

   Dividing through by 3 and then taking the 1/2 power of each side tells us that

   $$x_0 = (1/3)^{1/2}$$

   The 1/2 power of a quantity is the *positive square root* of that quantity. We're not interested in the negative square root here, because we're restricted to $0 \le x \le 1$. A calculator tells us that, to three decimal places,

   $$x_0 = 0.577$$

3. We can again apply the power rule. In this case, we get

$$f_5'(x) = 5x^4$$

To find x_0 in the interval $0 \le x \le 1$ such that $f_5'(x) = 1$, we set

$$5x_0^4 = 1$$

Dividing through by 5 and then taking the 1/4 power of each side tells us that

$$x_0 = (1/5)^{1/4}$$

The 1/4 power of a quantity is the *positive real fourth root* of that quantity. We're not interested in negative real roots or in nonreal complex roots in this case, because we're restricted to the positive real interval $0 \le x \le 1$. Using a calculator, we obtain

$$x_0 = 0.669$$

4. This time, we apply the power rule in a general sense to get

$$f_n'(x) = nx^{(n-1)}$$

To find x_0 in the interval $0 \le x \le 1$ such that $f_n'(x) = 1$, we set

$$nx_0^{(n-1)} = 1$$

Dividing through by n and then taking the $1/(n-1)$ power of each side tells us that

$$x_0 = (1/n)^{[1/(n-1)]}$$

The $1/(n-1)$ power of a quantity is the *positive real (n−1) root* of that quantity. As before, we're not interested in negative real roots or in nonreal complex roots here, because we're restricted to the positive real interval $0 \le x \le 1$.

5. Figure A-12 is a graph of the functions $f_n(x) = x^n$ in the interval $0 \le x \le 1$ for $n = 3$, $n = 4$, and $n = 5$. On each curve, the dot shows the point where the slope is 1. We didn't actually work out the case for $n = 4$, but the graph shows an educated guess!

6. We apply the power rule in a general sense to get

$$f_n'(x) = nx^{(n-1)}$$

To find the value $f_n'(1)$, we substitute 1 for x, getting

$$f_n'(1) = n \times 1^{(n-1)} = n$$

7. We again apply the power rule generally, obtaining

$$f_n'(x) = nx^{(n-1)}$$

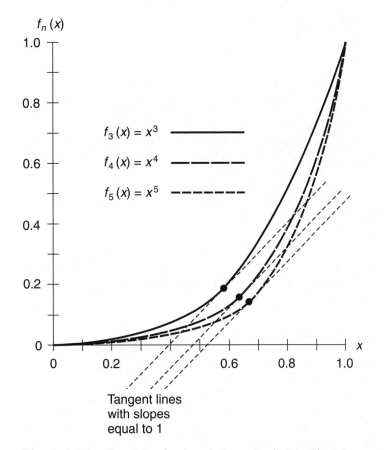

Figure A-12   Illustration for the solution to Prob. 5 in Chap. 5.

To find the value $f_n'(0)$, we substitute 0 for x, which gives us

$$f_n'(0) = n \times 0^{(n-1)} = 0^{(n-1)} = 0$$

8. To find the derivative of this polynomial function, we can apply the power rule to each term in the polynomial separately. That gives us

$$f'(x) = 56x^6 + 24x^5 - 15x^4 + 4x^3 + 2x - 0$$

The last term of 0 is useless, of course, so we can write

$$f'(x) = 56x^6 + 24x^5 - 15x^4 + 4x^3 + 2x$$

9. We again apply the power rule to each term individually, getting

$$f'(x) = 5a_5x^4 + 4a_4x^3 + 3a_3x^2 + 2a_2x + a_1 + 0$$

Taking off the last term of 0 gives us

$$f'(x) = 5a_5x^4 + 4a_4x^3 + 3a_3x^2 + 2a_2x + a_1$$

10. Once again, we can apply the power rule to each term. We know that the last term will be 0 so we don't have to write it. The derivative is

$$f'(x) = na_nx^{n-1} + (n-1)a_{n-1}x^{n-2} + (n-2)a_{n-2}x^{n-3} + \ldots + 2a_2x + a_1$$

Chapter 6

1. The original function f is

$$f(x) = -4x^4 + 2x^3 - x^2 - x + 1$$

Its derivative is

$$f'(x) = -16x^3 + 6x^2 - 2x - 1$$

When we multiply this function by 2, we get

$$2f'(x) = 2 \times (-16x^3 + 6x^2 - 2x - 1) = -32x^3 + 12x^2 - 4x - 2$$

The original function g is

$$g(x) = 2 \times (-4x^4 + 2x^3 - x^2 - x + 1) = -8x^4 + 4x^3 - 2x^2 - 2x + 2$$

Its derivative is

$$g'(x) = -32x^3 + 12x^2 - 4x - 2$$

which is exactly the same as $2f'(x)$.

2. The original function f is

$$f(x) = -40x^4 + 20x^3 - 10x^2 - 10x + 10$$

Its derivative is

$$f'(x) = -160x^3 + 60x^2 - 20x - 10$$

When we multiply this function by 1/5, we get

$$(1/5)\,f'(x) = (1/5) \times (-160x^3 + 60x^2 - 20x - 10) = -32x^3 + 12x^2 - 4x - 2$$

The original function *g* is

$$g(x) = (1/5) \times (-40x^4 + 20x^3 - 10x^2 - 10x + 10) = -8x^4 + 4x^3 - 2x^2 - 2x + 2$$

Its derivative is

$$g'(x) = -32x^3 + 12x^2 - 4x - 2$$

which is exactly the same as $(1/5) f'(x)$. It's worth noting here that the statement

$$g'(x) = (1/5) f'(x)$$

can also be written as

$$g'(x) = [f'(x)] / 5$$

Whenever we multiply a function by the reciprocal of a constant, it's the same as dividing that function by the constant. This gives us a way to define *division of a function by a constant*. Of course, the constant must be nonzero!

3. Here's the original statement of the two-function product rule for differentiation:

$$(fg)' = f'g + g'f$$

The commutative law for addition (from basic algebra) allows us to transpose the addends on the right side, getting

$$(fg)' = g'f + f'g$$

The expression $g'f + f'g$ is equal to $(gf)'$ according to the two-function product rule for differentiation "stated backward." Therefore, we can conclude that

$$(fg)' = (gf)'$$

4. Suppose we have three differentiable functions *f*, *g*, and *h*. The two-function product rule for differentiation tells us that

$$(fg)' = f'g + g'f$$

and

$$(gh)' = g'h + h'g$$

Let's rename *fg*, calling it *p* (for "product") instead. Then we can restate the first of the above two equations as

$$p' = f'g + g'f$$

We want to find $(fgh)'$, which is the same thing as finding $(ph)'$. The two-function product rule for differentiation can be applied here, giving us

$$(ph)' = p'h + h'p$$

Substituting fg back in for p gives us

$$(fgh)' = (fg)'h + h'fg$$

Once again taking advantage of the two-function product rule for differentiation, we can expand the first addend on the right side to get

$$(fgh)' = (f'g + g'f)\,h + h'fg$$

The distributive law for multiplication over addition (from basic algebra) allows us to reorganize the right side further, getting

$$(fgh)' = f'gh + g'fh + h'fg$$

This can be expressed another way, which some people find easier to remember:

$$(fgh)' = f'gh + fg'h + fgh'$$

When reading or writing something like this, we must remember that the *prime symbol,* indicating the derivative, applies only to the *single* function or parenthetical expression *immediately to its left.*

5. We want to find the derivative of the function

$$p(y) = 1 / (y^2 + 1)$$

which is the reciprocal of a function we can call g, like this:

$$g(y) = y^2 + 1$$

Taking the derivative of g, we get

$$g'(y) = 2y$$

Multiplying by -1 gives us

$$-g'(y) = -2y$$

The square of g is

$$[g(y)]^2 = (y^2 + 1)^2 = y^4 + 2y^2 + 1$$

Now we can apply the reciprocal rule for differentiation to obtain

$$p'(y) = \{1 \,/\, [g(y)]\}' = -g'(y) \,/\, [g(y)]^2 = -2y \,/\, (y^4 + 2y^2 + 1)$$

Our function p is defined for all reals because the original denominator function, g, never becomes 0. The function p is also continuous, and it doesn't turn any corners as we can see if we graph it. (The graph is not included here. If you want, feel free to draw it and see for yourself.) We can therefore say with confidence that p is differentiable over the entire set of reals.

6. This function looks a lot like the one we worked with in Prob. 5. The only difference is that there's a minus sign in place of the plus sign. But that makes a big difference in the extent to which the function is differentiable! We want to find the derivative of

$$r(y) = 1 \,/\, (y^2 - 1)$$

which is the reciprocal of a function we can call f, like this:

$$f(y) = y^2 - 1$$

Taking the derivative of f, we get

$$f'(y) = 2y$$

Multiplying by -1 gives us

$$-f'(y) = -2y$$

The square of f is

$$[f(y)]^2 = (y^2 - 1)^2 = y^4 - 2y^2 + 1$$

The reciprocal rule for differentiation tells us that

$$r'(y) = \{1 \,/\, [f(y)]\}' = -f'(y) \,/\, [f(y)]^2 = -2y \,/\, (y^4 - 2y^2 + 1)$$

If $y = 1$ or $y = -1$, then $r(y)$ is undefined, so $r'(y)$ is also undefined. Those values of y are the zeros of the denominator function $f(y)$. These discontinuities tell us that $r(y)$ is nondifferentiable at the points where $y = 1$ and $y = -1$.

7. We want to find the derivative of

$$s(z) = (z^4 - 1) \,/\, (z^2 + 1)$$

which is the quotient of two functions we can call f and g, as follows:

$$f(z) = z^4 - 1$$

and

$$g(z) = z^2 + 1$$

When we differentiate the quotient directly, we get

$$(f/g)'(z) = d/dz\,[(z^4 - 1) / (z^2 + 1)] = d/dz\,(z^2 - 1) = 2z$$

When we apply the quotient rule, we get

$$
\begin{aligned}
(f/g)'(z) &= \{[f'(z)][g(z)] - [g'(z)][f(z)]\} / [g(z)]^2 \\
&= \{[d/dz\,(z^4 - 1)](z^2 + 1) - [d/dz\,(z^2 + 1)]\,(z^4 - 1)\} / (z^2 + 1)^2 \\
&= [(4z^3)(z^2 + 1) - (2z)(z^4 - 1)] / (z^4 + 2z^2 + 1) \\
&= (4z^5 + 4z^3 - 2z^5 + 2z) / (z^4 + 2z^2 + 1) \\
&= (2z^5 + 4z^3 + 2z) / (z^4 + 2z^2 + 1) = 2z\,(z^4 + 2z^2 + 1) / (z^4 + 2z^2 + 1) = 2z
\end{aligned}
$$

This works for all real numbers, because the denominator function *g* never attains the value 0. The original function is therefore differentiable over the entire set of reals.

8. This time, we want to find the derivative of

$$t(z) = (z^4 - 1) / (z^2 - 1)$$

which is the quotient of two functions we can call *f* and *g*, as follows:

$$f(z) = z^4 - 1$$

and

$$g(z) = z^2 - 1$$

When we differentiate the quotient directly, we get

$$(f/g)'(z) = d/dz\,[(z^4 - 1) / (z^2 - 1)] = d/dz\,(z^2 + 1) = 2z$$

When we apply the quotient rule, we get

$$
\begin{aligned}
(f/g)'(z) &= \{[f'(z)][g(z)] - [g'(z)][f(z)]\} / [g(z)]^2 \\
&= \{[d/dz\,(z^4 - 1)](z^2 - 1) - [d/dz\,(z^2 - 1)]\,(z^4 - 1)\} / (z^2 - 1)^2 \\
&= [(4z^3)(z^2 - 1) - (2z)(z^4 - 1)] / (z^4 - 2z^2 + 1) \\
&= (4z^5 - 4z^3 - 2z^5 + 2z) / (z^4 - 2z^2 + 1) \\
&= (2z^5 - 4z^3 + 2z) / (z^4 - 2z^2 + 1) = 2z\,(z^4 - 2z^2 + 1) / (z^4 - 2z^2 + 1) = 2z
\end{aligned}
$$

The function $g(z)$ becomes 0 when $z = 1$ or $z = -1$. Nondifferentiable points in $t(z)$ therefore exist where $z = 1$ and $z = -1$, because $t(z)$ blows up at those values of z.

9. Here are the functions again, for reference:

$$f(x) = x^3 - 4x^2$$

and

$$g(y) = y^2 + 5y$$

The composite function we want to differentiate is

$$g[f(x)] = (x^3 - 4x^2)^2 + 5(x^3 - 4x^2) = x^6 - 8x^5 + 16x^4 + 5x^3 - 20x^2$$

When we differentiate this directly, we get

$$\{g[f(x)]\}' = d/dx\,(x^6 - 8x^5 + 16x^4 + 5x^3 - 20x^2) = 6x^5 - 40x^4 + 64x^3 + 15x^2 - 40x$$

Differentiating the functions individually, we obtain

$$f'(x) = 3x^2 - 8x$$

and

$$g'(y) = 2y + 5$$

When we apply the chain rule, we get

$$\{g[f(x)]\}' = g'[f(x)] \times f'(x) = [2 \times (x^3 - 4x^2) + 5]\,(3x^2 - 8x)$$
$$= (2x^3 - 8x^2 + 5)\,(3x^2 - 8x) = 6x^5 - 40x^4 + 64x^3 + 15x^2 - 40x$$

which agrees with the result we got when we differentiated the composite function directly.

10. Here are the functions again, for reference:

$$f(x) = x^2 - 4x$$

and

$$g(y) = 2y^2 + 7y$$

The composite function we want to differentiate is

$$g[f(x)] = 2 \times (x^2 - 4x)^2 + 7 \times (x^2 - 4x) = 2x^4 - 16x^3 + 39x^2 - 28x$$

When we differentiate this composite function directly, we get

$$\{g[f(x)]\}' = d/dx\,(2x^4 - 16x^3 + 39x^2 - 28x) = 8x^3 - 48x^2 + 78x - 28$$

When we differentiate the functions individually, we get

$$f'(x) = 2x - 4$$

and

$$g'(y) = 4y + 7$$

Applying the chain rule gives us

$$\{g[f(x)]\}' = g'[f(x)] \times f'(x) = [4 \times (x^2 - 4x) + 7](2x - 4)$$
$$= (4x^2 - 16x + 7)(2x - 4) = 8x^3 - 48x^2 + 78x - 28$$

which agrees with the result we got when we differentiated the composite function directly.

Chapter 7

1. Here's the function we want to differentiate:

    $$p(t) = 2t^2 - 4t + 5 + 4t^{-1} + 6t^{-2}$$

    We can differentiate each term individually according to the real-number power rule, and then add the differentiated terms up. The derivative of the middle term is 0, so we can skip it in the final sum, getting

    $$p'(t) = 4t - 4 - 4t^{-2} - 12t^{-3}$$

    The function $p(t)$ is nondifferentiable at $t = 0$, because that value generates fractions with denominators equal to 0. But $p(t)$ is differentiable for all nonzero values of t.

2. Here's the function we want to differentiate:

    $$q(w) = (w^{-1} + 1)(w^{-1} - 1)$$

    We can approach this in two different ways. We can treat the factors individually and use the product rule to find the derivative of q, or we can multiply the factors together first and then differentiate. Let's use the second method. We have

    $$(w^{-1} + 1)(w^{-1} - 1) = w^{-2} - w^{-1} + w^{-1} - 1 = w^{-2} - 1$$

    When we differentiate this, we obtain

    $$q'(w) = -2w^{-3}$$

    The function $q(w)$ is nondifferentiable at $w = 0$, because that value generates a fraction with a denominator of 0. But $q(w)$ is differentiable for all nonzero values of w.

3. We can use the reciprocal rule for differentiation, which tells us that if f is a differentiable function of a single variable, then

$$(1 / f)' = -f' / f^2$$

We are told to find the derivative of the cosecant, which is the reciprocal of the sine. Using the reciprocal rule, we get

$$(\csc x)' = [1 / (\sin x)]' = -(\sin x)' / \sin^2 x = -\cos x / \sin^2 x$$

Some derivative tables show this result in a different form. If you come across a table like that, try converting the expression in that table to the one shown here, or vice versa. You should be able to show that the two expressions are equivalent. The function $\csc x$ is nondifferentiable when $\sin x = 0$. That includes all integer multiples of π. Discontinuities occur at those values of x; the reciprocal of the sine is singular there.

4. Again, we can use the reciprocal rule for differentiation. We want to find the derivative of the secant, which is the reciprocal of the cosine. Using the reciprocal rule, we get

$$(\sec x)' = [1 / (\cos x)]' = -(\cos x)' / \cos^2 x = \sin x / \cos^2 x$$

As with the solution to Prob. 3, some derivative tables show this in a different form. You should be able to show that the two expressions are equivalent. The function $\sec x$ is nondifferentiable when $\cos x = 0$. That includes all odd integer multiples of $\pi/2$. Discontinuities occur at those values of x; the reciprocal of the cosine is singular there.

5. In this situation, we'll use the quotient rule for differentiation. This rule tells us that if f and g are differentiable functions of the same variable, then

$$(f/g)' = (f'g - g'f) / g^2$$

We want to find the derivative of the tangent function, which is the sine divided by the cosine. Let

$$f(x) = \sin x$$

and

$$g(x) = \cos x$$

The derivatives are

$$f'(x) = \cos x$$

and

$$g'(x) = -\sin x$$

Therefore

$$(\tan x)' = (\sin x / \cos x)' = [(\sin x)' \cos x - (\cos x)' \sin x] / \cos^2 x$$
$$= [\cos x \cos x - (-\sin x) \sin x] / \cos^2 x = (\cos^2 x + \sin^2 x) / \cos^2 x$$

Now it's time to remember one of the cardinal rules of trigonometry:

$$\cos^2 x + \sin^2 x = 1$$

We can therefore simplify the above equation to

$$(\tan x)' = 1 / \cos^2 x$$

By definition, we know that

$$1 / \cos x = \sec x$$

That means we can simplify our solution further to

$$(\tan x)' = \sec^2 x$$

The function $\sec^2 x$ is nondifferentiable whenever $\cos x = 0$. That includes all odd integer multiples of $\pi/2$. Discontinuities occur at those values of x; the reciprocal of the cosine is singular there.

6. This is a sort of "fraternal twin" to Prob. 5. We want to find the derivative of the cotangent function, which is the cosine divided by the sine. Again, let's use the derivative-of-a-quotient formula

$$(f/g)' = (f'g - g'f) / g^2$$

This time, we'll assign

$$f(x) = \cos x$$

and

$$g(x) = \sin x$$

The derivatives are

$$f'(x) = -\sin x$$

and

$$g'(x) = \cos x$$

Therefore

$$(\cot x)' = (\cos x / \sin x)' = [(\cos x)' \sin x - (\sin x)' \cos x] / \sin^2 x$$
$$= (-\sin x \sin x - \cos x \cos x) / \sin^2 x = (-\sin^2 x - \cos^2 x) / \sin^2 x$$
$$= -(\sin^2 x + \cos^2 x) / \sin^2 x$$

We can again use the trigonometry rule that helped us in Prob. 5, reversing the order of the addends:

$$\sin^2 x + \cos^2 x = 1$$

This allows us to reduce the above equation to

$$(\cot x)' = -1 / \sin^2 x$$

By definition, we know that

$$1 / \sin x = \csc x$$

That means we can simplify our solution further to

$$(\cot x)' = -\csc^2 x$$

The function $-\csc^2 x$ is nondifferentiable whenever $\sin x = 0$. That includes all integer multiples of π. Discontinuities occur at those values of x; the reciprocal of the sine is singular there.

7. Here's the function we want to differentiate:

$$p(x) = e^{ax}$$

Let's use the chain rule. We can break the function p down into two component functions f and g, such that

$$f(x) = ax$$

and

$$g(y) = e^y$$

Then the function p becomes

$$p = g(f)$$

Both f and g are differentiable, so

$$f'(x) = a$$

and

$$g'(y) = e^y$$

The chain rule says that if f and g are differentiable functions of the same variable, then

$$[g(f)]' = g'(f) \cdot f'$$

Substituting the values we've defined into this formula, we get

$$p'(x) = e^{ax} \cdot a = ae^{ax}$$

8. This time, we want to find an expression for the derivative of

$$q(x) = be^{ax}$$

Let's use the results of the solution to Prob. 7 to help us solve this problem. We can rewrite the function q as

$$q(x) = b \cdot p(x)$$

where p is the function defined in Prob. 7 and its and solution:

$$p(x) = e^{ax}$$

Therefore

$$q'(x) = [b \cdot p(x)]'$$

The multiplication-by-constant rule for differentiation allows us to rewrite this as

$$q'(x) = b \cdot p'(x)$$

We already know that

$$p'(x) = ae^{ax}$$

Therefore,

$$q'(x) = b \cdot ae^{ax} = abe^{ax}$$

9. We want to differentiate

$$r(x) = \ln ax$$

This function is defined and continuous if and only if $ax > 0$. That means we must restrict the domain to the positive reals if $a > 0$, and to the negative reals if $a < 0$. (We can't have $a = 0$, because in that case $ax = 0$, and ln 0 is not defined.) Our function never turns a corner at any point where it is defined. That means it's differentiable at all such points. Let's break r into components f and g, such that

$$f(x) = ax$$

and

$$g(y) = \ln y$$

Then we have

$$r = g(f)$$

The derivatives of the components are

$$f'(x) = a$$

and

$$g'(y) = 1/y$$

The chain rule tells us that

$$[g(f)]' = g'(f) \cdot f'$$

Substituting the values we've defined into this formula, we get

$$r'(x) = 1 / (ax) \cdot a = 1/x$$

10. We want to find an expression for the derivative of

$$s(x) = b \ln ax$$

Using the results of the solution to Prob. 9, we can rewrite the function s as

$$s(x) = b \cdot r(x)$$

where r is the function defined in Prob. 9 and its solution 9:

$$r(x) = \ln ax$$

Therefore

$$s'(x) = [b \cdot r(x)]'$$

The multiplication-by-constant rule lets us rewrite this as

$$s'(x) = b \cdot r'(x)$$

We already know that

$$r'(x) = 1/x$$

Therefore,

$$q'(x) = b \cdot (1/x) = b/x$$

The same constraints apply here as in the solution to Prob. 9. We must restrict the domain to the positive reals if $a > 0$, and to the negative reals if $a < 0$.

Chapter 8

1. The fifth derivative of this function is the derivative of its fourth derivative. We found the fourth derivative to be the constant function

$$f^{(4)}(x) = -96$$

Differentiating this, we get

$$f^{(5)}(x) = 0$$

because the derivative of a constant function is always the zero function.

2. The sixth derivative of our original function is the derivative of

$$f^{(5)}(x) = 0$$

This is also the zero function. We've reached "the end of the line" in this sequence of derivatives. If we keep differentiating further, nothing changes!

3. We found the fourth derivative to be

$$f^{(4)}(x) = 96 - 5 \cos x$$

The sum rule tells us that

$$f^{(5)}(x) = d/dx\, 96 - d/dx\,(5 \cos x)$$

Using the multiplication-by-constant rule, and knowing that the derivative of any constant is equal to 0, we get

$$f^{(5)}(x) = 0 - 5\, d/dx\,(\cos x)$$

The derivative of the cosine is the negative of the sine, and we get rid of the 0. That leaves us with

$$f^{(5)}(x) = -5 \cdot (-\sin x) = 5 \sin x$$

4. The sixth derivative is the result of differentiating the fifth derivative. That means

$$f^{(6)}(x) = d/dx\,(5 \sin x) = 5\,d/dx\,(\sin x) = 5 \cos x$$

5. The seventh derivative is the result of differentiating the sixth derivative. That means

$$f^{(7)}(x) = d/dx\,(5 \cos x) = 5\,d/dx\,(\cos x) = 5 \cdot (-\sin x) = -5 \sin x$$

As we keep on differentiating, we'll go through a four-way cycle. Looking all the way back to the original function and then listing its succeeding derivatives up to the seventh, we see:

$$f(x) = 4x^4 - 5 \cos x$$
$$f'(x) = 16x^3 + 5 \sin x$$
$$f''(x) = 48x^2 + 5 \cos x$$
$$f'''(x) = 96x - 5 \sin x$$
$$f^{(4)}(x) = 96 - 5 \cos x$$
$$f^{(5)}(x) = 5 \sin x$$
$$f^{(6)}(x) = 5 \cos x$$
$$f^{(7)} = -5 \sin x$$

From here, we can extrapolate the higher derivatives:

$$f^{(8)} = -5 \cos x$$
$$f^{(9)}(x) = 5 \sin x$$
$$f^{(10)}(x) = 5 \cos x$$
$$f^{(11)} = -5 \sin x$$
$$f^{(12)} = -5 \cos x$$
$$f^{(13)}(x) = 5 \sin x$$
$$f^{(14)}(x) = 5 \cos x$$
$$f^{(15)} = -5 \sin x$$
$$\downarrow$$

And so on, forever

6. When we calculated the fourth derivative, we got

$$g^{(4)}(t) = -6t^{-4} + 96t^{-5} - 120t^{-6} + 720t^{-7}$$

Finding the fifth derivative involves differentiating this function, one term at a time, and then adding the terms back up. When we do this, we get

$$g^{(5)}(t) = 24t^{-5} - 480t^{-6} + 720t^{-7} - 5{,}040t^{-8}$$

7. To find the sixth derivative, we differentiate the result of the solution to Prob. 6 to get

$$g^{(6)}(t) = -120t^{-6} + 2{,}880t^{-7} - 5{,}040t^{-8} + 40{,}320t^{-9}$$

8. To find the seventh derivative, we differentiate the result of the solution to Prob. 7 to get

$$g^{(7)}(t) = 720t^{-7} - 20{,}160t^{-8} + 40{,}320t^{-9} - 362{,}880t^{-10}$$

If we write down the original function and then list all the derivatives we've found so far, in succession, we see a general pattern:

$$g(t) = 2t^2 + 5t - 7 + 4t^{-1} - t^{-2} + 2t^{-3} + \ln t$$
$$g'(t) = 4t + 5 + t^{-1} - 4t^{-2} + 2t^{-3} - 6t^{-4}$$
$$g''(t) = 4 - t^{-2} + 8t^{-3} - 6t^{-4} + 24t^{-5}$$
$$g'''(t) = 2t^{-3} - 24t^{-4} + 24t^{-5} - 120t^{-6}$$
$$g^{(4)}(t) = -6t^{-4} + 96t^{-5} - 120t^{-6} + 720t^{-7}$$
$$g^{(5)}(t) = 24t^{-5} - 480t^{-6} + 720t^{-7} - 5{,}040t^{-8}$$
$$g^{(6)}(t) = -120t^{-6} + 2{,}880t^{-7} - 5{,}040t^{-8} + 40{,}320t^{-9}$$
$$g^{(7)}(t) = 720t^{-7} - 20{,}160t^{-8} + 40{,}320t^{-9} - 362{,}880t^{-10}$$

And so on, forever

9. We're told that the height of the cliff, h_c, is 50 m. We can find the descent time t by using the equation for total fallen distance vs. time, plugging in 50 for h_c to get

$$50 = 5t^2$$

Dividing through by 10, we obtain

$$10 = t^2$$

This solves to $t = 10^{1/2}$, which is approximately 3.16 s. The vertical speed v of a falling object at time t after its release is always

$$v = dh/dt = d/dt\,(5t^2) = 10t$$

We can calculate the vertical speed at impact, v_a, by plugging in 3.16 for t, getting

$$v_a = 10 \cdot 3.16 = 31.6 \text{ m/s}$$

The vertical acceleration a of a falling object at time t after its release is always

$$a = d^2/dt^2\,(5t^2) = d/dt\,(10t) = 10 \text{ m/s}^2$$

so the vertical acceleration a_a of the apple at impact was 10 m/s^2, the same as it was when the apple was thrown from the higher cliff.

10. We can calculate the height of the cliff, h_c, by plugging in 11 for t in the equation

$$h = 3t^2$$

that describes total fallen distance h vs. time t on the "alien planet." When we do that, we get

$$h_c = 3 \cdot 11^2 = 3 \cdot 121 = 363 \text{ m}$$

The vertical speed v of a falling object at time t after its release on this planet is always

$$v = dh/dt = d/dt\,(3t^2) = 6t$$

We can calculate the vertical speed of the "alien apple" at impact, v_a, by plugging in 11 for t, getting

$$v_a = 6 \cdot 11 = 66 \text{ m/s}$$

The vertical acceleration a of a falling object on this planet at time t after its release is always

$$a = d^2/dt^2(3t^2) = d/dt\,(6t) = 6 \text{ m/s}^2$$

The "alien apple" was therefore accelerating downward at 6 m/s^2 when it landed.

Chapter 9

1. Let's determine the x-value of the inflection point first. The original function is

$$y = 3x^3 + 3x^2 - x - 7$$

The first derivative is

$$dy/dx = 9x^2 + 6x - 1$$

and the second derivative is

$$d^2y/dx^2 = 18x + 6$$

The inflection point occurs where the second derivative is equal to 0. We must therefore solve the equation

$$18x + 6 = 0$$

Using algebra, we find that $x = -1/3$. Plugging this into the original cubic, we get

$$y = 3x^3 + 3x^2 - x - 7 = 3 \cdot (-1/3)^3 + 3 \cdot (-1/3)^2 - (-1/3) - 7 = -58/9$$

The coordinates of the inflection point are therefore $(x,y) = (-1/3, -58/9)$.

2. When we found the first derivative of the function stated in Prob. 1, we got

$$dy/dx = 9x^2 + 6x - 1$$

To find the slope of the graph at the inflection point, we plug the x-value of the inflection point into the above equation, getting

$$dy/dx = 9x^2 + 6x - 1 = 9 \cdot (-1/3)^2 + 6 \cdot (-1/3) - 1 = -2$$

3. From our knowledge of inflection points, we can be sure that curve of the function stated in Prob. 1 is concave upward on one side of the inflection point, and concave downward on the other side. A curve is concave upward at a point if and only if the second derivative is positive at that point. A curve is concave downward at a point if and only if the second derivative is negative at that point. Let's set $x = 0$, which is in the zone to the right (that is, on the positive side) of the inflection point where $x = -1/3$. Plugging this into the second derivative, we get

$$d^2y/dx^2 = 18x + 6 = 18 \cdot 0 + 6 = 6$$

The fact that this is a positive number tells us that the curve is concave upward to the right of the inflection point, where $x > -1/3$. The curve must therefore be concave downward to the left of the inflection point, where $x < -1/3$.

4. Here is the quadratic function again, for reference:

$$y = 4x^2 - 7$$

When we differentiate, we get

$$dy/dx = 8x$$

The extremum is the point at which the first derivative is equal to 0. That means we must solve the equation

$$8x = 0$$

for x. That's easy! We get $x = 0$, which is the x-value of the extremum. To find the y-value of the extremum, we substitute 0 for x in the original quadratic function, getting

$$y = 4x^2 - 7 = 4 \cdot 0^2 - 7 = -7$$

The extremum point is therefore $(x,y) = (0,-7)$. Because the leading coefficient is positive in the polynomial for the original function, we know that the parabola opens upward, so this extremum is an absolute minimum. We can also tell by taking the second derivative and noting that it's equal to 8, which is positive, indicating that the curve is concave upward.

5. In the chapter text, we found that the slope of the parabola of Fig. 9-4 is equal to 1 at the point (2,0). Now let's recall, from algebra, the point-slope form of a linear equation:

$$y - y_0 = m(x - x_0)$$

where (x_0, y_0) are the coordinates of the known point, and m is the slope of the line. We've determined that $x_0 = 2$, $y_0 = 0$, and $m = 1$, so we can plug in these numbers to get the equation

$$y - 0 = 1 \cdot (x - 2)$$

which simplifies to

$$y = x - 2$$

That's the equation of a line tangent to the parabola, passing through the point (2,0).

6. Here's the cubic function again, for reference:

$$y = x^3 + 3x^2 - 3x + 4$$

The first derivative is

$$dy/dx = 3x^2 + 6x - 3$$

The second derivative is

$$d^2y/dx^2 = 6x + 6$$

We can find the x-value of the inflection point by setting $d^2y/dx^2 = 0$. That gives us the linear equation

$$6x + 6 = 0$$

which solves to $x = -1$. Plugging this value into the original cubic and solving for y, we get

$$y = x^3 + 3x^2 - 3x + 4 = (-1)^3 + 3 \cdot (-1)^2 - 3 \cdot (-1) + 4 = 9$$

The coordinates of the inflection point are therefore $(x,y) = (-1,9)$. We can find the slope of the curve at this point by plugging the x-value into the first derivative, obtaining

$$dy/dx = 3x^2 + 6x - 3 = 3 \cdot (-1)^2 + 6 \cdot (-1) - 3 = -6$$

7. When we plug $x = 100$ into the original cubic function, we get

$$y = x^3 + 3x^2 - 3x + 4 = 100^3 + 3 \cdot 100^2 - 3 \cdot 100 + 4 = 1,029,704$$

That's "way up there" in the positive y direction. The inflection point, by comparison, is near the origin. This means that the curve, being a cubic, trends upward in the overall sense. But in the solution to Prob. 6, we found that the slope at the inflection point is negative, indicating that in the vicinity of that point, the curve trends downward! We must therefore conclude that the graph reverses itself in a small region near the inflection point. On the basis of this information, we can be sure that this curve matches the general profile of Fig. 9-2E.

8. To find the y-intercept, we plug $x = 0$ into the original function and then calculate the result. That gives us

$$y = x^3 + 3x^2 - 3x + 4 = 0^3 + 3 \cdot 0^2 - 3 \cdot 0 + 4 = 4$$

This tells us that the point $(x,y) = (0,4)$ is on the curve. To find the points where the slope is 0, we can set the first derivative equal to 0 and then solve the resulting equation. Remember that

$$dy/dx = 3x^2 + 6x - 3$$

so the quadratic equation is

$$3x^2 + 6x - 3 = 0$$

Let's use the quadratic formula to see if this equation has any real roots, and if so, to find them. If you don't remember that formula, here's a reminder. When we have a quadratic equation of the form

$$ax^2 + bx + c = 0$$

then the roots are

$$x = [-b \pm (b^2 - 4ac)^{1/2}] / (2a)$$

In this case, $a = 3$, $b = 6$, and $c = -3$. We therefore get

$$x = \{-6 \pm [6^2 - 4 \cdot 3 \cdot (-3)]^{1/2}\} / (2 \cdot 3) = -1 + 2^{1/2} \text{ or } -1 - 2^{1/2}$$

For graphing purposes, we can use a calculator to approximate these x-values as 0.414 and −2.414. Let's plug these into the cubic and do the arithmetic. For $x = 0.414$, we have

$$y = x^3 + 3x^2 - 3x + 4 = 0.414^3 + 3 \cdot 0.414^2 - 3 \cdot 0.414 + 4 = 3.343$$

The coordinates of one local extremum are therefore approximately

$$(x,y) = (0.414, 3.343)$$

That's good enough for graphing purposes! In the case of $x = -2.414$, we have

$$y = x^3 + 3x^2 - 3x + 4 = (-2.414)^3 + 3 \cdot (-2.414)^2 - 3 \cdot (-2.414) + 4 = 14.657$$

The coordinates of the other local extremum are approximately

$$(x,y) = (-2.414, 14.657)$$

We now have this information:

- The *y*-intercept point is (0,4)
- The inflection point is (−1,9)
- One local extremum point is approximately (0.414,3.343)
- The other local extremum point is approximately (−2.414,14.657)
- The function trends upward in the overall sense, but reverses between the local extrema

That's enough data to sketch a reasonably good graph, as shown in Fig. A-13. Each division on the *x* axis represents 1/2 unit. Each division on the *y* axis represents 4 units.

9. To find this slope, we plug the known *x*-value at the inflection point into the equation for the derivative. That's *x* = π. Remember that

$$d/dx \, (\sin x) = \cos x$$

The slope at the inflection point is therefore equal to cos π, or −1.

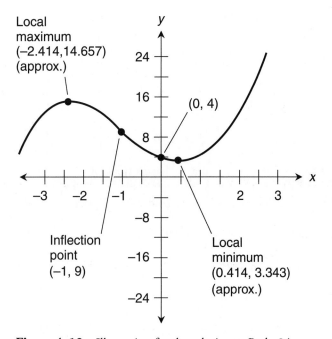

Figure A-13 Illustration for the solution to Prob. 8 in Chap. 9.

10. The *x*-values of the inflection points, which we figured out in the chapter text, are

$$x = \pi/4$$
$$x = 3\pi/4$$
$$x = 5\pi/4$$
$$x = 7\pi/4$$

The derivative of the square of the sine function, as we found in Chap. 7 and repeated in Chap. 9, is

$$d/dx \sin^2 x = 2 \sin x \cos x$$

Finding the slope at each of the inflection points involves plugging in the values and doing the arithmetic. Note that the *exact* values of the sines we'll encounter are

$$\sin \pi/4 = 1/(2^{1/2})$$
$$\sin 3\pi/4 = 1/(2^{1/2})$$
$$\sin 5\pi/4 = -1/(2^{1/2})$$
$$\sin 7\pi/4 = -1/(2^{1/2})$$

and the *exact* values of the cosines are

$$\cos \pi/4 = 1/(2^{1/2})$$
$$\cos 3\pi/4 = -1/(2^{1/2})$$
$$\cos 5\pi/4 = -1/(2^{1/2})$$
$$\cos 7\pi/4 = 1/(2^{1/2})$$

Now let's go through the calculations, taking care not to get the signs mixed up. At the first (left-most) inflection point in Fig. 9-7, we have a slope of

$$2 \sin \pi/4 \cos \pi/4 = 2 \cdot 1/(2^{1/2}) \cdot 1/(2^{1/2}) = 1$$

At the second inflection point, we have a slope of

$$2 \sin 3\pi/4 \cos 3\pi/4 = 2 \cdot 1/(2^{1/2}) \cdot [-1/(2^{1/2})] = -1$$

At the third inflection point, the slope is

$$2 \sin 5\pi/4 \cos 5\pi/4 = 2 \cdot [-1/(2^{1/2})] \cdot [(-1/(2^{1/2})] = 1$$

At the fourth (right-most) inflection point, the slope is

$$2 \sin 7\pi/4 \cos 7\pi/4 = 2 \cdot [-1/(2^{1/2})] \cdot 1/(2^{1/2}) = -1$$

B

Worked-Out Solutions to Exercises: Chapters 11 to 19

These solutions do not necessarily represent the only way a problem can be figured out. If you think you can solve a problem faster or better than the way it's done here, by all means try it! Always check your work to be sure your "alternative" answer is correct.

Chapter 11

1. To write the first n terms of a series in summation notation, we must figure out how it can be portrayed as a function of a steadily increasing sequence of integers, preferably

$$1, 2, 3, 4, 5, 6, 7, \ldots$$

Then we decide on a counting tag, give it a name such as i, and let it follow the above sequence. Here, our working series starts with 8, and we divide by 2 to get the second term. To get each succeeding term, we divide by 2. The series can be rewritten as

$$8/2^0 + 8/2^1 + 8/2^2 + 8/2^3 + \cdots$$

Alternatively, we can begin the series with a numerator of 16 and divide by increasing powers of 2 starting with the first power, like this:

$$16/2^1 + 16/2^2 + 16/2^3 + 16/2^4 + \cdots$$

That allows us to start indexing at $i = 1$, so the summation notation is

$$\sum_{i-1}^{n} 16/2^i$$

for the first n terms of the series.

2. In this series, we begin with 1/2, which is $i / (1 + i)$ if we let $i = 1$. We increase i by 1 to get the next term in the series, which is 2/3. Next, we increase the original numerator

and denominator by 2 to get 3/4. Then we add 3 to the original numerator and denominator. Then we add 4 to both, then we add 5 to both, and so on. The first n terms in the series can be expressed as

$$\sum_{i=1}^{n} i/(1+i)$$

3. To express the limit of the series stated in Probs. 1 and 2 as the number of terms grows endlessly, we write down the limits as n approaches infinity. In the first case, we have

$$\lim_{n\to\infty} \sum_{i=1}^{n} 16/2^i$$

and in the second case,

$$\lim_{n\to\infty} \sum_{i=1}^{n} i/(1+i)$$

4. The first limit in the solution to Prob. 3 is defined, because the series converges (approaches a specific limiting value) as the number of terms grows endlessly. The whole infinite series adds up to 16. Therefore

$$\lim_{n\to\infty} \sum_{i=1}^{n} 16/2^i = 16$$

The second limit is not defined because the series

$$1/2 + 2/3 + 3/4 + 4/5 + 5/6 + 6/7 + 7/8 \cdots$$

does not converge. When a series does not converge, it is said to *diverge*.

5. We want to find

$$\int_0^1 x^2 \, dx$$

This integral can be expressed in the limit form

$$\lim_{n\to\infty} \sum_{i=1}^{n} [(b-a)/n] \cdot f[a + i(b-a)/n]$$

Here, $a = 0$ and $b = 1$, so we can rewrite the above expression as

$$\lim_{n\to\infty} \sum_{i=1}^{n} [(1-0)/n] \cdot [0 + i(1-0)/n]^2$$

which simplifies to

$$\lim_{n\to\infty} \sum_{i=1}^{n} i^2/n^3$$

Let's look at the summation part alone:

$$\sum_{i=1}^{n} i^2/n^3$$

We can write this series out as

$$1^2/n^3 + 2^2/n^3 + 3^2/n^3 + \cdots + n^2/n^3$$

which simplifies to

$$(1^2 + 2^2 + 3^2 + \cdots + n^2) / n^3$$

From precalculus, recall that

$$\underset{n \to \infty}{Lim} \; (1^2 + 2^2 + 3^2 + \cdots + n^2) / n^3 = 1/3$$

Now we know that

$$\underset{n \to \infty}{Lim} \sum_{i=1}^{n} i^2/n^3 = 1/3$$

That's the integral we're looking for. We've figured out that

$$\int_{0}^{1} x^2 \, dx = 1/3$$

6. To determine the average value, we must find the height of a rectangle whose area is 1/3 square unit (the definite integral) and whose width is 1 unit (the distance from $x = 0$ to $x = 1$). If we call the rectangle's area *A*, its height *h*, and its width *w*, then

$$A = hw$$

from basic geometry. In this situation, $A = 1/3$ and $w = 1$, so

$$1/3 = h \cdot 1$$

Therefore, $h = 1/3$. That's the average value of the function

$$f(x) = x^2$$

over the interval from $x = 0$ to $x = 1$.

7. This time, we've been told to find

$$\int_{0}^{1} x^3 \, dx$$

This can be expressed in limit form as

$$\underset{n \to \infty}{Lim} \sum_{i=1}^{n} [(b-a)/n] \cdot f[a + i(b-a)/n]$$

In this situation, $a = 0$ and $b = 1$, so we can rewrite the above expression as

$$\text{Lim}_{n \to \infty} \sum_{i=1}^{n} [(1-0)/n] \cdot [0 + i(1-0)/n]^3$$

which simplifies to

$$\text{Lim}_{n \to \infty} \sum_{i=1}^{n} i^3/n^4$$

The summation alone is

$$\sum_{i=1}^{n} i^3/n^4$$

When we write this series out, we obtain

$$1^3/n^4 + 2^3/n^4 + 3^3/n^4 + \cdots + n^3/n^4$$

which simplifies to

$$(1^3 + 2^3 + 3^3 + \cdots + n^3) / n^4$$

Recall from precalculus that

$$\text{Lim}_{n \to \infty} (1^3 + 2^3 + 3^3 + \cdots + n^3) / n^4 = 1/4$$

Therefore

$$\text{Lim}_{n \to \infty} \sum_{i=1}^{n} i^3/n^4 = 1/4$$

That's the integral we want. We've worked out the fact that

$$\int_{0}^{1} x^3 \, dx = 1/4$$

8. We can again use the rectangle-area method to find the average value. This time, we must find the height of a rectangle whose area is 1/4 of a square unit and whose width is 1 unit. If we call the area A, the height h, and the width w as we did in the solution to Prob. 6, then again,

$$A = hw$$

This time, $A = 1/4$ and $w = 1$, so

$$1/4 = h \cdot 1$$

This solves to $h = 1/4$. The average value of the function

$$f(x) = x^3$$

over the interval from $x = 0$ to $x = 1$ is therefore equal to 1/4.

9. Remember the function that describes the car's speed vs. time:

$$s(t) = 2.4t$$

We want to find

$$\int_0^5 2.4t \, dt$$

This integral can be expressed as

$$\mathit{Lim}_{n \to \infty} \sum_{i=1}^{n} [(b-a)/n] \cdot s[a + i(b-a)/n]$$

In this scenario, $a = 0$ and $b = 5$, so we can rewrite the above expression as

$$\mathit{Lim}_{n \to \infty} \sum_{i=1}^{n} [(5-0)/n] \cdot 2.4 \cdot [0 + i(5-0)/n]$$

which simplifies to

$$\mathit{Lim}_{n \to \infty} \sum_{i=1}^{n} 60i/n^2$$

When we factor 60 out of the sum and the limit, we obtain

$$60 \; \mathit{Lim}_{n \to \infty} \sum_{i=1}^{n} i/n^2$$

The summation part by itself is

$$\sum_{i=1}^{n} i/n^2$$

We can write this series out as

$$1/n^2 + 2/n^2 + 3/n^2 + \cdots + n/n^2$$

which simplifies to

$$(1 + 2 + 3 + \cdots + n) / n^2$$

As we saw in the chapter text,

$$\mathit{Lim}_{n \to \infty} \sum_{i=1}^{n} (1 + 2 + 3 + \cdots + n) / n^2 = 1/2$$

Written in summation form, this is

$$\mathit{Lim}_{n \to \infty} \sum_{i=1}^{n} i/n^2 = 1/2$$

Therefore

$$60 \; \underset{n \to \infty}{Lim} \; \sum_{i=1}^{n} i/n^2 = 30$$

We've determined that

$$\int_{0}^{5} 2.4t \, dt = 30$$

This tells us that the car travels 30 meters in the first 5 seconds.

10. We can verify this result by finding the area of the triangle bounded on the top by the graph of the function, on the right by the line $t = 5$, and on the bottom by the t axis in Fig. 11-5. The height h of the triangle is equal to the value of the function when $t = 5$:

$$h = s(5) = 2.4 \cdot 5 = 12$$

The width w of the triangle is the distance between $t = 0$ and $t = 5$, which is 5 units. The area A of a triangle is half its width times its height. In this case,

$$A = hw/2 = 12 \cdot 5/2 = 30$$

This agrees with the result we got with the Riemann method.

Chapter 12

1. The general antiderivative is

$$F(x) = 2x + c$$

where c can be any real number. Figure B-1 shows the graphs of the three antiderivatives where $c = 1$, $c = 3$, and $c = -2$ respectively:

$$F_1(x) = 2x + 1$$
$$F_3(x) = 2x + 3$$
$$F_{-2}(x) = 2x - 2$$

All the antiderivative graphs are straight lines with slopes of 2. The constants correspond to the points where the lines pass through the dependent-variable axis.

2. The general antiderivative is

$$F(x) = x^2 + c$$

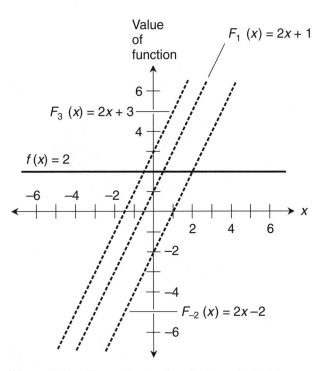

Figure B-1 Illustration for the solution to Prob. 1 in Chap. 12.

where c can be any real number. Figure B-2 shows the graphs of the three antiderivatives where $c=1$, $c=3$, and $c=-2$ respectively:

$$F_1(x) = x^2 + 1$$
$$F_3(x) = x^2 + 3$$
$$F_{-2}(x) = x^2 - 2$$

All the antiderivative graphs are parabolas that open upward. The constants correspond to the points where the curves pass through the dependent-variable axis. The curves all have identical orientations and contours; the only difference is their vertical positions.

3. The general antiderivative is

$$F(x) = x^3 + c$$

where c can be any real number. Figure B-3 shows the graphs of the three antiderivatives where $c = 5$, $c = 15$, and $c = -10$ respectively:

$$F_5(x) = x^3 + 5$$
$$F_{15}(x) = x^3 + 15$$
$$F_{-10}(x) = x^3 - 10$$

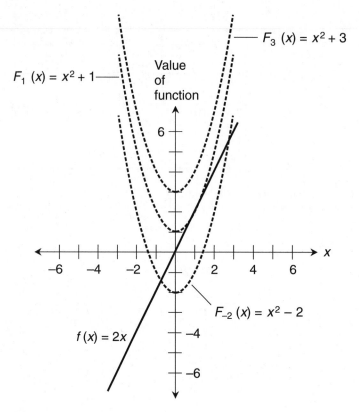

Figure B-2   Illustration for the solution to Prob. 2 in Chap. 12.

On this graph, each horizontal division represents 1 unit, and each vertical division represents 5 units. All the antiderivative graphs are cubic curves. The constants correspond to the points where the curves pass through the dependent-variable axis. The curves all have identical orientations and contours; the only difference is their vertical positions.

4. The general antiderivative is

$$F(x) = x^4 + c$$

where c can be any real number. Figure B-4 shows the graphs of the three antiderivatives where $c = 5$, $c = 15$, and $c = -10$ respectively:

$$F_5(x) = x^4 + 5$$
$$F_{15}(x) = x^4 + 15$$
$$F_{-10}(x) = x^4 - 10$$

On this graph, each horizontal division represents 1 unit, and each vertical division represents 5 units. All the antiderivative graphs are quartic (fourth-degree) curves.

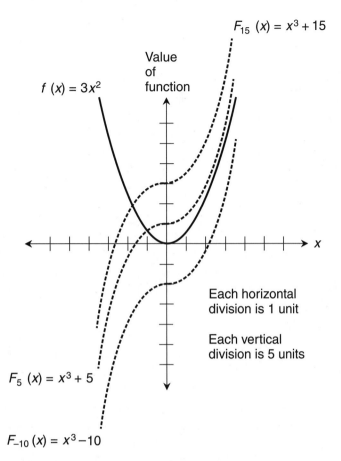

$F_{15}(x) = x^3 + 15$

$f(x) = 3x^2$

Value
of
function

Each horizontal
division is 1 unit

Each vertical
division is 5 units

x

$F_5(x) = x^3 + 5$

$F_{-10}(x) = x^3 - 10$

Figure B-3 Illustration for the solution to Prob. 3 in
Chap. 12. Each horizontal-axis division represents
1 unit. Each vertical-axis division represents
5 units.

The constants correspond to the points where the curves pass through the dependent-variable axis. The curves all have identical orientations and contours; the only difference is their vertical positions.

5. We've been given the following cubic function and told to find its definite integral from $x = 0$ to $x = 5$:

$$f(x) = 8x^3$$

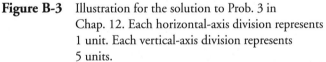

Let's remember the rule for differentiating basic nth-degree functions. If we have

$$f(x) = ax^n$$

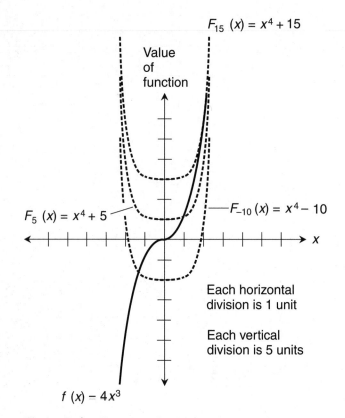

$F_{15}(x) = x^4 + 15$

Value of function

$F_5(x) = x^4 + 5$

$F_{-10}(x) = x^4 - 10$

x

Each horizontal division is 1 unit

Each vertical division is 5 units

$f(x) - 4x^3$

Figure B-4   Illustration for the solution to Prob. 4 in Chap. 12. Each horizontal-axis division represents 1 unit. Each vertical-axis division represents 5 units.

where *n* is a nonnegative integer, then the antiderivative in which the constant of integration is equal to 0 is

$$F(x) = ax^{(n+1)} / (n+1)$$

From this formula, the antiderivative of our cubic function is

$$F(x) = 8x^{(3+1)} / (3+1) = 8x^4/4 = 2x^4$$

We can omit the constant of integration whenever we use the Fundamental Theorem of Calculus. Let's use that theorem to evaluate

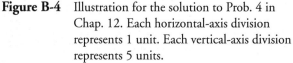

$$\int_0^5 8x^3 \, dx$$

Calculating the antiderivatives, we obtain

$$F(5) = 2 \cdot 5^4 = 1{,}250$$

and

$$F(0) = 2 \cdot (0)^4 = 0$$

Therefore

$$F(5) - F(0) = 1{,}250$$

so we get the final answer

$$\int_0^5 8x^3 \, dx = 1{,}250$$

6. We've been given the following quartic function and told to find its definite integral from $z = -6$ to $z = -3$:

$$g(z) = -2z^4$$

According to the rule for differentiating basic mth-degree functions, the antiderivative, not including the constant of integration, is

$$G(z) = -2z^{(4+1)} / (4 + 1) = -2z^5/5$$

We can use this antiderivative to evaluate the definite integral

$$\int_{-6}^{-3} -2z^4 \, dz$$

Calculating the antiderivatives, we obtain

$$G(-3) = [-2 \cdot (-3)^5] / 5 = [-2 \cdot (-243)] / 5 = 486/5 = 97.2$$

and

$$G(-6) = [-2 \cdot (-6)^5] / 5 = [-2 \cdot (-7{,}776)] / 5 = 15{,}552/5 = 3{,}110.4$$

Therefore

$$G(-3) - G(-6) = 97.2 - 3{,}110.4 = -3{,}013.2$$

so we get the final answer

$$\int_{-6}^{-3} -2z^4 \, dz = -3{,}013.2$$

7. Because the brick starts out having been tossed upward, its initial downward vertical speed is −5 meters per second. As always (on earth, at least), the downward vertical speed increases by 10 meters per second every second until the brick splashes down. If we call the acceleration function f and the time variable t, then, as before,

$$f(t) = 10$$

I start the timer when I toss the brick. When $t = 0$, the vertical downward speed is −5 meters per second, so the antiderivative is

$$F(t) = 10t - 5$$

This function F expresses the vertical *downward* speed of the brick vs. time, assuming that I toss the brick *straight up*. (That's why I had to lean out to toss it; otherwise it would've gone up into the balloon and then fallen back into the gondola!) Antidifferentiating again, remembering the rule for the sum or difference of indefinite integrals, and calling the result $\Phi(t)$, we get

$$\Phi(t) = 5t^2 - 5t$$

The function Φ expresses the "fallen distance" vs. the elapsed time. As before, the constant of integration is 0, because the initial "fallen distance" is 0. Let's calculate how long it will take the brick to hit the water after I've tossed it up. This, once again, is the time t at which the value of Φ is equal to the altitude of the balloon. We have

$$5t^2 - 5t = 1{,}000$$

We can get this equation into the standard quadratic form by subtracting 1,000 from each side, obtaining

$$5t^2 - 5t - 1{,}000 = 0$$

Dividing through by −5 gives us the simpler equation

$$-t^2 + t + 200 = 0$$

Using the quadratic formula to solve for t, we have

$$t = \{-1 \pm [1^2 - 4 \cdot (-1) \cdot 200]^{1/2}\} / [2 \cdot (-1)] = (-1 \pm 801^{1/2}) / (-2)$$

The positive square root of 801 is approximately 28.30. Therefore,

$$t \approx (-1 \pm 28.30) / (-2) \approx -13.65 \text{ or } 14.65$$

The only solution that makes sense is $t \approx 14.65$ seconds. The other solution is a "phantom negative."

8. To figure this out, we plug $t \approx 14.65$ into the function for downward vertical speed vs. time. That's the antiderivative of the acceleration function. We found that to be

$$F(t) = 10t - 5$$

Therefore, the vertical speed of the brick at splashdown is

$$F(t) = 10t - 5 \approx 10 \cdot 14.65 - 5 \approx 141.5$$

The brick is falling at 141.5 meters per second when it splashes down. It's a good thing there are no boats, rafts, or swimmers on the lake. We wouldn't want a brick to land on anybody at that speed!

9. As in the first experiment, the brick's initial speed is 0, and it accelerates downward by 10 meters per second. If the acceleration function is f and the time variable is t, then

$$f(t) = 10$$

We can leave out the constant of integration when we find the antiderivative

$$F(t) = 10t$$

This function expresses the vertical speed vs. time. When $t = 5$, we have

$$F(t) = 10 \cdot 5 = 50$$

After 5 seconds, the brick is falling at 50 meters per second. When $t = 6$,

$$F(t) = 10 \cdot 6 = 60$$

At the end of the 6th second, the brick is falling at 60 meters per second. Between $t = 5$ and $t = 6$, the brick gains 10 meters per second of downward vertical speed. It doesn't seem as if we've done definite integration here. But indirectly, we have. We subtracted one antiderivative (at $t = 5$) from another (at $t = 6$). According to the Fundamental Theorem of Calculus, we've just figured out that

$$\int_{5}^{6} 10 \, dt = 10$$

10. Let's take the antiderivative again and call the new function $\Phi(t)$, just as we did in the first experiment. Then

$$\Phi(t) = 5t^2$$

Once again, we can leave out the constant of integration. The function Φ tells us the vertical displacement in meters vs. the elapsed time in seconds. When $t = 5$, we have

$$\Phi(t) = 5 \cdot 5^2 = 5 \cdot 25 = 125$$

The brick falls 125 meters in the first 5 seconds. When $t = 6$, we have

$$\Phi\,(t) = 5 \cdot 6^2 = 5 \cdot 36 = 180$$

The brick falls 180 meters in the first 6 seconds. Therefore, it falls $180 - 125$, or 55, meters between $t = 5$ and $t = 6$. As in the solution to Prob. 9, it doesn't seem as if we've done formal integration here, but we've found a definite integral in a practical sense. We subtracted the antiderivative at the end of a time interval from the antiderivative at the beginning of the interval. According to the Fundamental Theorem of Calculus, we've determined that

$$\int_5^6 10t\, dt = 55$$

Chapter 13

1. According to the reversal rule for integration (which we've already proved),

$$\int_p^q h\,(v)\, dv = -\int_q^p h\,(v)\, dv$$

The right-hand side of this equation is the same as

$$-1 \int_q^p h\,(v)\, dv$$

The multiplication-by-constant rule tells us that we can rewrite this as

$$\int_q^p -1\, [h\,(v)]\, dv$$

which is the same as

$$\int_q^p -h\,(v)\, dv$$

We've just shown that if h is an integrable function of a variable v over an interval from $v = p$ to $v = q$ where p and q are real-number constants, then

$$\int_p^q h\,(v)\, dv = \int_q^p -h\,(v)\, dv$$

2. To solve this problem, we must merely do some calculations, being careful with the plus and minus signs. First, we want to evaluate the integral of a function over a specific interval going in the positive direction along the x axis:

$$\int_1^2 x^2\, dx$$

If we call our function $f\,(x) = x^2$, then we have the basic antiderivative

$$F\,(x) = x^3/3$$

We calculate

$$F(2) = 2^3/3 = 8/3$$

and

$$F(1) = 1^3/3 = 1/3$$

Therefore

$$F(2) - F(1) = 8/3 - 1/3 = 7/3$$

so we know that

$$\int_1^2 x^2 \, dx = 7/3$$

Next, we must evaluate the integral of the negative of the function, going in the negative direction along the x axis over the same interval as before:

$$\int_2^1 -x^2 \, dx$$

If we call the function $g(x) = -x^2$, then we have the basic antiderivative

$$G(x) = -x^3/3$$

We calculate

$$G(1) = -(1^3)/3 = -1/3$$

and

$$G(2) = -(2^3)/3 = -8/3$$

Therefore

$$G(1) - G(2) = -1/3 - (-8/3) = -1/3 + 8/3 = 7/3$$

So we know that

$$\int_2^{-1} -x^2 \, dx = 7/3$$

We've just demonstrated the following specific example of the theorem we proved in the solution to Prob. 1:

$$\int_1^2 x^2 \, dx = \int_2^1 -x^2 \, dx$$

3. Figure B-5 shows graphs of both functions, along with the regions defined by the curves in the interval between $x = 1$ and $x = 2$. When we integrate $f(x)$ over the interval in the positive direction, we get a positive result because the region lies above the x axis. When we integrate $g(x)$ over the same interval in the negative direction, we also get a positive result, because we're calculating the negative of a negative area.

4. We want to show that

$$\int_{-3}^{0} x^3 \, dx + \int_{0}^{3} x^3 \, dx = \int_{-3}^{3} x^3 \, dx$$

Let's call our function $f(x) = x^3$. We have the basic antiderivative

$$F(x) = x^4/4$$

First, we'll integrate over the interval from −3 to 0. We calculate

$$F(0) = 0^4/4 = 0$$

and

$$F(-3) = (-3)^4/4 = 81/4$$

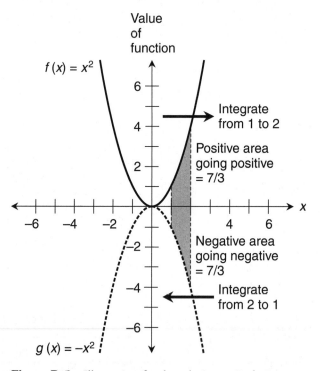

Figure B-5   Illustration for the solution to Prob. 3 in Chap. 13.

Therefore

$$F(0) - F(-3) = 0 - 81/4 = -81/4$$

so we know that

$$\int_{-3}^{0} x^3 \, dx = -81/4$$

Next, we integrate over the interval from 0 to 3. This time, we have

$$F(3) = 3^4/4 = 81/4$$

and

$$F(0) = 0^4/4 = 0$$

Therefore

$$F(3) - F(0) = 81/4 - 0 = 81/4$$

which tells us that

$$\int_{0}^{3} x^3 \, dx = 81/4$$

The sum of these two integrals is

$$\int_{-3}^{0} x^3 \, dx + \int_{0}^{3} x^3 \, dx = -81/4 + 81/4 = 0$$

Finally, let's integrate over the interval from −3 to 3. We calculate

$$F(3) = 3^4/4 = 81/4$$

and

$$F(-3) = (-3)^4/4 = 81/4$$

Therefore

$$F(3) - F(-3) = 81/4 - 81/4 = 0$$

which demonstrates that

$$\int_{-3}^{3} x^3 \, dx = 0$$

We've shown that the sum of the integrals from −3 to 0 and from 0 to 3 is equal to 0, and that the integral from −3 to 3 is also equal to 0. Mission accomplished!

5. Figure B-6 shows a graph of the function, along with the regions defined by the curve in the intervals between $x = -3$ and $x = 0$ and between $x = 0$ and $x = 3$. Each horizontal division represents 1 unit, and each vertical division represents 5 units. When we integrate from −3 to 0, we get a negative result, because the region lies below the x axis and we're integrating in the positive direction along that axis. When we integrate from 0 to 3, we get a positive result, because the region lies above the x axis and we're integrating in the positive direction along that axis. The two intervals give us equal but opposite integrals, so they add up to 0.

6. This time, we must go through calculations to show that

$$\int_{-3}^{-5} x^3 \, dx + \int_{-5}^{3} x^3 \, dx = \int_{-3}^{3} x^3 \, dx$$

We already know that

$$\int_{-3}^{3} x^3 \, dx = 0$$

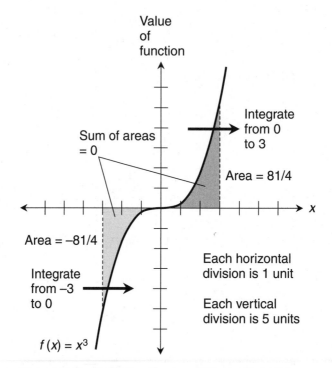

Figure B-6 Illustration for the solution to Prob. 5 in Chap. 13. Each horizontal-axis division represents 1 unit. Each vertical-axis division represents 5 units.

Now, our mission is to verify that

$$\int_{-3}^{-5} x^3\, dx + \int_{-5}^{3} x^3\, dx = 0$$

As before, let's call our function $f(x) = x^3$, giving us the basic antiderivative

$$F(x) = x^4/4$$

First, we integrate from -3 to -5. Calculating the antiderivatives, we get

$$F(-5) = (-5)^4/4 = 625/4$$

and

$$F(-3) = (-3)^4/4 = 81/4$$

Therefore

$$F(-5) - F(-3) = 625/4 - 81/4 = 544/4 = 136$$

so we know that

$$\int_{-3}^{-5} x^3\, dx = 136$$

Next, we integrate from -5 to 3. We have

$$F(3) = 3^4/4 = 81/4$$

and

$$F(-5) = (-5)^4/4 = 625/4$$

Therefore

$$F(3) - F(-5) = 81/4 - 625/4 = -544/4 = -136$$

which tells us that

$$\int_{-5}^{3} x^3\, dx = -136$$

The sum of these two integrals is

$$\int_{-3}^{-5} x^3\, dx + \int_{-5}^{3} x^3\, dx = 136 + (-136) = 0$$

7. From the Fundamental Theorem of Calculus, we know these four facts:

$$\int_a^b f(x)\,dx = F(b) - F(a)$$

$$\int_b^c f(x)\,dx = F(c) - F(b)$$

$$\int_c^d f(x)\,dx = F(d) - F(c)$$

$$\int_a^d f(x)\,dx = F(d) - F(a)$$

where F represents the antiderivative of f with the constant of integration equal to 0. Now let's rename the values of the functions at the points in the interval:

$$p = F(a)$$
$$q = F(b)$$
$$r = F(c)$$
$$s = F(d)$$

where p, q, r, and s are real numbers. All four of these numbers exist, because f is defined over the interval containing the x-values a, b, c, and d. By substitution, we can rewrite the first three of the above integrals as

$$\int_a^b f(x)\,dx = q - p$$

$$\int_b^c f(x)\,dx = r - q$$

$$\int_c^d f(x)\,dx = s - r$$

To find the sum of these three definite integrals, we can evaluate

$$(q - p) + (r - q) + (s - r)$$

By algebra, this simplifies to

$$s - p$$

Substituting back the original expressions for s and p, we get

$$F(d) - F(a)$$

According to the Fundamental Theorem of Calculus, that's

$$\int_a^d f(x)\, dx$$

8. We've been told to determine the definite integral

$$\int_{-1}^2 (4-x)^{-2}\, dx$$

We can consider this as a composite function in which

$$f(x) = 4 - x$$

and

$$g(y) = y^{-2}$$

In this situation,

$$f'(x) = -1$$

which means that

$$-f'(x) = 1$$

so we can rewrite the original integral in the form

$$\int_{-1}^2 g[f(x)] \cdot [-f'(x)]\, dx$$

The multiplication-by-constant rule for integration allows us to put the minus sign in front of the integral sign, giving us

$$-\int_{-1}^2 g[f(x)] \cdot f'(x)\, dx$$

We can get rid of the minus sign if we reverse the bounds of integration, because reversing the bounds multiplies a definite integral by -1. Now we have

$$\int_2^{-1} g[f(x)] \cdot f'(x)\, dx$$

Our integral is now in the form that's ready for substitution. We can calculate the "temporary bounds" as

$$f(2) = 4 - 2 = 2$$

and

$$f(-1) = 4 - (-1) = 5$$

We can write the integral as

$$\int_2^5 g(y)\, dy$$

which is

$$\int_2^5 y^{-2}\, dy$$

Remember that we're using y as a nickname for $f(x)$. Let's find the indefinite integral

$$\int y^{-2}\, dy$$

With the constant of integration set to 0, we get the antiderivative

$$G(y) = -y^{-1}$$

Now we calculate the values and subtract, getting

$$G(5) - G(2) = -(5^{-1}) - [-(2^{-1})] = -1/5 - (-1/2) = 3/10$$

We've determined that

$$\int_{-1}^2 (4-x)^{-2}\, dx = 3/10$$

9. In this situation, one of the bounds of our integration interval is at $x = 4$. When $x = 4$, the quantity $(4-x)^{-2}$ is a fraction with a denominator of 0. That's undefined. Therefore, this function blows up at one end of our interval. As we'll see in the next chapter, that presents a special challenge.

10. We've been told to determine, if possible, the definite integral

$$\int_{-1}^4 (4-x)^{-2}\, dx$$

Following the same process as in the solution to Prob. 8, we get

$$\int_4^{-1} g[f(x)] \cdot f'(x)\, dx$$

This is ready for substitution. We calculate the "temporary bounds" as

$$f(4) = 4 - 4 = 0$$

and

$$f(-1) = 4 - (-1) = 5$$

Now we can write the integral as

$$\int_0^5 g(y)\, dy$$

which is

$$\int_0^5 y^{-2}\, dy$$

As before, y is a nickname for $f(x)$. With the constant of integration set to 0, we get

$$G(y) = -y^{-1}$$

Now we calculate

$$G(5) - G(0) = -(5^{-1}) - (-0^{-1}) = -1/5 - (-1/0) = -1/5 + 1/0$$

The last expression is not defined, because its second term is $1/0$. We've determined that the definite integral

$$\int_{-1}^4 (4-x)^{-2}\, dx$$

is undefined.

Chapter 14

1. Let's call our function f, and call the independent variable x. Our improper integral is

$$\int_0^1 -7x^{-8}\, dx$$

Let's take a tiny positive δ, add it to the lower bound at $x = 0$, and get

$$\int_\delta^1 -7x^{-8}\, dx$$

Without the constant of integration, the basic antiderivative is

$$F(x) = x^{-7}$$

When we evaluate this from δ to 1, we get

$$F(1) - F(\delta) = 1^{-7} - \delta^{-7} = 1 - \delta^{-7}$$

Now we must work out the limit

$$\underset{\delta \to 0+}{Lim}\ 1 - \delta^{-7}$$

As δ approaches 0 from the positive side, the quantity δ^{-7} increases endlessly. When we subtract this increasing positive quantity from 1, we get a result that blows up negatively. Therefore

$$\underset{\delta \to 0+}{Lim}\ \int_\delta^1 -7x^{-8}\, dx$$

is infinite, telling us that

$$\int_0^1 -7x^{-8} \, dx$$

is undefined.

2. We want to evaluate

$$\int_{-\infty}^{-1} -7x^{-8} \, dx$$

Let's take a number p smaller than -1 and use it as the lower bound. Then we have

$$\int_p^{-1} -7x^{-8} \, dx$$

The antiderivative, as in Prob. 1, is

$$F(x) = x^{-7}$$

To find the value of the improper integral, we must determine

$$\underset{p \to -\infty}{Lim} \ F(-1) - F(p)$$

We can calculate that

$$F(-1) = (-1)^{-7} = -1$$

As p becomes large negatively, $F(p)$ approaches 0 from the negative direction, so

$$\underset{p \to -\infty}{Lim} \ F(-1) - F(p) = -1 - 0 = -1$$

Therefore

$$\underset{p \to -\infty}{Lim} \int_p^{-1} -7x^{-8} \, dx = -1$$

telling us that

$$\int_{-\infty}^{-1} -7x^{-8} \, dx = -1$$

3. Let's call our function f, and call the independent variable x. We want to evaluate

$$\int_0^2 x^{-3} \, dx$$

Let's take a tiny positive δ and add it to the lower bound at $x = 0$, getting

$$\int_\delta^2 x^{-3} \, dx$$

The antiderivative with the constant of integration set to 0 is

$$F(x) = -x^{-2}/2$$

When we evaluate this from δ to 2, we get

$$F(2) - F(\delta) = -2^{-2}/2 - (-\delta^{-2})/2 = -1/8 + \delta^{-2}/2 = \delta^{-2}/2 - 1/8$$

Now, we must look at the limit

$$\lim_{\delta \to 0+} \delta^{-2}/2 - 1/8$$

As δ becomes arbitrarily small while remaining positive, δ^{-2} increases without bound, so $\delta^{-2}/2 - 1/8$ also grows without bound. Therefore

$$\lim_{\delta \to 0+} \int_{\delta}^{2} x^{-3}\, dx$$

is infinite, telling us that

$$\int_{0}^{2} x^{-3}\, dx$$

is undefined.

4. We want to evaluate

$$\int_{2}^{\infty} x^{-3}\, dx$$

Let's consider a large positive number q as the upper bound, so the integral becomes

$$\int_{2}^{q} x^{-3}\, dx$$

The antiderivative, as we found in the solution to Prob. 3, is

$$F(x) = -x^{-2}/2$$

To find the integral, we must determine

$$\lim_{q \to \infty} F(q) - F(2)$$

We calculate

$$F(2) = -2^{-2}/2 = -1/8$$

As q becomes large positively, the value of $F(q)$ approaches 0, so

$$\lim_{q \to \infty} F(q) - F(2) = 0 - (-1/8) = 1/8$$

Therefore

$$\int_{2}^{\infty} x^{-3}\, dx = 1/8$$

5. We want to evaluate

$$\int_{-3}^{0} x^{-3/5}\, dx$$

Let's call this function $f(x)$. It has a singularity at $x = 0$. We can take a tiny positive number ε, subtract it from the upper bound at $x = 0$ where the singularity occurs, and obtain

$$\int_{-3}^{-\varepsilon} x^{-3/5}\, dx$$

Leaving out the constant of integration, the basic antiderivative is

$$F(x) = (5/2)x^{2/5}$$

When we evaluate F at $x = -3$, we get

$$F(-3) = (5/2) \cdot (-3)^{2/5} = (5/2) \cdot 9^{1/5}$$

When we evaluate F at $x = -\varepsilon$, we get

$$F(-\varepsilon) = (5/2) \cdot (-\varepsilon)^{2/5}$$

The difference is

$$F(-\varepsilon) - F(-3) = (5/2) \cdot (-\varepsilon)^{2/5} - (5/2) \cdot 9^{1/5}$$

Now consider

$$\operatorname*{Lim}_{\varepsilon \to 0+} \; (5/2) \cdot (-\varepsilon)^{2/5} - (5/2) \cdot 9^{1/5}$$

As ε approaches 0 from the positive direction, the value of $(-\varepsilon)^{2/5}$, which is the fifth root of the quantity $(-\varepsilon)^2$, approaches 0. If we multiply by 5/2, it still approaches 0. Now we know that

$$\operatorname*{Lim}_{\varepsilon \to 0+} \; (5/2) \cdot (-\varepsilon)^{2/5} - (5/2) \cdot 9^{1/5} = -(5/2) \cdot 9^{1/5}$$

so therefore

$$\operatorname*{Lim}_{\varepsilon \to 0+} \int_{-3}^{-\varepsilon} x^{-3/5}\, dx = -(5/2) \cdot 9^{1/5}$$

This tells us that

$$\int_{-3}^{0} x^{-3/5} \, dx = -(5/2) \cdot 9^{1/5}$$

This is an irrational number. With a calculator, we can approximate it to −3.880.

6. We want to evaluate

$$\int_{-3}^{-\infty} x^{-3/5} \, dx$$

This is the same function $f(x)$ as we worked with in Prob. 5. This time, we're integrating from −3 in the negative direction, indefinitely. Consider a large positive number q. Let's see what happens to

$$\int_{-3}^{-q} x^{-3/5} \, dx$$

as we make q increase without bound. (That means $-q$ grows larger negatively without bound). The basic antiderivative, as we found in the solution to Prob. 5, is

$$F(x) = (5/2)x^{2/5}$$

When we evaluate the antiderivative at $x = -3$, we get

$$F(-3) = (5/2) \cdot (-3)^{2/5} = (5/2) \cdot 9^{1/5}$$

When we evaluate the antiderivative at $x = -q$, we get

$$F(-q) = (5/2) \cdot (-q)^{2/5}$$

When we subtract the antiderivative at the "start" of our integration interval from the antiderivative at the "finish," we get

$$F(-q) - F(-3) = (5/2) \cdot (-q)^{2/5} - (5/2) \cdot 9^{1/5}$$

Now consider

$$\underset{q \to \infty}{Lim} \ (5/2) \cdot (-q)^{2/5} - (5/2) \cdot 9^{1/5}$$

As q grows positively without bound, the value of $(-q)^{2/5}$, which is the fifth root of the quantity $(-q)^2$, also increases positively without bound. When we multiply this quantity by 5/2, we make it grow even faster. Now we know that the above limit is infinite, so

$$\underset{q \to \infty}{Lim} \int_{-3}^{-q} x^{-3/5} \, dx$$

is also infinite. We must conclude that

$$\int_{-3}^{-\infty} x^{-3/5}\, dx$$

is undefined.

7. When we consider the graph of the function, the second of the two integrals is obviously undefined. That integral is

$$\int_{2}^{\infty} -5x\, dx$$

If we call the function $f(x)$, then

$$f(x) = -5x$$

The graph of $f(x)$ is a straight line with a slope of −5, passing through the origin. If we start integrating at $x = 2$ and proceed in the positive direction, the "curve" defines trapezoidal regions that grow endlessly larger, always with a left-hand edge −10 units long. We don't have to do any calculations to see this. It's obvious from the graph. (If you can't envision this graph in your "mental eye," you can sketch it.)

8. It seems that the following integral might be defined:

$$\int_{2}^{\infty} x^{-5}\, dx$$

Let's try to evaluate it. The antiderivative, not including the constant of integration, is

$$F(x) = -x^{-4}/4$$

Now consider the integral

$$\int_{2}^{q} x^{-5}\, dx$$

where q is some large positive number. To evaluate this, we must determine

$$\mathop{Lim}_{q \to \infty} F(q) - F(2)$$

where q increases without bound. We can easily calculate

$$F(2) = -(2^{-4})/4 = -1/64$$

As q becomes large, the value of $F(q)$ approaches 0, so

$$\mathop{Lim}_{q \to \infty} F(q) - F(2) = 0 - (-1/64) = 1/64$$

Therefore

$$\int_{2}^{\infty} x^{-5} \, dx = 1/64$$

9. Let's try to evaluate

$$\int_{-1}^{1} x^{-3} \, dx$$

using the Fundamental Theorem of Calculus directly, as if the interval contains no singularity. (It does, as we know, at $x = 0$.) The basic antiderivative is

$$F(x) = -x^{-2}/2$$

When we evaluate F straight through from -1 to 1, we get

$$F(1) - F(-1) = -1^{-2}/2 - [-(-1)^{-2}/2] = -1/2 - (-1/2) = 0$$

Let's look at this situation in the coordinate plane. Figure B-7 is a graph of

$$f(x) = x^{-3}$$

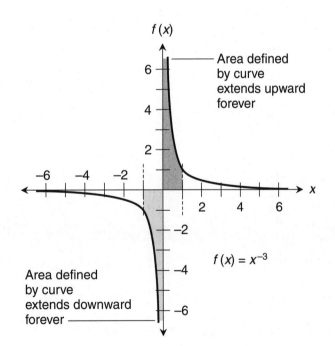

Figure B-7 Illustration for the solution to Probs. 9 and 10 in Chap. 14. This graph, while accurate in itself, might reinforce the mistaken idea that the integral of the function is equal to 0 over the interval shown.

with the boundaries $x = -1$ and $x = 1$ drawn in. The areas defined by the two parts of the curve are shaded. The negative area, which exists on the left-hand side of the vertical axis, is shaded lightly. The positive area, which appears on the right-hand side of the vertical axis, is shaded more darkly. It's tempting to imagine that these areas, even though both might be infinite, are equal and opposite. Figure B-7 can reinforce the mistaken notion that

$$\int_{-1}^{1} x^{-3}\, dx$$

is defined and equal to 0. In the next solution, we'll see that this is not the case.

10. Let's work out the part of the integral on the right-hand side of the singularity:

$$\int_{0}^{1} x^{-3}\, dx$$

We can take a tiny positive δ, add it to the lower bound at $x = 0$, and get

$$\int_{\delta}^{1} x^{-3}\, dx$$

The basic antiderivative is

$$F(x) = -x^{-2}/2$$

When we evaluate F from δ to 1, we get

$$F(1) - F(\delta) = -1^{-2}/2 - (-\delta^{-2}/2) = -1/2 + \delta^{-2}/2 = \delta^{-2}/2 - 1/2$$

Now, we must look at the limit

$$\operatorname*{Lim}_{\delta \to 0+} \delta^{-2}/2 - 1/2$$

The quantity $(\delta^{-2}/2)$ increases endlessly as δ approaches 0 from the positive direction, so

$$\operatorname*{Lim}_{\delta \to 0+} \int_{\delta}^{1} x^{-3}\, dx$$

is infinite. Therefore,

$$\int_{0}^{1} x^{-3}\, dx$$

is undefined. This fact alone is sufficient to tell us that

$$\int_{-1}^{1} x^{-3}\, dx$$

is undefined. We don't have to evaluate the left-hand part of the integral; no matter what we add to this undefined quantity, we'll get another undefined quantity.

Chapter 15

1. Here's the sum of integrals we've been told to evaluate:

$$\int_5^6 3x^2 \, dx + \int_5^6 -12x^2 \, dx$$

We can name the functions like this:

$$g(x) = 3x^2$$

and

$$h(x) = -12x^2$$

First, let's figure out the definite integral of $g(x)$ with respect to x from $x = 5$ to $x = 6$:

$$\int_5^6 3x^2 \, dx$$

The basic antiderivative, leaving out the constant of integration, is

$$G(x) = x^3$$

When we evaluate this from $x = 5$ to $x = 6$, we get

$$G(6) - G(5) = 6^3 - 5^3 = 216 - 125 = 91$$

Next, we figure out the definite integral of $h(x)$ with respect to x from $x = 5$ to $x = 6$:

$$\int_5^6 -12x^2 \, dx$$

The antiderivative, leaving out the constant of integration, is

$$H(x) = -4x^3$$

When we evaluate this from $x = 5$ to $x = 6$, we get

$$H(6) - H(5) = -4 \cdot 6^3 - (-4 \cdot 5^3) = -364$$

When we add these results, we get $91 - 364$, which is -273. Now, let's add the two functions together and then integrate their sum. Remember that

$$g(x) = 3x^2$$

and

$$h(x) = -12x^2$$

Therefore, if we call the sum function q, we have

$$g(x) + h(x) = q(x) = 3x^2 - 12x^2 = -9x^2$$

We want to work out the integral

$$\int_5^6 -9x^2 \, dx$$

The antiderivative, leaving out the constant of integration, is

$$Q(x) = -3x^3$$

When we evaluate this from $x = 5$ to $x = 6$, we get

$$Q(6) - Q(5) = -3 \cdot 6^3 - (-3 \cdot 5^3) = -273$$

This is the same as the result we got when we integrated the two functions separately and then added the integrals.

2. We've been told to evaluate this sum of integrals:

$$\int_5^6 3x^2 \, dx + \int_5^6 -12x^2 \, dx$$

Note the difference in the direction between the first and second integrals in this sum! Let's use the function names from the solution to Prob. 1:

$$g(x) = 3x^2$$

and

$$h(x) = -12x^2$$

In the solution to Prob. 1, we determined that

$$\int_5^6 3x^2 \, dx = 91$$

and

$$\int_5^6 -12x^2 \, dx = -364$$

According to the rule for reversing the direction of integration over a fixed interval,

$$\int_6^5 -12x^2 \, dx = -(-364) = 364$$

When we add these results, we get

$$\int_5^6 3x^2 \, dx + \int_6^5 -12x^2 \, dx = 91 + 364 = 455$$

Now, let's evaluate

$$\int_5^6 15x^2 \, dx$$

We can invent a new name $q(x)$ and assign it to the function we're integrating above:

$$q(x) = 15x^2$$

The antiderivative without the constant of integration is

$$Q(x) = 5x^3$$

When we evaluate Q from $x = 5$ to $x = 6$, we get

$$Q(6) - Q(5) = 5 \cdot 6^3 - 5 \cdot 5^3 = 455$$

This is the same as the result we got when we integrated the two functions separately and then added the integrals.

3. Here's the difference of integrals we've been told to evaluate:

$$\int_5^6 -12x^2 \, dx - \int_5^6 3x^2 \, dx$$

Let's keep using the same function names from the solutions to Probs. 1 and 2:

$$g(x) = 3x^2$$

and

$$h(x) = -12x^2$$

In the solution to Prob. 1, we found that

$$\int_5^6 -12x^2 \, dx = -364$$

and

$$\int_5^6 3x^2 \, dx = 91$$

The difference between these two definite integrals is

$$\int_5^6 -12x^2 \, dx - \int_5^6 3x^2 \, dx = -364 - 91 = -455$$

Our next step is to work out

$$\int_{5}^{6} -15x^2 \, dx$$

Let's invent a new name $p(x)$ and assign it to the function we're integrating:

$$p(x) = -15x^2$$

The *basic antiderivative* (that is, without the constant of integration) is

$$P(x) = -5x^3$$

When we evaluate this from $x = 5$ to $x = 6$, we get

$$P(6) - P(5) = -5 \cdot 6^3 - (-5 \cdot 5^3) = -455$$

This is the same as the result we got when we integrated the two functions separately and then subtracted the integrals.

4. We've been told to find

$$\int (x^{-2} + x^{-3} + x^{-4}) \, dx$$

This is the integral of a sum of three monomial functions of *x*. Let's call them

$$f_1(x) = x^{-2}$$
$$f_2(x) = x^{-3}$$
$$f_3(x) = x^{-4}$$

If we temporarily leave out the constants of integration, then the indefinite integrals of these monomial functions are

$$\int f_1(x) \, dx = -x^{-1}$$

$$\int f_2(x) \, dx = (-1/2)x^{-2}$$

$$\int f_3(x) \, dx = (-1/3)x^{-3}$$

The original integral without a constant of integration is

$$\int (x^{-2} + x^{-3} + x^{-4}) \, dx = \int x^{-2} \, dx + \int x^{-3} \, dx + \int x^{-4} \, dx$$
$$= -x^{-1} + (-1/2)x^{-2} + (-1/3)x^{-3}$$

Adding in a constant of integration *c*, we end up with

$$\int (x^{-2} + x^{-3} + x^{-4}) \, dx = -x^{-1} + (-1/2)x^{-2} + (-1/3)x^{-3} + c$$

5. We've been told to find

$$\int (7y)(6y-4)(-2y+3)\ dy$$

Let's multiply the factors to get a single polynomial:

$$(7y)(6y-4)(-2y+3) = -84y^3 + 182y^2 - 84y$$

The integral we want to evaluate is now

$$\int (-84y^3 + 182y^2 - 84y)\ dy$$

This is the integral of a sum of three monomial functions of y. We can call them

$$f_1(y) = -84y^3$$
$$f_2(y) = 182y^2$$
$$f_3(y) = -84y$$

If we temporarily leave out the constants of integration, then the indefinite integrals of these monomial functions are

$$\int f_1(y)\ dy = -21y^4$$

$$\int f_2(y)\ dy = (182/3)y^3$$

$$\int f_3(y)\ dy = -42y^2$$

The summed-up integral without a constant of integration is

$$\int (-84y^3 + 182y^2 - 84y)\ dy = -21y^4 + (182/3)y^3 + -42y^2$$

Adding in the consolidated constant of integration c, and writing the original integrand on the left-hand side of the equation, we have the final answer

$$\int (7y)(6y-4)(-2y+3)\ dy = -21y^4 + (182/3)y^3 + -42y^2 + c$$

6. We've been told to find

$$\int (z^8 - 1)(-3z^3 + 7z^2)\ dz$$

We begin by multiplying the factors together to get

$$(z^8 - 1)(-3z^3 + 7z^2) = -3z^{11} + 7z^{10} + 3z^3 - 7z^2$$

The integral we want to evaluate is now

$$\int (-3z^{11} + 7z^{10} + 3z^3 - 7z^2)\ dz$$

This is the integral of a sum of four monomial functions of z. We can call them

$$f_1(z) = -3z^{11}$$
$$f_2(z) = 7z^{10}$$
$$f_3(z) = 3z^3$$
$$f_4(z) = -7z^2$$

If we temporarily leave out the constants of integration, then the indefinite integrals of these monomial functions are

$$\int f_1(z)\ dz = (-1/4)z^{12}$$

$$\int f_2(z)\ dz = (7/11)z^{11}$$

$$\int f_3(z)\ dz = (3/4)z^4$$

$$\int f_4(z)\ dz = (-7/3)z^3$$

The complete integral, leaving out the constant of integration, is

$$\int (-3z^{11} + 7z^{10} + 3z^3 - 7z^2)\ dz = (-1/4)z^{12} + (7/11)z^{11} + (3/4)z^4 - (7/3)z^3$$

Adding in the constant of integration c, and writing the original integrand on the left-hand side, we have

$$\int (z^8 - 1)(-3z^3 + 7z^2)\ dz = (-1/4)z^{12} + (7/11)z^{11} + (3/4)z^4 - (7/3)z^3 + c$$

7. We've been told to find

$$\int_0^1 (4x^3 - 5x^2 + 7x - 4)\ dx$$

Let's start by working out the indefinite integral

$$\int (4x^3 - 5x^2 + 7x - 4)\ dx$$

The integrand is the sum of four monomial functions of x. Let's call them

$$f_1(x) = 4x^3$$
$$f_2(x) = -5x^2$$
$$f_3(x) = 7x$$
$$f_4(x) = -4$$

None of these functions has any singularities, so we can work out a definite integral for any of them over any interval. That means we can integrate the entire polynomial function (which is the sum of the monomial functions) over any interval. The basic antiderivatives of the monomial functions are

$$F_1 (x) = x^4$$
$$F_2 (x) = (-5/3)x^3$$
$$F_3 (x) = (7/2)x^2$$
$$F_4 (x) = -4x$$

The antiderivative of the original integrand is the sum of all these:

$$F (x) = F_1 (x) + F_2 (x) + F_3 (x) + F_4 (x) = x^4 - (5/3)x^3 + (7/2)x^2 - 4x$$

When we input 1 (the starting *x*-value of our interval) here, we get

$$F (1) = 1^4 - (5/3) \cdot 1^3 + (7/2) \cdot 1^2 - 4 \cdot 1 = -7/6$$

When we input 0 (the finishing *x*-value), we get

$$F (1) = 0^4 - (5/3) \cdot 0^3 + (7/2) \cdot 0^2 - 4 \cdot 0 = 0$$

The definite integral is therefore

$$F (1) - F (0) = -7/6 - 0 = -7/6$$

The complete answer to our problem is

$$\int_0^1 (4x^3 - 5x^2 + 7x - 4) \, dx = -7/6$$

8. We've been told to find

$$\int_{-1}^{-2} (x^{-2} + x^{-3} + x^{-4}) \, dx$$

We begin by working out

$$\int (x^{-2} + x^{-3} + x^{-4}) \, dx$$

This is the integral of the sum of the monomial functions

$$f_1 (x) = x^{-2}$$
$$f_2 (x) = x^{-3}$$
$$f_3 (x) = x^{-4}$$

Each of these functions has one singularity, which occurs at $x = 0$. But we're okay in the interval where $-2 < x < -1$. None of the three functions blows up anywhere in that span. The basic antiderivatives are

$$F_1(x) = -x^{-1}$$
$$F_2(x) = (-1/2)x^{-2}$$
$$F_3(x) = (-1/3)x^{-3}$$

The basic antiderivative of the original integrand is therefore

$$F(x) = F_1(x) + F_2(x) + F_3(x) = -x^{-1} - (1/2)x^{-2} - (1/3)x^{-3}$$

We must remember that we've been told to integrate over the interval in the negative direction, from $x = -1$ to $x = -2$! We input -2 (the finishing x-value) first, getting

$$F(-2) = -(-2)^{-1} - (1/2)(-2)^{-2} - (1/3)(-2)^{-3} = 5/12$$

When we input -1 (the starting x-value), we get

$$F(-1) = -(-1)^{-1} - (1/2)(-1)^{-2} - (1/3)(-1)^{-3} = 5/6$$

The definite integral is obtained by taking the finishing antiderivative minus the starting antiderivative. This gives us

$$F(-2) - F(-1) = 5/12 - 5/6 = -5/12$$

Therefore, the original definite integral is

$$\int_{-1}^{-2} (x^{-2} + x^{-3} + x^{-4}) \, dx = -5/12$$

9. We've been told to determine

$$\int_{0}^{1} (x - 2)^3 \, dx$$

Before we begin integrating, let's multiply out the cubed binomial, getting a straightforward polynomial. When we do that, we get

$$(x - 2)^3 = x^3 - 6x^2 + 12x - 8$$

Now we're ready to work out the indefinite integral

$$\int (x^3 - 6x^2 + 12x - 8) \, dx$$

This is the integral of the sum of four functions

$$f_1(x) = x^3$$
$$f_2(x) = -6x^2$$
$$f_3(x) = 12x$$
$$f_4(x) = -8$$

None of these functions has any singularities, so we're okay with any interval of integration. The basic antiderivatives are

$$F_1(x) = (1/4)x^4$$
$$F_2(x) = -2x^3$$
$$F_3(x) = 6x^2$$
$$F_4(x) = -8x$$

The basic antiderivative of the original integrand is

$$F(x) = F_1(x) + F_2(x) + F_3(x) + F_4(x) = (1/4)x^4 - 2x^3 + 6x^2 - 8x$$

When we input 1 here, we get

$$F(1) = (1/4) \cdot 1^4 - 2 \cdot 1^3 + 6 \cdot 1^2 - 8 \cdot 1 = -15/4$$

When we input 0, we get

$$F(0) = (1/4) \cdot 0^4 - 2 \cdot 0^3 + 6 \cdot 0^2 - 8 \cdot 0 = 0$$

The definite integral is

$$F(1) - F(0) = -15/4 - 0 = -15/4$$

The complete answer to our problem is

$$\int_0^1 (x-2)^3 \, dx = -15/4$$

10. We've been told to determine

$$\int_{-1}^1 (x^{1/3} + 4)^2 \, dx$$

To solve this problem, we should begin by multiplying out the squared binomial, getting a straightforward polynomial. That gives us

$$(x^{1/3} + 4)^2 = x^{2/3} + 8x^{1/3} + 16$$

Now we must work out

$$\int (x^{2/3} + 8x^{1/3} + 16) \, dx$$

Our integrand is the sum of the three monomial functions

$$f_1 (x) = x^{2/3}$$
$$f_2 (x) = 8x^{1/3}$$
$$f_3 (x) = 16$$

None of these functions has any singularities, so we don't have to be concerned about evaluating any improper integrals. The basic antiderivatives are

$$F_1 (x) = (3/5)x^{5/3}$$
$$F_2 (x) = 6x^{4/3}$$
$$F_3 (x) = 16x$$

The basic antiderivative of the original integrand is therefore

$$F (x) = F_1 (x) + F_2 (x) + F_3 (x) = (3/5)x^{5/3} + 6x^{4/3} + 16x$$

When we input 1 to this function, we get

$$F (1) = (3/5) \cdot 1^{5/3} + 6 \cdot 1^{4/3} + 16 \cdot 1 = 113/5$$

When we input −1, we get

$$F (-1) = (3/5) \cdot (-1)^{5/3} + 6 \cdot (-1)^{4/3} + 16 \cdot (-1) = -53/5$$

The definite integral is

$$F (1) - F (-1) = 113/5 - (-53/5) = 166/5$$

The complete answer to our problem is

$$\int_{-1}^{1} (x^{1/3} + 4)^2 \, dx = 166/5$$

Chapter 16

1. Figure B-8 illustrates the graphs of the two functions we now have. We want to find the area of the shaded region. The functions are

$$f (x) = 2x - 1$$

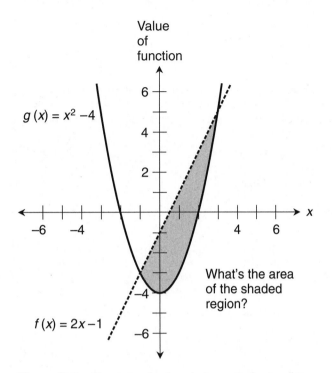

Figure B-8 Illustration for the solution to Probs. 1 and 2 in Chap. 16.

and

$$g\ (x) = x^2 - 4$$

To find this area, the first thing we should do is determine the x-values of the points where the line and the curve intersect. Then we should find the difference function

$$p\ (x) = f\ (x) - g\ (x)$$

Finally, we should integrate $p(x)$ with respect to x in the positive direction over the interval between the two intersection points.

2. To find the intersection points, let's set $f\ (x) = g\ (x)$, obtaining the single-variable equation

$$2x - 1 = x^2 - 4$$

Adding the quantity $(-2x + 1)$ to each side, transposing the left-hand and right-hand sides, and rearranging into standard quadratic form, we get

$$x^2 - 2x - 3 = 0$$

This factors into

$$(x+1)(x-3) = 0$$

The roots of this quadratic equation are $x = -1$ or $x = 3$. These are the x-values of the points where the line and the curve intersect, and are the bounds over which we should integrate the difference function. That difference function is

$$p(x) = f(x) - g(x) = (2x - 1) - (x^2 - 4) = -x^2 + 2x + 3$$

Now we must evaluate

$$\int_{-1}^{3} (-x^2 + 2x + 3) \, dx$$

The antiderivative, leaving out the constant of integration, is

$$P(x) = -x^3/3 + x^2 + 3x$$

When we evaluate this from $x = -1$ to $x = 3$, we obtain

$$P(3) - P(-1) = [-(3^3/3) + 3^2 + 3 \cdot 3] - [-(-1)^3/3 + (-1)^2 + 3 \cdot (-1)] = 32/3$$

3. First, we must find the intersection points between the line and the curve. Here are the functions again, for reference:

$$f(x) = x$$

and

$$g(x) = x^2$$

They're graphed in Fig. B-9, and the area we want to find is indicated by the shading. Let's set these two functions equal, so we get

$$x = x^2$$

To put this into standard quadratic form, we can subtract x from each side and then transpose the sides. That gives us

$$x^2 - x = 0$$

which factors into

$$x(x - 1) = 0$$

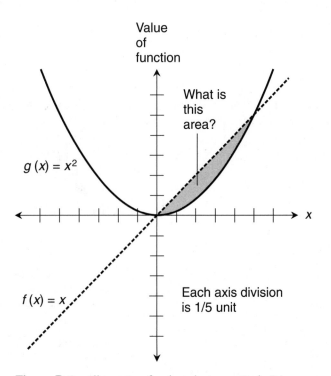

Figure B-9 Illustration for the solution to Prob. 3 in Chap. 16. Each axis division represents 1/5 unit.

This tells us that roots are $x = 0$ or $x = 1$. These are the bounds of the interval over which we must integrate the difference function

$$p(x) = f(x) - g(x) = x - x^2 = -x^2 + x$$

Remember that we subtract the "bottom function" from the "top function." Now we want to find

$$\int_0^1 (-x^2 + x) \, dx$$

The basic antiderivative is

$$P(x) = -x^3/3 + x^2/2$$

When we evaluate this from $x = 0$ to $x = 1$, we obtain

$$P(1) - P(0) - [-(1^3/3) + 1^2/2] - (-0^3/3 + 0^2/2) - 1/6$$

4. Refer to Fig. B-10. To begin, we determine the area of the shaded region, which is a triangle with a base length of 1 unit and a height of 1 unit. To find the area, we

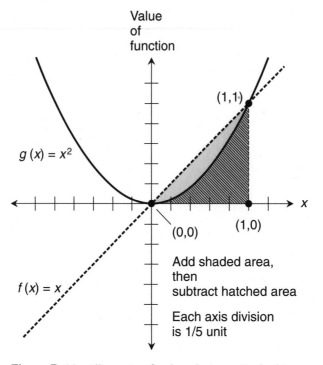

Figure B-10 Illustration for the solution to Prob. 4 in Chap. 16. Each axis division represents 1/5 unit.

multiply 1 by 1 and then divide by 2, getting 1/2 square unit. From this, we must subtract the area of the region defined by the parabola over the interval $0 < x < 1$. That's the hatched zone. Its area is

$$\int_0^1 x^2 \, dx$$

The antiderivative, leaving out the constant of integration, is

$$F(x) = x^3/3$$

When we evaluate this from $x = 0$ to $x = 1$, we obtain

$$F(1) - F(0) = 1^3/3 - 0^3/3 = 1/3$$

When we subtract this from the area of the triangle, we get $1/2 - 1/3$, which is 1/6 square unit. That's the same answer we got in the solution to Prob. 3.

5. Before we think about the integral, we must find the intersection points between the line and the curve. Here are the functions once again:

$$f(x) = x^2$$

and

$$g(x) = x^3$$

They're graphed in Fig. B-11. Setting the functions equal, we get

$$x^2 = x^3$$

To put this into standard cubic form, we can subtract x^2 from each side and then transpose the sides to get

$$x^3 - x^2 = 0$$

which factors into

$$x^2(x - 1) = 0$$

The real roots of this equation are $x = 0$ or $x = 1$. These x-values represent the bounds of our integration integral. The "top function" is f and the "bottom function" is g. The difference function is

$$p(x) = f(x) - g(x) = x^2 - x^3 = -x^3 + x^2$$

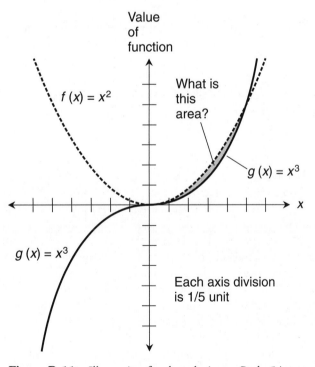

Figure B-11   Illustration for the solution to Prob. 5 in Chap. 16. Each axis division represents 1/5 unit.

Now we want to find

$$\int_0^1 (-x^3 + x^2)\, dx$$

The basic antiderivative is

$$P(x) = -x^4/4 + x^3/3$$

When we evaluate this from $x = 0$ to $x = 1$, we obtain

$$P(1) - P(0) = [-(1^4/4) + 1^3/3] - (-0^4/4 + 0^3/3) = 1/12$$

6. Once again, the first thing we must do is find the intersection points between the line and the curve. The functions are

$$f(x) = x$$

and

$$g(x) = x^3$$

Figure B-12 is a graph of this situation. Setting the functions equal, we get

$$x = x^3$$

To put this equation into standard cubic form, we subtract x from each side and then transpose the sides to get

$$x^3 - x = 0$$

which factors into

$$x(x - 1)(x + 1) = 0$$

The real roots of this equation are $x = -1$, $x = 0$, or $x = 1$. Bounding the region of interest for the interval $-1 < x < 0$, the "top function" is g and the "bottom function" is f. Bounding the region of interest for $0 < x < 1$, it's the other way around; the "top" function is f and the "bottom function" is g. When we look at Fig. B-12, it's tempting to suppose that the two regions have equal area, but we had better not assume it without proof! To be sure we get the right result, we must evaluate the areas of the two shaded regions separately.

Let's find the area of the shaded region in the third quadrant first. That's the region at the lower left. The difference function is

$$p(x) = g(x) - f(x) = x^3 - x$$

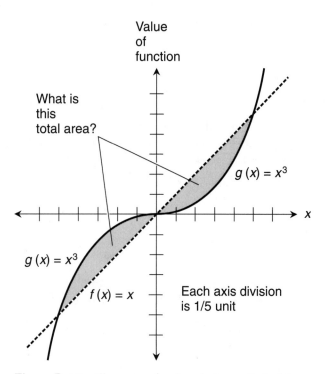

Figure B-12 Illustration for the solution to Prob. 6 in Chap. 16. Each axis division represents 1/5 unit.

To get a positive value for this integral, we must find the absolute value of

$$\int_{-1}^{0} (x^3 - x)\ dx$$

The basic antiderivative is

$$P(x) = x^4/4 - x^2/2$$

When we evaluate this from $x = -1$ to $x = 0$, we obtain

$$P(0) - P(-1) = (0^4/4 - 0^2/2) - [(-1)^4/4 - (-1)^2/2] = -1/4$$

The absolute value is 1/4. Now let's find the area of the shaded region in the first quadrant. The difference function in this case is

$$q(x) = f(x) - g(x) = x - x^3$$

This time, we must determine the definite integral

$$\int_{0}^{1} (x - x^3)\ dx$$

The basic antiderivative is

$$Q(x) = x^2/2 - x^4/4$$

When we evaluate this from $x = 0$ to $x = 1$, we obtain

$$Q(1) - Q(0) = (1^2/2 - 1^4/4) - (0^2/2 - 0^4/4) = 1/4$$

Now we are certain that the areas of the two shaded regions are equal. When we add them, we get the true geometric area between the line and the curve over the interval from $x = -1$ to $x - 1$. That area is $1/4 + 1/4 = 1/2$ square unit.

7. Here are the functions again, for reference:

$$f(x) = 1$$

and

$$g(x) = x^{-2/3}$$

Imagine the region bounded by the graph of f on the bottom, the graph of g on the right, and the dependent-variable axis on the left, as shown in Fig. B-13. The region has no top;

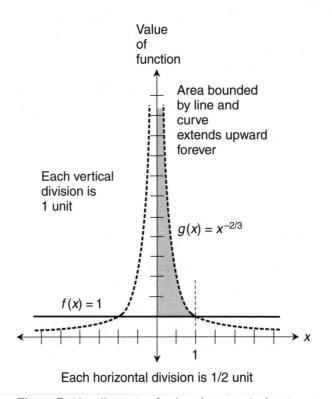

Figure B-13 Illustration for the solution to Prob. 7 in Chap. 16. Each horizontal-axis division represents 1/2 unit. Each vertical-axis division represents 1 unit.

it goes upward forever. We've been told to find the area of the shaded region, and that the interval of integration is $0 < x < 1$. The line and the curve intersect where $x = 1$ as the illustration suggests, because

$$f(1) = 1$$

and

$$g(1) = 1^{-2/3} = 1$$

The upper curve represents g and the lower curve represents f, so we subtract f from g to get the difference function

$$p(x) = g(x) - f(x) = x^{-2/3} - 1$$

We want to integrate p over the interval $0 < x < 1$, which is bounded on the left by the singularity. That integral is

$$\int_0^1 (x^{-2/3} - 1) \, dx$$

Let's take a tiny positive number δ and add it to the lower bound at $x = 0$ to obtain

$$\int_\delta^1 (x^{-2/3} - 1) \, dx$$

The basic antiderivative is

$$P(x) = 3x^{1/3} - x$$

When we evaluate this from δ to 1, we get

$$P(1) - P(\delta) = (3 \cdot 1^{1/3} - 1) - (3\delta^{1/3} - \delta) = 2 - (3\delta^{1/3} - \delta)$$

Now we must figure out

$$\lim_{\delta \to 0+} 2 - (3\delta^{1/3} - \delta)$$

As δ approaches 0 from the positive direction, the values of $3\delta^{1/3}$ and δ both approach 0. That means the quantity $(3\delta^{1/3} - \delta)$ approaches 0, so

$$\lim_{\delta \to 0+} 2 - (3\delta^{1/3} - \delta) = 2$$

It follows that

$$\lim_{\delta \to 0+} \int_\delta^1 (x^{-2/3} - 1) \, dx = 2$$

Therefore

$$\int_0^1 (x^{-2/3} - 1)\, dx = 2$$

The area of the shaded region in Fig. B-13 is 2 square units, even though it's infinitely tall.

8. The two functions in this situation are the same as the ones in Prob. 7. Here they are again, for reference:

$$f(x) = 1$$

and

$$g(x) = x^{-2/3}$$

This time, imagine the region bounded by the graph of g on the bottom, the graph of f on the top, and the line $x = 2$ on the right. We've been told to find the area of the shaded region in Fig. B-14, and that the interval of integration is $1 < x < 2$. In the

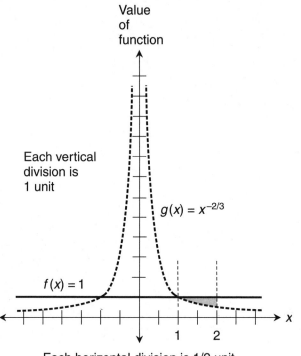

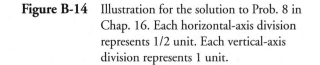

Figure B-14   Illustration for the solution to Prob. 8 in Chap. 16. Each horizontal-axis division represents 1/2 unit. Each vertical-axis division represents 1 unit.

solution to Prob. 7, we showed that the line and the curve intersect where $x = 1$. The upper curve represents f and the lower curve represents g, so we must subtract g from f to get the difference function. If we call that function q, then

$$q(x) = f(x) - g(x) = 1 - x^{-2/3}$$

We want to integrate q over the interval $1 < x < 2$. That integral is

$$\int_1^2 (1 - x^{-2/3}) \, dx$$

The basic antiderivative is

$$Q(x) = x - 3x^{1/3}$$

When we evaluate this from 1 to 2, we get

$$Q(2) - Q(1) = (2 - 3 \cdot 2^{1/3}) - (1 - 3 \cdot 1^{1/3}) = 4 - 3 \cdot 2^{1/3}$$

We have determined that

$$\int_1^2 (1 - x^{-2/3}) \, dx = 4 - 3 \cdot 2^{1/3}$$

That's the area of the shaded region in Fig. B-14. It's an irrational number; we can use a calculator to approximate it as 0.220 square units.

9. To find the total area in the interval $0 < x < 2$, we can add the areas we got in the solutions to Probs. 7 and 8. The situation is illustrated in Fig. B-15. The area that we found in the solution to Prob. 7 (and shown in Fig. B-13) was

$$\int_0^1 (x^{-2/3} - 1) \, dx = 2$$

The area that we found in the solution to Prob. 8 (and shown in Fig. B-14) was

$$\int_1^2 (1 - x^{-2/3}) \, dx = 4 - 3 \cdot 2^{1/3}$$

The total area is the sum of these, or $6 - 3 \cdot 2^{1/3}$ square units. This is an irrational number. A calculator tells us that it's equal to approximately 2.220 square units.

10. For reference, the two functions in this situation are

$$f(x) = x^{-2}$$

and

$$g(x) = -x^{-3}$$

We're interested in the shaded zone in the graph of Fig. B-16. This region is bounded by the curve for g on the bottom, the curve for f on the top, and the line $x = 1$ on the left.

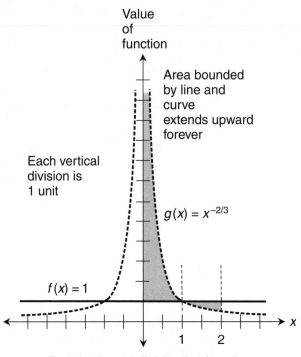

Figure B-15   Illustration for the solution to Prob. 9 in Chap. 16. Each horizontal-axis division represents 1/2 unit. Each vertical-axis division represents 1 unit.

The zone has no boundary on the right, because it extends in the positive x direction forever. We've been told to integrate over the infinitely wide interval $1 < x$, which can also be expressed as $(1,\infty)$. We must subtract g from f to get the difference function. If we call that function q, then

$$q(x) = f(x) - g(x) = x^{-2} - (-x^{-3}) = x^{-2} + x^{-3}$$

The area of the shaded region is equal to

$$\int_{1}^{\infty} (x^{-2} + x^{-3})\ dx$$

Imagine a positive real number s that we can make as large as we want. If we allow s to grow larger endlessly, the above integral can be expressed as

$$\underset{s \to \infty}{Lim} \int_{1}^{s} (x^{-2} + x^{-3})\ dx$$

The antiderivative of our difference function q without the constant of integration is

$$Q(x) = -x^{-1} + (-x^{-2}/2) = -x^{-1} - x^{-2}/2$$

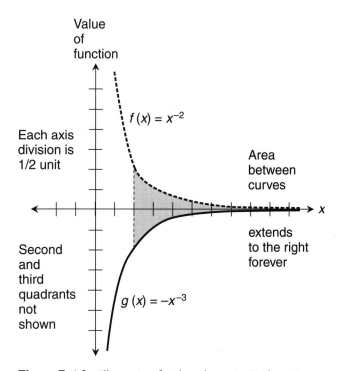

Figure B-16 Illustration for the solution to Prob. 10 in Chap. 16. Each axis division represents 1/2 unit. The second and third quadrants are not shown.

To find the limit of the integral stated above, we must figure out

$$\underset{s \to \infty}{Lim}\ Q(s) - Q(1)$$

As s grows larger endlessly, the value of $Q(s)$ approaches 0 from the negative side. We can see this by inputting some actual numbers and watching the trend:

$$Q(10) = -10^{-1} - 10^{-2}/2$$
$$Q(10^2) = -(10^2)^{-1} - (10^2)^{-2}/2$$
$$Q(10^3) = -(10^3)^{-1} - (10^3)^{-2}/2$$
$$Q(10^4) = -(10^4)^{-1} - (10^4)^{-2}/2$$

As s becomes larger without limit, $Q(s)$ converges on 0. We can easily calculate the value of our antiderivative when the input is 1:

$$Q(1) = -(1^{-1}) - (1^{-2})/2 = -1 - 1/2 = -3/2$$

Now we know that

$$\lim_{s \to \infty} \; Q(s) - Q(1) = 0 - (-3/2) = 0 + 3/2 = 3/2$$

Therefore

$$\lim_{s \to \infty} \int_1^s (x^{-2} + x^{-3}) \, dx = 3/2$$

which tells us that

$$\int_1^\infty (x^{-2} + x^{-3}) \, dx = 3/2$$

That's the area of the shaded zone in Fig. B-16. Despite the fact that the region is infinitely elongated toward the right, it encloses a finite area.

Chapter 17

1. Here's the integral we've been told to evaluate:

$$\int_0^{2\pi} -5 \sin x \, dx$$

We can pull out the constant 5 to get

$$5 \int_0^{2\pi} -\sin x \, dx$$

Let's call the integrand function

$$f(x) = -\sin x$$

The basic antiderivative is

$$F(x) = \cos x$$

When we evaluate this from $x = 0$ to $x = 2\pi$, we get

$$F(2\pi) - F(0) = \cos 2\pi - \cos 0 = 1 - 1 = 0$$

The definite integral we seek is five times this, which is also 0. We've found that

$$\int_0^{2\pi} -5 \sin x \, dx = 0$$

2. We want to see what happens to the definite integral of the cosine function over intervals whose lower bounds are always at $x = 0$ and whose upper bounds increase endlessly in the positive direction. That is, we want to evaluate

$$\lim_{s \to \infty} \int_0^s \cos x \, dx$$

Let's call the integrand function

$$g(x) = \cos x$$

The basic antiderivative is

$$G(x) = \sin x$$

When we evaluate this from $x = 0$ to $x = s$, we get

$$G(s) - G(0) = \sin s - \sin 0 = \sin s - 0 = \sin s$$

As s increases endlessly, $\sin s$ goes through a cycle that ranges over the entire set of reals between, and including, -1 and 1. The value of s never settles on any specific real number in this closed interval; it oscillates forever back and forth. That means

$$\underset{s \to \infty}{Lim} \int_{0}^{s} \cos x \, dx$$

is undefined. The integral never "settles down" as the interval grows. It oscillates like the function itself, but the *phase* (that is, the horizontal position of the "wave train") is not the same. Try graphing the sine and cosine functions together on the same coordinate plane, and see for yourself.

3. We want to evaluate the following definite integral, which involves the product of a composite function and another function:

$$\int_{0}^{1} (\cos x^2)(2x) \, dx$$

We can use the substitution rule here if we give the component functions the names

$$f(x) = x^2$$

and

$$g(y) = \cos y$$

When we differentiate $f(x)$, we get

$$f'(x) = 2x$$

This means we can use substitution to rewrite the above integral in the form

$$\int_{f(0)}^{f(1)} g(y) \, dy$$

The bounds of this integral are

$$f(0) = 0^2 = 0$$

and

$$f(1) = 1^2 = 1$$

so we can rewrite it as

$$\int_0^1 g(y)\, dy$$

which, in this case, is

$$\int_0^1 \cos y\, dy$$

We're using y as a nickname for $f(x)$. The basic antiderivative of g is

$$G(y) = \sin y$$

When we evaluate this from 0 to 1, we obtain

$$G(1) - G(0) = \sin 1 - \sin 0 = \sin 1 - 0 = \sin 1$$

This tells us that

$$\int_0^1 (\cos x^2)(2x)\, dx = \sin 1$$

A calculator tells us that $\sin 1 \approx 0.841$, rounded to three decimal places. (Remember that we're dealing with radians here, not degrees.)

4. Our task is to evaluate the definite integral

$$\int_0^\pi (4e^x + 4\sin x)\, dx$$

Let's start by factoring the constant 4 out of the integrand and putting it in front of the entire integral. That gives us

$$4\int_0^\pi (e^x + \sin x)\, dx$$

If we call the integrand function

$$f(x) = e^x + \sin x$$

then the basic antiderivative is

$$F(x) = e^x + (-\cos x) = e^x - \cos x$$

When we evaluate this from $x = 0$ to $x = \pi$, we get

$$F(\pi) - F(0) = (e^\pi - \cos \pi) - (e^0 - \cos 0) = [e^\pi - (-1)] - (1 - 1)$$
$$= (e^\pi + 1) - 0 = e^\pi + 1$$

The definite integral we seek is four times this, which is $4e^{\pi} + 4$. We've found that

$$\int_0^{\pi} (4e^x + 4 \sin x)\, dx = 4e^{\pi} + 4$$

Using a calculator and rounding to three decimal places, we get 96.563.

5. We want to evaluate

$$\int_0^1 2e^{2x}\, dx$$

Our integrand is a composite function. Let's give the component functions the names

$$f(x) = 2x$$

and

$$g(y) = e^y$$

The derivative of f is

$$f'(x) = 2$$

We can use substitution to rewrite the above integral in the form

$$\int_{f(0)}^{f(1)} g(y)\, dy$$

When we evaluate f at the bounds, we obtain

$$f(0) = 2 \cdot 0 = 0$$

and

$$f(1) = 2 \cdot 1 = 2$$

so the integral becomes

$$\int_0^2 g(y)\, dy$$

which is

$$\int_0^2 e^y\, dy$$

The basic antiderivative is

$$G(y) = e^y$$

When we evaluate this from $y = 0$ to $y = 2$, we get

$$G\,(2) - G\,(0) = e^2 - e^0 = e^2 - 1$$

which tells us that

$$\int_0^1 2e^{2x}\,dx = e^2 - 1$$

When we round to three decimal places with a calculator, we get 6.389.

6. We've been told to figure out the value of

$$\int_0^1 (\cos e^x)(e^x)\,dx$$

The integrand is a composite function. Using substitution, we can give the component functions the names

$$f\,(x) = e^x$$

and

$$g\,(y) = \cos y$$

When we differentiate f, we get

$$f'\,(x) = e^x$$

We can therefore rewrite the above integral in the form

$$\int_{f(0)}^{f(1)} g\,(y)\,dy$$

When we calculate the bounds of this integral, we get

$$f\,(0) = e^0 = 1$$

and

$$f\,(1) = e^1 = e$$

so it becomes

$$\int_1^e g\,(y)\,dy$$

which is

$$\int_1^e \cos y\,dy$$

the basic antiderivative is

$$G(y) = \sin y$$

When we evaluate this from $y = 1$ to $y = e$, we obtain

$$G(e) - G(1) = \sin e - \sin 1$$

which tells us that

$$\int_0^1 (\cos e^x)(e^x)\, dx = \sin e - \sin 1$$

We can use a calculator to work this out so we get a decimal expression. When we do that and round off to three decimal places, we get -0.431.

7. Let's call the function f and the independent variable x. Our integral is

$$\int_0^1 x^{-1}\, dx$$

This is an improper integral, because the reciprocal function of x is singular at $x = 0$. Let's take a tiny positive δ, and add it to the lower bound at $x = 0$. Then our integral becomes

$$\int_\delta^1 x^{-1}\, dx$$

The basic antiderivative of the integrand function is

$$F(x) = \ln |x|$$

The interval of integration doesn't include any negative values of x, so we can modify this to

$$F(x) = \ln x$$

When we evaluate this from δ to 1, we get

$$F(1) - F(\delta) = \ln 1 - \ln \delta = 0 - \ln \delta = -\ln \delta$$

Now, let's look at the limit

$$\lim_{\delta \to 0+} -\ln \delta$$

As δ approaches 0 from the positive side, the natural log of δ attains values that can become arbitrarily large in the negative direction. That means that the above limit blows up positively, so

$$\lim_{\delta \to 0+} \int_\delta^1 x^{-1}\, dx$$

is infinite. From this fact, we can conclude that

$$\int_0^1 x^{-1} \, dx$$

is undefined.

8. We want to evaluate

$$\int_{-\infty}^{-1} x^{-1} \, dx$$

The basic antiderivative of the integrand function is

$$F(x) = \ln |x|$$

To find the integral, we must determine

$$\underset{p \to -\infty}{Lim} \ F(-1) - F(p)$$

where p is a real-number constant that grows arbitrarily large in the negative direction. We can calculate that

$$F(-1) = \ln |-1| = \ln 1 = 0$$

As p becomes large negatively, $F(p)$ grows large to an unlimited extent while remaining positive. It's convenient to show this by inputting the negatives of increasing positive integer powers of e. Here are a few examples:

$$F(-e^2) = \ln |-e^2| = \ln e^2 = 2$$
$$F(-e^{20}) = \ln |-e^{20}| = \ln e^{20} = 20$$
$$F(-e^{200}) = \ln |-e^{200}| = \ln e^{200} = 200$$

There's no limit to how large $F(p)$ can become, if we make p large enough negatively. From that fact, we can conclude that

$$\underset{p \to -\infty}{Lim} \ \ln |-1| - \ln |p|$$

is infinite. Therefore

$$\int_{-\infty}^{-1} x^{-1} \, dx$$

is undefined.

9. Let's try to evaluate

$$\int_{-e}^{e} x^{-1} \, dx$$

as if the interval didn't contain a singularity. (It does, as we know, at $x = 0$. But we've been told to ignore this fact.) The basic antiderivative is

$$F(x) = \ln |x|$$

When we evaluate this directly from $-e$ to e, we get

$$F(e) - F(-e) = \ln |e| - \ln |-e| = \ln e - \ln e = 1 - 1 = 0$$

Let's look at this situation in the coordinate plane. Figure B-17 is a graph of

$$f(x) = x^{-1}$$

with the bounds $x = -e$ and $x = e$ drawn in. It looks as if the negative and positive areas defined by the curves cancel each other. They would if they were defined and finite. But they are not, as we'll discover when we solve the next problem.

10. Refer again to Fig. B-17. Let's work out the part of the integral on the right-hand side of the singularity. To do that, we must evaluate

$$\int_0^e x^{-1}\, dx$$

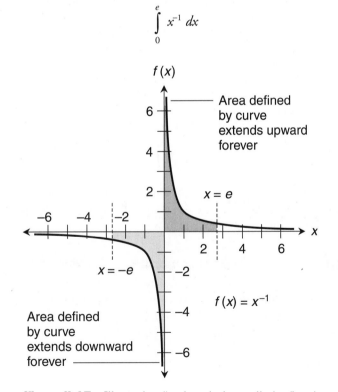

Figure B-17   Illustration for the solution to Probs. 9 and 10 in Chap. 17. This graph might suggest that the integral is 0 over the interval from $-e$ to e. But that isn't the case!

Let's take a tiny positive δ, add it to the lower bound where $x = 0$, and get

$$\int_{\delta}^{e} x^{-1} \, dx$$

The basic antiderivative is

$$F(x) = \ln |x|$$

Our interval does not span any negative values of x, so we can eliminate the absolute value symbols here. When we evaluate from δ to e, we get

$$F(e) - F(\delta) = \ln e - \ln \delta = 1 - \ln \delta$$

Now, we must look at the limit

$$\underset{\delta \to 0+}{Lim} \quad 1 - \ln \delta$$

As δ approaches 0 from the positive side, its natural log becomes arbitrarily large in the negative direction. Therefore, the quantity $(1 - \ln \delta)$ blows up, so

$$\underset{\delta \to 0+}{Lim} \int_{\delta}^{e} x^{-1} \, dx$$

is infinite, indicating that

$$\int_{0}^{e} x^{-1} \, dx$$

is undefined. This fact is all we need to conclude that

$$\int_{-e}^{e} x^{-1} \, dx$$

is undefined. Remember that we can't add an undefined quantity to anything and expect to get a sum that's defined! This is true even if both undefined quantities appear as if they ought to be additive inverses.

Chapter 18

1. The equation of the unit circle isn't written in a form that we usually see for a function. There's a good reason: The equation doesn't represent a function unless we restrict its range. Here's the equation again, for reference:

$$x^2 + y^2 = 1$$

If we consider only the first quadrant, representing that part of the circle where x and y are both positive, and if we let $y = f(x)$, then we can rearrange the equation to get

$$f(x) = (1 - x^2)^{1/2}$$

We want to find the arc length from the point where $x = 0$ to the point where $x = 1$. We begin by differentiating our function. We can do that using the chain rule, getting

$$f'(x) = -x(1 - x^2)^{-1/2}$$

If you've forgotten how the chain rule works, you might want to go back to Chap. 6 and review it. When we square the derivative, we get

$$[f'(x)]^2 = x^2(1 - x^2)^{-1}$$

Adding 1 gives us

$$1 + [f'(x)]^2 = 1 + x^2(1 - x^2)^{-1}$$

We can morph the right-hand side of this equation using algebra to get

$$1 + [f'(x)]^2 = (1 - x^2)^{-1}$$

The arc length L that we seek is the definite integral from $x = 0$ to $x = 1$, whose integrand is the positive square root of the above quantity. That's

$$L = \int_0^1 [(1 - x^2)^{-1}]^{1/2} \, dx$$

which can be simplified to

$$L = \int_0^1 (1 - x^2)^{-1/2} \, dx$$

Using an integral table such as App. G, we can see that the indefinite integral, leaving out the constant of integration, is

$$\int (1 - x^2)^{-1/2} \, dx = \text{Arcsin } x$$

That's the inverse sine function, which you'll learn more about in Chap. 21. (You'll also learn why the "A" in "Arcsin" is capitalized in equations!) When we evaluate this function from $x = 0$ to $x = 1$, we get

$$L = \text{Arcsin } 1 - \text{Arcsin } 0 = \pi/2 - 0 = \pi/2$$

A calculator will tell us that this is approximately 1.571 units, rounded to three decimal places.

2. Figure B-18 is a graph of the unit circle. Each axis division represents 1/5 unit. For the endpoint where $x = 1$,

$$f(1) = [1 - (-1)^2]^{1/2} = (1 - 1)^{1/2} = 0^{1/2} = 0$$

For the endpoint where $x = 0$,

$$f(0) = [1 - 0^2]^{1/2} = 1^{1/2} = 1$$

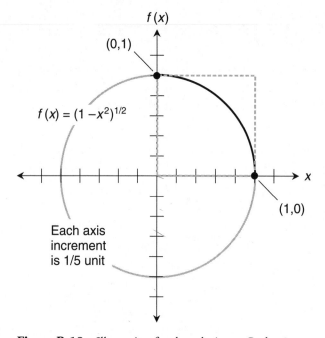

Figure B-18   Illustration for the solution to Probs. 2 and 3 in Chap. 18. Each axis division represents 1/5 unit.

The arc whose length we found is shown as a solid black curve. It's clearly equal to 1/4 of the circumference C of the circle. That means C is four times the arc length:

$$C = 4 \cdot \pi/2 = 2\pi$$

From basic Euclidean geometry, we remember that 2π units represents the circumference of a circle whose radius is 1 unit.

3. In Fig. B-18, the dashed gray square shows the guideline for using the arc-in-a-box method. If we say that $y = f(x)$, then we can call the arc's endpoints

$$(x_1, y_1) = (1,0)$$

and

$$(x_2, y_2) = (0,1)$$

The distance between these points, representing the diagonal measure of the box, is

$$[(y_2 - y_1)^2 + (x_2 - x_1)^2]^{1/2} = [(1 - 0)^2 + [(0 - 1)^2]^{1/2}$$
$$= [1^2 + (-1)^2]^{1/2} = (1 + 1)^{1/2} = 2^{1/2}$$

Using a calculator, we can see that this is approximately 1.414 units, rounded to three decimal places. Half the perimeter of the box is easily seen to be 1 + 1, or 2, units. In the solution to Prob. 1, we found $L \approx 1.571$, which is between these two limits.

4. As in Prob. 1, our equation is not written in the standard function form. Here it is again, for reference:

$$x^2 + y^2 = 4$$

If we consider only that part of the circle in the second quadrant, and if we let $y = f(x)$, then we can rearrange the equation to get

$$f(x) = (4 - x^2)^{1/2}$$

When we differentiate this using the chain rule, we get

$$f'(x) = -x(4 - x^2)^{-1/2}$$

When we square this derivative, we obtain

$$[f'(x)]^2 = x^2(4 - x^2)^{-1}$$

Adding 1 gives us

$$1 + [f'(x)]^2 = 1 + x^2(4 - x^2)^{-1}$$

We can morph the right-hand side using algebra to get

$$1 + [f'(x)]^2 = 4 \cdot (4 - x^2)^{-1}$$

The arc length L that we seek is the definite integral from $x = -2$ to $x = -1$ of the positive square root of the above quantity. That's

$$L = \int_{-2}^{-1} [4 \cdot (4 - x^2)^{-1}]^{1/2} \, dx$$

which can be simplified to

$$L = \int_{-2}^{-1} 2 \cdot (4 - x^2)^{-1/2} \, dx$$

and further to

$$L = 2 \int_{-2}^{-1} (4 - x^2)^{-1/2} \, dx$$

Using an integral table such as App. G, we can see that the indefinite integral, leaving out the constant of integration, is

$$\int (4 - x^2)^{-1/2} \, dx = \text{Arcsin} \, (x/2)$$

We want to evaluate this from $x = -2$ to $x = -1$. By convention, the domain of the sine function is restricted to the interval $[-\pi/2, \pi/2]$ to ensure that Arcsin is a true function. (Chap.21 will explain this.) When we do that, we get

$$L = \text{Arcsin}\,(-2/2) - \text{Arcsin}\,(-1/2) = \text{Arcsin}\,(-1) - \text{Arcsin}\,(-1/2)$$
$$= -\pi/2 - (-\pi/6) = -\pi/2 + \pi/6 = -\pi/3$$

Now we must remember that our arc length L is twice this, because the entire definite integral is supposed to be multiplied by 2. Therefore

$$L = -2\pi/3$$

The arc length comes out negative in this solution only because of the way we restrict Arcsin. We can take the absolute value to get the "real-world arc length," so

$$|L| = |-2\pi/3| = 2\pi/3$$

A calculator approximates this as 2.094 units, rounded to three decimal places.

5. Figure B-19 is a graph of a circle with radius 2, centered at the origin. Its equation is

$$x^2 + y^2 = 4$$

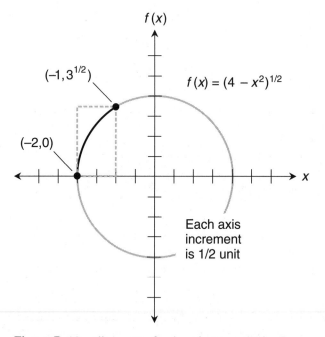

Figure B-19   Illustration for the solution to Probs. 5 and 6 in Chap. 18. Each axis division represents 1/2 unit.

Each axis division represents 1/2 unit. For the arc's endpoint where $x = -2$, we have

$$f(-2) = [4 - (-2)^2]^{1/2} = (4 - 4)^{1/2} = 0^{1/2} = 0$$

For the arc's endpoint at $x = -1$, we have

$$f(-1) = [4 - (-1)^2]^{1/2} = (4 - 1)^{1/2} = 3^{1/2}$$

The arc whose length we found is shown as a solid black curve.

6. In Fig. B-19, the dashed gray rectangle shows the arc-in-the-box guide. Suppose we say that $y = f(x)$. Then we can call the arc's endpoints

$$(x_1, y_1) = (-2, 0)$$

and

$$(x_2, y_2) = (-1, 3^{1/2})$$

The distance between these points, representing the diagonal measure of the box and the minimum possible arc length, is

$$[(y_2 - y_1)^2 + (x_2 - x_1)^2]^{1/2} = \{(3^{1/2} - 0)^2 + [(-1) - (-2)]^2\}^{1/2}$$
$$= [(3^{1/2})^2 + 1^2]^{1/2} = (3 + 1)^{1/2} = 4^{1/2} = 2$$

Half the perimeter of the box, is $1 + 3^{1/2}$ units. Using a calculator and rounding to three decimal places, we obtain 2.732 units as the maximum possible arc length. We found that the arc is $2\pi/3$ units long. That value is within the arc-in-the-box constraints.

7. Figure B-20 illustrates how we can show that the result of $L = 2\pi/3$ we got in the solution to Prob. 4 is exactly right. The radius r of the circle is 2 units, so its circumference C is

$$C = 2\pi r = 2\pi \cdot 2 = 4\pi$$

The ratio of the arc length to the circumference is therefore

$$L/C = (2\pi/3) / (4\pi) = 1/6$$

This tells us that the arc turns through an angle that is exactly 1/6 of a circle. A complete circle has 2π radians. If the arc length we found in the solution to Prob. 4 is correct, the arc should turn through exactly $2\pi/6$, or $\pi/3$, radians. We can verify this with trigonometry. We set up a right triangle as shown. The cosine of an angle θ inside a right triangle is equal to the length of the adjacent side (in this case 1 unit) divided by the length of the hypotenuse (in this case 2 units). That means we should have

$$\text{Arccos } (1/2) = \pi/3$$

This is a valid equation, which proves that the arc length we found is exactly right.

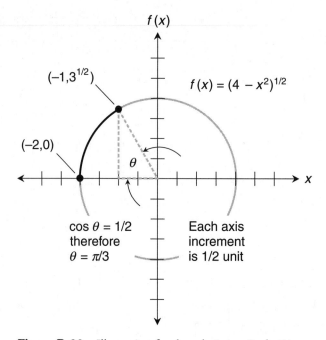

Figure B-20   Illustration for the solution to Prob. 7 in Chap. 18. Each axis division represents 1/2 unit.

8. Let's begin by differentiating the function

$$f(x) = x^2/2$$

The derivative is

$$f'(x) = x$$

When we plug x into the formula for the arc length in place of $f'(x)$, and if we also include the bounds of the integration interval, we get

$$L = \int_0^1 (1 + x^2)^{1/2} \, dx$$

Now let's consider the indefinite integral

$$\int (1 + x^2)^{1/2} \, dx$$

When we search through a table of integrals such as App. G at the back of this book, we'll find the form

$$\int (x^2 + a^2)^{1/2} = (x/2)(x^2 + a^2)^{1/2} + (1/2) \, a^2 \ln |x + (x^2 + a^2)^{1/2}| + c$$

where a is a general constant, and c is the constant of integration. Let's reverse the addends in the integrand. We can eliminate the absolute-value signs, because we won't encounter any negatives in the calculations to come. Now our equation is

$$\int (a^2 + x^2)^{1/2}\, dx = (x/2)(x^2 + a^2)^{1/2} + (1/2)\, a^2 \ln\, [x + (x^2 + a^2)^{1/2}] + c$$

If we substitute 1 in place of a and get rid of the constant of integration, we obtain

$$\int (1 + x^2)^{1/2}\, dx = (x/2)(x^2 + 1)^{1/2} + (1/2) \ln\, [x + (x^2 + 1)^{1/2}]$$

To determine the arc length, we must evaluate the following expression over the interval from $x = 0$ to $x = 1$:

$$(x/2)(x^2 + 1)^{1/2} + (1/2) \ln\, [x + (x^2 + 1)^{1/2}]$$

Plugging in $x = 1$, we get

$$(1/2) \cdot (1^2 + 1)^{1/2} + (1/2) \cdot \ln\, [1 + (1^2 + 1)^{1/2}] = (1/2) \cdot 2^{1/2} + (1/2) \cdot \ln\, (1 + 2^{1/2})$$
$$= (1/2) \cdot [2^{1/2} + \ln\, (1 + 2^{1/2})]$$

Plugging in $x = 0$, we get

$$(0/2) \cdot (0^2 + 1)^{1/2} + (1/2) \cdot \ln\, [0 + (0^2 + 1)^{1/2}]$$
$$= 0 + (1/2) \cdot \ln\, 1 = 0 + (1/2) \cdot 0 = 0 + 0 = 0$$

The arc length is the difference between these results, or simply the value for $x = 1$:

$$L = (1/2) \cdot [2^{1/2} + \ln\, (1 + 2^{1/2})]$$

Let's use a calculator to approximate this. When we round off to three decimal places, we get $L \approx 1.148$ units.

9. Figure B-21 is a graph of the parabola representing the function

$$f(x) = x^2/2$$

Each axis division represents 1/4 unit. For the endpoint at $x = 0$,

$$f(0) = 0^2/2 = 0/2 = 0$$

For the endpoint at $x = 1$,

$$f(1) = 1^2/2 = 1/2$$

The arc whose length we found is shown as a solid black curve.

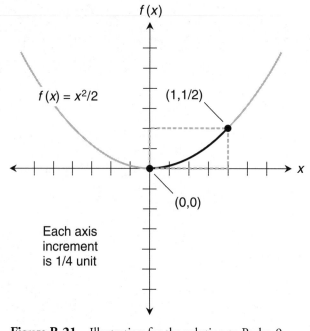

Figure B-21   Illustration for the solution to Probs. 9 and 10 in Chap. 18. Each axis division represents 1/4 unit.

10. In Fig. B-21, the dashed gray rectangle shows the arc-in-the-box guide. Suppose we say that $y = f(x)$. Then we can call the arc's endpoints

$$(x_1, y_1) = (0,0)$$

and

$$(x_2, y_2) = (1, 1/2)$$

The distance between these points, representing the diagonal of the box and the minimum possible arc length, is

$$[(y_2 - y_1)^2 + (x_2 - x_1)^2]^{1/2} = [(1/2 - 0)^2 + (1 - 0)^2]^{1/2} = (1/4 + 1)^{1/2} = (5/4)^{1/2}$$

Using a calculator, we find that this is approximately 1.118 units. Half the perimeter of the box, representing the maximum possible arc length, is $1 + 1/2$, or 1.5 units. In the solution to Prob. 8, we found that

$$L = (1/2) \cdot [2^{1/2} + \ln(1 + 2^{1/2})] \approx 1.148$$

This value lies between the arc-in-the-box constraints.

Chapter 19

1. We've been told to evaluate the definite integral

$$\int_0^\pi (3 \sin x - 5e^x) \, dx$$

The principle of linearity for subtraction allows us to rewrite this as

$$3 \int_0^\pi \sin x \, dx - 5 \int_0^\pi e^x \, dx$$

The basic antiderivative of the sine is the negative of the cosine. The exponential function is its own basic antiderivative. We can therefore evaluate the above sum of integrals as

$$3 \cdot [-\cos \pi - (-\cos 0)] - 5 \cdot (e^\pi - e^0)$$

which simplifies to

$$3 \cdot [\cos 0 - \cos \pi] - 5 \cdot (e^\pi - e^0)$$

We know that $\cos 0 = 1$, $\cos \pi = -1$, and $e^0 = 1$, so the above expression becomes

$$3 \cdot [1 - (-1)] - 5 \cdot (e^\pi - 1) = 3 \cdot 2 - 5e^\pi + 5 = 11 - 5e^\pi$$

The complete integral and its solution can be written out in full as

$$\int_0^\pi (3 \sin x - 5e^x) \, dx = 11 - 5e^\pi$$

2. We've been told to evaluate the definite integral

$$\int_1^2 (5e^x + 7x^{-1}) \, dx$$

The principle of linearity allows us to rewrite this as

$$5 \int_1^2 e^x \, dx + 7 \int_1^2 x^{-1} \, dx$$

The exponential function is its own basic antiderivative. The antiderivative of the reciprocal function is the absolute value of the natural logarithm. Because we're dealing entirely with positive values in our integration interval, we don't have to worry about absolute values. We can therefore express the above sum of integrals as

$$5 \cdot (e^2 - e^1) + 7 \cdot (\ln 2 - \ln 1)$$

Because $\ln 1 = 0$ and $e^1 = e$, we can simplify this to

$$5 \cdot (e^2 - e) + 7 \ln 2$$

We can get rid of the parentheses, rewriting the above expression as

$$5e^2 - 5e + 7 \ln 2$$

The complete integral and its solution can be written out in full as

$$\int_1^2 (5e^x + 7x^{-1})\, dx = 5e^2 - 5e + 7 \ln 2$$

3. Here's the integral we want to evaluate:

$$\int_{-\pi/2}^{\pi/2} (3 \sin x + 2 \cos x)\, dx$$

Using the principle of linearity, we can rearrange this to get

$$3 \int_{-\pi/2}^{\pi/2} \sin x\, dx \;+\; 2 \int_{-\pi/2}^{\pi/2} \cos x\, dx$$

The basic antiderivative of the sine is the negative of the cosine. The basic antiderivative of the cosine is the sine. Knowing these things, we can evaluate the above as

$$3 \cdot \{-\cos \pi/2 - [-\cos (-\pi/2)]\} + 2 \cdot [\sin \pi/2 - \sin (-\pi/2)]$$

which simplifies to

$$3 \cdot [\cos (-\pi/2) - \cos \pi/2] + 2 \cdot [\sin \pi/2 - \sin (-\pi/2)]$$

We know that $\cos \pi/2 = 0$, $\cos -\pi/2 = 0$, $\sin \pi/2 = 1$, and $\sin -\pi/2 = -1$, so the above becomes

$$3 \cdot (0 - 0) + 2 \cdot [1 - (-1)] = 3 \cdot 0 + 2 \cdot 2 = 4$$

The complete integral and its solution can be written out in full as

$$\int_{-\pi/2}^{\pi/2} (3 \sin x + 2 \cos x)\, dx = 4$$

4. Let's start with this:

$$\int_a^b \{k_1 [f_1 (x)] - k_2 [f_2 (x)]\}\, dx$$

According to the principle of linearity for subtraction, we can rewrite this as

$$k_1 \int_a^b f_1 (x)\, dx \;-\; k_2 \int_a^b f_2 (x)\, dx$$

Changing the subtraction to addition along with multiplication by -1, we get

$$k_1 \int_a^b f_1 (x)\, dx \;+\; (-1 \cdot k_2) \int_a^b f_2 (x)\, dx$$

which can be rewritten as

$$k_1 \int_a^b f_1(x) \ dx + k_2 \left(-\int_a^b f_2(x) \ dx \right)$$

The reversal rule lets us get rid of the minus sign and reverse the bounds of the second integral. That gives us the result we've been told to find:

$$k_1 \int_a^b f_1(x) \ dx + k_2 \int_b^a f_2(x) \ dx$$

5. Here's the integral we want to resolve:

$$\int (3 \cos x)(2x - 4) \ dx$$

To set up the parts, let's say that

$$f(x) = 2x - 4$$

and

$$g'(x) = 3 \cos x$$

The basic antiderivative of g', not including the constant of integration, is

$$g(x) = 3 \sin x$$

The derivative of f is

$$f'(x) = 2$$

For reference, the general formula for integration by parts is

$$\int f(x) \cdot g'(x) \ dx = f(x) \cdot g(x) - \int g(x) \cdot f'(x) \ dx$$

In this situation, we have

$$\int (3 \cos x)(2x - 4) \ dx = (2x - 4)(3 \sin x) - \int (3 \sin x) \cdot 2 \ dx$$

$$= (2x - 4)(3 \sin x) - 6 \int \sin x \ dx$$

The basic antiderivative of the sine is the negative of the cosine, so the last integral in the above expression becomes $-\cos x$ plus a constant of integration. Substituting back into the preceding equation, we get

$$\int (3 \cos x)(2x - 4) \ dx = (2x - 4)(3 \sin x) - 6 \cdot (-\cos x) + c$$

$$= 6x \sin x - 12 \sin x + 6 \cos x + c$$

where c is the constant of integration.

6. Here's the integral we want to resolve:

$$\int 7x^2 \cos x \, dx$$

To set up the parts, let's say that

$$f(x) = 7x^2$$

and

$$g'(x) = \cos x$$

The basic antiderivative of g' is

$$g(x) = \sin x$$

The derivative of f is

$$f'(x) = 14x$$

Again for reference, the general formula for integration by parts is

$$\int f(x) \cdot g'(x) \, dx = f(x) \cdot g(x) - \int g(x) \cdot f'(x) \, dx$$

In this case,

$$\int 7x^2 \cos x \, dx = 7x^2 \sin x - \int \sin x \cdot (14x) \, dx = 7x^2 \sin x - \int 14x \sin x \, dx$$

Now we've encountered another integral that requires us to apply the technique of integration by parts! Let's keep the above equation in mind as a "marker." A little later on, we'll want to come back to it. The new integral we must find is

$$\int 14x \sin x \, dx$$

Let's call the parts

$$f_*(x) = 14x$$

and

$$g_*'(x) = \sin x$$

The antiderivative of g_*', not including the constant of integration, is

$$g_*(x) = -\cos x$$

The derivative of f_* is

$$f_*'(x) = 14$$

An adaptation of the general formula for integration by parts tells us that

$$\int f_*(x) \cdot g'_*(x) \, dx = f_*(x) \cdot g_*(x) - \int g_*(x) \cdot f'_*(x) \, dx$$

In this situation, we have

$$\int 14\, x \sin x \, dx = 14x\, (-\cos x) - \int -\cos x \cdot 14 \, dx = -14x \cos x + 14 \int \cos x \, dx$$

The basic antiderivative of the cosine is the sine, so we can simplify the above to

$$= -14x \cos x + 14 \sin x$$

Remember that we're leaving out the constants of integration until we get to the end of this process. Now, the above expression can be substituted in place of the last integral in the equation we designated as a "marker" earlier. That equation (with the middle portion removed) is

$$\int 7x^2 \cos x \, dx = 7x^2 \sin x - \int 14x \sin x \, dx$$

Substitution gives us the complete solution to our problem:

$$\int 7x^2 \cos x \, dx = 7x^2 \sin x - (-14x \cos x + 14 \sin x) + c$$
$$= 7x^2 \sin x + 14x \cos x - 14 \sin x + c$$

7. Here's the integral we want to evaluate:

$$\int_0^\pi (\pi/2 - 2x)(5 \cos x) \, dx$$

First, let's find the indefinite integral. Then we can put the bounds back in and work out the definite integral. We want to resolve

$$\int (\pi/2 - 2x)(5 \cos x) \, dx$$

To set up the parts, let's say that

$$f(x) = \pi/2 - 2x$$

and

$$g'(x) = 5 \cos x$$

The basic antiderivative of g' is

$$g(x) = 5 \sin x$$

The derivative of f is

$$f'(x) = -2$$

Once again for reference, the general formula for integration by parts is

$$\int f(x) \cdot g'(x) \, dx = f(x) \cdot g(x) - \int g(x) \cdot f'(x) \, dx$$

In this case,

$$\int (\pi/2 - 2x)(5 \cos x) \, dx = (\pi/2 - 2x)(5 \sin x) - \int (5 \sin x) \cdot (-2) \, dx$$

$$= (\pi/2 - 2x)(5 \sin x) + 10 \int \sin x \, dx$$

The last integral in the above expression can be simplified to $-\cos x$, because the basic antiderivative of the sine is the negative of the cosine. We don't need the constant of integration, because we are going to evaluate a definite integral. Substituting back into the preceding equation, we get

$$\int (\pi/2 - 2x)(5 \cos x) \, dx = (\pi/2 - 2x)(5 \sin x) - 10 \cos x$$

$$= 5\pi/2 \sin x - 10x \sin x - 10 \cos x$$

Now we must evaluate

$$5\pi/2 \sin x - 10x \sin x - 10 \cos x \Big]_0^{\pi}$$

In this expression, the large square bracket with the values at the extreme right translates as "evaluated from 0 to π." First, let's plug π in to obtain

$$5\pi/2 \sin \pi - 10\pi \sin \pi - 10 \cos \pi = (5\pi/2) \cdot 0 - 10\pi \cdot 0 - 10 \cdot (-1)$$

$$= 0 - 0 - (-10) = 10$$

When we plug 0 into the same expression, we get

$$5\pi/2 \sin 0 - 10 \cdot 0 \sin 0 - 10 \cos 0 = (5\pi/2) \cdot 0 - 10\pi \cdot 0 - 10 \cdot 1$$

$$= 0 - 0 - 10 = -10$$

The definite integral is the difference between these results, which is

$$10 - (-10) = 10 + 10 = 20$$

We conclude that

$$\int_0^{\pi} (\pi/2 - 2x)(5 \cos x) \, dx = 20$$

8. Here's the integral we want to resolve:

$$\int -5 (2x - 7)^{-1} \, dx$$

To begin, we move the constant in front of the integral symbol, getting

$$-5 \int (2x-7)^{-1}\, dx$$

Then we use the indefinite integral formula from App. G to obtain

$$-5 \cdot 2^{-1} \ln |2x-7| + c$$

which can be simplified to

$$(-5/2) \ln |2x-7| + c$$

Therefore

$$\int -5\,(2x-7)^{-1}\, dx = (-5/2) \ln |2x-7| + c$$

9. Here's the integral we want to resolve:

$$\int 2x\,(x^2-1)^{-1}\, dx$$

With algebra, we can break the integrand into a sum of reciprocals of linear functions:

$$\int [(x+1)^{-1} + (x-1)^{-1}]\, dx$$

When we split this into a sum of integrals, we get

$$\int (x+1)^{-1}\, dx + \int (x-1)^{-1}\, dx$$

Adapting and applying the formula from App. G to each addend gives us

$$\ln |x+1| + \ln |x-1| + c$$

The complete statement of the integral and its resolution is

$$\int 2x\,(x^2-1)^{-1}\, dx = \ln |x+1| + \ln |x-1| + c$$

10. Here's the integral we want to resolve:

$$\int (5x-1)(x^2-x-2)^{-1}\, dx$$

With the help of algebra and intuition, the integrand breaks down into a sum of two constant multiples of reciprocal linear functions, like this:

$$\int [2\,(x+1)^{-1} + 3\,(x-2)^{-1}]\, dx$$

When we split this into a sum of two integrals and pull out the constants, we get

$$2 \int (x+1)^{-1}\,dx + 3 \int (x-2)^{-1}\,dx$$

Applying the usual formula from App. G to each addend gives us

$$2 \ln |x+1| + 3 \ln |x-2| + c$$

The complete statement of the integral and its resolution is

$$\int (5x-1)(x^2-x-2)^{-1}\,dx = 2 \ln |x+1| + 3 \ln |x-2| + c$$

Worked-Out Solutions to Exercises: Chapters 21 to 29

These solutions do not necessarily represent the only way a problem can be figured out. If you think you can solve a problem faster or better than the way it's done here, by all means try it! Always check your work to be sure your "alternative" answer is correct.

Chapter 21

1. We've been told to differentiate the inverse of

$$f(x) = x^2 + 2$$

for $x > 0$. (Do you know why we've restricted the domain this way?) Let

$$y = f(x)$$

and

$$x = f^{-1}(y)$$

We can derive x in terms of y using simple algebra:

$$f(x) = x^2 + 2$$
$$y = x^2 + 2$$
$$y - 2 = x^2$$
$$(y - 2)^{1/2} = x$$
$$x = (y - 2)^{1/7}$$
$$f^{-1}(y) = (y - 2)^{1/2}$$

Remember that if there's any ambiguity in a fractional power, it always denotes the positive value unless otherwise specified. The 1/2 power therefore means the positive square root, so we know that f^{-1} is a true function. When we differentiate directly using the chain rule, we get

$$f^{-1\prime}(y) = [(y-2)^{-1/2}] / 2$$

Now let's use the alternative formula we learned in the chapter text. The derivative of y with respect to x is

$$dy/dx = 2x$$

When we take the reciprocal of each side of this equation, we get

$$dx/dy = (2x)^{-1}$$

This is the derivative of the inverse function, but we want to state it in terms of y, not in terms of x. A moment ago, we found that

$$x = (y-2)^{1/2}$$

By substitution,

$$dx/dy = [2(y-2)^{1/2}]^{-1} = [(y-2)^{-1/2}]/2$$

which can also be written as

$$f^{-1\prime}(y) = [(y-2)^{-1/2}]/2$$

This agrees with the result we got when we differentiated the inverse function directly.

2. We want to differentiate the inverse of

$$f(x) = \ln x$$

for $x > 0$. (Do you know why we've restricted the domain this way?) Let's call

$$y = f(x)$$

and

$$x = f^{-1}(y)$$

We can derive x in terms of y from our knowledge of natural logarithms and exponentials:

$$f(x) = \ln x$$
$$y = \ln x$$
$$e^y = e^{(\ln x)}$$
$$e^y = x$$
$$x = e^y$$
$$f^{-1}(y) = e^y$$

When we differentiate this directly, we get

$$f^{-1\prime}(y) = e^y$$

because the exponential function is its own derivative. Now let's use the alternative method. We have

$$dy/dx = x^{-1}$$

When we take the reciprocal of both sides here, we get

$$dx/dy = (x^{-1})^{-1} = x$$

This is the derivative of the inverse function, but we want to state it in terms of *y*. A moment ago, we found that

$$x = e^y$$

By substitution,

$$dx/dy = e^y$$

which can also be written as

$$f^{-1\prime}(y) = e^y$$

This agrees with the result we got when we differentiated the inverse function directly.

3. Our task is to find the derivative of the inverse of

$$f(x) = x^5 + 4$$

As in the previous problems, let's call

$$y = f(x)$$

and

$$x = f^{-1}(y)$$

Deriving *x* in terms of *y*, we go through the algebra as follows:

$$\begin{aligned}
f(x) &= x^5 + 4 \\
y &= x^5 + 4 \\
y - 4 &= x^5 \\
(y-4)^{1/5} &= x \\
x &= (y-4)^{1/5} \\
f^{-1}(y) &= (y-4)^{1/5}
\end{aligned}$$

When we differentiate this directly, we get

$$f^{-1\prime}(y) = (y-4)^{-4/5}/5$$

Now let's use the alternative method. The derivative of y with respect to x is

$$dy/dx = 5x^4$$

When we take the reciprocal of both sides of this equation, we get

$$dx/dy = x^{-4}/5$$

We want to get the right-hand side in terms of y. A moment ago, we found that

$$x = (y-4)^{1/5}$$

By substitution,

$$dx/dy = [(y-4)^{1/5}]^{-4} / 5$$

which can also be written as

$$f^{-1\prime}(y) = (y-4)^{-4/5} / 5$$

This agrees with the result we got when we differentiated the inverse directly.

4. We've been told to find the domain and the derivative of the function

$$f(x) = 5 \text{ Arcsin } x$$

with respect to x. The domain of the Arcsine function is the set of all reals in $[-1,1]$. The domain of 5 times the Arcsine is also the set of all reals in $[-1,1]$. We can express the derivative of our function f as

$$d/dx \, (5 \text{ Arcsin } x)$$

Pulling out the constant, we get

$$5 \, d/dx \, (\text{Arcsin } x)$$

Using the formula we derived in the chapter text, this becomes

$$5 \, (1-x^2)^{-1/2}$$

5. We want to find the domain and the derivative of

$$g(t) = -6t^2 \text{ Arcsin } t$$

with respect to *t*. The domain of the Arcsine function is the set of all reals in [−1,1]. The domain of −6t^2 times the Arcsine is also the set of all reals in [−1,1]. Our function *g* is the product of two functions. Let's call them

$$p(t) = -6t^2$$

and

$$q(t) = \text{Arcsin } t$$

The derivatives are

$$p'(t) = -12t$$

and

$$q'(t) = (1 - t^2)^{-1/2}$$

According to the product rule for derivatives,

$$g' = p'q + q'p$$

When we substitute the expressions we've derived for *p*, *q*, *p'*, and *q'* in the above formula, we get

$$g'(t) = -12t \text{ Arcsin } t + (1 - t^2)^{-1/2} (-6t^2) = -12t \text{ Arcsin } t - 6t^2 (1 - t^2)^{-1/2}$$

6. We want to find the domain and the derivative of

$$h(z) = \text{Arccos } z^2$$

The domain is the set of all reals *z* such that −1 ≤ z^2 ≤ 1. That means *z* can be any real number in the closed interval [−1,1]. The half-open portion [−1,0) of the allowed interval for z^2 isn't "used" because that would require *z* to be imaginary, but this doesn't present a problem for us. We're concerned only with real-number values of *z*. Our function *h* is a composite function. Using the chain rule along with the formula for the derivative of the Arccosine, we get

$$h'(z) = -[1 - (z^2)^2]^{-1/2} (2z) = -2z (1 - z^4)^{-1/2}$$

7. We want to find the domain and the derivative of

$$f'(v) = 3v^3 \text{ Arccos } (v/3)$$

This function is the product of a cubic and the Arccosine. The domain of the cubic is the set of all reals. The argument of the Arccosine is *v*/3, so the domain is the set of reals

v such that $-1 \le v/3 \le 1$. Therefore, the input v for the Arccosine factor of f can be any real number in the interval $[-3,3]$. The domain of f is the intersection of the domains of its factors, or the set of all reals v such that $-3 \le v \le 3$. Our function f is the product of two functions. Let's call them

$$p(v) = 3v^3$$

and

$$q(v) = \mathrm{Arccos}\,(v/3)$$

The derivatives are

$$p'(v) = 9v^2$$

and

$$q'(v) = -[1 - (v/3)^2]^{-1/2}\,(1/3) = -(1 - v^2/9)^{-1/2}\,/\,3$$

Note that we've used the chain rule to differentiate q. According to the product rule for derivatives,

$$g' = p'\,q + q'\,p$$

When we substitute the expressions we've derived for p, q, p', and q' in the above formula, we get

$$
\begin{aligned}
f'(v) &= 9v^2\,\mathrm{Arccos}\,(v/3) + [-(1 - v^2/9)^{-1/2}\,/\,3]\,(3v^3)\\
&= 9v^2\,\mathrm{Arccos}\,(v/3) - v^3\,(1 - v^2/9)^{-1/2}
\end{aligned}
$$

8. We want to find the domain and derivative of

$$h(s) = \mathrm{Arcsin}\,s^3 - \mathrm{Arccos}\,2s$$

This is a difference of two functions. Let's call them p and q, and say that

$$p(s) = \mathrm{Arcsin}\,s^3$$

and

$$q(s) = \mathrm{Arccos}\,2s$$

The domain of p is the set of all real numbers s, such that s^3 is in the interval $[-1,1]$. That means we must have $-1 \le s \le 1$. The domain of q is the set of all real numbers s, such that $2s$ is in the interval $[-1,1]$. That means we must have $-1/2 \le s \le 1/2$. The domain of h is the set of all s in the intersection of the domains of p and q. In this case, that's the same as the domain of q, or $-1/2 \le s \le 1/2$. To find the derivative of h, we differentiate p and q separately, and then subtract. We must use the chain rule in both instances, getting

$$p'(s) = [1 - (s^3)^2]^{-1/2} (3s^2) = 3s^2 (1 - s^6)^{-1/2}$$

and

$$q'(s) = -[1 - (2s)^2]^{-1/2} (2) = -2 (1 - 4s^2)^{-1/2}$$

The derivative h' is the difference $p' - q'$, or

$$h'(s) = 3s^2 (1 - s^6)^{-1/2} - [-2 (1 - 4s^2)^{-1/2}] = 3s^2 (1 - s^6)^{-1/2} + 2 (1 - 4s^2)^{-1/2}$$

9. We want to find the domain and derivative of

$$g(w) = (\text{Arccos } w^2 - \text{Arcsin } w^2) / 2$$

This is a difference of two functions, divided by a constant. Let's say that

$$p(w) = \text{Arccos } w^2$$

and

$$q(w) = \text{Arcsin } w^2$$

The domain of p is the set of all real numbers w, such that w^2 is in the interval $[-1,1]$. That means we must have $-1 \le w \le 1$. The domain of q is the same. Therefore, the domain of g is the set of all w such that $-1 \le w \le 1$. The half-open portion $[-1,0)$ of the allowed interval for w^2 isn't "used" because that would require w to be imaginary, but that's okay; we're concerned only with real-number values of w. To find g', let's differentiate p and q separately, then subtract, and finally multiply by 1/2, which is the same as dividing by 2. We must use the chain rule to find the derivatives of p and q, getting

$$p'(w) = -[1 - (w^2)^2]^{-1/2} (2w) = -2w (1 - w^4)^{-1/2}$$

and

$$q'(w) = [1 - (w^2)^2]^{-1/2} (2w) = 2w (1 - w^4)^{-1/2}$$

The difference is

$$p'(w) - q'(w) = -2w (1 - w^4)^{-1/2} - 2w (1 - w^4)^{-1/2} = -4w (1 - w^4)^{-1/2}$$

Multiplying by 1/2 gives us the final result

$$g'(w) = -2w (1 - w^4)^{-1/2}$$

10. We want to find the domain and derivative of

$$f(x) = \text{Arcsin } e^x$$

The domain of f is the set of all real numbers x, such that $-1 \leq e^x \leq 1$. That means x can be anything less than or equal to 0. The half-open portion $[-1,0)$ of the allowed interval for e^x isn't "used" because e^x is never negative for any real x. Differentiation requires us to use the chain rule, but it's easy because the natural exponential function is its own derivative:

$$f'(x) = [1 - (e^x)^2]^{-1/2} (e^x) = e^x (1 - e^{2x})^{-1/2}$$

Chapter 22

1. Here's the equation we've been given:

$$3y^2 = 12x^2 - 48$$

We can subtract $12x^2$ from both sides, getting

$$-12x^2 + 3y^2 = -48$$

Multiplying through by -1 gives us

$$12x^2 - 3y^2 = 48$$

When we divide through by 48, we end up with an equation that's in the standard form for a hyperbola:

$$x^2/4 - y^2/16 = 1$$

The center is at $(0,0)$. The horizontal semi-axis is 2 units long (the positive square root of 4), and the vertical semi-axis is 4 units long (the positive square root of 16).

2. Here's the equation we've been given:

$$4x^2 + y^2 + 8x + 2y + 1 = 0$$

Let's rearrange this to get the monomials for each variable together:

$$4x^2 + 8x + y^2 + 2y + 1 = 0$$

The last three terms form a perfect square in y, so we can factor the trinomial they form into a binomial squared. When we do that, we get

$$4x^2 + 8x + (y + 1)^2 = 0$$

Dividing through by 4, we obtain

$$x^2 + 2x + (y+1)^2 / 4 = 0$$

Now let's add 1 to each side, inserting it as the third term on the left:

$$x^2 + 2x + 1 + (y+1)^2 / 4 = 1$$

We can factor the first three terms because they are a perfect square in x, getting an equation in the standard form for an ellipse:

$$(x+1)^2 + (y+1)^2 / 4 = 1$$

The center is at $(-1,-1)$. The horizontal semi-axis is 1 unit long (the positive square root of 1, the unwritten denominator of the first squared binomial), and the vertical semi-axis is 2 units long (the positive square root of 4).

3. Here's the equation we've been given:

$$2x^2 = 288 - 2y^2$$

We can add $2y^2$ to each side, getting

$$2x^2 + 2y^2 = 288$$

Dividing through by 2 gives us

$$x^2 + y^2 = 144$$

which can be rewritten as

$$x^2 + y^2 = 12^2$$

The standard form for the equation of a circle is

$$(x - x_0)^2 + (y - y_0)^2 = r^2$$

where x_0 and y_0 represent the coordinates of the center, and r is the radius. In this situation, $x_0 = 0$ and $y_0 = 0$, so the circle is centered at the origin. We have $r = 12$, so the radius is 12 units.

4. In the solution to Prob. 1, we found that the equation represents a hyperbola. In standard form, that equation is

$$x^2/4 - y^2/16 = 1$$

The x-coordinate of the center is $x_0 = 0$. The horizontal semi-axis is 2 units long. Therefore, the relation is defined for

$$x \leq 0 - 2 \quad \text{or} \quad x \geq 0 + 2$$

which means we must have

$$x \le -2 \quad \text{or} \quad x \ge 2$$

There are no restrictions on the domain when y is the independent variable. The relation that maps y into x is defined for all real numbers y.

To find y', let's get the equation into a more manageable form. Here's the original equation again:

$$x^2/4 - y^2/16 = 1$$

Multiplying through by 16 gives us

$$4x^2 - y^2 = 16$$

Differentiating with respect to x, we have

$$d/dx\,(4x^2) - d/dx\,(y^2) = d/dx\,(16)$$

This works out to

$$8x - 2yy' = 0$$

Subtracting $8x$ from each side produces

$$-2yy' = -8x$$

Dividing through by $-2y$ with the restriction that $y \ne 0$, we get

$$y' = (-8x)\,/\,(-2y) = 4x/y$$

This is defined only when $x < -2$ or $x > 2$. If $x = -2$ or $x = 2$, then the equation of the original hyperbola solves to $y = 0$, making y' undefined because the denominator, y, in the derivative is equal to 0. If $-2 < x < 2$, then y isn't part of the relation, so there is no derivative.

Now let's work out x'. We've already morphed the original equation into a form that's easy to contend with:

$$4x^2 - y^2 = 16$$

When we differentiate term by term with respect to y, we have

$$d/dy\,(4x^2) - d/dy\,(y^2) = d/dy\,(16)$$

Working it out, we get

$$8xx' - 2y = 0$$

We can add $2y$ to each side, producing

$$8xx' = 2y$$

Dividing through by $8x$ with the restriction that $x \neq 0$, we end up with

$$x' = (2y) / (8x) = y / (4x)$$

We've found dx/dy, and it's defined over the entire set of real numbers y. (If you aren't sure why this is true, sketch a graph of the hyperbola. There's no point on the curve for which its slope is parallel to the x axis. But that would have to occur to make dx/dy undefined.)

Checking back on our results, we can see that that y' and x' are reciprocals of each other, as we should expect. We've found that

$$y' = 4x/y$$

and

$$x' = y / (4x)$$

5. In the solution to Prob. 2, we found that the equation represents an ellipse. In standard form, that equation is

$$(x+1)^2 + (y+1)^2 / 4 = 1$$

The x-coordinate of the center is $x_0 = -1$. The horizontal semi-axis is 1 unit long. Therefore, the relation is defined for

$$-1 - 1 \leq x \leq -1 + 1$$

which means we must have

$$-2 \leq x \leq 0$$

The y-coordinate of the center is $y_0 = -1$. The vertical semi-axis is 2 units long. Therefore, the relation is defined for

$$-1 - 2 \leq y \leq -1 + 2$$

which means we must have

$$-3 \leq y \leq 1$$

To find y', let's go back to the original equation, because it's easier to differentiate than the equation for the ellipse in standard form. Here it is again:

$$4x^2 + y^2 + 8x + 2y + 1 = 0$$

Differentiating with respect to *x*, we get

$$d/dx\,(4x^2) + d/dx\,(y^2) + d/dx\,(8x) + d/dx\,(2y) + d/dx\,(1) = d/dx\,(0)$$

This works out to

$$8x + 2yy' + 8 + 2y' + 0 = 0$$

Dividing through by 2 and getting rid of the extraneous term 0, we get

$$4x + yy' + 4 + y' = 0$$

Adding the quantity $(-4x - 4)$ to each side yields

$$yy' + y' = -4x - 4$$

which can be rewritten as

$$(y + 1)\,y' = -4x - 4$$

Dividing through by the quantity $(y + 1)$ with the understanding that $y \neq -1$, we get

$$y' = (-4x - 4)\,/\,(y + 1)$$

This is defined only when $-2 < x < 0$. If $x = -2$ or $x = 0$, then the equation of the original ellipse tells us that $y = -1$, making y' undefined because the denominator of the ratio becomes 0. If $x < -2$ or $x > 0$, then y isn't part of the relation, so there can be no derivative.

Now let's figure out x'. We'll work again with the original equation, because it's easy to differentiate term-by-term:

$$4x^2 + y^2 + 8x + 2y + 1 = 0$$

Taking the derivatives of the terms with respect to *y*, we get

$$d/dy\,(4x^2) + d/dy\,(y^2) + d/dy\,(8x) + d/dy\,(2y) + d/dy\,(1) = d/dy\,(0)$$

This works out to

$$8xx' + 2y + 8x' + 2 + 0 = 0$$

Dividing through by 2 and getting rid of the extraneous term 0, we get

$$4xx' + y + 4x' + 1 = 0$$

Adding the quantity $(-y - 1)$ to each side yields

$$4xx' + 4x' = -y - 1$$

which can be rewritten as

$$(4x + 4)\, x' = -y - 1$$

Dividing through by the quantity $(4x + 4)$ with the understanding that $x \neq -1$, we get

$$x' = (-y - 1)\, /\, (4x + 4) = (y + 1)\, /\, (-4x - 4)$$

This is defined only when $-3 < y < 1$. If $y = -3$ or $y = 1$, then the equation of the original ellipse tells us that $x = -1$, making x' undefined because the denominator of the ratio becomes 0. If $y < -3$ or $y > 1$, then x isn't part of the relation, so x' can't exist.

   Checking back on our results, we can see that that y' and x' are reciprocals of each other. We've found that

$$y' = (-4x - 4)\, /\, (y + 1)$$

and

$$x' = (y + 1)\, /\, (-4x - 4)$$

6. In the solution to Prob. 3, we found that the equation represents a circle. In standard form, it's

$$x^2 + y^2 = 144$$

The circle is centered at the origin, and the radius is 12 units. That means the two-way relation is defined for

$$-12 \leq x \leq 12$$

and

$$-12 \leq y \leq 12$$

   To find y', let's use the standard form directly. When we differentiate with respect to x term-by-term, we get

$$d/dx\, (x^2) + d/dx\, (y^2) = d/dx\, (144)$$

which works out to

$$2x + 2yy' = 0$$

Dividing through by 2, we get

$$x + yy' = 0$$

Subtracting x from each side produces

$$yy' = -x$$

Dividing through by y with the restriction that $y \neq 0$, we obtain

$$y' = -x/y$$

We've found the derivative dy/dx of the two-way relation. This derivative exists only for values of x such that $-12 < x < 12$. If $x = -12$ or $x = 12$, then we get $y = 0$ when we solve for it in the original equation, making y' undefined. If $x < -12$ or $x > 12$, then y isn't part of the relation, so again, y' is undefined.

Now let's find x' by differentiating both sides with respect to y, and then solving the result for x'. We have

$$d/dy\,(x^2) + d/dy\,(y^2) = d/dy\,(144)$$

When we differentiate each term, we get

$$2xx' + 2y = 0$$

Dividing through by 2, we get

$$xx' + y = 0$$

When we subtract y from each side, we obtain

$$xx' = -y$$

We can divide each side by x while insisting that $x \neq 0$ to obtain

$$x' = -y/x$$

We've found the derivative dx/dy. But there's a restriction, just as there was with the other derivative. The value of x' is defined only for $-12 < y < 12$. When $y = -12$ or $y = 12$, then we get $x = 0$ when we solve for it in the original equation. That makes x' undefined. When we have $y < -12$ or $y > 12$, then x isn't in the relation, so x' is undefined.

Checking back on our results, we can see that that y' and x' are reciprocals of each other, as we should expect. We've found that

$$y' = -x/y$$

and

$$x' = -y/x$$

Have you noticed that the formulas we got for y' and x' are the same for this circle as for the unit circle? This happens when we implicitly differentiate the equation for *any* circle centered at the origin, regardless of its radius. If that seems strange to you, remember that we're taking a ratio. As things work out, the ratio is independent of the circle's radius, even though the denominator and numerator change with the radius.

7. In the chapter text, we found these general derivatives for the unit circle:

$$y' = -x/y$$

and

$$x' = -y/x$$

To calculate either derivative at a point, we must know both the x and the y values at that point. We want to find y' when $x = 1/2$. Let's solve for y by plugging $x = 1/2$ into the equation for the unit circle. For reference, that equation is

$$x^2 + y^2 = 1$$

Substituting, we obtain

$$(1/2)^2 + y^2 = 1$$

Morphing this equation produces

$$y^2 = 1 - (1/2)^2 = 1 - 1/4 = 3/4$$

Solving for y gives us the two values

$$y = \pm(3/4)^{1/2} = \pm 3^{1/2}/2$$

We finish by calculating

$$y' = -x/y = -(1/2) / [\pm 3^{1/2}/2] = \pm 3^{-1/2}$$

Now we want to find x' when $y = 2^{-1/2}$. First we solve for x by plugging the given value for y into the equation for the unit circle. That gives us

$$x^2 + (2^{-1/2})^2 = 1$$

Morphing, we get

$$x^2 = 1 - (2^{-1/2})^2 = 1 - 2^{-1} = 1 - 1/2 = 1/2$$

Solving for x, we obtain the two values

$$x = \pm(1/2)^{1/2} = \pm 1/2^{1/2} = \pm 2^{-1/2}$$

To finish, we calculate

$$x' = -y/x = (-2^{-1/2}) / (\pm 2^{-1/2}) = \pm 1$$

8. Here's the original equation representing the ellipse shown in Fig. 22-4:

$$(x-1)^2 / 4 + (y+1)^2 / 9 = 1$$

We found the derivatives

$$y' = (9 - 9x) / (4y + 4)$$

and

$$x' = (4y + 4) / (9 - 9x)$$

When $x = 1$, we get

$$y' = (9 - 9 \cdot 1) / (4y + 4) = 0 / (4y - 4)$$

This is defined and equal to 0, as long as $x = 1$ is part of the relation. In the chapter text, we found that the requirement is

$$-1 \le x \le 3$$

We're within the allowed interval, so when $x = 1$, we have $y' = 0$. Now let's look at what happens to x' when $y = -1$. We obtain

$$x' = [4 \cdot (-1) + 4] / (9 - 9x) = 0 / (9 - 9x)$$

This is defined and equal to 0, as long as $y = -1$ is part of the relation. We found that we must have

$$-4 \le y \le 2$$

Our value of $y = -1$ is in the relation, so when $y = -1$, we have $x' = 0$.

9. Here's the standard-form equation for the hyperbola shown in Fig. 22-5:

$$(x-1)^2 / 4 - (y+1)^2 / 9 = 1$$

We found these derivatives:

$$y' = (9x - 9) / (4y + 4)$$

and

$$x' = (4y + 4) / (9x - 9)$$

When $x = -3$, we get

$$y' = [9 \cdot (-3) - 9] / (4y + 4) = -36 / (4y + 4) = -9 / (y + 1)$$

In the chapter text, we found that y' is defined for $x < -1$ or $x > 3$. Our value $x = -3$ is therefore within the "left-hand" allowed zone. Let's find the actual numbers for y'. Plugging $x = -3$ into the original equation for the hyperbola, we get

$$(-3 - 1)^2 / 4 - (y + 1)^2 / 9 = 1$$

which simplifies to

$$4 - (y + 1)^2 / 9 = 1$$

Subtracting 4 from each side and then multiplying through by -9, we get

$$(y + 1)^2 = 27$$

We can take the plus-or-minus square root of each side to obtain

$$y + 1 = \pm 27^{1/2}$$

A minute ago, we found that when $x = -3$, we have

$$y' = -9 / (y + 1)$$

Substituting $\pm 27^{1/2}$ for the quantity $(y + 1)$, we get

$$y' = -9 / (y + 1) = -9 / (\pm 27^{1/2}) = \pm 9 / 27^{1/2} = \pm 3^{1/2}$$

10. Let's start with the general equation for a circle centered at the origin in Cartesian xy-coordinates. That equation is

$$x^2 + y^2 = r^2$$

where r is the radius. When we differentiate each term with respect to x, we get

$$d/dx\,(x^2) + d/dx\,(y^2) = d/dx\,(r^2)$$

Because r is a constant, r^2 is also a constant. When we work out the derivatives of each term, we get

$$2x + 2yy' = 0$$

Dividing through by 2 produces

$$x + yy' = 0$$

Subtracting x from each side, we get

$$yy' = -x$$

We can divide through by y if we insist that $y \neq 0$. That gives us the familiar ratio

$$y' = -x/y$$

Now let's solve for y in terms of x, using the original equation

$$x^2 + y^2 = r^2$$

We can subtract x^2 from each side to obtain

$$y^2 = r^2 - x^2$$

We can take the square root of both sides of this equation, but we must include both the positive and the negative values to be sure that we account for the complete circle. That gives us

$$y = \pm(r^2 - x^2)^{1/2}$$

Substituting into the equation we obtained for the derivative, we get

$$y' = -x/y = -x / [\pm(r^2 - x^2)^{1/2}] = \pm x / (r^2 - x^2)^{1/2}$$

The expression at the right-hand end describes two quantities that are exact negatives of each other for any x in the relation:

$$x / (r^2 - x^2)^{1/2}$$

and

$$-x / (r^2 - x^2)^{1/2}$$

Chapter 23

1. We've been told to evaluate the following limit, where the numerator and denominator both approach 0 as x approaches 0 (from either side):

$$\lim_{x \to 0} (\sin x) / (8x)$$

If we name the functions

$$f(x) = \sin x$$

and

$$g(x) = 8x$$

then the derivatives are

$$f'(x) = \cos x$$

and

$$g'(x) = 8$$

According to the l'Hôpital rule,

$$\lim_{x \to 0} f(x) \,/\, g(x) = \lim_{x \to 0} f'(x) \,/\, g'(x) = \lim_{x \to 0} (\cos x) \,/\, 8$$

The numerator of our last expression, cos x, approaches 1 as x approaches 0. The denominator is always 8. Therefore,

$$\lim_{x \to 0} (\sin x) \,/\, (8x) = 1/8$$

2. We've been told to evaluate the following limit, where the numerator and denominator both approach 0 as x approaches 1 (from either side):

$$\lim_{x \to 1} (6x^2 - 12x + 6) \,/\, (x^2 - 2x + 1)$$

If we name the functions

$$f(x) = 6x^2 - 12x + 6$$

and

$$g(x) = x^2 - 2x + 1$$

then we have

$$f'(x) = 12x - 12$$

and

$$g'(x) = 2x - 2$$

The l'Hôpital rule tells us that

$$\lim_{x \to 1} f(x) \,/\, g(x) = \lim_{x \to 1} f'(x) \,/\, g'(x) = \lim_{x \to 1} (12x - 12) \,/\, (2x - 2)$$

This expression, like the first one, tends toward 0/0 as x approaches 1. We can differentiate the numerator and denominator again to obtain

$$f''(x) = 12$$

and

$$g''(x) = 2$$

The l'Hôpital rule now indicates that

$$\lim_{x \to 1} f'(x) \,/\, g'(x) = \lim_{x \to 1} f''(x) \,/\, g''(x) = \lim_{x \to 1} 12/2$$

Our last expression is equal to 6, no matter what the value of x. We've found that

$$\lim_{x \to 1} (6x^2 - 12x + 6) \,/\, (x^2 - 2x + 1) = 6$$

3. Our mission is to evaluate the following limit, where the numerator and denominator both approach 0 as x approaches 0 (from either side):

$$\lim_{x \to 0} (12 \sin x - 12x) \,/\, x^3$$

If we name the functions

$$f(x) = 12 \sin x - 12x$$

and

$$g(x) = x^3$$

then the derivatives are

$$f'(x) = 12 \cos x - 12$$

and

$$g'(x) = 3x^2$$

According to l'Hôpital,

$$\lim_{x \to 0} f(x) \,/\, g(x) = \lim_{x \to 0} f'(x) \,/\, g'(x) = \lim_{x \to 0} (12 \cos x - 12) \,/\, (3x^2)$$

This expression approaches 0/0 as x approaches 0, just as the first one does. Let's apply the l'Hôpital rule again. Differentiating a second time, we get

$$f''(x) = -12 \sin x$$

and

$$g''(x) = 6x$$

The rule now indicates that

$$\underset{x \to 0}{Lim}\ f'(x)\ /\ g'(x) = \underset{x \to 0}{Lim}\ f''(x)\ /\ g''(x) = \underset{x \to 0}{Lim}\ (-12 \sin x)\ /\ (6x)$$

We still have an expression that tends toward 0/0 as x approaches 0, so we can apply the l'Hôpital rule a third time. We obtain

$$f'''(x) = -12 \cos x$$

and

$$g'''(x) = 6$$

Therefore,

$$\underset{x \to 0}{Lim}\ f''(x)\ /\ g''(x) = \underset{x \to 0}{Lim}\ f'''(x)\ /\ g'''(x) = \underset{x \to 0}{Lim}\ (-12 \cos x)\ /\ 6$$

As x approaches 0, the numerator in the rightmost expression approaches -12, because $\cos x$ approaches 1. The denominator is always 6. We can therefore conclude that our original limit is defined and is equal to $-12/6$, or -2. That is,

$$\underset{x \to 0}{Lim}\ (12 \sin x - 12x)\ /\ x^3 = -2$$

4. We want to evaluate the limit, as x is positive and increases without end, of $7x$ divided by the natural logarithm of x:

$$\underset{x \to +\infty}{Lim}\ 7x\ /\ (\ln x)$$

Let's name the functions

$$f(x) = 7x$$

and

$$g(x) = \ln x$$

Both $f(x)$ and $g(x)$ tend toward positive infinity as x increases endlessly. The derivatives are

$$f'(x) = 7$$

and

$$g'(x) = x^{-1}$$

Applying l'Hôpital's rule, we find that

$$\underset{x \to +\infty}{Lim}\ f(x) / g(x) = \underset{x \to +\infty}{Lim}\ f'(x) / g'(x) = \underset{x \to +\infty}{Lim}\ 7 / (x^{-1})$$

The expression in the last limit above can be rewritten as

$$7 / (x^{-1}) = 7x$$

It's easy to see that $7x$ tends toward positive infinity as x increases without end. We've found that

$$\underset{x \to +\infty}{Lim}\ 7x / (\ln x) = +\infty$$

5. We've been told to evaluate the following limit, where the numerator and denominator both tend toward positive infinity as x increases endlessly:

$$\underset{x \to +\infty}{Lim}\ e^x / x^2$$

Let's name the functions

$$f(x) = e^x$$

and

$$g(x) = x^2$$

When we differentiate them, we obtain

$$f'(x) = e^x$$

and

$$g'(x) = 2x$$

The l'Hôpital rule tells us that

$$\underset{x \to +\infty}{Lim}\ f(x) / g(x) = \underset{x \to +\infty}{Lim}\ f'(x) / g'(x) = \underset{x \to +\infty}{Lim}\ e^x / (2x)$$

We have another expression where the numerator and denominator both tend toward positive infinity as x increases without end. Let's apply l'Hôpital's rule again. The second derivatives of our functions are

$$f''(x) = e^x$$

and

$$g''(x) = 2$$

The l'Hôpital rule says that

$$\lim_{x \to +\infty} f'(x) / g'(x) = \lim_{x \to +\infty} f''(x) / g''(x) = \lim_{x \to +\infty} e^x/2$$

The numerator of the final expression in the above equation tends toward positive infinity as x increases forever, but the denominator is a constant 2. That means

$$\lim_{x \to +\infty} e^x / x^2 = +\infty$$

6. We want to work out the following limit, where the numerator and denominator both tend toward positive infinity as x approaches 0 from the right:

$$\lim_{x \to 0+} x^{-3} / x^{-2}$$

Let's simplify the expression before we seek the limit. With algebra, we can manipulate it like this:

$$x^{-3} / x^{-2} = (1/x^3) / (1/x^2) = (1/x^3) (x^2) = x^2 / x^3 = x^{-1}$$

Now we have reduced the problem to

$$\lim_{x \to 0+} x^{-1}$$

The value of x^{-1} tends toward positive infinity as x approaches 0 from the right, so

$$\lim_{x \to 0+} x^{-3} / x^{-2} = +\infty$$

7. We've been told to evaluate the limit, as x approaches 0 from the right, of the product of the sine function and the reciprocal function:

$$\lim_{x \to 0+} (\sin x) \, x^{-1}$$

When we rewrite the expression as a ratio, we have

$$\lim_{x \to 0+} (\sin x) / x$$

As x approaches 0 from the right, the numerator and denominator of this expression both approach 0 from the right. We can therefore apply the l'Hôpital rule for limits whose expressions take the form 0/0. Let's call the functions

$$f(x) = \sin x$$

and

$$g(x) = x$$

The derivatives are

$$f'(x) = \cos x$$

and

$$g'(x) = 1$$

so we have

$$\lim_{x \to 0+} f(x) / g(x) = \lim_{x \to 0+} f'(x) / g'(x) = \lim_{x \to 0+} (\cos x) / 1$$

The rightmost expression approaches 1 as x approaches 0 from the right, so

$$\lim_{x \to 0+} (\sin x) \, x^{-1} = 1$$

This is the same as the limit we got when we approached 0 from the left in the chapter text. In that case, both $\sin x$ and x were negative, producing a positive ratio (minus over minus equals plus). In the example here, both $\sin x$ and x are positive, again producing a positive ratio (plus over plus equals plus).

8. We want to evaluate the following limit, where the first factor tends toward 0 and the second factor tends toward negative infinity as x approaches 0 from the left:

$$\lim_{x \to 0-} x^5 \ln |x|$$

Let's rewrite this as the limit of a ratio by putting x^{-5} in the denominator, getting

$$\lim_{x \to 0-} (\ln |x|) / x^{-5}$$

Both the numerator and the denominator in this ratio tend toward negative infinity as x approaches 0 from the left, so we can apply the l'Hôpital rule for limits of expressions of the form $-\infty/(-\infty)$. Let's call the functions

$$f(x) = \ln |x|$$

and

$$g(x) = x^{-5}$$

The derivatives are

$$f'(x) = x^{-1}$$

and

$$g'(x) = -5x^{-6}$$

Therefore

$$\text{Lim}_{x\to0-} f(x) / g(x) = \text{Lim}_{x\to0-} f'(x) / g'(x) = \text{Lim}_{x\to0-} x^{-1} / (-5x^{-6})$$

The rightmost expression can be simplified with algebra:

$$x^{-1} / (-5x^{-6}) = (1/x) / (-5/x^6) = (1/x) [x^6 / (-5)] = -x^6 / (5x) = -x^5 / 5$$

Now we have the manageable limit

$$\text{Lim}_{x\to0-} -x^5 / 5$$

This expression approaches 0 as x approaches 0 from the left. We've determined that

$$\text{Lim}_{x\to0-} x^5 \ln |x| = 0$$

9. We've been told to evaluate the following limit, where both terms tend toward positive infinity as x approaches 0 from the right:

$$\text{Lim}_{x\to0+} 2x^{-1} - 3(e^x - 1)^{-1}$$

We can rewrite the expression in this limit as a ratio, getting

$$\text{Lim}_{x\to0+} (2e^x - 2 - 3x) / (xe^x - x)$$

As x approaches 0 from the right, the numerator and denominator of the above ratio both approach 0, so the l'Hôpital rule for limits of expressions tending toward 0/0 applies. Let

$$f(x) = 2e^x - 2 - 3x$$

and

$$g(x) = xe^x - x$$

The derivatives are

$$f'(x) = 2e^x - 3$$

and

$$g'(x) = e^x + xe^x - 1$$

Therefore

$$\text{Lim}_{x\to0+} f(x) / g(x) = \text{Lim}_{x\to0+} f'(x) / g'(x) = \text{Lim}_{x\to0+} (2e^x - 3) / (e^x + xe^x - 1)$$

As x approaches 0 from the right, the numerator in the above ratio approaches -1, while the denominator approaches 0 from the right. The ratio grows large endlessly in the negative direction, so the limit is negative-infinite. We can conclude that the original limit is

$$\operatorname*{Lim}_{x\to0+} \; 2x^{-1} - 3(e^x - 1)^{-1} = -\infty$$

10. We've been told to evaluate the following limit, where both terms tend toward positive infinity as x approaches $\pi/2$ from the left:

$$\operatorname*{Lim}_{x\to\pi/2-} \; \sec x - \tan x$$

We know from trigonometry that

$$\sec x = (\cos x)^{-1}$$

and

$$\tan x = (\sin x) \, / \, (\cos x)$$

so we can rewrite the limit as

$$\operatorname*{Lim}_{x\to\pi/2-} \; (\cos x)^{-1} - (\sin x) \, / \, (\cos x)$$

Let's manipulate the expression to obtain a ratio:

$$(\cos x)^{-1} - (\sin x) \, / \, (\cos x) = (\cos x - \cos x \sin x) \, / \, (\cos x)^2$$
$$= (1 - \sin x) \, / \, (\cos x)$$

Our limit is now

$$\operatorname*{Lim}_{x\to\pi/2-} \; (1 - \sin x) \, / \, (\cos x)$$

As x approaches $\pi/2$ from the left, the numerator and denominator both approach 0. We can use the l'Hôpital rule for ratios of the form 0/0. Let's call the functions

$$f(x) = 1 - \sin x$$

and

$$g(x) = \cos x$$

The derivatives are

$$f'(x) = -\cos x$$

and

$$g'(x) = -\sin x$$

Therefore

$$\lim_{x \to \pi/2-} f(x) / g(x) = \lim_{x \to \pi/2-} f'(x) / g'(x) = \lim_{x \to \pi/2-} (-\cos x) / (-\sin x)$$

$$= \lim_{x \to \pi/2-} (\cos x) / (\sin x)$$

The cosine over the sine equals the cotangent, so the above limit is

$$\lim_{x \to \pi/2-} \cot x$$

From trigonometry, we know that $\cot \pi/2 = 0$. Our original limit is therefore

$$\lim_{x \to \pi/2-} \sec x - \tan x = 0$$

Chapter 24

1. If we consider y to be the dependent variable in the situation of Fig. 24-1, the surface defines a relation

$$y = g(x,z)$$

where x and z are the independent variables. To see if this surface represents a true function of x and z, we can imagine a movable straight line that's always parallel to the y axis, which is the dependent-variable axis. If this line never intersects the surface at more than one point, then g is a true function. Otherwise, g is not a true function. It's visually apparent that any such line passing through the surface must intersect it twice, unless the line is tangent to the surface. Therefore,

$$y = g(x,z)$$

is not a true function of x and z.

2. This time, x is our dependent variable, so the surface defines a relation

$$x = h(y,z)$$

where y and z are the independent variables. To test this graph to see if it represents a true function of y and z, we can imagine a movable straight line parallel to the x axis. We can see that any such line passing through the paraboloid will go through twice, unless the line is tangent to the surface. This tells us that

$$x = h(y,z)$$

is not a true function of y and z.

3. For reference, our original function is

$$z = f(x,y) = -x^2 + 2xy + 4y^3$$

To differentiate f with respect to x, we treat y as a constant and let x be the independent variable. The derivative of the first term with respect to x is $-2x$. The second term contains the constant $2y$. Its derivative with respect to x is equal to $2y$. The third term is the constant we get by cubing y and then multiplying by 4. Its derivative with respect to x is equal to 0, because the derivative of a constant is always 0. Therefore,

$$\partial z/\partial x = -2x + 2y + 0 = -2x + 2y$$

4. Once more, our original function is

$$z = f(x,y) = -x^2 + 2xy + 4y^3$$

To differentiate f with respect to y, we treat x as a constant, so the independent variable is y. The first term is the constant we get when we square x and then take the negative, so its derivative with respect to y is 0. In second term, $2x$ is a constant. Its derivative with respect to y is equal to that constant $2x$. The derivative of the third term with respect to y is $12y^2$. Adding these results, we get

$$\partial z/\partial y = 0 + 2x + 12y^2 = 2x + 12y^2$$

5. To find $\partial/\partial x \, (1,1)$, we plug in the values $x = 1$ and $y = 1$ to the general equation for $\partial z/\partial x$ we got in the solution to Prob. 3. That gives us

$$\partial/\partial x \, (1,1) = -2 \cdot 1 + 2 \cdot 1 = -2 + 2 = 0$$

To find $\partial/\partial y \, (1,1)$, we plug in $x = 1$ and $y = 1$ to the general equation for $\partial z/\partial y$ we got in the solution to Prob. 4, getting

$$\partial/\partial y \, (1,1) = 2 \cdot 1 + 12 \cdot 1^2 = 2 + 12 = 14$$

6. To find $\partial/\partial x \, (-2,-3)$, we plug in $x = -2$ and $y = -3$ to the general equation for $\partial z/\partial x$ we got in the solution to Prob. 3 to get

$$\partial/\partial x \, (-2,-3) = -2 \cdot (-2) + 2 \cdot (-3) = 4 + (-6) = -2$$

To find $\partial/\partial y \, (-2,-3)$, we plug in the values $x = -2$ and $y = -3$ to the general equation for $\partial z/\partial y$ we got in the solution to Prob. 4, obtaining

$$\partial/\partial y \, (-2,-3) = 2 \cdot (-2) + 12 \cdot (-3)^2 = -4 + 12 \cdot 9 = -4 + 108 = 104$$

7. The original function is

$$w = g(x,y,z) = xy \ln |z| - 2x^2 y^3 z^4$$

provided $z \neq 0$, because $\ln |0|$ is undefined. To figure out $\partial w / \partial x$, we hold y and z constant and think of x as the independent variable. In the first term, $y \ln |z|$ is a constant, so the derivative with respect to x is $y \ln |z|$. In the second term, $y^3 z^4$ is a constant, so the derivative with respect to x is $4xy^3 z^4$. Subtracting these two results gives us

$$\partial w / \partial x = y \ln |z| - 4xy^3 z^4$$

8. Again, our original function is

$$w = g(x,y,z) = xy \ln |z| - 2x^2 y^3 z^4$$

provided $z \neq 0$. To find $\partial w / \partial y$, we keep x and z constant and let y be the independent variable. The first term contains the constant $x \ln |z|$, so its derivative with respect to y is $x \ln |z|$. In the second term, $2x^2 z^4$ is the constant, and the variable y is cubed; that means the derivative with respect to y is $6x^2 y^2 z^4$. Subtracting, we get

$$\partial w / \partial y = x \ln |z| - 6x^2 y^2 z^4$$

9. Once more for reference, the original function is

$$w = g(x,y,z) = xy \ln |z| - 2x^2 y^3 z^4$$

provided $z \neq 0$. To figure out $\partial w / \partial z$, we hold x and y constant and let z be the independent variable. In the first term, xy is a constant, so the derivative with respect to z is equal to xy times the derivative of $\ln |z|$. That's xyz^{-1}. In the second term, $2x^2 y^3$ is a constant, so the derivative with respect to z is $8x^2 y^3 z^3$. Subtracting these derivatives yields

$$\partial w / \partial z = xyz^{-1} - 8x^2 y^3 z^3$$

10. To work out $\partial / \partial x \,(-1,2,e)$, we put $x = -1$, $y = 2$, and $z = e$ into $\partial w / \partial x$ from the solution to Prob. 7. The general formula is

$$\partial w / \partial x = y \ln |z| - 4xy^3 z^4$$

Working out the arithmetic, we get

$$\partial / \partial x \,(-1,2,e) = 2 \cdot \ln |e| - 4 \cdot (-1) \cdot 2^3 \cdot e^4 = 2 + 32e^4$$

To find $\partial / \partial y \,(-2,-1,2e)$, we plug in $x = -2$, $y = -1$, and $z = 2e$ to $\partial w / \partial y$ we got in the solution to Prob. 8. The general formula is

$$\partial w / \partial y = x \ln |z| - 6x^2 y^2 z^4$$

When we do the arithmetic, we obtain

$$\partial / \partial y \,(-2,-1,2e) = -2 \cdot \ln |2e| - 6 \cdot (-2)^2 \cdot (-1)^2 \cdot (2e)^4 = -2 \cdot \ln (2e) - 384e^4$$

To calculate $\partial/\partial z\,(1,-2,0)$, we plug in $x=1$, $y=-2$, and $z=0$ to $\partial w/\partial z$ from the solution to Prob. 9. The general formula is

$$\partial w/\partial z = xyz^{-1} - 8x^2y^3z^3$$

The arithmetic yields

$$\partial/\partial z\,(1,-2,0) = 1 \cdot (-2) \cdot 0^{-1} - 8 \cdot 1^2 \cdot (-2)^3 \cdot 0^3$$

We can stop here! The first term is undefined, because it contains the reciprocal of 0. We must conclude that $\partial/\partial z\,(1,-2,0)$ is not defined. The original function is singular at any point where $z=0$, so we should not be surprised at this result.

Chapter 25

1. We want to find the second partial of the following function with respect to *x*:

$$z = f(x,y) = -x^2 + 2xy + 4y^3$$

The first partial $\partial z/\partial x$ appears in the solution to Prob. 3 in Chap. 24. That partial is

$$\partial z/\partial x = -2x + 2y$$

To find $\partial^2 z/\partial x^2$, we differentiate again with respect to *x*, continuing to hold *y* constant. When we do that, we get

$$\partial^2 z/\partial x^2 = -2 + 0 = -2$$

2. We found the first partial with respect to *y* in the solution to Prob. 4 in Chap. 24. That derivative is

$$\partial z/\partial y = 2x + 12y^2$$

To obtain the second partial with respect to *y*, we hold *x* constant and differentiate the above function relative to *y*, getting

$$\partial^2 z/\partial y^2 = 0 + 24y = 24y$$

3. To find the second partial with respect to *x* at (3,2), we input $x=3$ and $y=2$ to

$$\partial^2 z/\partial x^2 = -2$$

We have no calculations to do here! The specific second partial is equal to -2, no matter what *x* or *y* values we input. To find $\partial^2/\partial y^2\,(3,2)$, we plug in $x=3$ and $y=2$ to

$$\partial^2 z/\partial y^2 = 24y$$

The arithmetic gives us

$$\partial^2/\partial y^2\,(3,2) = 24 \cdot 2 = 48$$

In this case, the specific second partial derivative is affected only by the value of y. It doesn't matter what happens to x.

4. We want to find $\partial^2 z/\partial y\partial x$ for the function

$$z = f(x,y) = -x^2 + 2xy + 4y^3$$

We must differentiate f with respect to x, and then differentiate $\partial z/\partial x$ with respect to y. The first partial relative to x, which we've seen before, is

$$\partial z/\partial x = -2x + 2y$$

Holding x constant and differentiating with respect to y, we get

$$\partial^2 z/\partial y\partial x = 0 + 2 = 2$$

5. To find $\partial^2 z/\partial x\partial y$, we differentiate with respect to y and then with respect to x. We've already seen that

$$\partial z/\partial y = 2x + 12y^2$$

To differentiate with respect to x, we hold y constant, getting

$$\partial^2 z/\partial x\partial y = 2 + 0 = 2$$

6. The mixed second partials are both constant functions with a value of 2. The values of the input variables are irrelevant, so we simply have

$$\partial^2/\partial y\partial x\,(3,2) = 2$$

and

$$\partial^2/\partial x\partial y\,(3,2) = 2$$

7. We found the first partials for this function in the solutions to Probs. 7, 8, and 9 in Chap. 24. For reference, the original function is

$$w = g(x,y,z) = xy \ln |z| - 2x^2 y^3 z^4$$

We must insist that $z \neq 0$ no matter what we do with this function. That's because, if we let $z = 0$, then g contains the factor $\ln |0|$, which is undefined. The first partial of the original function relative to x is

$$\partial w/\partial x = y \ln |z| - 4xy^3 z^4$$

Differentiating against x again, we get

$$\partial^2 w / \partial x^2 = 0 - 4y^3 z^4 = -4y^3 z^4$$

The first partial of the original function relative to y is

$$\partial w / \partial y = x \ln |z| - 6x^2 y^2 z^4$$

Differentiating against y again, we get

$$\partial^2 w / \partial y^2 = 0 - 12x^2 yz^4 = -12x^2 yz^4$$

The first partial of the original function relative to z is

$$\partial w / \partial z = xyz^{-1} - 8x^2 y^3 z^3$$

Differentiating against z again, we get

$$\partial^2 w / \partial z^2 = -xyz^{-2} - 24x^2 y^3 z^2$$

8. To work out $\partial^2 / \partial x^2 \, (-1,2,e)$, we input $x = -1$, $y = 2$, and $z = e$ to

$$\partial^2 w / \partial x^2 = -4y^3 z^4$$

Doing the arithmetic, we get

$$\partial^2 / \partial x^2 \, (-1,2,e) = -4 \cdot 2^3 \cdot e^4 = -32e^4$$

To find $\partial^2 / \partial y^2 \, (-2,-1,2e)$, we input $x = -2$, $y = -1$, and $z = 2e$ to

$$\partial^2 w / \partial y^2 = -12x^2 yz^4$$

Calculating, we get

$$\partial^2 / \partial y^2 \, (-2,-1,2e) = -12 \cdot (-2)^2 \cdot (-1) \cdot (2e)^4 = 768e^4$$

To determine $\partial^2 / \partial z^2 \, (1,-2,0)$, we input $x = 1$, $y = -2$, and $z = 0$ to

$$\partial^2 w / \partial z^2 = -xyz^{-2} - 24x^2 y^3 z^2$$

The first term here contains z^{-2}, which is not defined for $z = 0$. We remember that the original function g is singular at any point where $z = 0$, so we conclude that g is not differentiable at $(1,-2,0)$. This specific second partial is undefined.

9. We've been told to find all six mixed second partials for the function

$$w = g\,(x,y,z) = xy \ln |z| - 2x^2 y^3 z^4$$

provided that $z \neq 0$, because ln |0| is not defined. From the solution to Prob. 7, the first partial of the original function relative to x is

$$\partial w / \partial x = y \ln |z| - 4xy^3z^4$$

To differentiate $\partial w / \partial x$ against y, we hold x and z constant to get

$$\partial^2 w / \partial y \partial x = \ln |z| - 12xy^2z^4$$

To differentiate $\partial w / \partial x$ against z, we hold x and y constant to get

$$\partial^2 w / \partial z \partial x = yz^{-1} - 16xy^3z^3$$

From the solution to Prob. 7, the first partial of the original function relative to y is

$$\partial w / \partial y = x \ln |z| - 6x^2y^2z^4$$

To differentiate $\partial w / \partial y$ against x, we hold y and z constant to get

$$\partial^2 w / \partial x \partial y = \ln |z| - 12xy^2z^4$$

To differentiate $\partial w / \partial y$ against z, we hold x and y constant to get

$$\partial^2 w / \partial z \partial y = xz^{-1} - 24x^2y^2z^3$$

From the solution to Prob. 7, the first partial of the original function relative to z is

$$\partial w / \partial z = xyz^{-1} - 8x^2y^3z^3$$

To differentiate $\partial w / \partial z$ against x, we hold y and z constant to get

$$\partial^2 w / \partial x \partial z = yz^{-1} - 16xy^3z^3$$

To differentiate $\partial w / \partial z$ against y, we hold x and z constant to get

$$\partial^2 w / \partial y \partial z = xz^{-1} - 24x^2y^2z^3$$

10. Solving this problem requires some tricky arithmetic! To find $\partial^2 / \partial y \partial x \, (1,2,e)$, we input $x = 1$, $y = 2$, and $z = e$ to

$$\partial^2 w / \partial y \partial x = \ln |z| - 12xy^2z^4$$

Calculating, we get

$$\partial^2 / \partial y \partial x \, (1,2,e) = \ln e - 12 \cdot 1 \cdot 2^2 \cdot e^4 = 1 - 48e^4$$

To work out $\partial^2 / \partial z \partial x \, (1,2,e)$, we input $x = 1$, $y = 2$, and $z = e$ to

$$\partial^2 w / \partial z \partial x = yz^{-1} - 16xy^3z^3$$

Calculating, we get

$$\partial^2/\partial z \partial x \, (1,2,e) = 2e^{-1} - 16 \cdot 1 \cdot 2^3 \cdot e^3 = 2e^{-1} - 128e^3$$

To work out $\partial^2/\partial x \partial y \, (1,2,e)$, we input $x = 1$, $y = 2$, and $z = e$ to

$$\partial^2 w/\partial x \partial y = \ln |z| - 12xy^2z^4$$

Calculating, we get

$$\partial^2/\partial x \partial y \, (1,2,e) = \ln e - 12 \cdot 1 \cdot 2^2 \cdot e^4 = 1 - 48e^4$$

To work out $\partial^2/\partial z \partial y \, (1,2,e)$, we input $x = 1$, $y = 2$, and $z = e$ to

$$\partial^2 w/\partial z \partial y = xz^{-1} - 24x^2y^2z^3$$

Calculating, we get

$$\partial^2/\partial z \partial y \, (1,2,e) = e^{-1} - 24 \cdot 1^2 \cdot 2^2 \cdot e^3 = e^{-1} - 96e^3$$

To work out $\partial^2/\partial x \partial z \, (1,2,e)$, we input $x = 1$, $y = 2$, and $z = e$ to

$$\partial^2 w/\partial x \partial z = yz^{-1} - 16xy^3z^3$$

Calculating, we get

$$\partial^2/\partial x \partial z \, (1,2,e) = 2e^{-1} - 16 \cdot 1 \cdot 2^3 \cdot e^3 = 2e^{-1} - 128e^3$$

To work out $\partial^2/\partial y \partial z \, (1,2,e)$, we input $x = 1$, $y = 2$, and $z = e$ to

$$\partial^2 w/\partial y \partial z = xz^{-1} - 24x^2y^2z^3$$

Calculating, we get

$$\partial^2/\partial y \partial z \, (1,2,e) = e^{-1} - 24 \cdot 1^2 \cdot 2^2 \cdot e^3 = e^{-1} - 96e^3$$

Chapter 26

1. We've been told that the radius of the cylinder is 2 units and its length is 5 units. Therefore, $r = 2$ and $h = 5$. The circumference of a cross-sectional circle is a function f of the displacement x to the right of the cylinder's left-hand face:

$$f(x) = 2\pi r = 2\pi \cdot 2 = 4\pi$$

To find the lateral-surface area A, we integrate f along the cylinder from $x = 0$ to $x = 5$, getting

$$A = \int_0^5 4\pi \, dx$$

The basic antiderivative is

$$F(x) = 4\pi x$$

When we evaluate this antiderivative from $x = 0$ to $x = 5$, we obtain

$$A = 4\pi x \bigg]_0^5 = 4\pi \cdot 5 - 4\pi \cdot 0 = 20\pi$$

The formula from solid geometry tells us that

$$A = 2\pi r h = 2\pi \cdot 2 \cdot 5 = 20\pi$$

2. We know that the radius of the cylinder is 2 units and the length is 5 units. The area of a cross-sectional disk, cut perpendicular to the cylinder's axis and also perpendicular to the x axis, is a constant function

$$g(x) = \pi r^2 = \pi \cdot 2^2 = 4\pi$$

To find the volume V, we integrate g along the cylinder from $x = 0$ to $x = 5$. That gives us

$$V = \int_0^5 4\pi \, dx$$

The basic antiderivative is

$$G(x) = 4\pi x$$

When we evaluate this antiderivative from $x = 0$ to $x = 5$, we get

$$V = 4\pi x \bigg]_0^5 = 4\pi \cdot 5 - 4\pi \cdot 0 = 20\pi$$

This is the same numerical value as we got for the lateral-surface area, but it's *not* the same quantity! Remember that volume is expressed in *cubic* units, while area is expressed in *square* units. The fact that they're both represented by the same number is a coincidence for this particular cylinder. When we use the formula from 3D geometry for the volume, we get

$$V = \pi r^2 h = \pi \cdot 2^2 \cdot 5 = 20\pi$$

3. We're told that the radius of the cone is 12 units and the slant height is 13 units. Therefore, $r = 12$ and $s = 13$. The circumference of a circular slice as a function f of the distance t between it and the cone's apex is

$$f(t) = (2\pi r/s)t = (2\pi \cdot 12/13)t = (24\pi/13)t$$

To find the slant-surface area A, we evaluate the definite integral

$$A = \int_0^{13} (24\pi/13)t \, dt$$

The basic antiderivative is

$$F(t) = [(24\pi/13)t^2] / 2 = (12\pi/13)t^2$$

When we evaluate F from $t = 0$ to $t = 13$, we get

$$A = (12\pi/13)t^2 \Big]_0^{13} = (12\pi/13) \cdot 13^2 - (12\pi/13) \cdot 0^2 = (12\pi/13) \cdot 169 = 156\pi$$

Using the geometric formula, we get

$$A = \pi r s = \pi \cdot 12 \cdot 13 = 156\pi$$

4. Again, we have $r = 12$ and $s = 13$. We haven't been told the true height h of the cone. The Pythagorean theorem from geometry tells us that

$$s^2 = r^2 + h^2$$

so therefore

$$h = (s^2 - r^2)^{1/2} = (13^2 - 12^2)^{1/2} = (169 - 144)^{1/2} = 25^{1/2} = 5$$

The area of a cross-sectional disk as a function g of the distance x between its center and the cone's apex is

$$g(x) = (\pi r^2/h^2)x^2 = (\pi \cdot 12^2/5^2)x^2 = (144\pi/25)x^2$$

To find the cone's volume V, we integrate g along the axis of the cone from its apex, where $x = 0$, to its base, where $x = 5$, getting

$$V = \int_0^5 (144\pi/25)x^2 \, dx$$

The basic antiderivative is

$$G(x) = [(144\pi/25)x^3] / 3 = (48\pi/25)x^3$$

Evaluating G from $x = 0$ to $x = 5$, we get

$$V = [(48\pi/25)x^3 \Big]_0^5 = (48\pi/25) \cdot 5^3 - (48\pi/25) \cdot 0^3 = (48\pi/25) \cdot 125 = 240\pi$$

The formula from solid geometry tells us that

$$V = \pi r^2 h/3 = \pi \cdot 12^2 \cdot 5/3 = \pi \cdot 144 \cdot 5/3 = 240\pi$$

5. The circumference of a cross-sectional circle as a function of the arc displacement t is described by

$$f(t) = 2\pi r \sin(t/r) = 2\pi \cdot 10 \sin(t/10) = 20\pi \sin(t/10)$$

To find the surface area A, we integrate f from $t = 0$ to $t = \pi r$. We know that $r = 10$, so

$$A = \int_{0}^{10\pi} 20\pi \sin(t/10) \, dt = 20\pi \int_{0}^{10\pi} \sin(t/10) \, dt$$

The table of integrals, App. G, gives us the formula

$$\int \sin ax \, dx = -a^{-1} \cos ax + c$$

Letting $a = 1/10$ and leaving out the constant of integration, we obtain

$$\int \sin(t/10) \, dt = -10 \cos(t/10)$$

The basic antiderivative F of the complete function f is 20π times this, or

$$F(t) = -200\pi \cos(t/10)$$

When we evaluate F from $t = 0$ to $t = 10\pi$, we get

$$A = -200\pi \cos(t/10) \Big]_{0}^{10\pi} = -200\pi \cos \pi - (-200\pi \cos 0)$$

$$= 200\pi + 200\pi = 400\pi$$

Using the formula from solid geometry, we get

$$A = 4\pi r^2 = 4\pi \cdot 10^2 = 400\pi$$

6. The area of any particular cross-sectional disk cutting through the sphere, based on the location of the disk's center along the x axis, is

$$g(x) = \pi r^2 - \pi x^2$$

We've been told that $r = 10$, so this function becomes

$$g(x) = 100\pi - \pi x^2$$

The sphere's volume V is obtained by integrating the function g along the x axis from the point where $x = -10$ to the point where $x - 10$. That gives us

$$V = \int_{-10}^{10} (100\pi - \pi x^2) \, dx$$

The basic antiderivative is

$$G(x) = 100\pi x - \pi x^3/3$$

When we evaluate G from $x = -10$ to $x = 10$, we get

$$V = (100\pi x - \pi x^3/3) \Big]_{-10}^{10} = (100\pi \cdot 10 - \pi \cdot 10^3/3) - [100\pi \cdot (-10) - \pi \cdot (-10)^3/3]$$

$$= (1{,}000\pi - 1{,}000\pi/3) - [-1{,}000\pi - (-1{,}000\pi/3)]$$

$$= 2{,}000\pi/3 + 2{,}000\pi/3 = 4{,}000\pi/3$$

From solid geometry, we can calculate the volume as

$$V = 4\pi r^3/3 = 4\pi \cdot 10^3/3 = 4{,}000\pi/3$$

7. As is shown in Fig. 26-5 and reproduced here in Fig. C-1, the left-hand face of the prism is in the yz-plane. The right-hand face is h units to the right of the left-hand face. The faces are rectangles measuring a units tall and b units deep. Imagine that we cut the outer shell into cross-sectional slices, and then increase the number of slices indefinitely. The slices are all rectangles parallel to the end faces. Like those faces, they all measure a units tall and b units deep. The perimeter of any particular cross-sectional rectangle is a constant function f of the displacement x to the right of the cylinder's left-hand face:

$$f(x) = 2a + 2b$$

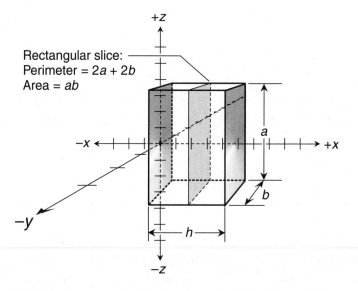

Figure C-1   Illustration for the solution to Probs. 7 through 10 in Chap. 26.

Remember that we're treating all three prism dimensions *a, b,* and *h* as fixed constants. To find the lateral-surface area *A* (not including either end face), we integrate the constant function *f* along the prism from its left-hand face, where *x* = 0, to its right-hand face, where *x* = *h*. The definite integral is

$$A = \int_0^h (2a + 2b)\ dx$$

The basic antiderivative is

$$F(x) = (2a + 2b)x$$

When we evaluate this expression from *x* = 0 to *x* = *h*, we get

$$A = (2a + 2b)x \Big]_0^h = (2a + 2b)h - (2a + 2b) \cdot 0 = (2a + 2b)h = 2ah + 2bh$$

8. We can derive a formula for the lateral-surface area of a prism in terms of *a, b,* and *h* as shown in Fig. C-1 by adding up the areas of each of the four rectangles that form the prism's sleeve. Two of the rectangles, which appear at the front and back of the prism as shown in Fig. C-1, measure *a* units tall by *h* units long, so they both have area *ah*. The other two rectangles, which appear at the top and bottom of the prism, measure *b* units deep by *h* units long. (The lateral-surface area doesn't include the areas of the end faces, which appear at the left and right.) When we add up the areas of the faces making up the sleeve, we get

$$A = ah + ah + bh + bh = 2ah + 2bh$$

This agrees with the formula we got by integration.

9. Let's keep referring to Fig. C-1. We can cut our prism into cross-sectional slices again, and make the number of slices approach infinity. But instead of slicing through only the outer shell, we cut through the interior, getting filled-in rectangles. These regions are all parallel to the prism's end faces, and they all measure *a* units tall by *b* units deep. The area of any particular cross-sectional rectangle is a constant function *g* of the displacement *x* to the right of the cylinder's left-hand face:

$$g(x) = ab$$

To find the volume *V*, we integrate *g* along the prism from *x* = 0 to *x* = *h*. That gives us

$$V = \int_0^h ab\ dx$$

The basic antiderivative is

$$G(x) = abx$$

When we evaluate this antiderivative from *x* = 0 to *x* = *h*, we get

$$V = abx \Big]_0^h = abh - ab \cdot 0 = abh$$

10. In solid geometry, we learned that the volume V of a rectangular prism is equal to the product of the measures of its edges in each dimension. In this situation (Fig. C-1), those measures are a, b, and h. Therefore

$$V = abh$$

This agrees with the formula we derived by integration.

Chapter 27

1. The acceleration of gravity on the planet is 24 feet per second per second. This is a constant. Let's integrate the acceleration function twice with respect to time, from $t = 0$ seconds (when we drop the brick) to $t = 14$ seconds (when the brick lands). That gives us

$$\int_0^{14} \int_0^{14} 24 \, dt \, dt$$

The first basic antiderivative is

$$\int 24 \, dt = 24t$$

Antidifferentiating again, we get

$$\int 24t \, dt = 12t^2$$

The repeated integral tells us that our altitude, in feet, is

$$\int_0^{14} \int_0^{14} 24 \, dt \, dt = 12t^2 \Big]_0^{14} = 12 \cdot 14^2 - 12 \cdot 0^2 = 2{,}352$$

2. Our acceleration rate depends on time. If we call the acceleration function f, then

$$f(t) = 6t$$

where t is the time, in seconds, after we start moving. The first basic antiderivative is

$$\int 6t \, dt = 3t^2$$

Antidifferentiating again, we get

$$\int 3t^2 \, dt = t^3$$

This function describes the distance we travel from the starting point, in meters, as a function of the time t, in seconds, after the starting instant. If we call the function h, then

$$h(t) = t^3$$

3. To find the distance that the car travels in any particular length of time after it starts moving, we plug in the time values to the function $h(t)$ that we derived in the solution to Prob. 2. After 1 second, we travel

$$h(1) = 1^3 = 1 \text{ meter}$$

After 2 seconds, our total distance is

$$h(2) = 2^3 = 8 \text{ meters}$$

After 3 seconds, our total distance is

$$h(3) = 3^3 = 27 \text{ meters}$$

In the first 4 seconds, we cover a distance of

$$h(4) = 4^3 = 64 \text{ meters}$$

If we can keep this acceleration-increase rate steady for 10 seconds, we will have gone

$$h(10) = 10^3 = 1{,}000 \text{ meters}$$

4. We have a flat surface described by

$$f(x,y) = 4$$

Let's use ordinary geometry to find the mathematical volume of the box defined by f relative to the region whose edges are at $x = -3$, $x = 5$, $y = -5$, and $y = 3$ in the xy-plane. Figure C-2 is a simplified graph of this situation. The base and the top of the box both

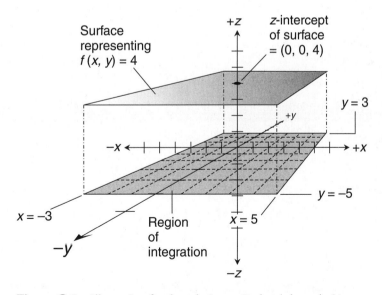

Figure C-2   Illustration for the solution to Probs. 4 through 6 in Chap. 27.

measure 8 units by 8 units. The height of the box is 4 units. The mathematical volume *V* is therefore

$$V = 8 \cdot 8 \cdot 4 = 256$$

5. Here's the two-variable constant function *f* in Cartesian *xyz*-space that describes the surface:

$$f(x,y) = 4$$

We want to calculate the mathematical volume of the solid box relative to the region in the *xy*-plane whose edges are at $x = -3$, $x = 5$, $y = -5$, and $y = 3$ by integrating with respect to *x* and then with respect to *y*. In shorthand, our double integral is

$$\iint_R 4 \; dx \; dy$$

Written out in full, the iterated integral is

$$\int_{-5}^{3} \left(\int_{-3}^{5} 4 \; dx \right) dy$$

The integral inside the large parentheses is

$$\int_{-3}^{5} 4 \; dx$$

which resolves to

$$4x \Big]_{-3}^{5}$$

Evaluating, we get

$$4 \cdot 5 - 4 \cdot (-3) = 32$$

We substitute this in place of the integral inside the parentheses above, obtaining

$$\int_{-5}^{3} 32 \; dy$$

Integrating against *y*, we get

$$32y \Big]_{-5}^{3}$$

Doing the arithmetic, we get:

$$32 \cdot 3 - 32 \cdot (-5) = 256$$

6. This time, we'll integrate the constant function with respect to y first, and then with respect to x. Written in the shorthand form, our double integral is

$$\iint_R 4 \, dy \, dx$$

The iterated integral is

$$\int_{-3}^{5} \left(\int_{-5}^{3} 4 \, dy \right) dx$$

The inside integral is

$$\int_{-5}^{3} 4 \, dy$$

which resolves to

$$4y \Big]_{-5}^{3}$$

Evaluating from -5 to 3 yields

$$4 \cdot 3 - 4 \cdot (-5) = 32$$

Now we have the definite integral

$$\int_{-3}^{5} 32 \, dx$$

which resolves to

$$32x \Big]_{-3}^{5}$$

Working out the arithmetic, we get the same mathematical volume as we did in the solutions to Probs. 4 and 5:

$$32 \cdot 5 - 32 \cdot (-3) = 256$$

7. Here's the function in Cartesian xyz-space that describes the surface of interest:

$$f(x,y) = 4x + 4y$$

We want to calculate the mathematical volume of the solid relative to the region in the xy-plane whose edges are at $x = 1$, $x = 3$, $y = 0$, and $y = 5$ by integrating with respect to x and then with respect to y. In shorthand, our double integral is

$$\iint_R (4x + 4y) \, dx \, dy$$

Written out in full, the iterated integral is

$$\int_0^5 \left[\int_1^3 (4x + 4y) \, dx \right] dy$$

The integral inside the large square brackets is

$$\int_1^3 (4x + 4y) \, dx$$

Integrating with respect to x gives us

$$2x^2 + 4xy \, \Big]_1^3$$

Evaluating this expression from $x = 1$ to $x = 3$, we get

$$(2 \cdot 3^2 + 4 \cdot 3y) - (2 \cdot 1^2 + 4 \cdot y) = 16 + 8y$$

We substitute this for the integral inside the large square brackets above, obtaining

$$\int_0^5 (16 + 8y) \, dy$$

Integrating against y, we get

$$16y + 4y^2 \, \Big]_0^5$$

Working out the arithmetic yields

$$(16 \cdot 5 + 4 \cdot 5^2) - (16 \cdot 0 + 4 \cdot 0^2) = 180$$

8. Let's calculate the mathematical volume again, but this time we'll integrate with respect to y and then with respect to x. The shorthand integral is

$$\iint_R (4x + 4y) \, dy \, dx$$

Written out in full, the iterated integral is

$$\int_1^3 \left[\int_0^5 (4x + 4y) \, dy \right] dx$$

The integral inside the square brackets is

$$\int_0^5 (4x + 4y) \, dy$$

which resolves to

$$4xy + 2y^2 \, \Big]_0^5$$

Evaluating this expression from $y = 0$ to $y = 5$ yields

$$(4x \cdot 5 + 2 \cdot 5^2) - (4x \cdot 0 + 2 \cdot 0^2) = 20x + 50$$

We substitute this in place of the integral inside the large square brackets above to get

$$\int_1^3 (20x + 50) \, dx$$

which resolves to

$$10x^2 + 50x \Big]_1^3$$

Evaluating, we obtain

$$(10 \cdot 3^2 + 50 \cdot 3) - (10 \cdot 1^2 + 50 \cdot 1) = 180$$

9. Here's the function in Cartesian *xyz*-space that describes the surface of interest:

$$f(x,y) = x^2 + 2xy + y^2$$

Let's find the mathematical volume of the solid relative to the region in the *xy*-plane whose edges are at $x = 1$, $x = 3$, $y = 0$, and $y = 5$ by integrating with respect to x and then with respect to y. In shorthand, our double integral is

$$\iint_R (x^2 + 2xy + y^2) \, dx \, dy$$

Written out in full, the iterated integral is

$$\int_0^5 \left[\int_1^3 (x^2 + 2xy + y^2) \, dx \right] dy$$

The integral inside the large brackets is

$$\int_1^3 (x^2 + 2xy + y^2) \, dx$$

Integrating with respect to x gives us

$$x^3/3 + x^2 y + xy^2 \Big]_1^3$$

Evaluating from $x = 1$ to $x = 3$, we get

$$(3^3/3 + 3^2 y + 3y^2) - (1^3/3 + 1^2 y + y^2) = 2y^2 + 8y + 26/3$$

We substitute this for the integral inside the large square brackets above, obtaining

$$\int_0^5 (2y^2 + 8y + 26/3) \, dy$$

which works out to

$$2y^3/3 + 4y^2 + 26y/3 \,\Big]_0^5$$

When we evaluate this, we get

$$(2 \cdot 5^3/3 + 4 \cdot 5^2 + 26 \cdot 5/3) - (2 \cdot 0^3/3 + 4 \cdot 0^2 + 26 \cdot 0/3) = 680/3$$

10. Now we'll figure out the mathematical volume by integrating with respect to y and then with respect to x. The shorthand integral is

$$\iint\limits_R (x^2 + 2xy + y^2) \, dy \, dx$$

The iterated integral is

$$\int_1^3 \left[\int_0^5 (x^2 + 2xy + y^2) \, dy \right] dx$$

The integral inside the square brackets is

$$\int_0^5 (x^2 + 2xy + y^2) \, dy$$

Integrating with respect to y, we get

$$x^2y + xy^2 + y^3/3 \,\Big]_0^5$$

Evaluating this expression from $y = 0$ to $y = 5$ yields

$$(x^2 \cdot 5 + x \cdot 5^2 + 5^3/3) - (x^2 \cdot 0 + x \cdot 0^2 + 0^3/3) = 5x^2 + 25x + 125/3$$

We substitute this for the integral inside the large square brackets above, obtaining

$$\int_1^3 (5x^2 + 25x + 125/3) \, dx$$

which resolves to

$$5x^3/3 + 25x^2/2 + 125x/3 \,\Big]_1^3$$

Grinding out the arithmetic to evaluate this expression, we end up with

$$(5 \cdot 3^3/3 + 25 \cdot 3^2/2 + 125 \cdot 3/3) - (5 \cdot 1^3/3 + 25 \cdot 1^2/2 + 125 \cdot 1/3) = 680/3$$

Chapter 28

1. Imagine that we're at the point where $x = a$ in the situation of Fig. 28-6. Here, the cross-sectional slab is a line segment parallel to the z axis and "above" the xy-plane. As such, it has no area, either geometric or mathematical. As we start moving toward the right along the x axis, the slab starts out slender and tall, and becomes progressively

wider but shorter (less tall). As we keep moving, the mathematical area of the slab increases until it reaches a maximum at some value of x where $a < x < b$. But as the slab keeps getting shorter, its area decreases until, at the line where the surface and the region intersect, its mathematical area becomes 0. As we keep moving toward the right along the x axis, the slab acquires negative mathematical area, which becomes more negative until, at another value of x between a and b, the slab attains its maximum negative mathematical area. As we get close to $x = b$, the slab becomes "negatively taller" but slimmer until, when we reach $x = b$, it collapses into a line segment parallel to the z-axis but below the xy-plane, and, as in the beginning, it has no area, either geometric or mathematical.

2. Figure C-3 shows the region in the xy-plane for which we want to find the true geometric area. We can do this by integrating the difference function

$$q(x) = h(x) - g(x)$$

over the interval from $x = 0$ to $x = 2$. Because $g(x) = 0$, we have

$$q(x) = h(x)$$

The definite integral representing the area A of the region is therefore

$$A = \int_0^2 h(x)\, dx = \int_0^2 3x^2\, dx$$

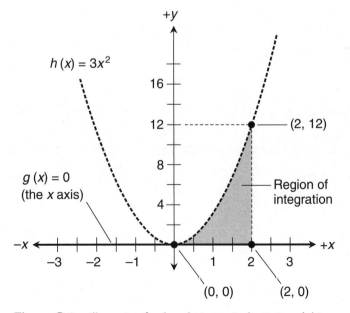

Figure C-3   Illustration for the solution to Probs. 2, 3, and 4 in Chap. 28.

which resolves to

$$A = x^3 \Big]_0^2$$

Evaluating, we get

$$A = 2^3 - 0^3 = 8$$

The height of the solid is 4, so the total volume V is

$$V = 4A = 4 \cdot 8 = 32$$

This agrees with the result we got by double integration.

3. We've been told to find the mathematical volume of a solid with respect to the region of integration shown in Fig. 28-2 (simplified here in Fig. C-3) as defined by

$$f(x,y) = 2x - 6y$$

In shorthand form, the double integral is

$$\iint_R (2x - 6y) \, dy \, dx$$

The x-value interval enclosing the region is (0,2), so the iterated integral is

$$\int_0^2 \left[\int_{g(x)}^{h(x)} (2x - 6y) \, dy \right] dx$$

The integral inside the large square brackets is

$$\int_{g(x)}^{h(x)} (2x - 6y) \, dy$$

When we integrate against y while holding x constant, we obtain

$$2xy - 3y^2 \Big]_{g(x)}^{h(x)}$$

The xy-plane functions whose graphs define the shaded region are

$$g(x) = 0$$

and

$$h(x) = 3x^2$$

Evaluating the quantity $(2xy - 3y^2)$ from $y = 0$ to $y = 3x^2$ yields

$$[2x \cdot 3x^2 - 3(3x^2)^2] - (2x \cdot 0 - 3 \cdot 0^2) = 6x^3 - 27x^4$$

Substituting this for the integral inside the large square brackets above, we get

$$\int_0^2 (6x^3 - 27x^4)\, dx$$

which resolves to

$$3x^4/2 - 27x^5/5 \ \Big]_0^2$$

Working out the arithmetic to get the solid's mathematical volume, we obtain

$$V = (3 \cdot 2^4/2 - 27 \cdot 2^5/5) - (3 \cdot 0^4/2 - 27 \cdot 0^5/5) = -744/5$$

The negative result tells us that the surface, on the average, lies "below" the *xy*-plane where $z < 0$. If we could see the surface in Fig. C-3, it would lie mostly "behind" the region of integration.

4. We want to find the mathematical volume of a solid defined by the following function with respect to the region of integration shown in Fig. 28-2 (simplified here in Fig. C-3) as defined by

$$f(x,y) = 3x^2 - 4y$$

Written in shorthand, our double integral is

$$\iint_R (3x^2 - 4y)\, dy\, dx$$

Again, the span of *x* values is (0,2). The iterated integral is therefore

$$\int_0^2 \Big[\int_{g(x)}^{h(x)} (3x^2 - 4y)\, dy \Big]\, dx$$

The integral inside the large square brackets is

$$\int_{g(x)}^{h(x)} (3x^2 - 4y)\, dy$$

Holding *x* constant and integrating against *y*, we get

$$3x^2y - 2y^2 \ \Big]_{g(x)}^{h(x)}$$

The *xy*-plane functions whose graphs define the region of integration are

$$g(x) = 0$$

and

$$h(x) = 3x^2$$

Evaluating the quantity $(3x^2y - 2y^2)$ from $y = 0$ to $y = 3x^2$, we get

$$[3x^2 \cdot 3x^2 - 2 \cdot (3x^2)^2] - (3x^2 \cdot 0 - 2 \cdot 0^2) = -9x^4$$

We substitute this for the integral inside the large square brackets above to obtain

$$\int_0^2 -9x^4 \, dx$$

which resolves to

$$-9x^5/5 \, \Big]_0^2$$

Working out the arithmetic, we get a mathematical volume of

$$V = -9 \cdot 2^5/5 - (-9 \cdot 0^5/5) = -288/5$$

5. First, we must find the true geometric area A, between the line and the curve shown in Fig. 28-3 (simplified here in Fig. C-4) by subtracting the lower function g from the upper function h, and then integrating over the interval between the points where the

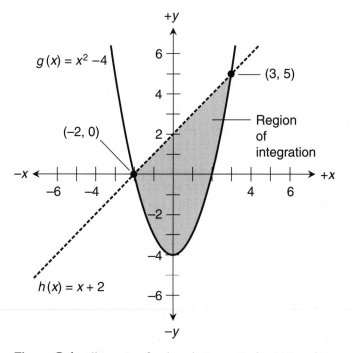

Figure C-4   Illustration for the solution to Probs. 5, 6, and 7 in Chap. 28.

graphs intersect. When we subtract the function for the parabola from the function for the straight line, we get

$$p(x) = h(x) - g(x) = (x + 2) - (x^2 - 4) = -x^2 + x + 6$$

We integrate $p(x)$ over the interval between $x = -2$ and $x = 3$, which are the x values of the points where the line and the curve intersect:

$$A = \int_{-2}^{3} (-x^2 + x + 6) \, dx$$

which resolves to

$$A = -x^3/3 + x^2/2 + 6x \Big]_{-2}^{3}$$

Working out the arithmetic, we get

$$A = [-(3^3/3) + 3^2/2 + 6 \cdot 3] - [-(-2)^3/3 + (-2)^2/2 + 6 \cdot (-2)] = 125/6$$

To find the volume of the solid, we must multiply this area by 4, which is the height of the solid. That gives us

$$V = 4A = 4 \cdot 125/6 = 250/3$$

This agrees with the result we got by double integration.

6. We want to calculate the mathematical volume of a solid with respect to the region in Fig. 28-3 (simplified here in Fig. C-4) as defined by the following function, which represents a flat surface oriented at a slant:

$$f(x,y) = -4x$$

In shorthand, we have

$$\iint_{R} -4x \, dy \, dx$$

The x-value interval enclosing the region of integration is $(-2,3)$, so we can write the iterated integral as

$$\int_{-2}^{3} \left[\int_{g(x)}^{h(x)} -4x \, dy \right] dx$$

The integral inside the large square brackets is

$$\int_{g(x)}^{h(x)} -4x \, dy$$

When we integrate against y, we obtain

$$-4xy \Big]_{g(x)}^{h(x)}$$

The *xy*-plane functions whose graphs define the region of integration are

$$g(x) = x^2 - 4$$

and

$$h(x) = x + 2$$

We evaluate $-4xy$ from $y = x^2 - 4$ to $y = x + 2$ to get

$$-4x(x+2) - [-4x(x^2 - 4)] = 4x^3 - 4x^2 - 24x$$

Substituting this for the integral inside the large square brackets above, we get

$$\int_{-2}^{3} (4x^3 - 4x^2 - 24x)\, dx$$

which resolves to

$$x^4 - 4x^3/3 - 12x^2 \Big]_{-2}^{3}$$

Calculating to get the mathematical volume of the solid, we obtain

$$V = (3^4 - 4 \cdot 3^3/3 - 12 \cdot 3^2) - [(-2)^4 - 4 \cdot (-2)^3/3 - 12 \cdot (-2)^2] = -125/3$$

7. Imagine a warped surface in *xyz*-space that represents the function

$$f(x,y) = -x^2$$

We want to find the mathematical volume of the solid defined by this surface and the region illustrated in Fig. 28-3 (simplified here in Fig. C-4). The shorthand form of the double integral is

$$\iint_R -x^2 \, dy \, dx$$

The interval representing the span of *x* values is $(-2,3)$, so the iterated integral is

$$\int_{-2}^{3} \left[\int_{g(x)}^{h(x)} -x^2 \, dy \right] dx$$

The integral inside the large square brackets is

$$\int_{g(x)}^{h(x)} -x^2 \, dy$$

Integrating with respect to *y*, we get

$$-x^2 y \Big]_{g(x)}^{h(x)}$$

The functions whose graphs define the region in the *xy*-plane are

$$g(x) = x^2 - 4$$

and

$$h(x) = x + 2$$

Evaluating $-x^2 y$ from $y = x^2 - 4$ to $y = x + 2$ produces

$$-x^2(x+2) - [-x^2(x^2-4)] = x^4 - x^3 - 6x^2$$

When we substitute this for the integral inside the large square brackets above, we get

$$\int_{-2}^{3} (x^4 - x^3 - 6x^2)\, dx$$

which resolves to

$$x^5/5 - x^4/4 - 2x^3 \Big]_{-2}^{3}$$

Working out the arithmetic to derive the mathematical volume, we get

$$V = (3^5/5 - 3^4/4 - 2 \cdot 3^3) - [(-2)^5/5 - (-2)^4/4 - 2 \cdot (-2)^3] = -125/4$$

8. First, we must find the true geometric area *A*, between the curves shown in Fig. 28-4 (and simplified here in Fig. C-5) by subtracting the lower function *g* from the upper function *h*, and then integrating over the interval between the points where the graphs intersect. We must integrate the difference function

$$p(x) = h(x) - g(x) = (-x^2/2 + 2) - (x^2/2 - 2) = -x^2 + 4$$

The interval of interest is (−2,2), so the integral we want to evaluate is

$$A = \int_{-2}^{2} (-x^2 + 4)\, dx$$

which resolves to

$$A = -x^3/3 + 4x \Big]_{-2}^{2}$$

When we evaluate this, we obtain

$$A = [-(2^3/3) + 4 \cdot 2] - [-(-2)^3/3 + 4 \cdot (-2)] = 32/3$$

To find the mathematical volume, we multiply this area by 4, which is the mathematical height of the solid. That gives us

$$V = 4A = 4 \cdot 32/3 = 128/3$$

This agrees with the result we got by double integration.

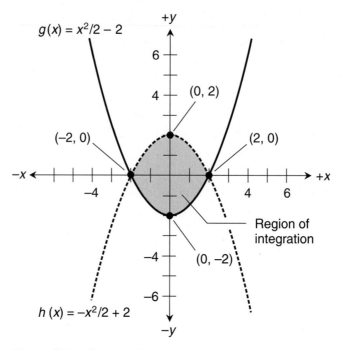

Figure C-5   Illustration for the solution to Probs. 8, 9, and 10 in Chap. 28.

9. We've been told to figure out the mathematical volume of a solid defined by the region shown in Fig. 28-4 (simplified here in Fig. C-5) and the function

$$f(x,y) = 2x + 1$$

In shorthand form, the double integral is

$$\iint\limits_{R} (2x + 1)\ dy\ dx$$

The span of x values is $(-2, 2)$, so the iterated integral is

$$\int_{-2}^{2} \left[\int_{g(x)}^{h(x)} (2x + 1)\ dy \right] dx$$

The integral inside the large square brackets is

$$\int_{g(x)}^{h(x)} (2x + 1)\ dy$$

Integrating against y, we hold x constant to get

$$2xy + y \ \Bigg]_{g(x)}^{h(x)}$$

The functions whose graphs define our *xy*-plane region are

$$g(x) = x^2/2 - 2$$

and

$$h(x) = -x^2/2 + 2$$

Evaluating the quantity $(2xy + y)$ from $y = x^2/2 - 2$ to $y = -x^2/2 + 2$, we get

$$[2x(-x^2/2 + 2) + (-x^2/2 + 2)] - [2x(x^2/2 - 2) + (x^2/2 - 2)] = -2x^3 - x^2 + 8x + 4$$

Substituting this for the integral inside the large square brackets above, we get

$$\int_{-2}^{2} (-2x^3 - x^2 + 8x + 4)\, dx$$

which resolves to

$$-x^4/2 - x^3/3 + 4x^2 + 4x \Big]_{-2}^{2}$$

Working out the arithmetic, we obtain

$$V = (-2^4/2 - 2^3/3 + 4 \cdot 2^2 + 4 \cdot 2) - [-(-2)^4/2 - (-2)^3/3 + 4 \cdot (-2)^2 + 4 \cdot (-2)] = 32/3$$

10. We want find the mathematical volume of a solid relative to the region in Fig. 28-4 (simplified here in Fig. C-5), as defined by the following function:

$$f(x,y) = 2y - 1$$

In shorthand form, the double integral is

$$\iint_{R} (2y - 1)\, dy\, dx$$

The span of *x*-values is (−2,2). The iterated integral is therefore

$$\int_{-2}^{2} \left[\int_{g(x)}^{h(x)} (2y - 1)\, dy \right] dx$$

The integral inside the large square brackets is

$$\int_{g(x)}^{h(x)} (2y - 1)\, dy$$

Integrating with respect to *y*, we get

$$y^2 - y \Big]_{g(x)}^{h(x)}$$

The functions whose graphs define our *xy*-plane region are

$$g(x) = x^2/2 - 2$$

and

$$h(x) = -x^2/2 + 2$$

Evaluating the quantity $(y^2 - y)$ from $y = x^2/2 - 2$ to $y = -x^2/2 + 2$, we get

$$[(-x^2/2 + 2)^2 - (-x^2/2 + 2)] - [(x^2/2 - 2)^2 - (x^2/2 - 2)] = x^2 - 4$$

Substituting this for the integral inside the large square brackets above, we get

$$\int_{-2}^{2} (x^2 - 4)\ dx$$

which resolves to

$$x^3/3 - 4x \ \Big]_{-2}^{2}$$

The arithmetic yields

$$V = (2^3/3 - 4 \cdot 2) - [(-2)^3/3 - 4 \cdot (-2)] = -32/3$$

Chapter 29

1. Here's the differential equation we want to solve:

$$dy/dx = \sin x + 3$$

Integrating both sides with respect to x, we get

$$\int (dy/dx)\ dx = \int (\sin x + 3)\ dx$$

Working out the integrals gives us

$$\int (dy/dx)\ dx = y + c_1$$

and

$$\int (\sin x + 3)\ dx = -\cos x + 3x + c_2$$

where c_1 and c_2 are constants. Combining the general antiderivatives into a single equation, we get

$$y + c_1 = -\cos x + 3x + c_2$$

Subtracting c_1 from each side produces

$$y = -\cos x + 3x - c_1 + c_2$$

Combining the constants of integration into a single constant c yields the solution

$$y = H(x) = -\cos x + 3x + c$$

where H is the family of solution functions.

2. Here's the differential equation we want to solve:

$$dy/dx + \sin x = \cos x$$

First, we subtract $\sin x$ from both sides to obtain

$$dy/dx = \cos x - \sin x$$

When we integrate this equation through, we obtain

$$\int (dy/dx)\ dx = \int (\cos x - \sin x)\ dx$$

The general antiderivatives are

$$\int (dy/dx)\ dx = y + c_1$$

and

$$\int (\cos x - \sin x)\ dx = \sin x + \cos x + c_2$$

Combining the general antiderivatives into a single equation, we get

$$y + c_1 = \sin x + \cos x + c_2$$

Subtracting c_1 from each side gives us

$$y = \sin x + \cos x - c_1 + c_2$$

Combining the constants of integration yields the solution

$$y = H(x) = \sin x + \cos x + c$$

3. Here's the differential equation we want to solve:

$$2\,dy/dx - 4e^x = 16x^3$$

Dividing through by 2 gives us

$$dy/dx - 2e^x = 8x^3$$

Adding $2e^x$ to both sides, we get

$$dy/dx = 2e^x + 8x^3$$

Integrating both sides produces

$$\int (dy/dx)\ dx = \int (2e^x + 8x^3)\ dx$$

The individual indefinite integrals are

$$\int (dy/dx)\ dx = y + c_1$$

and

$$\int (2e^x + 8x^3)\ dx = 2e^x + 2x^4 + c_2$$

Combining the general antiderivatives into a single equation, we get

$$y + c_1 = 2e^x + 2x^4 + c_2$$

Subtracting c_1 from each side gives us

$$y = 2e^x + 2x^4 - c_1 + c_2$$

Combining the constants of integration, we get the solution

$$y = H(x) = 2e^x + 2x^4 + c$$

4. We've been told to solve this differential equation:

$$d^2y/dx^2 = \cos x + 5x$$

When we take the indefinite integrals of both sides, we get

$$\int (d^2y/dx^2)\ dx = \int (\cos x + 5x)\ dx$$

The general antiderivatives are

$$\int (d^2y/dx^2)\ dx = dy/dx + c_1$$

and

$$\int (\cos x + 5x)\ dx = \sin x + 5x^2/2 + c_2$$

where c_1 and c_2 are constants of integration. Combining the general antiderivatives into a single equation, we get

$$dy/dx + c_1 = \sin x + 5x^2/2 + c_2$$

We can rearrange this as

$$dy/dx = \sin x + 5x^2/2 - c_1 + c_2$$

Let's consolidate the constants, adding $-c_1$ to c_2 and calling the sum p. That gives us

$$dy/dx = \sin x + 5x^2/2 + p$$

Integrating through, we get

$$\int (dy/dx)\ dx = \int (\sin x + 5x^2/2 + p)\ dx$$

The general antiderivatives are

$$\int (dy/dx)\ dx = y + c_3$$

and

$$\int (\sin x + 5x^2/2 + p)\ dx = -\cos x + 5x^3/6 + px + c_4$$

where c_3 and c_4 are new constants. Combining the general antiderivatives into a single equation, we get

$$y + c_3 = -\cos x + 5x^3/6 + px + c_4$$

Subtracting c_3 from each side, we get

$$y = -\cos x + 5x^3/6 + px - c_3 + c_4$$

Adding $-c_3$ to c_4 and calling the sum q, the solution simplifies to

$$y = H(x) = -\cos x + 5x^3/6 + px + q$$

where H is the family of solution functions, and p and q are constants whose values we don't necessarily know.

5. We want to solve this differential equation:

$$d^2y/dx^2 + 2 \sin x = 3 \cos x$$

Subtracting the quantity $(2 \sin x)$ from both sides, we get

$$d^2y/dx^2 = 3 \cos x - 2 \sin x$$

Integrating both sides gives us

$$\int (d^2y/dx^2)\ dx = \int (3 \cos x - 2 \sin x)\ dx$$

The general antiderivatives are

$$\int (d^2y/dx^2)\ dx = dy/dx + c_1$$

and

$$\int (3 \cos x - 2 \sin x) \, dx = 3 \sin x + 2 \cos x + c_2$$

Combining the general antiderivatives into a single equation, we get

$$dy/dx + c_1 = 3 \sin x + 2 \cos x + c_2$$

which can be rearranged to get

$$dy/dx = 3 \sin x + 2 \cos x - c_1 + c_2$$

Letting $-c_1 + c_2 = p$, we can simplify this to

$$dy/dx = 3 \sin x + 2 \cos x + p$$

When we integrate both sides of this equation, we get

$$\int (dy/dx) \, dx = \int (3 \sin x + 2 \cos x + p) \, dx$$

The general antiderivatives are

$$\int (dy/dx) \, dx = y + c_3$$

and

$$\int (3 \sin x + 2 \cos x + p) \, dx = -3 \cos x + 2 \sin x + px + c_4$$

Combining the general antiderivatives into a single equation, we get

$$y + c_3 = -3 \cos x + 2 \sin x + px + c_4$$

Subtracting c_3 from each side yields

$$y = -3 \cos x + 2 \sin x + px - c_3 + c_4$$

When we add $-c_3$ to c_4 and call the combination q, we can simplify our solution to

$$y = H(x) = -3 \cos x + 2 \sin x + px + q$$

6. Here's the differential equation we've been told to solve:

$$2d^2y/dx^2 - 2e^x = 24x^2$$

Adding $2e^x$ to both sides, we get

$$2d^2y/dx^2 = 2e^x + 24x^2$$

Dividing through by 2 gives us

$$d^2y/dx^2 = e^x + 12x^2$$

Integrating both sides, we get

$$\int (d^2y/dx^2)\, dx = \int (e^x + 12x^2)\, dx$$

The general antiderivatives are

$$\int (d^2y/dx^2)\, dx = dy/dx + c_1$$

and

$$\int (e^x + 12x^2)\, dx = e^x + 4x^3 + c_2$$

Combining the general antiderivatives into a single equation, we get

$$dy/dx + c_1 = e^x + 4x^3 + c_2$$

Subtracting c_1 from each side yields

$$dy/dx = e^x + 4x^3 - c_1 + c_2$$

After adding $-c_1$ to c_2 to get a single constant p, we have

$$dy/dx = e^x + 4x^3 + p$$

Integrating this equation through gives us

$$\int (dy/dx)\, dx = \int (e^x + 4x^3 + p)\, dx$$

These integrals work out as

$$\int (dy/dx)\, dx = y + c_3$$

and

$$\int (e^x + 4x^3 + p)\, dx = e^x + x^4 + px + c_4$$

Combining the general antiderivatives into a single equation, we get

$$y + c_3 = e^x + x^4 + px + c_4$$

When we subtract c_3 from each side, we get

$$y = e^x + x^4 + px - c_3 + c_4$$

We can let $-c_3 + c_4 = q$ to produce the solution

$$y = H(x) = e^x + x^4 + px + q$$

7. We've been told to solve this differential equation:

$$dy/(3\,dx) + 6x^{-2} = e^x$$

Multiplying through by 3, we get

$$dy/dx + 18x^{-2} = 3e^x$$

Subtracting $18x^{-2}$ from both sides gives us

$$dy/dx = 3e^x - 18x^{-2}$$

Integrating both sides produces

$$\int (dy/dx)\ dx = \int (3e^x - 18x^{-2})\ dx$$

The general antiderivatives are

$$\int (dy/dx)\ dx = y + c_1$$

and

$$\int (3e^x - 18x^{-2})\ dx = 3e^x + 18x^{-1} + c_2$$

Combining the general antiderivatives into a single equation, we get

$$y + c_1 = 3e^x + 18x^{-1} + c_2$$

Subtracting c_1 from each side produces

$$y = 3e^x + 18x^{-1} - c_1 + c_2$$

We can let $-c_1 + c_2 = c$ to produce the solution

$$y = H(x) = 3e^x + 18x^{-1} + c$$

8. Here's the differential equation we've been told to solve:

$$dy - 4e^x\ dx = x^2\ dx + 2x\ dx$$

Dividing through by dx, we obtain

$$dy/dx - 4e^x = x^2 + 2x$$

Adding $4e^x$ to both sides, we get

$$dy/dx = x^2 + 2x + 4e^x$$

When we integrate through with respect to x, we get

$$\int (dy/dx)\, dx = \int (x^2 + 2x + 4e^x)\, dx$$

The general antiderivatives are

$$\int (dy/dx)\, dx = y + c_1$$

and

$$\int (x^2 + 2x + 4e^x)\, dx = x^3/3 + x^2 + 4e^x + c_2$$

Combining the general antiderivatives into a single equation, we get

$$y + c_1 = x^3/3 + x^2 + 4e^x + c_2$$

Subtracting c_1 from each side yields

$$y = x^3/3 + x^2 + 4e^x - c_1 + c_2$$

Combining the constants of integration gives us the solution

$$y = H(x) = x^3/3 + x^2 + 4e^x + c$$

9. Once again, here's the differential equation we want to solve:

$$dy - 4e^x\, dx = x^2\, dx + 2x\, dx$$

Adding the quantity $(4e^x\, dx)$ to both sides, we obtain

$$dy = x^2\, dx + 2x\, dx + 4e^x\, dx$$

Factoring out dx on the right-hand side, we get

$$dy = (x^2 + 2x + 4e^x)\, dx$$

When we integrate straight through, we get

$$\int dy = \int (x^2 + 2x + 4e^x)\, dx$$

The general antiderivatives are

$$\int dy = y + c_1$$

and

$$\int (x^2 + 2x + 4e^x)\ dx = x^3/3 + x^2 + 4e^x + c_2$$

Combining the general antiderivatives into a single equation, we get

$$y + c_1 = x^3/3 + x^2 + 4e^x + c_2$$

Subtracting c_1 from each side yields

$$y = x^3/3 + x^2 + 4e^x - c_1 + c_2$$

Combining the constants of integration gives us the solution

$$y = H(x) = x^3/3 + x^2 + 4e^x + c$$

10. Let's check the solutions for Exercises 1 through 8 in order. In some cases we must differentiate once; in other cases we must differentiate twice. Our goal is always to get back the original differential equation.

Checking solution 1. We finished with

$$y = -\cos x + 3x + c$$

Differentiating both sides gives us the original differential equation

$$dy/dx = \sin x + 3$$

Checking solution 2. We finished with

$$y = \sin x + \cos x + c$$

Differentiating both sides, we get

$$dy/dx = \cos x - \sin x$$

We can add $\sin x$ to both sides, getting the original differential equation

$$dy/dx + \sin x = \cos x$$

Checking solution 3. We finished with

$$y = 2e^x + 2x^4 + c$$

Differentiating through produces the equation

$$dy/dx = 2e^x + 8x^3$$

We can subtract $2e^x$ from both sides to obtain

$$dy/dx - 2e^x = 8x^3$$

Multiplying through by 2 gets us back to the original differential equation

$$2dy/dx - 4e^x = 16x^3$$

Checking solution 4. We finished with

$$y = -\cos x + 5x^3/6 + px + q$$

Differentiating each side, we get

$$dy/dx = \sin x + 5x^2/2 + p$$

Differentiating through again, we get the original differential equation

$$d^2y/dx^2 = \cos x + 5x$$

Checking solution 5. We finished with

$$y = -3 \cos x + 2 \sin x + px + q$$

Differentiating each side, we get

$$dy/dx = 3 \sin x + 2 \cos x + p$$

Differentiating through again, we get

$$d^2y/dx^2 = 3 \cos x - 2 \sin x$$

Adding the quantity $(2 \sin x)$ to both sides, we get the original differential equation

$$d^2y/dx^2 + 2 \sin x = 3 \cos x$$

Checking solution 6. We finished with

$$y = e^x + x^4 + px + q$$

Differentiating each side, we get

$$dy/dx = e^x + 4x^3 + p$$

Differentiating through again, we get

$$d^2y/dx^2 = e^x + 12x^2$$

^{x} from both sides gives us

$$d^2y/dx^2 - e^x = 12x^2$$

differential equation comes back when we multiply through by 2, yielding

$$2d^2y/dx^2 - 2e^x = 24x^2$$

...ecking solution 7. We finished with

$$y = 3e^x + 18x^{-1} + c$$

Differentiating each side, we get

$$dy/dx = 3e^x - 18x^{-2}$$

Adding $18x^{-2}$ to each side gives us

$$dy/dx + 18x^{-2} = 3e^x$$

Dividing through by 3, we get back the original differential equation

$$dy/(3dx) + 6x^{-2} = e^x$$

Checking solutions 8 and 9. We finished with

$$y = x^3/3 + x^2 + 4e^x + c$$

Differentiating both sides, we get

$$dy/dx = x^2 + 2x + 4e^x$$

Subtracting $4e^x$ from each side produces

$$dy/dx - 4e^x = x^2 + 2x$$

Multiplying through by dx, we get back the original equation

$$dy - 4e^x\, dx = x^2\, dx + 2x\, dx$$

D

Answers to Final Exam Questions

1. c	2. d	3. a	4. a	5. e
6. a	7. a	8. c	9. a	10. c
11. c	12. b	13. a	14. a	15. d
16. d	17. c	18. d	19. c	20. b
21. d	22. a	23. a	24. c	25. d
26. e	27. d	28. e	29. b	30. c
31. b	32. a	33. a	34. b	35. c
36. e	37. b	38. a	39. b	40. b
41. e	42. e	43. e	44. d	45. b
46. d	47. c	48. b	49. d	50. c
51. c	52. a	53. d	54. d	55. b
56. e	57. b	58. a	59. c	60. e
61. c	62. a	63. b	64. a	65. b
66. c	67. a	68. e	69. e	70. b
71. d	72. c	73. b	74. c	75. d
76. d	77. c	78. d	79. e	80. a
81. c	82. a	83. e	84. a	85. c
86. a	87. e	88. d	89. e	90. e
91. e	92. a	93. e	94. b	95. b
96. c	97. d	98. d	99. c	100. e
101. d	102. a	103. b	104. e	105. d
106. a	107. b	108. d	109. a	110. c
111. d	112. e	113. e	114. c	115. a
116. c	117. a	118. d	119. c	120. c
121. b	122. c	123. a	124. b	125. e
126. d	127. a	128. d	129. a	130. c
131. a	132. b	133. b	134. e	135. a
136. d	137. a	138. b	139. c	140. c
141. e	142. a	143. c	144. d	145. a
146. d	147. a	148. c	149. b	150. a

E

Special Characters in Order of Appearance

Symbol	First use	Meaning
$\{a, b\}$	Chapter 1	Set containing elements a and b
(a,b)	Chapter 1	Ordered pair of elements a and b
$f(x)$	Chapter 1	Function or relation f with independent variable x
$f^{-1}(x)$	Chapter 1	Inverse of a function or relation $f(x)$
Δx	Chapter 1	Increment in x; difference in x-values
Δy	Chapter 1	Increment in y; difference in y-values
$\pm$	Chapter 1	Plus-or-minus
Lim	Chapter 2	Limit
$\rightarrow$	Chapter 2	Symbol indicating that one quantity approaches another
dy/dx	Chapter 3	Derivative of y with respect to x
y'	Chapter 3	Derivative of y with respect to independent variable
$df(x)/dx$	Chapter 3	Derivative of function $f(x)$ with respect to x
$d/dx\, f(x)$		
df/dx		
$f'(x)$		
f'	Chapter 3	Derivative of function f with respect to independent variable
$\rightarrow +\infty$	Chapter 4	Increases endlessly or "approaches positive infinity"
$\rightarrow 0+$	Chapter 4	Approaches 0 from the positive side
$\rightarrow -\infty$	Chapter 4	Decreases endlessly or "approaches negative infinity"
$\rightarrow 0-$	Chapter 4	Approaches 0 from the negative side
$|x|$	Chapter 4	Absolute value of x
δ	Chapter 5	Alternative notation for small values of Δx
o	Chapter 6	Alternative notation for composite function or a function of another function
·	Chapter 7	Alternative notation for multiplication
$\approx$	Chapter 7	Approximate equality
. . .	Chapter 11	Stand-in for terms in a set, sequence, or series

Σ	Chapter 11	Sum or summation
$\int$	Chapter 11	Indefinite integral
$\int_a^b$	Chapter 11	Definite integral over interval from a to b
μ	Chapter 11	Mean of a probability distribution
σ	Chapter 11	Standard deviation of a probability distribution
$\Big]_a^b$	Chapter 12	Symbol for evaluation over interval from a to b
∂	Chapter 24	Symbol for partial derivative
$\int_a^b \int_a^b$	Chapter 27	Repeated integral of single-variable function from a to b
$\iint_R$	Chapter 27	Double integral of two-variable function over a region R

F

Table of Derivatives

The letter a denotes a general constant, f and g denote functions, x denotes a variable, and e represents the exponential constant (approximately 2.71828).

Function $f(x)$	Derivative $f'(x)$
a	0
ax	a
ax^n	nax^{n-1}
$\ln|x|$	x^{-1}
$\ln|g(x)|$	$g^{-1}(x)\,g'(x)$
x^{-a}	$-ax^{(-a-1)}$
e^x	e^x
a^x	$a^x \ln|a|$
$a^{g(x)}$	$[a^{g(x)}]\,[g'(x)]\ln|a|$
e^{ax}	ae^{ax}
$e^{g(x)}$	$[e^{g(x)}]\,[g'(x)]$
xe^x	$e^x + xe^x$
$\sin x$	$\cos x$
$\cos x$	$-\sin x$
$\tan x$	$\sec^2 x$
$\csc x$	$-\csc x \cot x$
$\sec x$	$\sec x \tan x$
$\cot x$	$-\csc^2 x$
$\text{Arcsin } x$	$(1-x^2)^{-1/2}$
$\text{Arccos } x$	$-(1-x^2)^{-1/2}$
$\text{Arctan } x$	$(1+x^2)^{-1}$

G

Table of Integrals

Letters a and b denote general constants, c denotes a constant of integration, f denotes a function, x denotes a variable, and e represents the exponential constant (approximately 2.71828).

Function	**Indefinite integral**
$f(x)$	$\int f(x)\, dx$
0	c
1	$x + c$
a	$ax + c$
x	$(1/2)\, x^2 + c$
ax	$(1/2)\, ax^2 + c$
ax^2	$(1/3)\, ax^3 + c$
ax^3	$(1/4)\, ax^4 + c$
ax^4	$(1/5)\, ax^5 + c$
ax^{-1}	$a \ln |x| + c$
ax^{-2}	$-ax^{-1} + c$
ax^{-3}	$(-1/2)\, ax^{-2} + c$
ax^{-4}	$(-1/3)\, ax^{-3} + c$
ax^n for $n \neq -1$	$(n+1)^{-1}\, ax^{n+1} + c$
$(ax + b)^{-1}$	$a^{-1} \ln |ax + b| + c$
$(ax + b)^n$ for $n \neq -1$	$(an + a)^{-1}\, (ax + b)^{n+1} + c$
$x\,(ax + b)^{1/2}$	$(1/15)\, a^{-2}\,(6ax - 4b)(ax + b)^{3/2} + c$
$x\,(ax + b)^{-1/2}$	$(1/3)\, a^{-2}\,(2ax - 4b)(ax + b)^{1/2} + c$
$x\,(ax + b)^{-2}$	$b\,(a^3x + a^2b)^{-1} + a^{-2} \ln |ax + b| + c$
$(x^2 + a^2)^{1/2}$	$(x/2)\,(x^2 + a^2)^{1/2} + (1/2)\, a^2 \ln |x + (x^2 + a^2)^{1/2}| + c$
$(x^2 - a^2)^{1/2}$	$(x/2)\,(x^2 - a^2)^{1/2} - (1/2)\, a^2 \ln |x + (x^2 - a^2)^{1/2}| + c$
$(a^2 - x^2)^{1/2}$	$(x/2)\,(a^2 - x^2)^{1/2} + (1/2)\, a^2 \operatorname{Arcsin}(a^{-1}x) + c$
$(x^2 + a^2)^{-1/2}$	$\ln |x + (x^2 + a^2)^{1/2}| + c$
$(x^2 - a^2)^{-1/2}$	$\ln |x + (x^2 - a^2)^{1/2}| + c$
$(a^2 - x^2)^{-1/2}$	$\operatorname{Arcsin}(a^{-1}x) + c$

$(x^2 + a^2)^{-1}$ for $a > 0$	$a^{-1} \operatorname{Arctan} (a^{-1}x) + c$
$(x^2 - a^2)^{-1}$	$(1/2) \, a^{-1} \ln |(x + a)^{-1} \, (x - a)| + c$
$(a^2 - x^2)^{-1}$ for $|a| > |x|$	$(1/2) \, a^{-1} \ln |(a - x)^{-1} \, (a + x)| + c$
$(x^2 + a^2)^{-2}$	$(2a^2x^2 + 2a^4)^{-1} \, x + (1/2) \, a^{-3} \operatorname{Arctan} (a^{-1}x) + c$
$(x^2 - a^2)^{-2}$	$(-x) \, (2a^2x^2 - 2a^4)^{-1} - (1/4) \, a^{-3} \ln |(x + a)^{-1} \, (x - a)| + c$
$(a^2 - x^2)^{-2}$ for $|a| > |x|$	$-x \, (2a^4 - 2a^2x^2)^{-1} + (1/4) \, a^{-3} \ln |(a - x)^{-1} \, (a + x)| + c$
$x \, (x^2 + a^2)^{1/2}$	$(1/3) \, (x^2 + a^2)^{3/2} + c$
$x \, (x^2 - a^2)^{1/2}$	$(1/3) \, (x^2 - a^2)^{3/2} + c$
$x \, (a^2 - x^2)^{1/2}$	$(-1/3) \, (a^2 - x^2)^{3/2} + c$
$x \, (x^2 + a^2)^{-1/2}$	$(x^2 + a^2)^{1/2} + c$
$x \, (x^2 - a^2)^{-1/2}$	$(x^2 - a^2)^{1/2} + c$
$x \, (a^2 - x^2)^{-1/2}$	$-(a^2 - x^2)^{1/2} + c$
$x \, (x^2 + a^2)^{-1}$	$(1/2) \ln |x^2 + a^2| + c$
$x \, (x^2 - a^2)^{-1}$	$(1/2) \ln |x^2 - a^2| + c$
$x \, (a^2 - x^2)^{-1}$ for $|a| > |x|$	$-(1/2) \ln |a^2 - x^2| + c$
$x \, (x^2 + a^2)^{-2}$	$(-2x^2 - 2a^2)^{-1} + c$
$x \, (x^2 - a^2)^{-2}$	$-(1/2) \, (x^2 - a^2)^{-1} + c$
$x \, (a^2 - x^2)^{-2}$ for $|a| > |x|$	$(1/2) \, (a^2 - x^2)^{-1} + c$
e^x	$e^x + c$
e^{ax}	$a^{-1} \, e^{ax} + c$
$a \, e^{bx}$	$b^{-1} a \, e^{bx} + c$
$x \, e^{ax}$	$a^{-1}x \, e^{ax} - a^{-2} \, e^{ax} + c$
$x^2 \, e^{bx}$	$b^{-1}x^2 \, e^{bx} - 2b^{-2}x \, e^{bx} + 2b^{-3} \, e^{bx} + c$
$\ln |x|$	$x \ln |x| - x + c$
$x^n \ln |x|$ for $n \neq -1$	$(n + 1)^{-1} \, x^{(n+1)} \ln |x| - (n + 1)^{-2} \, x^{(n+1)} + c$
$x^{-1} \ln x$	$(1/2) \ln^2 |x| + c$
$\sin x$	$-\cos x + c$
$\cos x$	$\sin x + c$
$\tan x$	$\ln |\sec x| + c$
$\csc x$	$\ln |\tan (x/2)| + c$
$\sec x$	$\ln |\sec x + \tan x| + c$
$\cot x$	$\ln |\sin x| + c$
$\sin ax$	$-a^{-1} \cos ax + c$
$\cos ax$	$a^{-1} \sin ax + c$
$\tan ax$	$a^{-1} \ln |\sec ax| + c$
$\csc ax$	$a^{-1} \ln |\tan (ax/2)| + c$
$\sec ax$	$a^{-1} \ln |\tan (\pi/4 + ax/2)| + c$
$\cot ax$	$a^{-1} \ln |\sin ax| + c$
$\sin^2 x$	$(1/2) \{x - [(1/2) \sin (2x)]\} + c$
$\cos^2 x$	$(1/2) \{x + [(1/2) \sin (2x)]\} + c$
$\tan^2 x$	$\tan x - x + c$
$\csc^2 x$	$-\cot x + c$
$\sec^2 x$	$\tan x + c$
$\cot^2 x$	$-\cot x - x + c$
$\sin^2 ax$	$(1/2) \, x - (1/4) \, a^{-1} \, (\sin 2ax) + c$
$\cos^2 ax$	$(1/2) \, x + (1/4) \, a^{-1} \, (\sin 2ax) + c$
$\tan^2 ax$	$a^{-1} \tan ax - x + c$

$\csc^2 ax$	$-a^{-1} \cot ax + c$
$\sec^2 ax$	$a^{-1} \tan ax + c$
$\cot^2 ax$	$-a^{-1} \cot ax - x + c$
$x \sin ax$	$a^{-2} \sin ax - a^{-1}x \cos ax + c$
$x \cos ax$	$a^{-2} \cos ax + a^{-1}x \sin ax + c$
$x^2 \sin ax$	$2a^{-2}x \sin ax + (2a^{-3} - a^{-1}x^2) \cos ax + c$
$x^2 \cos ax$	$2a^{-2}x \cos ax + (a^{-1}x^2 - 2a^{-3}) \sin ax + c$
$(\sin x \cos x)^{-2}$	$-2 \cot 2x + c$
$(\sin x \cos x)^{-1}$	$\ln |\tan x| + c$
$\sin x \cos x$	$(1/2) \sin^2 x + c$
$\sin^2 x \cos^2 x$	$(1/8) x - (1/32) \sin 4x + c$
$(\sin ax \cos ax)^{-2}$	$-2a^{-1} \cot 2ax + c$
$(\sin ax \cos ax)^{-1}$	$a^{-1} \ln |\tan ax| + c$
$\sin ax \cos ax$	$(1/2) a^{-1} \sin^2 ax + c$
$\sin^2 ax \cos^2 ax$	$(1/8) x - (1/32) (a^{-1}) \sin 4ax + c$
$\sec x \tan x$	$\sec x + c$
$\text{Arcsin } x$	$x \text{ Arcsin } x + (1 - x^2)^{1/2} + c$
$\text{Arccos } x$	$x \text{ Arccos } x - (1 - x^2)^{1/2} + c$
$\text{Arctan } x$	$x \text{ Arctan } x - (1/2) \ln |1 + x^2| + c$

Suggested Additional Reading

- Bachman, D. *Advanced Calculus Demystified*. New York: McGraw-Hill, 2007.
- Gibilisco, S. *Mastering Technical Mathematics,* 3rd ed. New York: McGraw-Hill, 2008.
- Gibilisco, S. *Technical Math Demystified*. New York: McGraw-Hill, 2006.
- Huettenmueller, R. *Algebra Demystified*. New York: McGraw-Hill, 2003.
- Huettenmueller, R. *College Algebra Demystified*. New York: McGraw-Hill, 2004.
- Huettenmueller, R. *Pre-Calculus Demystified*. New York: McGraw-Hill, 2005.
- Krantz, S. *Calculus Demystified*. New York: McGraw-Hill, 2003.
- Krantz, S. *Differential Equations Demystified*. New York: McGraw-Hill, 2005.
- Olive, J. *Maths: A Student's Survival Guide,* 2d ed. Cambridge, England: Cambridge University Press, 2003.

Index